Gene Cloning

Gene Cloning

Edited by Patrick Faraday

SYRAWOOD
PUBLISHING HOUSE

New York

Published by Syrawood Publishing House,
750 Third Avenue, 9th Floor,
New York, NY 10017, USA
www.syrawoodpublishinghouse.com

Gene Cloning
Edited by Patrick Faraday

© 2018 Syrawood Publishing House

International Standard Book Number: 978-1-68286-515-6 (Hardback)

Cataloging-in-Publication Data

Gene cloning / edited by Patrick Faraday.
    p. cm.
Includes bibliographical references and index.
ISBN 978-1-68286-515-6
1. Molecular cloning. 2. Genetic engineering. I. Faraday, Patrick.
QH442.2 .G46 2018
660.65--dc23

TABLE OF CONTENTS

PREFACE

The main aim of this book is to educate learners and enhance their research focus by presenting diverse topics covering this vast field. This is an advanced book which compiles significant studies by distinguished experts in the area of analysis. This book addresses successive solutions to the challenges arising in the area of application, along with it; the book provides scope for future developments.

Gene cloning is the method by which a gene is copied and encoded for modification. It is very important in processes related to cellular biology such as gene sequencing, mutagenesis, protein expression and genotyping. Gene cloning has various industrial applications such as in the creation of genetically modified crops, gene therapy, tissue engineering, hormone creation in the field of medicine, etc. This book elucidates the concepts and innovative models around prospective developments with respect to gene cloning. It strives to provide a fair idea about this discipline and to help develop a better understanding of the latest advances within this field. The book will prove to be of great help to students and researchers in the fields of molecular biology, genetics and engineering.

It was a great honour to edit this book, though there were challenges, as it involved a lot of communication and networking between me and the editorial team. However, the end result was this all-inclusive book covering diverse themes in the field.

Finally, it is important to acknowledge the efforts of the contributors for their excellent chapters, through which a wide variety of issues have been addressed. I would also like to thank my colleagues for their valuable feedback during the making of this book.

Editor

A Modified Recombineering Protocol for the Genetic Manipulation of Gene Clusters in *Aspergillus fumigatus*

Laura Alcazar-Fuoli[1¤a], Timothy Cairns[1¤b], Jordi F. Lopez[2], Bozo Zonja[3], Sandra Pérez[3], Damià Barceló[3,4], Yasuhiro Igarashi[5], Paul Bowyer[1], Elaine Bignell[1]*

1 Manchester Fungal Infection Group, Institute for Inflammation and Repair, Faculty of Medicine and Human Sciences, Manchester Academic Health Science Centre, The University of Manchester, Manchester, United Kingdom, 2 Department of Environmental Chemistry, Institute of Environmental Assessment and Water Research (IDÆA), Consejo Superior de Investigaciones Científicas, c/Jordi Girona, Barcelona, Spain, 3 Water and Soil Quality Research Group, Institute of Environmental Assessment and Water Research (IDÆA), Consejo Superior de Investigaciones Científicas, c/Jordi Girona, Barcelona, Spain, 4 Catalan Institute of Water Research, ICRA, C/Emili Grahit, 101, edifici H2O Parc Científic i Tecnològic de la Universitat de Girona, Girona, Spain, 5 Biotechnology Research Center, Toyama Prefectural University 5180 Kurokawa, Imizu, Toyama, Japan

Abstract

Genomic analyses of fungal genome structure have revealed the presence of physically-linked groups of genes, termed gene clusters, where collective functionality of encoded gene products serves a common biosynthetic purpose. In multiple fungal pathogens of humans and plants gene clusters have been shown to encode pathways for biosynthesis of secondary metabolites including metabolites required for pathogenicity. In the major mould pathogen of humans *Aspergillus fumigatus*, multiple clusters of co-ordinately upregulated genes were identified as having heightened transcript abundances, relative to laboratory cultured equivalents, during the early stages of murine infection. The aim of this study was to develop and optimise a methodology for manipulation of gene cluster architecture, thereby providing the means to assess their relevance to fungal pathogenicity. To this end we adapted a recombineering methodology which exploits lambda phage-mediated recombination of DNA in bacteria, for the generation of gene cluster deletion cassettes. By exploiting a pre-existing bacterial artificial chromosome (BAC) library of *A. fumigatus* genomic clones we were able to implement single or multiple intra-cluster gene replacement events at both subtelomeric and telomere distal chromosomal locations, in both wild type and highly recombinogenic *A. fumigatus* isolates. We then applied the methodology to address the boundaries of a gene cluster producing a nematocidal secondary metabolite, pseurotin A, and to address the role of this secondary metabolite in insect and mammalian responses to *A. fumigatus* challenge.

Editor: Robert A. Cramer, Geisel School of Medicine at Dartmouth, United States of America

Funding: This work was supported in part by grants to EB from the Medical Research Council (G0501164 and MR/L00822/1; http://www.mrc.ac.uk); and BBSRC (BB/G009619/1; http://www.bbsrc.ac.uk). LAF is funded by Fondo de Investigación Sanitaria with a Miguel Servet fellowship (FIS:CP11/00026; http://www.isciii.es). TC was funded by a BBSRC PhD Studentship (BB/D526396/1; http://www.bbsrc.ac.uk). BZ acknowledges the Marie Curie Actions ITN CSI: Environment PITNGA-2010-264329 for the Early Stage Researcher contract and funding. SP acknowledges the contract from the Ramón y Cajal Program of the Spanish Ministry of Economy and Competitiveness (http://www.mineco.gob.es). The funders had no role in study design, data collection and analysis, decision to publish, or preparation of the manuscript.

Competing Interests: EB, LAF, TC PB, JFL, SP, BZ, DB do not have a financial relationship with any commercial entities that have an interest in the subject of this manuscript.

* Email: elaine.bignell@manchester.ac.uk

¤a Current address: Mycology Reference laboratory, National Centre for Microbiology, Instituto de Salud Carlos III. Crta. Majadahonda-Pozuelo, Madrid, Spain
¤b Current address: Biosciences, Geoffrey Pope Building, University of Exeter, Exeter, United Kingdom

Introduction

In the genomic era of fungal molecular genetics, the context and/or spatial organisation of genes is emerging as an important regulatory determinant [1]. In some instances the mechanistic significance of such organisational structures remains unclear but it is now widely accepted that genes involved in the biosynthesis of certain secondary metabolites are co-localised, in series, as gene clusters [2]. Secondary metabolites (SMs) can be produced by most fungal species [2,3] and in some cases, such as the biosyntheses of penicillin, sterigmatocystin and aflatoxin by *Aspergillus* species, the genetic regulation of cluster activities has been well characterised [4–6]. Many putative SM gene clusters

have been inferred by genome sequencing and comparative genomics or by transcriptional analyses where co-regulation of neighbouring genes is in evidence [3,4,7–9]. Lack of clearly defined biosynthetic pathways for many secondary metabolites means that the boundaries and number of genes comprising each gene cluster are often poorly defined, although common features can be identified including the involvement of polyketide synthases (PKSs) and nonribosomal peptide synthetases (NRPSs), and hybrids thereof [10]. In addition it has been demonstrated that the collective functionality of such gene products is ensured by their chromosomal colocalisation [11,12]. Noteworthy is the fact that the majority of known and putative SM gene clusters are located at subtelomeric regions of the chromosomes, [8] most likely facilitating their epigenetic regulation by chromatin-based mechanisms [13]. This epigenetic control of secondary metabolism might provide a means by which SM biosynthesis can be tailored to specific growth conditions while remaining otherwise silent.

In the major mould pathogen of humans, *A. fumigatus*, transcriptional upregulation of 70 *A. fumigatus* genes involved in SM biosynthesis was found during initiation of infection in the mammalian lung relative to laboratory cultures [9]. The direct relevance of SM biosynthesis to disease outcomes in whole animals is evidenced by a crucial role for the epipolythiodipiperazine toxin, gliotoxin, in pathogenicity in corticosteroid-treated hosts [14], however, the role of most individual secondary metabolites in pathogenicity of *A. fumigatus* remains a major unanswered question. A clue to the potential relevance of secondary metabolites during mammalian infection is provided by the putative methyltransferase LaeA, which in *Aspergillus* spp. is a major regulator of SM biosynthesis. In *A. fumigatus* a *ΔlaeA* mutant is hypovirulent in mouse models of invasive aspergillosis [15,16] and transcriptional analysis of a *ΔlaeA* mutant compared to the parental strain showed that LaeA influenced expression of 13 out of 22 secondary metabolite gene clusters [17].

In order to derive functional insight on both gene cluster organisation and the role of the *A. fumigatus* biosynthetic products in fungal pathogenicity, we sought the means to delete and/or re-organise groups of genes. Genetic manipulation of *A. fumigatus* has been fraught with difficulties due to relatively low efficiencies of homologous recombination. Several advances have augmented the success of gene replacements in *A. fumigatus* including the disablement of non-homologous end joining and the exploitation of split-marker strategies to facilitate the direct selection of appropriately mutated transformants [18–22]. Specific PCR-based gene targeting strategies have been gainfully employed to achieve deletion of gene clusters in *Ustilago maydis* [23]. In *A. nidulans* the deletion and regulatable expression of gene clusters has been achieved by exploiting highly recombinogenic strains [24].

Chaveroche et al. first exploited recombineering for *Aspergillus* gene knockouts by using the large insert sizes of cosmid gDNA clones to maximise homologous integration frequencies in *A. nidulans* [25]. This strategy provided a much-needed solution to the bottleneck then associated with low rates of homologous recombination in *A. fumigatus* and was more widely adopted for deletion of single *A. fumigatus* genes [26] but was limited to DNA insert sizes amenable to cosmid cloning (~37–52 kb), and reliant upon plasmid-mediated induction of recombinogenic functions. Recent availability of new recombineering reagents, and refinement of culturing and recombineering protocols, has elevated recombineering efficiency and practicability [27]. We have exploited these advances to expand the repertoire of tools available for *A. fumigatus* manipulation. Relative to the previously-used methodology [25,26] the new reagents promote, via one-step λ-infection of BAC-harbouring *E. coli* clones, a means for higher throughput construction of large recombinant *A. fumigatus* DNA fragments and critically for this study, the ability to work with larger inserts, thereby enabling multiple manipulations of gene cluster architecture from a single BAC clone. A key refinement is the use of a lambda phage which is replication-defective in *E. coli* cells harbouring bacterial artificial chromosomes (BACs), but retains heat-inducible homologous recombination functions. This allows users to render BACs competent for recombineering by a simple lambda infection and to induce recombination in *E. coli* via a simple temperature shift, thereby permitting high throughput manipulations of BAC clones. We used clones from a pre-existing BAC library of *A. fumigatus* genomic clones [28] to delete single genes and gene clusters in *A. fumigatus* by using a modification of

this recombineering approach. We standardized the methodology by targeting two, physically unlinked, individual genes: a telomere distal pH-responsive transcription factor-encoding gene *pacC* [18,29] and a telomere-proximal putative transcription factor-encoding gene *regA*. We then applied the methodology to address the boundaries of a gene cluster producing a nematocidal secondary metabolite, pseurotin A, and to address the role of this secondary metabolite in insect viability and during interactions between *A. fumigatus* and mammalian phagocytic, or respiratory epithelial cells.

Materials and Methods

Strains, media and culture conditions

Aspergillus fumigatus strains used in this study are presented in Table 1. Fungal strains were routinely grown at 37°C on *Aspergillus* complete medium (ACM) according to Pontecorvo et al. [30] containing 1% (w/v) glucose as carbon source and 5 mM ammonium tartrate as nitrogen source. For solid media 1% (w/v) agar was added. Minimal media (MM) containing 5 mM ammonium tartrate and 1% (w/v) glucose [31] was used for phenotypic testing. For *Aspergillus* transformation MM was supplemented with 1 M sucrose to produce regeneration medium (RM). Liquid cultures were agitated by orbital shaking at 150 rpm unless otherwise stated. For propagation of plasmids, *E. coli* strain XL-10 (Agilent technologies) was grown in Luria-Bertani (LB medium) supplemented with ampicillin (100 μg/ml). The *A. fumigatus* BAC library was maintained in *E. coli* DH10B (Invitrogen, UK). Propagation of BAC clones was performed in LB supplemented with chloramphenicol (12.5 μg/ml). Reagents for recombineering were kindly provided by Donald L Court (Wellcome Trust Sanger Institute, Hinxton, Cambridge, UK). The replication deficient λ phage (λ cI_{857} $ind1$ $Cro_{TYR26amber}$ $P_{GLN59amber}$ $rex< >tetra$) [27] was maintained in *E. coli* LE392 (Promega, UK). The BAC library as well as the reagents for recombineering in *A. fumigatus*, are available on request.

Cloning and BAC recombineering procedures

For standard cloning and sub-cloning procedures the vectors pUC19 [32] or pGEM-T Easy (Promega, UK) were used. All DNA oligonucleotides used in this study were purchased from Sigma (UK) (Table 2). PCRs were performed and optimised according to the manufacturer's guidelines.

Plasmids expressing biselectable markers BSM-Z/P and BSM-A/H respectively conferring Zeocin (Z) and pyrithiamine (P), or ampicillin (A) and hygromycin (H) resistances were constructed as follows. To construct pBSM-Z/P a gene conferring resistance to zeomycin was amplified by PCR from the plasmid pCDA21 [25] using the primers Zeo1F and Zeo1R (Table 2). The amplicon was blunt ended and cloned into the SmaI site of pUC19 to produce the plasmid pZ3. The *ptrA* gene was obtained by PCR amplification from plasmid pPTRII [33] using primers PtrAF and PtrAR (Table 2) and was cloned into the SalI site of pZ3. pBSM-A/H was constructed by PCR amplification of a gene conferring ampicillin resistance from the plasmid pSK379 [34] using primers AmpR-F and AmpRHyg-R (Table 2); and PCR amplification of a gene (*hph*) conferring hygromycin resistance from pID621 [35] using primers AmpRHyg-F and Hyg-R (Table 2). Fusion of the two amplicons was performed by an overlapping PCR procedure, using PrimeSTAR DNA polymerase enzyme (Clontech, UK) and the primers AmpR-F and Hyg-R (Table 2). The resulting PCR product was cloned into the pGEM-T Easy vector (promega, UK) according to the manufacturer's guidelines.

Table 1. Strains used in this study.

Strain	Genotype	Source
CEA17_ΔakuB^KU80	CEA17akuB^KU80:: pyrG	[19]
CM237	Wild type	[41]
Af293	Wild type	[47]
PsoA	Af293 pyrG⁻ psoA:: pyrG	[47]
H515	CM237 pabaA: hph	[35]
ATCC46645	Wild type	American Type Culture Collection
ΔpacC	CEA17akuB^KU80:: pyrG pacC::BSM-A/H	This study
ΔPsoAcluster	CEA17akuB^KU80:: pyrG PsoAcluster:: BSM-A/H	This study
ΔAFUA_8G00520	CEA17akuB^KU80:: pyrG AFUA_8G00520:: BSM-A/H	This study
ΔAFUA_8G00550	CEA17akuB^KU80:: pyrG AFUA_8G00550:: BSM-A/H	This study
ΔregA	CEA17akuB^KU80:: pyrG regA:: BSM-Z/P	This study

pyrG encodes an *A. niger* orotidine-5-monophosphate decarboxylase conferring prototrophy to uracil and uridine; BSM-A/H is a biselectable marker constructed during this study which includes the *hph* gene encoding an *E. coli* hygromycin phosphotransferase conferring hygromycin B resistance in *A. fumigatus*; BSM-Z/P is a biselectable marker constructed during this study which includes the *ptrA* gene conferring pyrithiamine resistance in *A. fumigatus*.

Construction of recombinant BACs by recombineering

For construction of recombinant BAC clones a library of end-sequenced, indexed, Af293 *A. fumigatus* BAC clones generated at the Wellcome Trust Sanger Institute (in collaboration with the University of Manchester) [28] was utilised (Table S1). In order to construct gene replacement cassettes for recombineering, biselect-able markers (BSMs) were amplified by PCR using tailed oligonucleotide primers. To avoid contamination by BSM-Z/P or BSM-H/P the plasmids were linearized prior to PCR and absence of circular DNA was confirmed by *E. coli* transformation. Primer tail sequences were designed to introduce 80 bp of homology to the target genetic locus, at both of the 5′ and 3′ extremities of the BSM (Figure 1 and Table 2). PCR amplicons for recombineering were generated using the high fidelity DNA polymerase PrimeSTAR (Clontech, UK). Amplicons were purified gel extracted using the purification kit NucleoSpin for gel extraction (Macherey-Nagel, Germany). To generate recombinant BACs, 1 ng of the appropriately tailed PCR amplicon was used as a template for PCR. PCR products were precipitated with 100 μl ethanol and 2 μl 5 M NaCl per 50 μl of PCR reaction. Air-dried PCR products (~3 μg) were dissolved in 100 mM CaCl₂ and stored at 4°C until needed.

BACs containing the *A. fumigatus* DNA sequences of interest were selected from the library (Table S1) and recombinant BACs were generated by transformation and heat induction according to the method described by Chan et al. [27,36]. A precise protocol for this procedure is provided as Protocol S1 in supporting information.

After recombineering BAC DNA extraction was performed with Qiagen reagents for plasmid isolation [36] and verification of insertion or deletion in targeted BACs was obtained by PCR using appropriate primers (Table 2).

Genetic manipulation of *A. fumigatus*

For fungal transformation BAC DNA was extracted from 50 ml LB cultures. DNA was resuspended in 200 μl of water and 60 μl of recombined BACs (~1 μg) were linearized overnight with the appropriate restriction enzyme in a total volume of 70 μl. Restriction enzymes were heat inactivated before *A. fumigatus* transformation. Only freshly prepared BAC DNA was used for transformation.

Protoplast transformation was based on the protocol described by Szewczyk and co-workers [37]. Selection of transformants was performed in RM media containing 150 ug/ml of hygromycin or 0.5 ug/ml of pyrithiamine. Plates were incubated at room temperature for 24 hours and then incubated at 37°C for 72–144 hours. Gene targeting was first verified by PCR using primers targeting the corresponding gene, selected outside of the flanking regions and/or in combination with primers targeting the resistance gene marker. For PCR verification DNA was extracted from spores [38]. Single homologous recombination into the *A. fumigatus* genome was confirmed by Southern blot analysis [39] using digoxigenin-labeled probes and the DIG system (Roche, UK) for hybridization and detection.

Aspergillus phenotypic analysis

Aspergillus fumigatus strains were evaluated by spotting 4 sequential 10 fold dilutions of candidate *A. fumigatus* spores, starting at a concentration of 2.5×10^4 per spot, onto MM pH 6.5. Alkaline tolerance was assessed by spotting 2.5×10^4 spores onto MM pH 8. Plates were incubated at 37°C for 48 hours. Images were captured using a Nikon Coolpix 990 digital camera.

Quantification of pseurotin A in fungal culture filtrates

Fungal strains were grown at 37°C on *Aspergillus* complete agar (ACM) for 4 days prior to conidial harvest. 200 ml ACM liquid media was inoculated with *A. fumigatus* strains at a concentration of 10^6 spores per ml. Cultures were incubated at 30°C for 1 week with agitation at 180 rpm. After the incubation period, chloroform (180 ml) was added to each bottle, and the contents were mixed on a rotary shaker at 180 rpm for 15 minutes. After filtration through Miracloth (Calbiochem, USA), the chloroform phase was separated and evaporated to dryness on a rotary evaporator at 40°C. The residue was dissolved in chloroform and filtered. The extracts were evaporated under nitrogen, then reconstituted in 1 ml of methanol and filtered through a 0.22 μm filter (Millipore, UK). The extracts were analyzed for pseurotin A levels using ultra high pressure liquid chromatography coupled to a triple quadruple mass spectrometry (UPLC-(ESI)-QqQ-MS), in positive mode. Pseurotin A (CAS 58523-30-1) purchased from ENZOlife sciences was used as standard while 2-fluoro and 3-fluoro-pseurotin A, obtained in high

Table 2. Oligonucleotides used in this study.

Name	Sequence (5′–>3′)	Purpose in this study
AmpRF	GGTCTGACAGTTACCAATGC	Plasmid construction
Hyg-R	GCTTGATATCGAATTCGTCG	Plasmid construction
Zeo1F	GAATTCTCAGTCCTGCTCCT	Plasmid construction
Zeo1R	CGGGGGATCCACTAGTTCT	Plasmid construction
PtrAF	GGCCAATTGATTACGGGAT	Plasmid construction
PtrAR	ATGGCCTCTTGCATCTTTG	Plasmid construction
pacC BSM-A/H -F	GCGAGCGTCACGTCGGTCG AAAGAGCACAAATAACCTT AACCTGACATGCCAGTGGG GAAGCTGCCGCACCACGACT GTTGGTCTGACAGTTACCAATGC	5′-flanking amplification and fusion
BSM-A/H -R	TGCTCCTTCAATATCAGTTA ACGTCGACGAGGAAATGTG CGCGGAACCCCTATTTGTTTA	5′-flanking amplification
pacC BSM-A/H -R	TGGATGGAGGGGCGACGCTC TTCGGGGCGAGCACGCTGTAA AGTGCCTCCAGTGTACCGGCG ACGATCATCGTCAAAGATGCTTG ATATCGAATTCGTCG	3′-flanking amplification and fusion
BSM-A/H -F	TAAACAAATAGGGGTTCCGCGCAC ATTTCCTCGTCGACGTTAACTG ATATTGAAGGAGCA	3′-flanking amplification
pacCzeo-F	GCGAGCGTCACGTCGGTCGAA AGAGCACAAATAACCTTAACCT GACATGCCAGTGGGGAAGCTGC CGCACCACGACTGTTTTCTAGAG CGGCCGCGATAT	BSM-Z/P biselectable marker *amplification*
pacCPtrA-R	TGGATGGAGGGGCGACGCTCT TCGGGGCGAGCACGCTGTAAA GTGCCTCCAGTGTACCGGCGA CGATCATCGTCAAAGATGGCC TAGATGGCCTCTTGCA	BSM-Z/P biselectable marker amplification
CF5R	ATAAGGTTAGCCGAGATGCG	*pacC* replacement verification
CF5F	CTAGCACTTCCATGAGCAAC	*pacC* replacement verification
520A-F	CAAAGCCACATCGACCCTT GCCCTCTGGCCGGATCACACCT GGACAAGCTACCCTCCTCTATC GCGTCCTCTTCTTCCTGGGTCT GACAGTTACCAATGC	BSM-A/H biselectable marker amplification
520H-R	GCCGCATCCATATCCAAGCG TGATCTGTAGCTATGTCCCAC TAGGTCTACTCAATGGCATTG TCAGGTCCAGTCCGCCTTGCT TGATATCGAATTCGTCG	BSM-A/H biselectable marker amplification
520-1F	GCCCTCTGGCCGGATCACAC	AFUA_8G00520 deletion verification
520-1R	CATGGGGACTGGCCGCATCC	AFUA_8G00520 deletion verification
550A-F	CGGTCATAGACAAGAGGAATCT CTACATAAAGGCGCTATCCCTGC TATTAAATGACGGGCATGGGATT GGTAGTTCGTAGGGTCTGACA GTTACCAATGC	BSM-A/H biselectable marker amplification
550H-R	CGATTGTACATGCTCACACGTA GAATCGGCACAGTCTTGGGAA AAGTATGTGGTTTAACTAGCC TTGTCGACCTTGCTTTGCTTGA TATCGAATTCGTCG	BSM-A/H biselectable marker amplification
550-1F	CGGGCATGGGATTGGTAGTTCGT	AFUA_8G00550 deletion verification
550-1R	GGCCTAACCGGGTTCCAGCG	AFUA_8G00550 deletion verification

Table 2. Cont.

Name	Sequence (5'->3')	Purpose in this study
PsoACA-F	AAGGCGGACTGGACCTGACA ATGCCATTGAGTAGACCTAGT GGGACATAGCTACAGATCAC GCTTGGATATGGATGCGGCG GTCTGACAGTTACCAATGC	BSM-A/H biselectable marker amplification
PsoACH-R	TCAGCACTAGGGAAGTCGGT GTAATGGTGTCAGCCTACTCA GTCACGTGCAGGACATAATCC TCCATCCCCCGAACGACAGCTT GATATCGAATTCGTCG	BSM-A/H biselectable marker amplification
PSOAClustF1	CAGCCTGTGGCTCGCTGGTC	PsoA cluster deletion verification
PSOAClustR1	TCCCCGCGTCCACACTCGAT	PsoA cluster deletion verification
520SB-F	TTCAGGTGCTGCAAGATGTC	Southern Blot probe
520SB-R	CTTCATGGCCGTTCTGGTAT	Southern Blot probe
550SB-F	GGCCTGATCTACCTTCACCA	Southern Blot probe
550SB-R	TAGCAGGGATAGCGCCTTTA	Southern Blot probe
640_F	CCCCTTGACATAGGGTAATAAT GTGCTTTCGCATTGTTCCACCC ATGGCCCCCCCGCGTTCGGAG CTGCGTTAGCTAGGCTTCT AGAGCGGCCGCGATAT	BSM-Z/P biselectable marker amplification
640_R	TTATTTGCTGGCATCTCGCAA CTTCTCAAGAAGATGGACCAA GTTATCCACCAGTGGCGGAGT CTGTTTAGAAGTTTCATATGGCC TCTTGCATCTTTG	BSM-Z/P biselectable marker amplification
640_INT_F	ATTCGGCTCTGCATATCACC	regA disruption verification
640_INT_R	TGAATGATAGGCGTCCTTCC	regA disruption verification
5'_640	TCCAGGATCTTCGCATAGGT	regA disruption verification
3'_640	GACCGAGTTGACTCGGATCT	regA disruption verification

purity by directed biosynthesis by the group of Dr. Igarashi were used as internal standards [40]. The chromatographic separation of the compounds was performed on a Waters Acquity UPLC BEH C_{18} column (50 mm×2.1 mm, 1.7 um) (Waters, Milford, USA) preceeded by a pre-column of the same packing material (5 mm×2.1 mm, 1.7 um). The mobile phases employed were: (A) acetonitrile and (B) water (20 mM Ammonium Acetate). Elution was accomplished with the following solvent gradient: 0 min (20% A) – 1 min (20% A) – 6 min (30% A), – 6.5 min (95% A) – 6.8 min (20% A) and stabilizing until 8 min. The flow rate was 300 μL min^{-1} and the column temperature was held at 35°C. The injection volume was 10 μL.

Macrophage phagocytosis assays

Phagocytosis of *A. fumigatus* spores was measured in the murine macrophage cell line RAW 264.7. Macrophages were grown in RPMI media supplemented with 10% fetal bovine serum (containing L-glutamine (200 mM), penicillin (10,000 units/ml), streptomycin (10 mg/ml), Sigma, UK) at 37°C in an atmosphere of 5% CO_2. Prior to infection, macrophages were adjusted to a cell density of 5×10^5 cells per ml. 1 ml of cells was added to 24 well Multiwell (BD Falcon) plates and allowed to adhere overnight at 37°C in an atmosphere of 5% CO_2. Macrophages were challenged with *A. fumigatus* spores at an effector to target (E:T) ratio of 1:1 and incubated for 2 hours at 37°C in an atmosphere of 5% CO_2.

Next, the residual spores in the culture medium were recovered, washed three times in PBS and colony forming units (CFUs) were enumerated and expressed as a percentage of infectious dose to reflect the proportion of internalised spores.

Quantification of lactate dehydrogenase (LDH) release from A549 lung epithelial cells

LDH release from A549 monolayers was quantified following co-incubation with *A. fumigatus* conidia at an E:T ratio of 1:0.1. LDH release was determined using the CytoTox 96 Non-Radioactive Cytotoxicity Assay (Promega). A549 cells were cultured at 5×10^5 cells/well in 24-well plates for 24 hours in MEM media supplemented with 10% fetal bovine serum and containing L-glutamine (200 mM), penicillin (10,000 units/ml), streptomycin (10 mg/ml) (Sigma, UK) at 37°C in an atmosphere of 5% CO_2. The assays were performed according to the manufacturer's instructions and the measurements from three biological replications were evaluated.

Galleria mellonella survival assay

Wax-moth larvae were infected with *A. fumigatus* CEA17_Δa-kuB^{KU80} or transformants (Table 1). CM237 parental [41] and a para-aminobenzoic acid (PABA) auxotroph of CM237 (referred to as *A. fumigatus* H515) [35] were used as positive controls for attenuated and non-attenuated virulence respectively. Wax moth

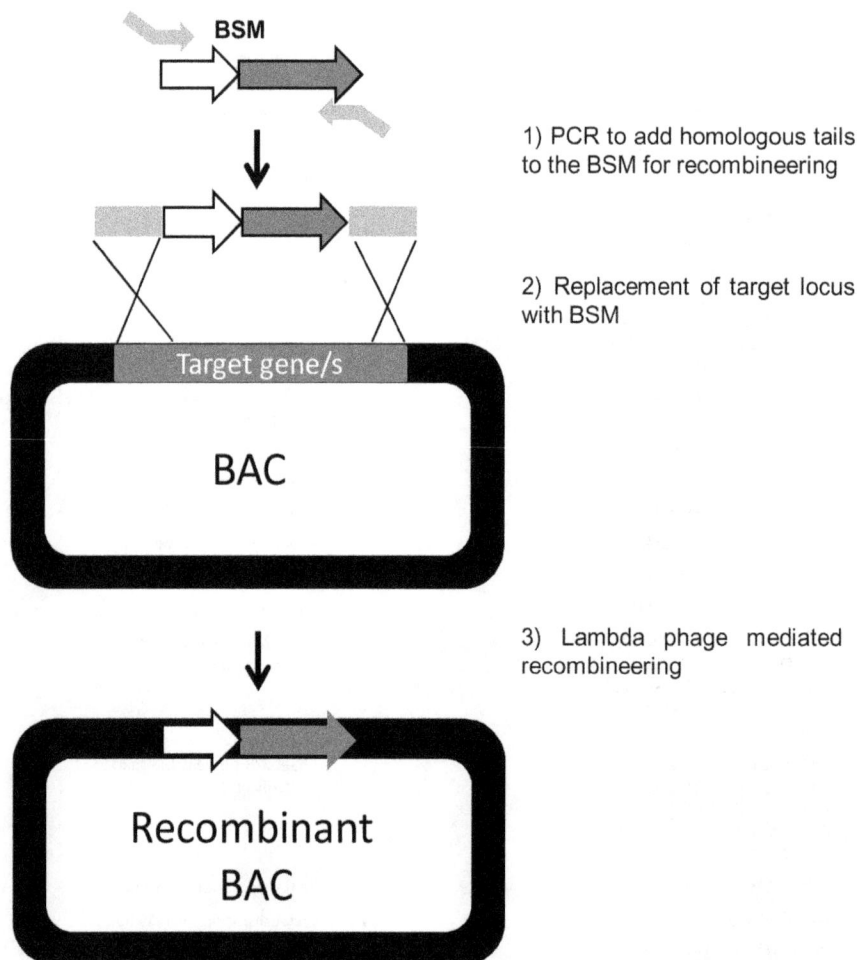

Figure 1. Overview of BAC recombineering in *E. coli*. BSMs were amplified by PCR using tailed oligonucleotide primers: 1) Primer tail sequences were designed to introduce 80 bp of homology to the target locus, at both of the 5′ and 3′ extremities of the BSM. 2) Replacement of target locus with BSM. 3) Heat-induction of homologous recombination functions mediate by lambda phage in *E. coli*.

larvae killing assays were carried out as described previously [42,43]. Briefly, groups of 10 larvae (0.3–0.5 grams, R.J. Mous Livebait, The Netherlands) were inoculated into the haemocoel with 10 µL of a 10^7 conidia/ml suspension in water for injection so the final inoculum in each group was 10^5 conidia per larva. Additionally, non-infected larvae, injected with 10 µl of water were included in parallel in every infection. 4 µg/ml of para-aminobenzoic acid (PABA) was used to inject *G. mellonela* infected with the H515 strain. Mortality, defined by lack of movement in response to stimulation and discoloration (melanization) of the cuticle, was recorded daily.

Statistical analyses

Statistical analyses were performed in GraphPad Prism, version 5. The statistical significance of variances between phagocytosis and cell cytotoxicity was calculated by using a nonparametric Mann-Whitney t test. A p value<0.05 was considered significant. Kaplan-Meier survival curves were analysed by using a log-rank (Mantel-Cox) test for significance. A p value<0.01 was considered significant.

Results and Discussion

Optimisation of BAC-mediated gene replacement in *A. fumigatus*

In order to commence a functional genomic analysis of gene clusters and virulence in the human fungal pathogen *A. fumigatus* we developed the means to manipulate gene content of complex genetic loci using BAC-mediated recombineering. A recombineering approach was previously successfully applied in *A. nidulans* [25] where recombinant cosmids were generated in *E. coli* following transformation with a plasmid carrying the λ phage *red*γαβ operon. In our study a higher throughput approach was adopted whereby recombineering functions are transiently supplied, via phage infection and a simple temperature shift, to BAC-harbouring *E. coli* cells [27]. In this manner BAC-cloned genomic regions of interest are replaced with a biselectable marker (BSM) which confers selectable tolerance to antibiotics and/or toxic metabolites in both *E. coli* and *A. fumigatus*. We used a previously constructed library of end-sequenced, indexed *A. fumigatus* BAC clones (Table S1) to facilitate our analysis [28]. This *A. fumigatus* BAC library provides 10X genome coverage and contains 8380 clones having an average insert size of 75 Kb (Table S1 and File S1).

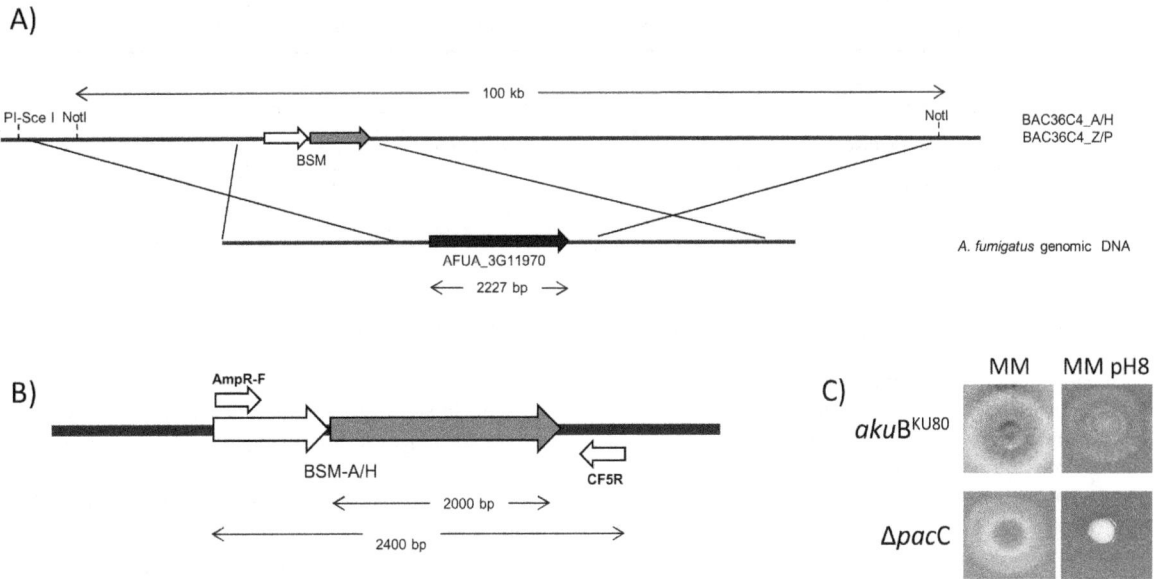

Figure 2. Deletion of the *pacC* gene in *A. fumigatus* CEA17_Δ*akuB*[KU80] (referred as *akuB*[KU80]). A) Schematic view of *pacC* gene deletion. B) Primers used to check gene replacement at the *pacC* locus by PCR. C) Phenotypic analysis of Δ*pacC* mutants compared with the wild type. 2.5×10^4 spores were point inoculated onto MM pH 6.5 and MM pH 8. Plates were incubated at 37°C for 48 hours.

In order to establish and optimise the methodology we first elected to replace single genes, selecting two, physically unlinked, individual genes AFUA_1G17640 and AFUA_3G11970. AFUA_3G11970 is a telomere distal gene encoding the transcription factor, PacC which is involved in alkaline signal transduction [18]. AFUA_1G17640 is a telomere-proximal gene encoding a putative transcription factor, RegA, which resides in a cluster of genes upregulated during murine infection and has a possible role in melanin biosynthesis [44,45]. For our initial experiments we exploited the well-characterised alkaline sensitivity of PacC null mutants to permit rapid assessment of homologous gene replacements amongst transformants, and utilised two newly constructed biselectable marker plasmids (pBSM-Z/P and pBSM-A/H) to permit comparative assessment of achievable gene replacement frequencies. Schematic overviews of recombinant BAC construction in *E. coli* and BAC-mediated *pacC* gene deletion are provided in Figures 1 and 2A, respectively. A BAC clone (AfB28-mq1_36C04) having appropriate coverage of the *pacC* AFUA_3G11970 genomic locus was retrieved from the library (Table S1). The BAC insert spanned the entire AFUA_3G11970 gene incorporating 24 kb and 74 kb of 5' and 3' flanking regions respectively. Appropriately recombined BAC clones were identified by PCR with primers AmpR-F and CF5R (Figure 2B) or PtR-F and CF5R. Recombinant BACs were denoted as BAC36C4-Z/P and BAC36C4-A/H where the AFUA_3G11970 gene had been

replaced with an ampicillin/hygromycin or zeocin/pyrithiamine biselectable marker respectively.

In order to linearize the recombinant BACs prior to *A. fumigatus* transformation two methods were used. For transformations using BAC36C4-A/H two different approaches were tested; (i) NotI digestion, which cuts twice in the polylinker of the pBACe3.6 vector and also excises a linear DNA fragment of 100 kb containing 24 kb and 74 kb of *pacC*-flanking 5' and 3' sequences respectively, and (ii) PI-SceI digestion which simply linearises the recombinant BAC clone. NotI digestion was not an option prior to transformation with BAC36C4-Z/P, because the zeocin resistance cassette contains a NotI restriction site. Transformants were analysed by PCR using DNA extracted from spores, and primers indicated in Figure 2B and Table 2. First, loss of the *pacC* gene was determined using primers CF5F and CF5R (data not shown). Appropriate insertion of the BSM was subsequently identified with primers AmpR-F and CF5R (Figure 2B) or PtR-F and CF5R. Transformants which had taken up the exogenous DNA were screened for growth at alkaline pH (Figure 2C) to determine the frequency of gene replacement amongst transformants analysed. Table 3 shows the efficiency of allelic replacement at the *pacC* locus when recombinant BACs are linearised according to these two strategies. Regardless of the strategy undertaken, and independent of the BSM utilised, we reproducibly obtained a minimum of 19% of total transformants having undergone gene replacements. We found gene replace-

Table 3. Efficiency of allelic replacement at the AFUA_3G11970 locus when vectors were linearized with different enzymes.

DNA	Enzyme	Frequency (%) of gene replacement at the AFUA_3G11970 locus/total of transformants
BAC36C4-A/H	NotI	7/17 (41%)
BAC36C4-A/H	PI-SceI	2/6 (33%)
BAC36C4-Z/P	PI-SceI	4/21 (19%)

Table 4. Efficiency of allelic replacement at the pseurotin A locus using different *A. fumigatus* strains.

A.fumigatus strain	Percentage of appropriate recombinant transformants/total transformants tested by PCR		
	ΔAFUA_8G00520	ΔAFUA_8G00550	ΔPsoAcluster
CEA17_ΔakuB^KU80	8/9 (88%)	6/6 (100%)	6/7 (85%)
ATCC46645	2/63 (3.1%)	1/8 (12.5%)	1/8 (12.5%)

ments utilising the BSM-A/H biselectable marker to be most favourable due to superior efficiencies of homologous recombination and an easier restriction digestion strategy.

Keller et al identified that telomere position effect (TPE) can influence the expression of selectable markers targeted to telomeric loci [46]. In order to assess the impact of such effects upon our methodology, and to verify that telomeric loci are amenable to BAC-mediated gene replacements, we constructed a recombinant BAC to replace the *reg*A (AFUA_1G17640) gene, which is positioned 80 kb from the right terminus of Chromosome 1, according to the genome of the sequenced isolate Af293. A BAC clone (AfB28_mq1_17e12) whose insert spanned the entire AFUA_1G17640 gene with 12326 and 1442 bp of 5 and 3 flanking regions respectively, was selected as the recombineering substrate. The zeocin and pyrithiamine biselectable marker was amplified from the pBSM-Z/P plasmid with primers 640_F and 640_R (Table 2) and recombinant BACs were sourced via diagnostic PCR (Figure S1). The recombinant BAC was digested with SacI, liberating a 12.5 kb deletion cassette (Figure S1). SacI digests were heat inactivated and used for subsequent *A. fumigatus* transformations. Homologous integrants were identified by PCR (Figure S1), exploiting the presence of a NotI site in the BSM. *reg*A deletion was verified by Southern blot, probing with a 600 bp fragment of the zeocin cassette, which produced a single fragment of the expected size (Figure S1).

BAC mediated deletion of the entire pseurotin biosynthetic gene cluster, or associated genes

Pseurotin A is a cyclic peptide putatively biosynthesised by a cluster of five genes housed on the left subtelomeric arm of chromosome 8. Genes in the cluster encode two putative hydrolases (AFUA_8G00530, AFUA_8G00570) a putative methyltransferase (AFUA_8G00550), a putative P450 monooxygenase (AFUA_8G00560) and the hybrid PKS/NRPS *psoA* (AFUA_8G00540) [47]. It has been demonstrated, via gene replacement analyses that integrity of the *psoA* gene is required for the biosynthesis of pseurotin A in *A. fumigatus* [47]. Pseurotin A is a compound that has been reported as a competitive inhibitor of chitin synthase, inducer of nerve-cell differentiation [48] and a suppressor of immunoglobulin E production [49]. Additionally, recent transcriptional, proteomic and metabolic analyses have demonstrated pseurotin A biosynthesis in hypoxic, but not normoxic culture [50] suggesting the production of a toxic and/or immunomodulatory secondary metabolite in hypoxic microenvironments encountered during pulmonary infection [51,52]. To further understand the regulation of pseurotin A production we

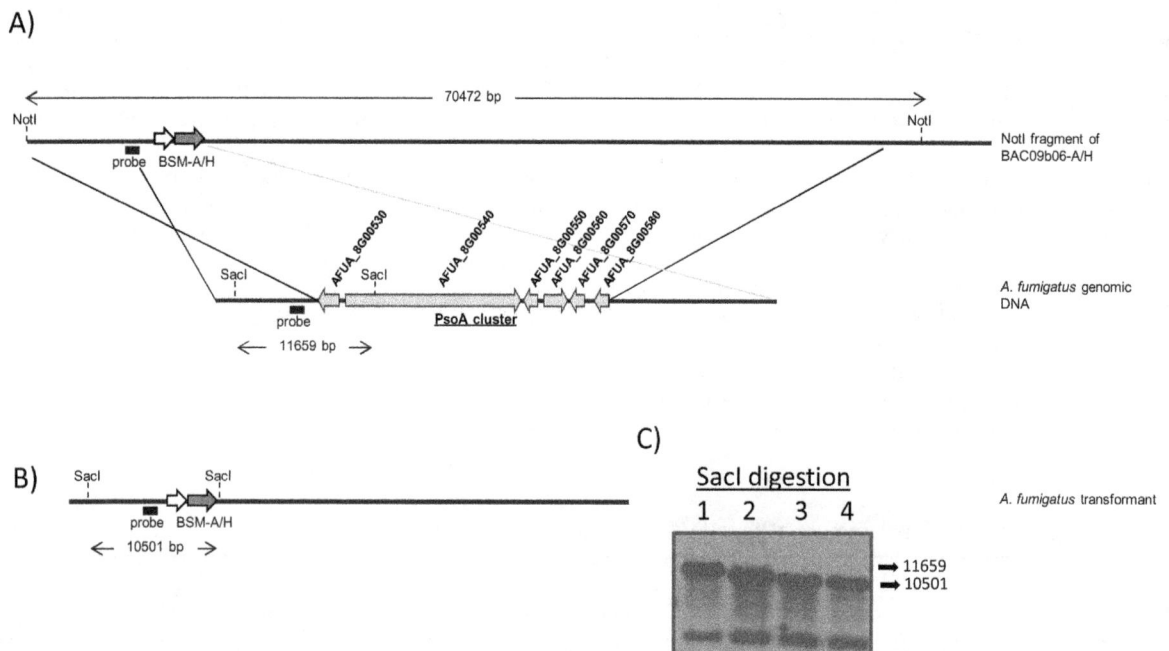

Figure 3. Deletion of the PsoA cluster in *A. fumigatus* CEA17_Δ*akuB*^KU80. A) Schematic representation of PsoA cluster replacement by BSM-A/H cassette in *A. fumigatus* CEA17_Δ*akuB*^KU80. B) Expected structure of the replacement locus and C) Southern blot analysis of PsoAcluster deleted mutant and wild type (WT) strains. Expected hybridization band pattern: (1) 11659 bp for WT, and (2, 3, 4) 10501 bp for ΔPsoAcluster mutants.

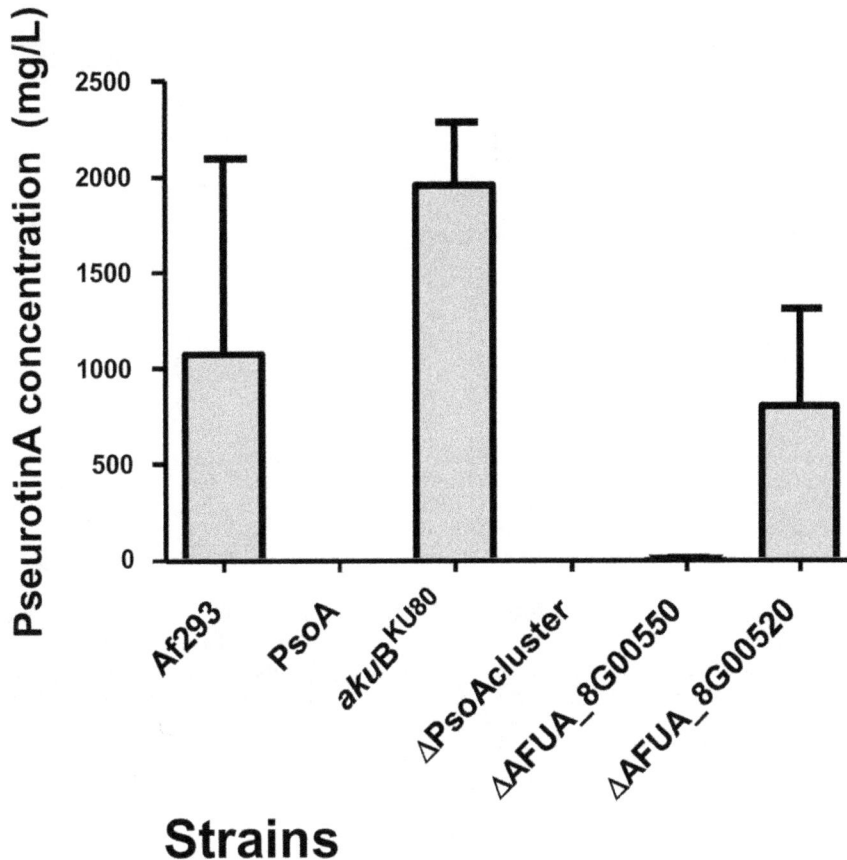

Figure 4. Quantification of pseurotin A by UPLC-ESI coupled to mass spectrometry in wild type (*A. fumigatus* CEA17_Δ*akuB*[KU80] referred as *akuB*[KU80]) and mutant strains.

constructed *A. fumigatus* mutants lacking the entire pseurotin gene cluster (AFUA_8G00530 – AFUA_8G00580) or other genes within or surrounding the cluster limits. We focused upon AFUA_8G00520, which encodes an integral membrane protein which lies beyond the cluster boundaries but is co-regulated, during murine infection [9], with the genes within the cluster. We also studied the AFUA_8G00550 gene, which encodes a SirN-like methyltransferase predicted to transform an intermediate of the pseurotin A biosynthetic pathway [47,53].

The pseurotin gene cluster is located 115 kb from the left arm of chromosome 8. Gene sequences conforming to those included within the PsoA gene cluster [47] matched with three different

clones from the *A. fumigatus* BAC library (AfB46-09f02, AfB46-09a06 and AfB46-09b06 as indicated in Table S1, and abbreviated in this study to BAC09f02, BAC09a06 and BAC09b06, respectively). Regions within these BAC clones were targeted by recombineering with BSM-A/H′ and used for *A. fumigatus* transformation. We worked in parallel with all three clones to demonstrate the versatility of our BAC-mediated approach. Thus, the BAC09f02 clone was utilised to delete AFUA_8G00520, encoding an integral membrane protein. The BAC09a06 clone was utilised to delete AFUA_8G00550, encoding a methyltransferase, and the BAC09b06 clone was utilised to delete the entire cluster of genes (AFUA_8G00530 – AFUA_8G00580). BAC

Table 5. Mass spectrometry conditions for Pseurotin A and internal standards.

Type	Compound	Retention time (min)	Transitions	Cone voltage	Collision energy
Target	Pseurotin A	3.95	**432>316**	**30**	**10**
			432>348	15	10
Internal Standard (1)	3-F Pseurotin A	3.59	450>334	15	5
			450>366	15	5
Internal Standard (2)	2-F Pseurotin A	4.63	450>344	15	5
			450>366	15	5

*In bold is the quantification transition.

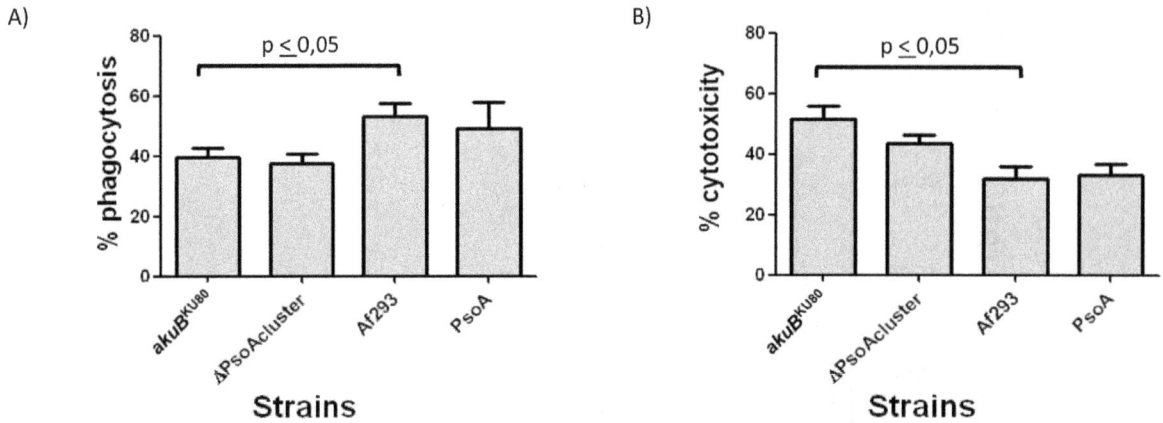

Figure 5. Effect of pseurotin A production upon interaction of *A. fumigatus* **strains with mammalian cells.** A) Percentage conidial phagocytosis following 2 h incubation with murine macrophages (RAW 264.7). B) Relative cytotoxicity (LDH release) after 24 h of co-incubation of *A. fumigatus* and human alveolar epithelial cells (A549). The statistical significance was calculated by using a nonparametric Mann-Whitney t test. A p value<0.05 was considered significant. *A. fumigatus* CEA17_$\Delta akuB^{KU80}$ referred as $akuB^{KU80}$.

AfB46-09f02 (insert size 56305 bp) contained sequence spanning genes AFUA_8G00420 to AFUA_8G00590 of the Af293 *A. fumigatus* genome, BAC09a06 (insert size 98501 bp) contained sequence spanning genes AFUA_8G00360 to AFUA_8G00740 and BAC09b06 (insert size 119804 bp) contained sequence spanning genes (AFUA_8G00490 to AFUA_8G00900).

Allelic replacements were first tested by PCR and single integration was confirmed by Southern blot (Figure 3, S2 and S3). Based on the PCR results frequency of homologous recombination in the *A. fumigatus* CEA17_$\Delta akuB^{KU80}$ genetic background was high with more than 85% of tested transformants (n = 6–9) undergoing allelic replacement at the correct genomic locus. Although lower frequencies of gene and gene cluster deletions were obtained when the clinical isolate ATCC46645 was used (Table 4), we obtained relevant mutants within the first 8 transformants tested, indicating practically useful, if not heightened, frequencies of gene replacement in non-mutated clinical isolates.

Our data demonstrate the utility of this method for deletion of single genes, gene clusters and neighbouring genes. Although outside the objectives of this study, our method would also facilitate analysis of specific protein domains of *A. fumigatus* PKS and NRPSs. For example, the hybrid PKS/NRPS gene *psoA* is

12024 bp in length, and encodes a protein having multiple functional domains including those conferring putative acyltransferase, dehydratase, methyltransferase, ketoreductase, acyl carrier, thiolation and reductase activities [47]. Phage based recombineering of BAC09b06, which spans the entire *psoA* gene, could therefore facilitate the rapid deletion of DNA regions which encode distinct activities. This type of targeted mutational approach was used by Hahn and Stachelhaus to generate multiple mutations in the C-terminus of the prokaryotic PKS TycA in order to demonstrate the presence of short communication-mediating (COM) domains [54].

An important caveat to consider with this method is that the BAC clones from the *A. fumigatus* library are derived from the Af293 strain, so genetic replacements in other genetic backgrounds will carry any sequence polymorphisms from the original strain. In fact, comparative genomic analysis of A1163 and Af293 *A. fumigatus* isolates identified the number of unique genes in each genome as up to 2% of total genomic cohorts [55]. It is therefore important, when working in alternative genetic backgrounds to scrutinise/moderate the region of replaced sequence to mitigate the introduction of polymorphisms.

In our study phenotypic analyses of the newly constructed mutants was performed in parallel with a previously constructed

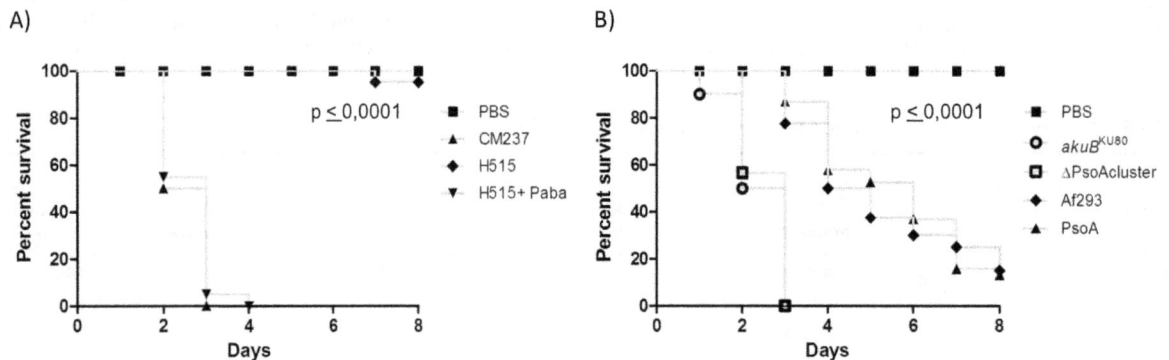

Figure 6. Survival of *G. mellonella* **infected with** *A. fumigatus.* A) Survival following infection with CM237 or H515, +4 µg/ml PABA. P value corresponds to comparison of survival rate between CM237 and H515 infected larvae. B) Survival following infection with wild type (CEA17_$\Delta akuB^{KU80}$ ($akuB^{KU80}$) and Af293) or mutant strains (PsoA and ΔPsoAcluster); P value corresponds to comparison of survival rate between CEA17_$\Delta akuB^{KU80}$ and Af293. A p value<0.01 was considered significant.

ΔpsoA strain [47] which lacks the hybrid non-ribosomal-polyketide synthase PsoA (AFUA_8G00540), and its progenitor the Af293 parental strain (Table 1). Radial growth analyses revealed no growth defects amongst the PsoA cluster mutants (data not shown). Analysis of pseurotin production revealed that pseurotin biosynthesis was completely abrogated in mutants lacking the biosynthetic gene cluster (Figure 4) and further, that integrity of AFUA_8G00550 is required for pseurotin biosynthesis. Mass spectrometry conditions for Pseurotin A and internal standards are listed in Table 5. This result agrees with a recent study in which a role for the AFUA_8G00550 gene product in O-methylation of an intermediate metabolite has been demonstrated, and where elimination of this methylating activity limited the synthesis of pseurotin A [53]. In contrast, deletion of the AFUA_8G00520 gene did not eliminate pseurotin A biosynthesis compared to a congenic parental isolate.

Macrophage-mediated phagocytosis of *A. fumigatus,* and host cell damage

Genes of the pseurotin A biosynthetic gene cluster are co-ordinately upregulated, relative to laboratory culture, during initiation of murine infection [9]. This observation, coupled with reported activity of pseurotin A as an inducer of nerve-cell differentiation [48] and suppressor of immunoglobulin E production [49] prompted us to determine whether the absence of pseurotin A had any impact on the mammalian host response to *A. fumigatus* challenge. Since macrophages and lung epithelial cells constitute the main initial immunological barriers to *A. fumigatus* infection [56] the role of pseurotin A as a cytotoxic molecule was tested in these two cell types. Murine macrophages (RAW 264.7) and human lung epithelial cells (A549) were co-incubated with spores from either CEA17_$\Delta akuB^{KU80}$ or pseurotin-deficient strains. For analysis of phagocytosis by macrophages, the proportion of unphagocytosed *A. fumigatus* spores was calculated, after two hours of host and pathogen co-incubation (Figure 5A). This analysis revealed no differences between pseurotin A-producing and non-producing isolates in either of the *A. fumigatus* CEA17_$\Delta akuB^{KU80}$ or Af293 genetic backgrounds, however a significant difference in phagocytosis of isolates derived from CEA17_$\Delta akuB^{KU80}$ and Af293 genetic backgrounds was discernable (Figure 5A). Relative cytotoxicity, to A549 epithelial cells, of pseurotin A-producing and non-producing isolates *A. fumigatus* was determined by release of LDH (Figure 5B). Again, no significant impact of pseurotin A upon epithelial cell lysis was measurable in our assays; however, we observed that *A. fumigatus* CEA17_$\Delta akuB^{KU80}$ isolate was reproducibly more resistant than Af293 to phagocytosis by macrophages, and more cytotoxic to epithelial cells (Figure 5B).

Screening of pathogenicity of *A. fumigatus* strains in *G. mellonella*

In keeping with the diverse bioactivity of fungal secondary metabolites, nematocidal properties of pseurotin A have previously been reported [57]. A precedent for the biosynthetic products of *A. fumigatus* gene clusters to moderate the outcome of host-pathogen interactions has recently been demonstrated, where deletion of the NRP synthetase Pes3 (AFUA_5G12730) significantly increases the virulence of *A. fumigatus* in wax moth larvae [58]. To test the role of pseurotin A in pathogenicity towards invertebrate hosts we utilized an established *G. mellonella* wax moth infection assay. This model has emerged as a useful alternative to mammalian infection assays for the study of fungal

virulence and pathogenesis, fungus-host interactions, and antifungal drug efficacy [42,59].

All *A. fumigatus* strains, with the exception of the H515 PABA auxotroph (Figure 6A) were able to kill the larvae at 37°C following injection of conidia into the larval haemocoel (Figure 6). In keeping with observations in murine models of infection, the H515 PABA auxotroph became pathogenic when administered in combination with exogenous PABA. Relative to the CEA17_$\Delta akuB^{KU80}$ strain or Af293 strain, no significant differences in larval mortality were found following injection of pseurotin non-producing isolates (Figure 6B). Our data suggest that the pseurotin A gene cluster is not required for *A. fumigatus* virulence in *G. mellonella*. A full analysis in the mammalian host (under way in our laboratory) will be required to dismiss the involvement of this secondary metabolite in mammalian virulence.

We observed differential pathogenicity traits in Af293 and CEA17_$\Delta akuB^{KU80}$ which extend to pathogenicity in invertebrate hosts (Figure 6B). Such strain-dependent variance of the host response to *A. fumigatus* has been also recently reported for CEA10 (which is the CEA17_$\Delta akuB^{KU80}$ progenitor) which elicits a stronger inflammatory response, based on the cytokine secretion profile of *Aspergillus*-stimulated dentritic cells, compared to Af293 [60].

Conclusions

We have developed a new protocol based on *E. coli* recombineering methodology to efficiently target genes and gene clusters in *A. fumigatus*. Advantages of this system.

include: (i) a single PCR is required for construction of gene replacement cassettes; (ii) maximisation of flanking regions promotes efficient sequence replacement in *A. fumigatus;* (iii) the approach works well in wild-type clinical isolates. Our methodology significantly expands the toolkit available for manipulation of *A. fumigatus* gene clusters.

Supporting Information

Figure   S1 A) Schematic representation of gene AFUA_1G17640 (*reg*A) replacement by BSM-Z/P cassette in *A. fumigatus* CEA17_$\Delta akuB^{KU80}$. (B) Schematic representation of the *reg*A region following a homologous recombination event with the BAC CC_e12.9 disruption cassette. A NotI restriction site is introduced by the BSM-Z/P cassette. (C) Southern blot analyses of CEA17_$\Delta akuB^{KU80}$ (lane 1) and *Δreg*A transformant (lane 2) gDNA which was digested with SacI. Blots were probed with a 600 bp fragment of the zeocin cassette. *reg*A deletion is indicated by a single SacI fragment of 12.5 Kb observed for *Δreg*A gDNA. (D) Gel electrophoresis image of diagnostic PCR to confirm *reg*A gene deletion. Primers 5′_640 and 3′_640 were used to amplify the *reg*A region from CEA17_$\Delta akuB^{KU80}$ and putative *Δreg*A transformant gDNA. PCR amplicons from CEA17_$\Delta akuB^{KU80}$ and *Δreg*A were NotI digested and analysed using gel electrophoresis, which demonstrated wild-type banding patterns in the CEA17_$\Delta akuB^{KU80}$ (lane 1) and introduction of the NotI site by the zeocin cassette in the transformant strain (lane 2). PCR using primers internal to the *reg*A coding sequence demonstrated a product of the expected 1.55 kb size from CEA17_$\Delta akuB^{KU80}$ gDNA template (lane 3) but no product from *Δreg*A transformant (lane 4), indicating *Δreg*A gene replacement. (TIF)

Figure S2 A) Schematic view of AFUA_8G00520 replacement by BSM-A/H cassette in *A. fumigatus* CEA17_$\Delta akuB^{KU80}$. B and C) Southern blot analysis of AFUA_8G00520 deleted mutant and wild type (WT) strains. Expected hybridization band pattern: (1) 5042 bp for WT, and (2, 3, 4) 6724 bp for the mutants. (TIF)

Figure S3   A) Schematic view of AFUA_8G00550 replacement by BSM-A/H cassette in *A. fumigatus* CEA17_*ΔakuB*^{KU80}. B and C) Southern blot analysis of AFUA_8G00550 deleted mutant and wild type (WT) strains. Expected hybridization band pattern: (1, 2, 3) 3772 bp for the mutants and (4) 756 bp for WT. (TIF)

Table S1   BAC library. (XLS)

File S1   BAC clones sequences. (GZ)

Protocol S1   Protocol for recombineering. (DOCX)

Author Contributions

Conceived and designed the experiments: LAF EB. Performed the experiments: LAF TC BZ SP. Analyzed the data: LAF TC BZ SP EB. Contributed reagents/materials/analysis tools: PB YI JFL DB. Contributed to the writing of the manuscript: LAF TC EB.

References

1. Dean RA (2007) Fungal gene clusters. Nat Biotechnol 25: 67. nbt0107-67 [pii];10.1038/nbt0107-67 [doi].
2. Keller NP, Hohn TM (1997) Metabolic Pathway Gene Clusters in Filamentous Fungi. Fungal Genet Biol 21: 17–29. FG970970 [pii].
3. Keller NP, Turner G, Bennett JW (2005) Fungal secondary metabolism - from biochemistry to genomics. Nat Rev Microbiol 3: 937–947. nrmicro1286 [pii];10.1038/nrmicro1286 [doi].
4. MacCabe AP, van LH, Palissa H, Unkles SE, Riach MB, et al. (1991) Delta-(L-alpha-aminoadipyl)-L-cysteinyl-D-valine synthetase from *Aspergillus nidulans*. Molecular characterization of the acvA gene encoding the first enzyme of the penicillin biosynthetic pathway. J Biol Chem 266: 12646–12654.
5. Brakhage AA (2013) Regulation of fungal secondary metabolism. Nat Rev Microbiol 11: 21–32. nrmicro2916 [pii];10.1038/nrmicro2916 [doi].
6. Yu JH, Leonard TJ (1995) Sterigmatocystin biosynthesis in *Aspergillus nidulans* requires a novel type I polyketide synthase. J Bacteriol 177: 4792–4800.
7. Brakhage AA, Schroeckh V (2011) Fungal secondary metabolites - strategies to activate silent gene clusters. Fungal Genet Biol 48: 15–22. S1087-1845(10)00068-X [pii];10.1016/j.fgb.2010.04.004 [doi].
8. Palmer JM, Keller NP (2010) Secondary metabolism in fungi: does chromosomal location matter? Curr Opin Microbiol 13: 431–436. S1369-5274(10)00060-3 [pii];10.1016/j.mib.2010.04.008 [doi].
9. McDonagh A, Fedorova ND, Crabtree J, Yu Y, Kim S, et al. (2008) Sub-telomere directed gene expression during initiation of invasive aspergillosis. PLoS Pathog 4: e1000154. 10.1371/journal.ppat.1000154 [doi].
10. Bumpus SB, Evans BS, Thomas PM, Ntai I, Kelleher NL (2009) A proteomics approach to discovering natural products and their biosynthetic pathways. Nature Biotechnology 27: 951–U120.
11. Chiou CH, Miller M, Wilson DL, Trail F, Linz JE (2002) Chromosomal location plays a role in regulation of aflatoxin gene expression in *Aspergillus parasiticus*. Applied and Environmental Microbiology 68: 306–315.
12. Liang SH, Wu TS, Lee R, Chu FS, Linz JE (1997) Analysis of mechanisms regulating expression of the ver-1 gene, involved in aflatoxin biosynthesis. Applied and Environmental Microbiology 63: 1058–1065.
13. Gacek A, Strauss J (2012) The chromatin code of fungal secondary metabolite gene clusters. Applied Microbiology and Biotechnology 95: 1389–1404.
14. Bok JW, Chung D, Balajee SA, Marr KA, Andes D, et al. (2006) GliZ, a transcriptional regulator of gliotoxin biosynthesis, contributes to *Aspergillus fumigatus* virulence. Infect Immun 74: 6761–6768. IAI.00780-06 [pii];10.1128/IAI.00780-06 [doi].
15. Sugui JA, Pardo J, Chang YC, Mullbacher A, Zarember KA, et al. (2007) Role of laeA in the Regulation of alb1, gliP, Conidial Morphology, and Virulence in *Aspergillus fumigatus*. Eukaryot Cell 6: 1552–1561. EC.00140-07 [pii];10.1128/EC.00140-07 [doi].
16. Bok JW, Balajee SA, Marr KA, Andes D, Nielsen KF, et al. (2005) LaeA, a regulator of morphogenetic fungal virulence factors. Eukaryot Cell 4: 1574–1582. 4/9/1574 [pii];10.1128/EC.4.9.1574-1582.2005 [doi].
17. Perrin RM, Fedorova ND, Bok JW, Cramer RA, Wortman JR, et al. (2007) Transcriptional regulation of chemical diversity in *Aspergillus fumigatus* by LaeA. PLoS Pathog 3: e50. 06-PLPA-RA-0390R2 [pii];10.1371/journal.ppat.0030050 [doi].
18. Amich J, Leal F, Calera JA (2009) Repression of the acid ZrfA/ZrfB zinc-uptake system of *Aspergillus fumigatus* mediated by PacC under neutral, zinc-limiting conditions. Int Microbiol 12: 39–47. im2306120 [pii].
19. da Silva Ferreira ME, Kress MR, Savoldi M, Goldman MH, Hartl A, et al. (2006) The akuB(KU80) mutant deficient for nonhomologous end joining is a powerful tool for analyzing pathogenicity in *Aspergillus fumigatus*. Eukaryot Cell 5: 207–211. 5/1/207 [pii];10.1128/EC.5.1.207-211.2006 [doi].
20. Hartmann T, Dumig M, Jaber BM, Szewczyk E, Olbermann P, et al. (2010) Validation of a self-excising marker in the human pathogen *Aspergillus fumigatus* by employing the beta-rec/six site-specific recombination system. Appl Environ Microbiol 76: 6313–6317. AEM.00882-10 [pii];10.1128/AEM.00882-10 [doi].
21. Krappmann S, Sasse C, Braus GH (2006) Gene targeting in *Aspergillus fumigatus* by homologous recombination is facilitated in a nonhomologous end-joining-deficient genetic background. Eukaryot Cell 5: 212–215. 5/1/212 [pii];10.1128/EC.5.1.212-215.2006 [doi].
22. Krappmann S (2006) Tools to study molecular mechanisms of *Aspergillus* pathogenicity. Trends Microbiol 14: 356–364. S0966-842X(06)00150-8 [pii];10.1016/j.tim.2006.06.005 [doi].
23. Kamper J (2004) A PCR-based system for highly efficient generation of gene replacement mutants in *Ustilago maydis*. Mol Genet Genomics 271: 103–110. 10.1007/s00438-003-0962-8 [doi].
24. Chiang YM, Oakley CE, Ahuja M, Entwistle R, Schultz A, et al. (2013) An Efficient System for Heterologous Expression of Secondary Metabolite Genes in *Aspergillus nidulans*. J Am Chem Soc 135: 7720–7731. 10.1021/ja401945a [doi].
25. Chaveroche MK, Ghigo JM, d'Enfert C (2000) A rapid method for efficient gene replacement in the filamentous fungus *Aspergillus nidulans*. Nucleic Acids Res 28: E97.
26. Langfelder K, Gattung S, Brakhage AA (2002) A novel method used to delete a new *Aspergillus fumigatus* ABC transporter-encoding gene. Curr Genet 41: 268–274. 10.1007/s00294-002-0313-z [doi].
27. Chan W, Costantino N, Li R, Lee SC, Su Q, et al. (2007) A recombineering based approach for high-throughput conditional knockout targeting vector construction. Nucleic Acids Res 35: e64. gkm163 [pii];10.1093/nar/gkm163 [doi].
28. Pain A, Woodward J, Quail MA, Anderson MJ, Clark R, et al. (2004) Insight into the genome of *Aspergillus fumigatus*: analysis of a 922 kb region encompassing the nitrate assimilation gene cluster. Fungal Genet Biol 41: 443–453. 10.1016/j.fgb.2003.12.003 [doi];S1087184503002111 [pii].
29. Penalva MA, Tilburn J, Bignell E, Arst HN Jr (2008) Ambient pH gene regulation in fungi: making connections. Trends Microbiol 16: 291–300. S0966-842X(08)00090-5 [pii];10.1016/j.tim.2008.03.006 [doi].
30. Pontecorvo G, Roper JA, Hemmons LM, Macdonald KD, Bufton AW (1953) The genetics of *Aspergillus nidulans*. Adv Genet 5: 141–238.
31. Cove DJ (1966) The induction and repression of nitrate reductase in the fungus *Aspergillus nidulans*. Biochim Biophys Acta 113: 51–56.
32. Vieira J, Messing J (1982) The pUC plasmids, an M13mp7-derived system for insertion mutagenesis and sequencing with synthetic universal primers. Gene 19: 259–268.
33. Kubodera T, Yamashita N, Nishimura A (2000) Pyrithiamine resistance gene (ptrA) of *Aspergillus oryzae*: cloning, characterization and application as a dominant selectable marker for transformation. Biosci Biotechnol Biochem 64: 1416–1421.
34. Szewczyk E, Krappmann S (2010) Conserved regulators of mating are essential for *Aspergillus fumigatus* cleistothecium formation. Eukaryot Cell 9: 774–783. EC.00375-09 [pii];10.1128/EC.00375-09 [doi].
35. Brown JS, Aufauvre-Brown A, Brown J, Jennings JM, Arst H Jr, et al. (2000) Signature-tagged and directed mutagenesis identify PABA synthetase as essential for *Aspergillus fumigatus* pathogenicity. Mol Microbiol 36: 1371–1380.
36. Liu P, Jenkins NA, Copeland NG (2003) A highly efficient recombineering-based method for generating conditional knockout mutations. Genome Res 13: 476–484. 10.1101/gr.749203 [doi].
37. Szewczyk E, Nayak T, Oakley CE, Edgerton H, Xiong Y, et al. (2006) Fusion PCR and gene targeting in *Aspergillus nidulans*. Nat Protoc 1: 3111–3120. nprot.2006.405 [pii];10.1038/nprot.2006.405 [doi].
38. Hervas-Aguilar A, Rodriguez JM, Tilburn J, Arst HN Jr, Penalva MA (2007) Evidence for the direct involvement of the proteasome in the proteolytic processing of the *Aspergillus nidulans* zinc finger transcription factor PacC. J Biol Chem 282: 34735–34747. M706723200 [pii];10.1074/jbc.M706723200 [doi].
39. Southern EM (1975) Detection of specific sequences among DNA fragments separated by gel electrophoresis. J Mol Biol 98: 503–517.
40. Igarashi Y, Yabuta Y, Sekine A, Fujii K, Harada K, et al. (2004) Directed biosynthesis of fluorinated pseurotin A, synerazol and gliotoxin. J Antibiot (Tokyo) 57: 748–754.
41. Tang CM, Cohen J, Krausz T, Van NS, Holden DW (1993) The alkaline protease of *Aspergillus fumigatus* is not a virulence determinant in two murine models of invasive pulmonary aspergillosis. Infect Immun 61: 1650–1656.

42. Mesa-Arango AC, Forastiero A, Bernal-Martinez L, Cuenca-Estrella M, Mellado E, et al. (2013) The non-mammalian host *Galleria mellonella* can be used to study the virulence of the fungal pathogen *Candida tropicalis* and the efficacy of antifungal drugs during infection by this pathogenic yeast. Med Mycol 51: 461–472. 10.3109/13693786.2012.737031 [doi].

43. Slater JL, Gregson L, Denning DW, Warn PA (2011) Pathogenicity of *Aspergillus fumigatus* mutants assessed in *Galleria mellonella* matches that in mice. Med Mycol 49 Suppl 1: S107–S113. 10.3109/13693786.2010.523852 [doi].

44. Eliahu N, Igbaria A, Rose MS, Horwitz BA, Lev S (2007) Melanin biosynthesis in the maize pathogen *Cochliobolus heterostrophus* depends on two mitogen-activated protein kinases, Chk1 and Mps1, and the transcription factor Cmr1. Eukaryot Cell 6: 421–429. EC.00264-06 [pii];10.1128/EC.00264-06 [doi].

45. Tsuji G, Kenmochi Y, Takano Y, Sweigard J, Farrall L, et al. (2000) Novel fungal transcriptional activators, Cmr1p of *Colletotrichum lagenarium* and pig1p of *Magnaporthe grisea*, contain Cys2His2 zinc finger and Zn(II)2Cys6 binuclear cluster DNA-binding motifs and regulate transcription of melanin biosynthesis genes in a developmentally specific manner. Mol Microbiol 38: 940–954. mmi2181 [pii].

46. Palmer JM, Mallaredy S, Perry DW, Sanchez JF, Theisen JM, et al. (2010) Telomere position effect is regulated by heterochromatin-associated proteins and NkuA in *Aspergillus nidulans*. Microbiology 156: 3522–3531. mic.0.039255-0 [pii];10.1099/mic.0.039255-0 [doi].

47. Maiya S, Grundmann A, Li X, Li SM, Turner G (2007) Identification of a hybrid PKS/NRPS required for pseurotin A biosynthesis in the human pathogen *Aspergillus fumigatus*. Chembiochem 8: 1736–1743. 10.1002/cbic.200700202 [doi].

48. Komagata D, Fujita S, Yamashita N, Saito S, Morino T (1996) Novel neuritogenic activities of pseurotin A and penicillic acid. Journal of Antibiotics 49: 958–959.

49. Ishikawa M, Ninomiya T, Akabane H, Kushida N, Tsujiuchi G, et al. (2009) Pseurotin A and its analogues as inhibitors of immunoglobuline E production. Bioorganic & Medicinal Chemistry Letters 19: 1457–1460.

50. Vodisch M, Scherlach K, Winkler R, Hertweck C, Braun HP, et al. (2011) Analysis of the *Aspergillus fumigatus* Proteome Reveals Metabolic Changes and the Activation of the Pseurotin A Biosynthesis Gene Cluster in Response to Hypoxia. Journal of Proteome Research 10: 2508–2524.

51. Brock M, Jouvion G, Droin-Bergere S, Dussurget O, Nicola MA, et al. (2008) Bioluminescent *Aspergillus fumigatus*, a new tool for drug efficiency testing and in vivo monitoring of invasive aspergillosis. Appl Environ Microbiol 74: 7023–7035. AEM.01288-08 [pii];10.1128/AEM.01288-08 [doi].

52. Grahl N, Puttikamonkul S, Macdonald JM, Gamcsik MP, Ngo LY, et al. (2011) In vivo hypoxia and a fungal alcohol dehydrogenase influence the pathogenesis of invasive pulmonary aspergillosis. PLoS Pathog 7: e1002145. 10.1371/journal.ppat.1002145 [doi];PPATHOGENS-D-11-00223 [pii].

53. Tsunematsu Y, Fukutomi M, Saruwatari T, Noguchi H, Hotta K, et al. (2014) Elucidation of Pseurotin Biosynthetic Pathway Points to Trans-Acting C-Methyltransferase: Generation of Chemical Diversity. Angew Chem Int Ed Engl 53: 1–6.

54. Hahn M, Stachelhaus T (2004) Selective interaction between nonribosomal peptide synthetases is facilitated by short communication-mediating domains. Proceedings of the National Academy of Sciences of the United States of America 101: 15585–15590.

55. Fedorova ND, Khaldi N, Joardar VS, Maiti R, Amedeo P, et al. (2008) Genomic islands in the pathogenic filamentous fungus *Aspergillus fumigatus*. PLoS Genet 4: e1000046. 10.1371/journal.pgen.1000046 [doi].

56. Latge JP (1999) *Aspergillus fumigatus* and aspergillosis. Clinical Microbiology Reviews 12: 310–350.

57. Hayashi A, Fujioka S, Nukina M, Kawano T, Shimada A, et al. (2007) Fumiquinones A and B, nematicidal quinones produced by *Aspergillus fumigatus*. Biosci Biotechnol Biochem 71(7): 1697–1702.

58. O'Hanlon KA, Cairns T, Stack D, Schrettl M, Bignell EM, et al. (2011) Targeted Disruption of Nonribosomal Peptide Synthetase pes3 Augments the Virulence of *Aspergillus fumigatus*. Infection and Immunity 79: 3978–3992.

59. Gomez-Lopez A, Forastiero A, Cendejas-Bueno E, Gregson L, Mellado E, et al. (2014) An invertebrate model to evaluate virulence in *Aspergillus fumigatus*: The role of azole resistance. Med Mycol 52(3): 311–9. myt022 [pii];10.1093/mmy/myt022 [doi].

60. Rizzetto L, Giovannini G, Bromley M, Bowyer P, Romani L, et al. (2013) Strain Dependent Variation of Immune Responses to *A-fumigatus*: Definition of Pathogenic Species. Plos One 8(2): e56651.

2

The Relevance of External Quality Assessment for Molecular Testing for *ALK* Positive Non-Small Cell Lung Cancer: Results from Two Pilot Rounds Show Room for Optimization

Lien Tembuyser[1], Véronique Tack[1], Karen Zwaenepoel[2], Patrick Pauwels[2], Keith Miller[3], Lukas Bubendorf[4], Keith Kerr[5], Ed Schuuring[6], Erik Thunnissen[7], Elisabeth M. C. Dequeker[1]*

1 Department of Public Health and Primary Care, Biomedical Quality Assurance Research Unit, KU Leuven – University of Leuven, Leuven, Belgium, 2 Department of Pathology, Antwerp University Hospital, Edegem, Belgium, 3 UK NEQAS ICC & ISH, London, United Kingdom, 4 Institute for Pathology, Basel University Hospital, Basel, Switzerland, 5 Department of Pathology, Aberdeen Royal Infirmary, Aberdeen, United Kingdom, 6 Department of Pathology and Medical Biology, University of Groningen, Groningen, the Netherlands, 7 Department of Pathology, VU University Medical Center, Amsterdam, The Netherlands

Abstract

Background and Purpose: Molecular profiling should be performed on all advanced non-small cell lung cancer with non-squamous histology to allow treatment selection. Currently, this should include *EGFR* mutation testing and testing for *ALK* rearrangements. *ROS1* is another emerging target. *ALK* rearrangement status is a critical biomarker to predict response to tyrosine kinase inhibitors such as crizotinib. To promote high quality testing in non-small cell lung cancer, the European Society of Pathology has introduced an external quality assessment scheme. This article summarizes the results of the first two pilot rounds organized in 2012–2013.

Materials and Methods: Tissue microarray slides consisting of cell-lines and resection specimens were distributed with the request for routine *ALK* testing using IHC or FISH. Participation in *ALK* FISH testing included the interpretation of four digital FISH images.

Results: Data from 173 different laboratories was obtained. Results demonstrate decreased error rates in the second round for both *ALK* FISH and *ALK* IHC, although the error rates were still high and the need for external quality assessment in laboratories performing *ALK* testing is evident. Error rates obtained by FISH were lower than by IHC. The lowest error rates were observed for the interpretation of digital FISH images.

Conclusion: There was a large variety in FISH enumeration practices. Based on the results from this study, recommendations for the methodology, analysis, interpretation and result reporting were issued. External quality assessment is a crucial element to improve the quality of molecular testing.

Editor: Ramon A. de Mello, University of Algarve, Portugal

Funding: This work was supported by an unrestricted grant from Pfizer: grant number ZL520506, www.pfizer.com/. The funders had no role in study design, data collection and analysis, decision to publish, or preparation of the manuscript.

Competing Interests: Patrick Pauwels received a scientific grant from Pfizer and receives occasional speaker fees from Pfizer, Ventana (Roche) and Abbott. Lukas Bubendorf receives consultation and lecture fees from Roche and Pfizer. Keith Kerr receives consultation and lecture fees from Pfizer, Novartis and Abbott. Ed Schuuring has board membership (Roche, Pfizer and Novartis) and receives lecture fees (Roche and Abbott). Erik Thunnissen received an unrestricted grant from Pfizer and travel and speaker's fee for presentation (Pfizer). Elisabeth M. C. Dequeker received an unrestricted grant from Pfizer and remuneration from AstraZeneca. The other authors have no competing interests to declare.

* Email: els.dequeker@med.kuleuven.be

Introduction

Lung cancer is amongst the leading causes of cancer related mortality worldwide [1]. Approximately 85% of lung cancers are non-small cell lung cancers (NSCLC), traditionally divided into three major cell types: adenocarcinoma, squamous cell carcinoma and large cell carcinoma [2]. Over the past decade, the availability of molecular targeted therapies has increased the progression-free survival for patients with NSCLC, adenocarcinoma in particular [3–6].

The approach of using biomarkers to select treatments that are tailored to individual patient profiles is referred to as precision medicine. In advanced NSCLC, *EGFR* gene mutations and *ALK* rearrangements are currently critical biomarkers to predict treatment response. The fusion protein from *ROS1* rearrangement is an emerging target.

In 2007, it was first reported that an inversion on chromosome 2p resulted in the creation of an *EML4-ALK* fusion gene in lung cancer [7]. Multiple *EML4-ALK* variants, represented by different *EML4* breakpoints, have been identified, as well as other fusion partners for *ALK*, such as *KIF5B* and *TFG* [8–10]. *ALK* rearrangements result in oncogenic fusions which lead to constitutive activity of the *ALK* tyrosine kinase with subsequent effects on proliferation, migration and survival [11]. Lung cancers harboring *ALK* rearrangements represent a unique subpopulation of lung cancer patients. The frequency of the *EML4-ALK* rearrangement ranges from 2% to 7% in unselected NSCLC patients [3,12]. The frequency is higher in NSCLC patients with adenocarcinoma histology, non or light cigarette smoking history, and younger age, regardless of ethnicity [3,12,13]. However, these clinical characteristics are not shared by all carriers and molecular characterization is necessary to determine treatment eligibility [3,14,15].

ALK rearrangements are pharmacologically targetable with the small molecule tyrosine kinase inhibitor (TKI) crizotinib. In 2011, the FDA granted accelerated approval of crizotinib in response to the manifested clinical benefit.

Routine molecular diagnostics need to include evaluations for both *EGFR* mutations and *ALK* rearrangements [13,15,16]. It is expected that testing for *ROS1* rearrangements will be included soon. *ROS1* is another receptor tyrosine kinase that forms fusions in NSCLC and has shown responsiveness to crizotinib [17]. Diagnostic testing laboratories have been expected to rapidly introduce and perform molecular testing for NSCLC. For successful patient treatment, it is of great importance that molecular test results are accurate, highly reliable, and presented in a timely fashion. In 2012, the European Society of Pathology (ESP) proposed an external quality assessment (EQA) scheme to promote high quality biomarker testing in NSCLC for *EGFR* mutation analysis and *ALK* rearrangement detection. From 2014 on, *ROS1* testing is also included. The scheme aims to assess and improve the current status of molecular testing in NSCLC, to provide education and remedial measures, to permit inter-laboratory comparison and to allow validation of test methods by distributing validated material harboring well-defined aberrations. For *EGFR*, EQA results have been reported [18]. This article summarizes the results of the two *ALK* testing pilot rounds of the ESP Lung EQA scheme, organized in 2012–2013 with the purpose to reflect the current status of *ALK* rearrangement testing practices and to issue recommendations for the improvement of testing quality.

Materials and Methods

A pilot EQA scheme consisting of two rounds was set up. Tissue microarray (TMA) slides that consisted of NSCLC cell-lines and resection specimens were distributed. Three expert laboratories (University of Groningen, the Netherlands, UK NEQAS ICC & ISH, United Kingdom and VU University Medical Center, Amsterdam) provided material for this EQA program. All patient samples were leftover tissues that were obtained as part of routine care and testing from the three laboratories mentioned above and then handed over to the researchers anonymously. These laboratories signed a statement that the patient material was obtained according to the national legal requirements for the use of patient samples. Informed consent is not a mandatory prerequisite for the use of patient derived material, since samples for test validation are exempt from research regulations requiring informed consent. The treating physician was responsible to obtain informed consent from the patients to use their tissues and

data for research purposes and this consent is kept in the patient's medical file. The authors had no contact with the patients or received any patient identifying information. The samples were to be analyzed for the presence of *ALK* rearrangements using IHC or FISH. In addition to the TMA slides, participation in *ALK* FISH included the interpretation of four digital FISH images which were accessible online. The *ALK* FISH digital cases were provided in close collaboration with UK NEQAS ICC & ISH.

In both rounds, mock clinical information was provided for several cases, for which the delivery of a full written report was requested. Report content was assessed in agreement with established standards/guidelines on reporting [19–21].

A central database was used for the submission of the results, accessible through the ESP Lung EQA Scheme website. Through their personal account, participants could access their results, scheme documentation and assessor feedback. Upon registration, each laboratory was assigned a unique EQA identity number to guarantee anonymity. A team of medical and technical experts supported the validation of the samples and the evaluation of the scheme results. Results were discussed during an assessment meeting to obtain final consensus scores. Participants received individual feedback and a general report with aggregated scheme results.

The set-up of both rounds slightly differed and the scheme was a pilot for the development and standardization of homogeneous testing material. Eight samples (four resection specimens and four cell lines) and twelve samples (six resection specimens and six cell lines) were prepared and send to the participants for respectively the first and second round. Different cell lines either with or without *ALK* break were routinely fixed with neutral-buffered formalin, mixed with agar and embedded in paraffin (reflecting routine pathology tissue block) were included. Results from samples for which less than 75% of the participants were able to obtain a result were not taken into account to assess performance [22]. Consequently, for *ALK* FISH, 3/8 and 7/12 samples were regarded as educational samples for respectively the first and second round. For the assessment, the accepted cases were two resection specimens and three cell lines for the first round of *ALK* FISH, and for the second round, five resection specimens were included. For *ALK* IHC, all samples were approved in both rounds.

For *ALK* FISH, it was requested for each case to report the number of neoplastic cell nuclei without hybridization signals, the number of neoplastic nuclei with fused signal, with split signal and with a single red signal. An algorithm automatically generated the number of neoplastic nuclei with FISH signal, the number of neoplastic nuclei with split or single red, and the fraction of FISH positive and negative nuclei. Participants were asked to then determine the outcome of the *ALK* FISH test (positive/negative). Samples for which a laboratory did not obtain results due to sample quality or technical failures were not assessed. For *ALK* FISH TMA and *ALK* FISH Digital, error rates were calculated based on the samples for which a minimum of 50 nuclei were enumerated, in order to exclude uninformative cases. This paper does not emphasize the marking criteria and assigned scores, as the scoring criteria slightly differed for the pilot rounds, but the study aims to reflect the current status of *ALK* rearrangement testing practices in molecular pathology laboratories.

For *ALK* IHC, it was requested to use the H-score procedure as described by Ruschoff et al. [23]. This modified H-score procedure is educational as it gives a better understanding of the *ALK* IHC sensitivity and reliability [14]. In the first round, a cut-off IHC score of 32 was determined by the mean score plus standard deviation from the laboratories on the IHC negative

Table 1. Laboratory characteristics.

Laboratory characteristics		Round 1		Round 2	
		N	% (n = 67)	N	% (n = 149)
Laboratory settings	Community Hospital	30	48%	80	54%
	University Hospital	24	36%	30	20%
	University	6	9%	17	11%
	Private	3	4%	16	11%
	Industry	2	3%	3	2%
	Private Hospital	0	0%	2	1%
	Other	2	3%	1	1%
Clinical tests offered on lung biopsy samples in routine diagnostics	ALK rearrangement detection	56	84%	110	74%
	EGFR mutation detection	/	/	128	86%
	KRAS mutation detection	/	/	112	75%
Analysis performed under the authority of the department of pathology	Yes	61	91%	116	78%
	No	6	9%	33	22%
Tests most frequently ordered on lung biopsies	EGFR	/	/	132	89%
	ALK	/	/	49	33%
	KRAS	/	/	44	30%
	None of the above	/	/	7	5%

specimens (except outliers with H-score >100). The same threshold for positivity/negativity was applied in the second round.

For the statistical analysis, scheme error rates from both rounds for FISH digital and FISH TMA were compared using the Mann-Whitney U test. Scheme error rates for IHC were compared using an unpaired t-test. The level of significance was set at $\alpha = 0,05$.

Results

In total, 173 different laboratories (primarily from European Union countries) participated in the pilot rounds. In the first round, 29 laboratories submitted results for *ALK* IHC, 55 for *ALK* FISH TMA, and 67 laboratories performed the interpretation of the digital *ALK* FISH images. In the second round, 58 laboratories submitted results for *ALK* IHC, 104 for *ALK* FISH TMA, and 106 for the *ALK* FISH digital cases. For the data-analysis, missing values were ignored and only valid answers to questions were included, which explains why sample sizes slightly differ. Laboratory characteristics are listed in Table 1. The total number of laboratories that provided information was used as the denominator to calculate percentages. Because it was sometimes possible to indicate more than one answer, percentages may not add up to 100%.

The majority of the participants were set in a community hospital or university hospital environment. The analysis was mostly performed under the authority of the department of pathology. Regarding the interpretation of *ALK* FISH, a pathologist was most frequently involved (23% and 27% of laboratories in the first and second round, respectively), in some cases assisted by a scientist (18% and 27%) or a technician (20%

and 13%). A scientist alone performed the FISH reading in 15% of the laboratories in both the first and second round. The final reading conclusion was the responsibility of a pathologist alone in more than half of the laboratories (52% and 59% in the first and second round). A pathologist in cooperation with a scientist was responsible in 13% and 14% of the laboratories. A scientist alone was responsible for the final reading conclusion in 16% and 12% of the laboratories in the first and second round.

ALK FISH digital results

Results for both rounds of the *ALK* FISH digital subscheme are summarized in Table 2. There were no clear differences in the error rates depending on the number of nuclei enumerated. Enumeration practices were evaluated on sample level and on laboratory level. In both rounds, the bulk of the participants enumerated 50–100 nuclei for each case. At laboratory level, in the first round, 34/67 laboratories (51%) counted ≥50 cells for each sample. In the second round, an increase was observed to 77/106 (73%). Table 3 illustrates the performance of the labs that participated in both rounds for each subscheme. Improvement in enumeration practices was defined as enumeration of ≥50 cells in a larger number of samples in the second round compared to the first round.

A decrease was observed in the error rates between both rounds (error rates were calculated taking only the samples for which ≥50 nuclei were enumerated into account). In the first round, 7 out of 195 scored samples were incorrectly assigned (3,6%), while in the second round, 4 errors out of 366 scored samples (1,1%) occurred. The comparison of the number of errors made for laboratories

Table 2. Results for the interpretation of digital *ALK* FISH images.

Round	Sample ID	*ALK* rearrangement status	Number of labs that counted <50 nuclei	Number of labs that counted 50–100 nuclei	Number of labs that counted >100 nuclei	Number of FP/FN from labs that counted <50 nuclei	Number of FP/FN from labs that counted 50–100 nuclei	Number of FP/FN from labs that counted >100 nuclei
Round 1	LUNG12.109	negative	12 (18%)	38 (58%)	16 (24%)	1 (8%)	0 (0%)	1 (6%)
	LUNG12.110	negative	21 (32%)	43 (65%)	2 (3%)	0 (0%)	0 (0%)	0 (0%)
	LUNG12.111	positive	15 (23%)	45 (70%)	4 (6%)	0 (0%)	0 (0%)	0 (0%)
	LUNG12.112	positive	13 (22%)	37 (62%)	10 (17%)	3 (23%)	6 (16%)	0 (0%)
	Average		15,3 (23,8%)	40,8 (63,8%)	8,0 (12,5%)	1,0 (7,75%)	1,5 (4,0%)	0,3 (1,5%)
Round 2	LUNG12.213	negative	19 (18%)	74 (70%)	11 (10%)	0 (0%)	0 (0%)	0 (0%)
	LUNG12.214	positive	11 (11%)	69 (65%)	24 (23%)	0 (0%)	0 (0%)	0 (0%)
	LUNG12.215	negative	10 (10%)	66 (62%)	27 (25%)	0 (0%)	1 (2%)	2 (7%)
	LUNG12.216	positive	9 (9%)	68 (64%)	27 (25%)	0 (0%)	1 (1%)	0 (0%)
	Average		12,3 (12,0%)	69,3 (65,3%)	22,3 (20,75%)	0 (0%)	0,5 (0,8%)	0,5 (1,8%)

FP, false positives; FN, false negatives.

that participated in both rounds and that counted ≥50 nuclei for each case can be found in Table 3.

ALK FISH TMA results

Table 4 summarizes the *ALK* FISH TMA results for both rounds. In both rounds, the majority of the participants enumerated 50–100 nuclei for each case. Again, there were only small differences in the error rates depending on the number of nuclei enumerated. In the first round, 30/55 laboratories (55%) counted ≥50 cells for each sample; in the second round there was an increase to 81/104 (78%).

For the TMA cases there was also a decrease in the error rates between the two rounds. In the first round, 14 out of 193 scored samples were incorrectly assigned (7,3%), while in the second round, 22 errors out of 423 scored samples (5,2%) occurred. Comparison of the enumeration performance and number of errors made for laboratories participating in both rounds can be found in Table 3.

ALK IHC results

Results for both *ALK* IHC rounds are provided in Table 5. In the first round, 30/230 scored cases (13,0%) were incorrectly called (false positive or false negative). In the second round, a decrease to 44/540 (8,2%) was observed. Table 3 illustrates the performance for the labs that participated in both IHC rounds.

Summary scheme error rates

The Mann-Whitney U test revealed no significant differences between both rounds for Digital FISH (U = 7, z = −0.308, p = 0.758) or FISH TMA (U = 9, z = −0.731, p = 0.465). For IHC, an unpaired t-test showed no significant difference between the first (M = 0.13, SD = 0.06) and second round (M = 0.08, SD = 0.05); t(18) = 1.845, p = 0.082. Although not statistically significant, comparing the error rates in both rounds suggests a learning effect (Table 6). The smallest error rates were observed for the digital cases, assessing only the post-analytical interpretation phase. In both rounds, the error rate for *ALK* FISH TMA was lower than the error rate for *ALK* IHC TMA.

Methods used

The most frequently used method for FISH analysis was the Vysis *ALK* break apart FISH probe kit (Abbott Molecular, Illinois, USA), used by over 70% of the participants. For IHC, the most frequently used antibodies were clone 5A4 and clone D5F3 for the first and second round, respectively. An overview of the used methods and the error rate per method is provided in Tables 7 and 8. For the percentage of laboratories that used a certain method, the total number of laboratories that provided information was used as the denominator. Because it was possible to indicate more than one used method, percentages may not add up to 100%.

For *ALK* FISH, the Repeat-Free Poseidon ALK/EML4 t(2;2) inv(2) Fusion Probe (Kreatech Diagnostics, Amsterdam, the Netherlands) revealed a high error rate of 50% in the second round (Table 7). For IHC, the smallest error rates were observed for clones 5A4 and D5F3 (Table 8).

Clinical result reporting

Evaluation of the written reports for the second round (n = 102) showed that a case-specific clinical interpretation was missing in 74% and 79% of the reports for an *ALK* positive and *ALK* negative case, respectively. Patient name and date of birth were correctly present in the majority of the reports, as well as a

Table 3. Performance of laboratories that participated in both rounds.

Subscheme	Nr of labs with improved enumeration performance	Nr of labs with equal enumeration performance	Nr of labs with worse enumeration performance	Nr of labs with a decrease in nr of errors	Nr of labs with equal nr of errors	Nr of labs with higher nr of errors
ALK FISH Digital	14/39 (36%)	22/39 (56%)	3/39 (8%)	1/37 (3%)	36/37 (97%)	0/37 (0%)
ALK FISH TMA	13/35 (37%)	17/35 (49%)	5/35 (14%)	7/29 (24%)	20/29 (69%)	2/29 (7%)
ALK IHC	/	/	/	3/14 (21%)	9/14 (64%)	2/14 (14%)

specification of the methods used (FISH kit information or IHC antibody). However, a specification of the aberrations tested and the threshold of the method were not mentioned in 46% and 47% of the FISH reports. The total number of neoplastic cells analyzed and the number of cells with split and/or single signal were missing in 23% of the FISH reports. For IHC, the threshold for positivity/negativity was not defined in 81% of the reports, and the staining intensity was missing in 39% of the reports.

Discussion

Major advances have been made in the management of patients with NSCLC, with improved treatment response and survival following the introduction of molecular targeted TKI therapies focusing on *EGFR* mutations and *ALK* rearrangements. The increasing importance of morphology-based studies such as IHC or FISH has made the pathologist's involvement a key element in precision medicine for NSCLC [2,24].

In reply to the growing demands, laboratories have introduced molecular testing for NSCLC in routine diagnostics. Regular participation in quality assurance programs is crucial to ensure a high quality of testing service and to warrant patient safety [15,18,19].

Our results show that in the majority of the participating laboratories, *ALK* testing is performed under the authority of the pathology department. This is a necessity as FISH and IHC are both histological tests. Pathology review and assessment of section quality is essential considering the diversity and heterogeneity of tumor tissue [19], as false negatives may be due to poor fixation or insufficient neoplastic cell content [14,18].

Three methods are generally used in routine diagnostics for *ALK* rearrangement detection: FISH, RT-PCR, and immunohistochemistry for aberrant expression of ALK protein [12,14,25]. Importantly, every assay should undergo validation in the laboratory before clinical interpretation and should be subject to regular internal and external quality controls [14,19,24]. The FDA approved test to determine *ALK* status is the Vysis LSI *ALK* dual color, break apart rearrangement probe (Abbott Molecular, Illinois, USA) [12,14]. Although other IVD-CE labeled kits are available in Europe, this kit was by far the most frequently used method in both pilot rounds. *ALK* break apart (or split-signal) probes detect disruption of the *ALK* 2p23 locus but do not identify the partner fusion gene [3,25]. Surprisingly, the *ALK/EML4* fusion probe is still occasionally used, although these probes miss the translocation of *ALK* with partners other than *EML4*. In the second round, the fusion probe revealed a high error rate of 50%. The cut-off values used during the clinical trials to prove the efficacy of crizotinib can be transferred from the Vysis probe to other break apart probes since the design (size + location) is highly similar [26]. The ZytoLight TriCheck (ZytoVision, Bremerhaven, Germany), used by approximately 7% of the participants in both

rounds, can identify the presence of an *ALK* rearrangement and if the rearrangement partner is *EML4*. Today it is still under discussion whether it is important to know the fusion partner of *ALK* in relation to expected response to *ALK* TKIs [27,28].

According to Abbott Molecular scoring criteria, a nucleus is considered positive if it contains at least one split signal or one isolated red signal. A first enumerator should count 50 nuclei. Cases with >50% and <10% positive nuclei are considered positive and negative respectively. If a sample shows between 10–50% positive nuclei, a second enumerator should also count 50 nuclei. If the average of the two readings contains at least 15% positive cells, the sample is considered positive. The kit specifies uninformative specimens as those in which fewer than 50 nuclei within the scribed area can be enumerated. In our evaluation these cases were therefore not included to calculate and compare the scheme error rates. It has been shown that the sensitivity and specificity of the kit increase as the number of tumor areas and number of nuclei scored increase [29,30]. Our results showed that *ALK* rearrangement status was often determined on the evaluation of less than 50 nuclei by many participants. The percentage of false positive and false negative results upon enumeration of <50 nuclei did not reveal a clear difference compared to the percentage upon enumeration of ≥50 tumor nuclei. These findings correlate with the fact that the *ALK* rearrangement appears to be a homogenous event in the tumor population [26,29], Enumeration of <50 nuclei is not advisable because this number is based on the minimal number that is statistically needed to be able to reliably define a sample without FISH break signals (<15% of nuclei) as a case without *ALK* rearrangement. In addition, the predictive value of phase III trial is based on this. Remarkably, some participants enumerated a large number of nuclei (e.g. >600 evaluated nuclei for case 12.215), which is not a requirement for daily practice. FISH interpretation should be performed in areas of the slide with clear signals, which are clearly distinct from the nuclear fluorescent 'noise' as well as from the background [15]. Importantly, selection of neoplastic nuclei is essential, and to this end sufficient morphological knowledge in FISH stained slides is obligatory, which stresses involvement of a pathologist.

It is not a surprise that the TMA FISH error rates were substantially higher than those of the digital FISH images. The FISH digital subscheme specifically assesses the interpretation of identical digital images whereas the TMA FISH error rate also incorporates variation in serial TMA sections, technical execution, and reading. Suboptimal *ALK* FISH procedure may lead to a low signal versus background ratio, increasing the chance for interpretation errors.

Although FISH is used as a standard test, it demonstrates considerable inter-observer variability. Therefore, experienced (> 100 cases/year) and well-trained FISH reviewers/enumerators are necessary. If the clinical scientist is well trained and experienced in histo- and cytomorphology with specialized training in solid tumor

Table 4. ALK FISH TMA results.

Round	Sample ID	ALK rearrangement status	Number of labs that counted <50 nuclei	Number of labs that counted 50–100 nuclei	Number of labs that counted >100	Number of FP/FN from labs that counted <50 nuclei	Number of FP/FN from labs that counted 50–100 nuclei	Number of FP/FN from labs that counted >100 nuclei
Round 1	LUNG12.101	positive	11(22%)	37 (74%)	2 (4%)	2 (18%)	4 (11%)	0 (0%)
	LUNG12.105	positive	11 (26%)	26 (62%)	5 (12%)	2 (18%)	6 (23%)	1 (20%)
	LUNG12.106	negative	10 (19%)	34 (64%)	9 (17%)	1 (10%)	1 (3%)	0 (0%)
	LUNG12.107	positive	11 (21%)	34 (64%)	8 (15%)	0 (0%)	2 (6%)	0 (0%)
	LUNG12.108	negative	10 (20%)	34 (68%)	6 (12%)	0 (0%)	0 (0%)	0 (0%)
	Average		10,6 (21,6%)	33,0 (66,4%)	6,0 (12,0%)	1 (9,2%)	2,6 (8,6%)	0,2 (4,0%)
Round 2	LUNG12.202	positive	10 (11%)	46 (51%)	34 (38%)	1 (10%)	14 (30%)	1 (3%)
	LUNG12.204	negative	8 (9%)	48 (52%)	37 (40%)	0 (0%)	1 (2%)	2 (5%)
	LUNG12.206	negative	7 (8%)	56 (60%)	30 (32%)	0 (0%)	0 (0%)	1 (3%)
	LUNG12.208	positive	6 (6%)	55 (57%)	36 (37%)	0 (0%)	1 (2%)	1 (3%)
	LUNG12.210	negative	6 (7%)	48 (53%)	36 (40%)	0 (0%)	0 (0%)	1 (3%)
	Average		7,4% (8,2%)	50,6 (54,6%)	34,6 (37,4%)	0,2 (2,0%)	3,2 (6,8%)	1,2 (3,4%)

FP, false positives; FN, false negatives.

Table 5. *ALK* IHC results.

Round	Sample ID	ALK rearrangement status	Average IHC score	Standard deviation IHC score	Number of FP/FN
Round 1	LUNG 12.101	positive	154	53	2 (7%)
	LUNG 12.102	positive	113	73	6 (22%)
	LUNG 12.103	negative	15	51	2 (7%)
	LUNG 12.104	negative	24	49	5 (17%)
	LUNG 12.105	positive	118	61	3 (10%)
	LUNG 12.106	negative	25	53	6 (21%)
	LUNG 12.107	positive	154	62	2 (7%)
	LUNG 12.108	negative	13	32	4 (14%)
	Average	/	/	/	3,8 (13,1%)
Round 2	LUNG12.201	negative	18	52	5 (12%)
	LUNG12.202	negative	16	51	5 (11%)
	LUNG12.203	negative	5	22	2 (5%)
	LUNG12.204	negative	6	23	2 (4%)
	LUNG12.205	negative	18	47	7 (16%)
	LUNG12.206	negative	1	4	0 (0%)
	LUNG12.207	negative	9	34	4 (9%)
	LUNG12.208	positive	163	58	4 (9%)
	LUNG12.209	positive	137	69	7 (16%)
	LUNG12.210	negative	2	11	1 (2%)
	LUNG12.211	positive	143	68	6 (13%)
	LUNG12.212	negative	1	6	1 (2%)
	Average	/	/	/	3,7 (8,3%)

FP, false positives; FN, false negatives.

FISH analysis, he/she can be responsible for the technical performance and molecular interpretation. A pathologist should at least be responsible for the selection of the right cells, the review of the interpretation and the authorization of the pathology report [14,15]. Our data demonstrate that a pathologist was responsible for the final conclusion in the majority of the laboratories. Participating laboratories indicated that scientists and technicians were often involved in FISH enumeration. In this set-up it is important that the clinical scientist can consult a pathologist at any time in case of doubt concerning the location of the tumor cell area.

ALK IHC, if carefully clinically validated according to ISO 15189, may be considered as a screening method to select specimens for *ALK* FISH testing [15]. It is a cost-effective screening tool which correlated significantly with *ALK* FISH, using a number of antibodies including the 5A4 and D5F3 [25,31–33]. However, discrepancies are reported also and need to be elucidated [34]. The 5A4 and D5F3 antibodies were the most frequently used clones in our study and revealed the smallest error rates, which is in accordance with literature [32,33] and the findings of a recent NordiQC assessment [35]. Not surprisingly however, the error rates for IHC were greater than for FISH. Recently different validation projects for *ALK* IHC tests were done in collaboration with a lot of laboratories [36]. Moreover, on the website of NORDIQC (http://www.nordiqc.org/), advice on IHC staining protocols is given for several antibody clones.

Table 6. Scheme error rates per subscheme for both pilot rounds.

Subscheme	Error rate round 1*		Error rate round 2*	
Digital *ALK* FISH	3,6%	0,5% due to FP	1,1%	0,8% due to FP
		3,1% due to FN		0,3% due to FN
ALK FISH TMA	7,3%	0,5% due to FP	5,2%	4,0% due to FP
		6,7% due to FN		1,2% due to FN
ALK IHC	13,0%	7,4% due to FP	8,2%	5,0% due to FP
		5,6% due to FN		3,2% due to FN

*Error rate = (number of FP + number of FN)/total number of informative results FP, false positives; FN, false negatives.

Table 7. *ALK* FISH methods used in the scheme and error rate per probe.

ALK FISH rearrangement detection method	Round 1				Round 2			
	Percentage of labs that used the probe (n=60)	Number of FP/FN obtained with the probe	Total number of samples analysed with the probe	Percentage of samples resulting in a FP/FN	Percentage of labs that used the probe (n=103)	Number of FP/FN obtained with the probe	Total number of samples analysed with the probe	Percentage of samples resulting in a FP/FN
ALK FISH DNA Probe, Split Signal (Dako, Glostrup, Denmark)	12%	3	29	10,3%	11%	0	38	0,0%
ZytoVision (Bremerhaven, Germany), not further specified	/	/	/	/	2%	1	9	11,1%
Master Diagnostica (Granada, Spain)	/	/	/	/	1%	/	/	/
Repeat-Free Poseidon ALK (2p23) Break Probe (Kreatech Diagnostics, Amsterdam, the Netherlands)	2%	1	5	20,0%	3%	0	10	0,0%
Repeat-Free Poseidon ALK/EML4 t(2;2) inv(2) Fusion Probe (Kreatech Diagnostics, Amsterdam, the Netherlands)	7%	0	16	0,0%	1%	2	4	50,0%
Vysis ALK break apart FISH probe kit (Abbott Molecular, Illinois, USA)	82%	11	157	7,0%	78%	16	311	5,1%
ZytoLight SPEC ALK Dual Color Break Apart Probe (ZytoVision, Bremerhaven, Germany)	/	/	/	/	1%	0	4	0,0%
ZytoLight SPEC ALK/EML4 TriChec Probe (ZytoVision, Bremerhaven, Germany)	7%	0	14	0,0%	7%	2	30	6,7%

FP, false positives; FN, false negatives.

Table 8. Antibodies used for *ALK* IHC in the scheme and error rate per antibody.

ALK IHC antibody	Round 1				Round 2			
	Percentage of labs that used the antibody (n = 22)	Number of FP/FN obtained with the antibody	Total number of samples analysed with the antibody	Percentage of samples resulting in a FP/FN	Percentage of labs that used the antibody (n = 62)	Number of FP/FN obtained with the antibody	Total number of samples analysed with the antibody	Percentage of samples resulting in a FP/FN
5A4	64%	6	104	5,8%	40%	5	185	2,7%
D5F3	14%	1	16	6,3%	44%	30	291	10,3%
ALK1	18%	9	32	28,1%	15%	7	35	20,0%
SP8	9%	8	16	50,0%	2%	1	11	9,1%
4C5B8	/	/	/	/	2%	1	6	16,7%

FP, false positives; FN, false negatives.

Our study demonstrates improvement of *ALK* testing after only two EQA rounds. This suggests that laboratories constructively use the assessors' feedback from the previous round to enhance their performance. Participation in EQA facilitates rapid exposure of errors and the timely implementation of corrective and preventive actions. However, other factors such as increased expertise and experience may play a part. It is expected that larger datasets, spanning a larger number of EQA participations will demonstrate a statistically significant improvement. On scheme level, the error rates for both *ALK* FISH and *ALK* IHC were lower in the second round and the *ALK* FISH digital scheme demonstrated an error rate of only 1,1%. Error rates for *ALK* FISH TMA and *ALK* IHC were still high (>5%), which stresses the need for continued education through EQA. Progress was also seen on individual laboratory level. For FISH analysis, improvements were observed both in the number of errors made and in enumeration practices.

Reporting of test results should take into account sample adequacy relative to the assay performance characteristics and limitations, and clinical reports should be readily interpretable by non-expert clinicians [19,21]. Previous EQA schemes have exposed existing deficiencies in clinical reporting [18,37]. Our results show that the content of reports for *ALK* rearrangement detection should be improved. Especially, a case-specific clinical interpretation, predicting the effect of the rearrangement status on therapy response, should be integrated in each report since a clear and concise assessment of the clinical implications of the result is crucial to fully inform treatment options.

Maintenance of quality assurance measures, including stringent internal quality controls and continued education by repeated EQA participations is essential to ensure high testing quality and rapid exposure of errors in order to warrant appropriate treatment choices. This article has demonstrated improvement in the performance of *ALK* FISH and *ALK* IHC in two consecutive EQA rounds. Several recommendations were made to improve the quality of *ALK* testing.

Acknowledgments

The following laboratories were responsible for the preparation and validation of the samples: Carola Andersson (Sweden), Lukas Bubendorf (Switzerland), Keith Kerr (United Kingdom), Keith Miller (United Kingdom), Patrick Pauwels (Belgium), Ed Schuuring, Lorian Slagter and Rianne Pelgrim (The Netherlands), Erik Thunnissen (The Netherlands). We thank Sofie Delen for her help in the assessment of the scheme results. We also thank the laboratories that participated in the pilot rounds and we thank the administrative office of the European Society of Pathology for their assistance.

Author Contributions

Conceived and designed the experiments: LT KM ES ET ED. Performed the experiments: LT KM ES ET ED. Analyzed the data: LT VT KZ PP KK KM LB ES ET ED. Contributed reagents/materials/analysis tools: LT KM ES ET ED. Contributed to the writing of the manuscript: LT VT PP KZ ED.

References

1. Jemal A, Bray F, Center MM, Ferlay J, Ward E, et al. (2011) Global cancer statistics. CA Cancer J Clin 61: 69–90.
2. Cagle PT, Allen TC, Olsen RJ (2013) Lung cancer biomarkers: present status and future developments. Arch Pathol Lab Med 137: 1191–1198.
3. Kwak EL, Bang YJ, Camidge DR, Shaw AT, Solomon B, et al. (2010) Anaplastic lymphoma kinase inhibition in non-small-cell lung cancer. N Engl J Med 363: 1693–1703.
4. Stinchcombe TE, Socinski MA (2008) Gefitinib in advanced non-small cell lung cancer: does it deserve a second chance? Oncologist 13: 933–944.
5. Shepherd FA, Rodrigues Pereira J, Ciuleanu T, Tan EH, Hirsh V, et al. (2005) Erlotinib in previously treated non-small-cell lung cancer. N Engl J Med 353: 123–132.

6. Reck M, Heigener DF, Mok T, Soria JC, Rabe KF (2013) Management of non-small-cell lung cancer: recent developments. Lancet 382: 709–719.

7. Soda M, Choi YL, Enomoto M, Takada S, Yamashita Y, et al. (2007) Identification of the transforming EML4-ALK fusion gene in non-small-cell lung cancer. Nature 448: 561–566.

8. Choi YL, Takeuchi K, Soda M, Inamura K, Togashi Y, et al. (2008) Identification of novel isoforms of the EML4-ALK transforming gene in non-small cell lung cancer. Cancer Res 68: 4971–4976.

9. Takeuchi K, Choi YL, Togashi Y, Soda M, Hatano S, et al. (2009) KIF5B-ALK, a novel fusion oncokinase identified by an immunohistochemistry-based diagnostic system for ALK-positive lung cancer. Clin Cancer Res 15: 3143–3149.

10. Rikova K, Guo A, Zeng Q, Possemato A, Yu J, et al. (2007) Global survey of phosphotyrosine signaling identifies oncogenic kinases in lung cancer. Cell 131: 1190–1203.

11. Dacic S (2013) Molecular genetic testing for lung adenocarcinomas: a practical approach to clinically relevant mutations and translocations. J Clin Pathol 66: 870–874.

12. Mino-Kenudson M, Mark EJ (2011) Reflex testing for epidermal growth factor receptor mutation and anaplastic lymphoma kinase fluorescence in situ hybridization in non-small cell lung cancer. Arch Pathol Lab Med 135: 655–664.

13. Sasaki T, Janne PA (2011) New strategies for treatment of ALK-rearranged non-small cell lung cancers. Clin Cancer Res 17: 7213–7218.

14. Thunnissen E, Bubendorf L, Dietel M, Elmberger G, Kerr K, et al. (2012) EML4-ALK testing in non-small cell carcinomas of the lung: a review with recommendations. Virchows Arch 461: 245–257.

15. Lindeman NI, Cagle PT, Beasley MB, Chitale DA, Dacic S, et al. (2013) Molecular testing guideline for selection of lung cancer patients for EGFR and ALK tyrosine kinase inhibitors: guideline from the College of American Pathologists, International Association for the Study of Lung Cancer, and Association for Molecular Pathology. J Mol Diagn 15: 415–453.

16. Mescam-Mancini L, Lantuejoul S, Moro-Sibilot D, Rouquette I, Souquet PJ, et al. (2014) On the relevance of a testing algorithm for the detection of ROS1-rearranged lung adenocarcinomas. Lung Cancer 83: 168–173.

17. Thunnissen E, van der Oord K, den Bakker M (2014) Prognostic and predictive biomarkers in lung cancer. A review. Virchows Arch 464: 347–358.

18. Deans ZC, Bilbe N, O'Sullivan B, Lazarou LP, de Castro DG, et al. (2013) Improvement in the quality of molecular analysis of EGFR in non-small-cell lung cancer detected by three rounds of external quality assessment. J Clin Pathol 66: 319–325.

19. van Krieken JH, Normanno N, Blackhall F, Boone E, Botti G, et al. (2013) Guideline on the requirements of external quality assessment programs in molecular pathology. Virchows Arch 462: 27–37.

20. International Organization for Standardization (2012) ISO 15189: 2012 Medical laboratories - Particular requirements for quality and competence. Geneva, ISO.

21. Gulley ML, Braziel RM, Halling KC, Hsi ED, Kant JA, et al. (2007) Clinical laboratory reports in molecular pathology. Arch Pathol Lab Med 131: 852–863.

22. Thunnissen E, Bovee JV, Bruinsma H, van den Brule AJ, Dinjens W, et al. (2011) EGFR and KRAS quality assurance schemes in pathology: generating normative data for molecular predictive marker analysis in targeted therapy. J Clin Pathol 64: 884–892.

23. Ruschoff J, Dietel M, Baretton G, Arbogast S, Walch A, et al. (2010) HER2 diagnostics in gastric cancer-guideline validation and development of standardized immunohistochemical testing. Virchows Arch 457: 299–307.

24. Garrido P, de Castro J, Concha A, Felip E, Isla D, et al. (2012) Guidelines for biomarker testing in advanced non-small-cell lung cancer. A national consensus of the Spanish Society of Medical Oncology (SEOM) and the Spanish Society of Pathology (SEAP). Clin Transl Oncol 14: 338–349.

25. Selinger CI, Rogers TM, Russell PA, O'Toole S, Yip P, et al. (2013) Testing for ALK rearrangement in lung adenocarcinoma: a multicenter comparison of immunohistochemistry and fluorescent in situ hybridization. Mod Pathol 26: 1545–1553.

26. Zwaenepoel K, Van Dongen A, Lambin S, Weyn C, Pauwels P (2014) Detection of ALK expression in NSCLC with ALK-gene rearrangements - comparison of multiple IHC methods. Histopathology. In press.

27. Heuckmann JM, Balke-Want H, Malchers F, Peifer M, Sos ML, et al. (2012) Differential protein stability and ALK inhibitor sensitivity of EML4-ALK fusion variants. Clin Cancer Res 18: 4682–4690.

28. Crystal AS, Shaw AT (2012) Variants on a theme: a biomarker of crizotinib response in ALK-positive non-small cell lung cancer? Clin Cancer Res 18: 4479–4481.

29. Camidge DR, Kono SA, Flacco A, Tan AC, Doebele RC, et al. (2010) Optimizing the detection of lung cancer patients harboring anaplastic lymphoma kinase (ALK) gene rearrangements potentially suitable for ALK inhibitor treatment. Clin Cancer Res 16: 5581–5590.

30. Yoshida A, Tsuta K, Nitta H, Hatanaka Y, Asamura H, et al. (2011) Bright-field dual-color chromogenic in situ hybridization for diagnosing echinoderm microtubule-associated protein-like 4-anaplastic lymphoma kinase-positive lung adenocarcinomas. J Thorac Oncol 6: 1677–1686.

31. Han XH, Zhang NN, Ma L, Lin DM, Hao XZ, et al. (2013) Immunohistochemistry reliably detects ALK rearrangements in patients with advanced non-small-cell lung cancer. Virchows Arch 463: 583–591.

32. Conklin CM, Craddock KJ, Have C, Laskin J, Couture C, et al. (2013) Immunohistochemistry is a reliable screening tool for identification of ALK rearrangement in non-small-cell lung carcinoma and is antibody dependent. J Thorac Oncol 8: 45–51.

33. Mino-Kenudson M, Chirieac LR, Law K, Hornick JL, Lindeman N, et al. (2010) A novel, highly sensitive antibody allows for the routine detection of ALK-rearranged lung adenocarcinomas by standard immunohistochemistry. Clin Cancer Res 16: 1561–1571.

34. Cabillic F, Gros A, Dugay F, Begueret H, Mesuroux L, et al. (2014) Parallel FISH and Immunohistochemical Studies of ALK Status in 3244 Non-Small-Cell Lung Cancers Reveal Major Discordances. J Thorac Oncol 9: 295–306.

35. Nordic Immunohistochemical Quality Control. Lung Anaplastic Lymphoma Kinase Assessment Run 39 2013. Available: http://www.nordiqc.org/Run-39-B16-H4/Assessment/Run39_ALK.pdf. Accessed 2014 Apr 1.

36. Blackhall FH, Peters S, Bubendorf L, Dafni U, Hager H, et al. (2014) Prevalence and Clinical Outcomes for Patients With ALK-Positive Resected Stage I to III Adenocarcinoma: Results From the European Thoracic Oncology Platform Lungscape Project. J Clin Oncol 32: 2780–2787.

37. Tembuyser L, Ligtenberg MJ, Normanno N, Delen S, van Krieken JH, et al. (2014) Higher Quality of Molecular Testing, an Unfulfilled Priority: Results from External Quality Assessment for KRAS Mutation Testing in Colorectal Cancer. J Mol Diagn 16: 371–377.

3

Regulation of the Orphan Nuclear Receptor *Nr2f2* by the DFNA15 Deafness Gene *Pou4f3*

Chrysostomos Tornari, Emily R. Towers, Jonathan E. Gale, Sally J. Dawson*

UCL Ear Institute, University College London, London, United Kingdom

Abstract

Hair cells are the mechanotransducing cells of the inner ear that are essential for hearing and balance. POU4F3 – a POU-domain transcription factor selectively expressed by these cells – has been shown to be essential for hair cell differentiation and survival in mice and its mutation in humans underlies late-onset progressive hearing loss (DFNA15). The downstream targets of POU4F3 are required for hair cell differentiation and survival. We aimed to identify such targets in order to elucidate the molecular pathways involved in hair cell production and maintenance. The orphan thyroid nuclear receptor *Nr2f2* was identified as a POU4F3 target using a subtractive hybridization strategy and EMSA analysis showed that POU4F3 binds to two sites in the *Nr2f2* 5′ flanking region. These sites were shown to be required for POU4F3 activation as their mutation leads to a reduction in the response of an *Nr2f2* 5′ flanking region reporter construct to POU4F3. Immunocytochemistry was carried out in the developing and adult inner ear in order to investigate the relevance of this interaction in hearing. NR2F2 expression in the postnatal mouse organ of Corti was shown to be detectable in all sensory epithelia examined and characterised. These data demonstrate that *Nr2f2* is a direct target of POU4F3 *in vitro* and that this regulatory relationship may be relevant to hair cell development and survival.

Editor: Berta Alsina, Universitat Pompeu Fabra, Spain

Funding: This work was funded by grants from: Action on hearing loss, http://actiononhearingloss.org.uk/ (grant number 483 to JEG and SJD); the Wellcome Trust, http://www.wellcome.ac.uk/ (grant number 091092/Z/09/Z to JEG and SJD); and the original subtractive hybridisation screen was supported by a grant from the Wellcome Trust (grant number 064599 to SJD). CT's research was undertaken as part of a studentship that was funded by GlaxoSmithKline (http://www.gsk.com/uk) via the UCL MB PhD programme (http://www.ucl.ac.uk/mbphd). The funders had no role in study design, data collection and analysis, decision to publish, or preparation of the manuscript.

Competing Interests: The authors have read the journal's policy and the authors of this manuscript have the following competing interests: CT's research was undertaken as part of a studentship that was funded by a commercial funder, GlaxoSmithKline (http://www.gsk.com/uk) via the UCL MB PhD programme (http://www.ucl.ac.uk/mbphd). GlaxoSmithKline had no role in study design, data collection and analysis, decision to publish, or preparation of the manuscript and have no intellectual property involved in the research.

* Email: sally.dawson@ucl.ac.uk

Introduction

POU4F3 is a member of the POU family of transcription factors that regulate a wide array of neuroendocrine developmental pathways [1]. This family is characterised by the presence of a bipartite DNA binding domain known as the POU domain which comprises a POU-homeodomain and a POU-specific domain separated by a linker [2]. All of these components are required for sequence-specific DNA binding [3,4].

Though known to be expressed in the retina, spinal cord, dorsal root ganglia and Merkel cells, the investigation of POU4F3 expression and function has focused on its role in the development and maintenance of hair cells – the sensory cells of the inner ear [5–8]. Homozygous *Pou4f3* mutant mice demonstrate deafness and balance deficits due to the failure of nascent hair cells – cells that express hair cell markers – to develop into morphologically recognisable hair cells and subsequent death of these cells [6–9]. Furthermore, heterozygous mutation of human *POU4F3* is the cause of autosomal dominant late-onset progressive hearing loss (DFNA15) in several families [10–13].

As the molecular pathways involved in differentiation and maintenance of hair cells are largely unknown, identification of the downstream targets of POU4F3 would greatly improve current understanding of the transcription factors and target genes that mediate hair cell differentiation and maturation. Such an understanding may also aid in the explanation of the mechanism by which *POU4F3* mutation causes hearing loss in humans.

To date, the only known direct targets of POU4F3 are BDNF, NT-3 and Caprin-1 [14,15], though the expression of other genes is strongly associated with POU4F3 expression [14,16,17]. In previous work we identified Caprin-1 as a direct target of POU4F3 in a subtractive hybridization screen that was carried out in UB/OC-2 cells – a conditionally immortal cell line, derived from the developing mouse inner ear sensory epithelium, that displays expression of hair cell markers [18]– that were manipulated to differentially express POU4F3 [15]. The Caprin-1 5′ flanking sequence contains POU4F3 binding sites which mediate repression of Caprin-1 expression by POU4F3 [15]. We showed that Caprin-1, which is known to promote stress granule formation [19–21], is expressed in both the hair cells and supporting cells of the organ of Corti as well as being involved in the hair cell response to stress [15].

The same subtractive hybridization screen also returned the orphan thyroid nuclear receptor *Nr2f2* as a putative target of

POU4F3. *Nr2f2* is an orphan steroid/thyroid hormone nuclear receptor which is expressed in a range of organs in the developing embryo [22] as well as adult tissues [23]. Its most essential roles in development are in angiogenesis, heart development and remodelling the primitive capillary plexus into large and small microcapillaries; as demonstrated by knocking out the *Nr2f2* gene in mice. This change results in lethality at around E10 that is likely to be due to haemorrhage and oedema in the brain and heart [24]. Further study of the function of NR2F2 in the vascular system has revealed its importance in designating venous identity via its suppression of the Notch signalling pathway. In addition to its previously reported roles in generation of venous identity, these studies show a role for NR2F2 in cell fate determination [25].

Despite having a general role in embryogenesis, NR2F2 also demonstrates more specific actions in organogenesis. In the brain, it regulates cell migration [26]; it is essential for correct stomach patterning [27]; and has been proposed to be involved in dorso-ventral patterning of the mammalian retina [28]. Both NR2F2 and its closely related family member NR2F1 have been shown to be variably expressed in the developing mouse inner ear. NR2F2 is expressed in the distal tip of the elongating cochlear epithelium at E10 to E13.5; homogeneous expression is seen across cell types of both the greater and lesser epithelial ridges (GER & LER) in the apical coil at E15.5; and predominantly LER expression is seen in the apical-to-middle coil at the same age [29].

In contrast to NR2F2, NR2F1 expression in the developing inner ear between E14.5 and E15.5 becomes more confined to the GER in the apical cochlea. In the basal cochlea, its expression extends to the LER and the level of expression appears to decrease [29]. This decrease in basal expression correlates with the wave of hair cell differentiation from base to apex [30]. Consistent with a role in cochlear development, $Nr2f1^{-/-}$ mice display shortened cochlear ducts with an increased number of hair cells in the mid-to-apical cochlear turns; an anomaly which appears consistent with a role in Notch regulation of organ of Corti differentiation [31].

Though NR2F2 has been implicated in a number of developmental processes, its expression is known to be maintained in postnatal vertebrates. Post-natal NR2F2 expression is perhaps best-characterised in the liver where it is known to downregulate the apolipoprotein A1 (apoA1) gene [32]. However, its expression is maintained in the adult mouse brain and in a wide array of postnatal human tissue types, suggesting additional uncharacterised roles in cell maintenance [23,33].

In this paper, we present evidence for the direct regulation of *Nr2f2* by POU4F3. Furthermore, we show that NR2F2 expression is maintained postnatally in the mouse cochlear sensory epithelium. These data add to our understanding of the downstream effects of POU4F3 signalling and widen the array of possible functions of NR2F2 to the postnatal inner ear.

Materials and Methods

Subtractive hybridization

POU4F3-regulated genes were identified as previously described [15]. Briefly, two populations of UB/OC-2 cells [18] were created. In one population, POU4F3 expression was increased by stable transfection of a *Pou4f3* expression construct. In the other, POU4F3 expression was reduced by stable transfection of an antisense *Pou4f3* construct. cDNA was prepared from these two cell populations and used in a subtractive hybridization screen using the PCR-Select cDNA Subtraction kit (Clontech). Differentially expressed cDNA sequences identified in this analysis were verified by a series of hybridization experiments with cDNA from the original analysis and virtual northern blot experiments with cDNA from transiently transfected cells to increase the stringency of the analysis [15]. Clones that displayed differential expression were selected for further analysis.

BLAST analysis

Following identification of differentially expressed transcripts by subtractive hybridization, cloned cDNA was subjected to Sanger sequencing. This was followed by BLASTN analysis of transcript sequences against the ENSEMBL mouse cDNA database (NCBI m35) with a 'near-exact match' stringency. Matches were confirmed by alignment of clone sequence to the cDNA sequence in the ENSEMBL database using the BioEdit software package [34].

Nr2f2 5' flanking region Genomatix analysis

The Genomatix Gene2Promoter software [35] was used to identify predicted *Nr2f2* promoters for analysis. The promoter that corresponded to the transcript that best matched the *Nr2f2* transcript identified in the BLAST analysis was selected for further investigation (GXP_158616). To identify putative POU4F3 binding sites in this predicted promoter, the selected region was interrogated using the Genomatix MatInspector [36,37] and ModelInspector [38] programs. Binding sites for analysis were selected on the basis of search stringency, proximity to the predicted transcriptional start site and promoter architecture at the relevant site. Two predicted binding sites were selected for functional analysis and were named POU recognition element-1 (PRE1) and POU recognition element-2 (PRE2).

Electrophoretic mobility-shift assay (EMSA) analysis

Both *in vitro* translated protein and UB/OC-2 cell nuclear protein extract were used. *Pou4f3* was cloned into the pGEM-T Easy vector (Promega) under the control of the T7 or SP6 promoter and *in vitro* translated protein was generated using either the TNT-T7 Coupled Reticulocyte Lysate System (Promega) or the TNT-SP6 Quick Coupled Transcription/Translation System (Promega).

Double-stranded probe oligonucleotides were labelled in a standard T4 kinase reaction with γ^{32}P γATP (GE Healthcare) using 50 ng probe per reaction. Assay reactions containing 10 μl 2x Parker buffer (16% Ficoll, 40 mM HEPES at pH 7.9, 100 mM KCl, 2 mM EDTA at pH 8.0 and 1 mM DTT), 3 μg poly(dI·dC), 2–4 μl nuclear protein extract or *in vitro* translated protein and 0–1000 ng non-radiolabelled competition probe were made up to a total volume of 20 μl and incubated on ice for ten minutes. Assay reactions were subsequently incubated at room temperature for 15–30 minutes following the addition of 1 ng labelled probe.

Reactions were loaded onto a 4% polyacrylamide (29:1), 0.25x TBE gel and electrophoresed in 0.25x TBE at 200 V for one to three hours at 4°C. The polyacrylamide gel was subsequently dried and autoradiographs were produced by exposure of dried gels to X-Ray film at −80°C or room temperature for varying durations.

Luciferase assays

The 4.2 kb-*Nr2f2*-Luc *Nr2f2* 5' flanking region luciferase reporter construct was kindly provided by Dr M Vasseur-Cognet [39]. The PRE1-Luc and PRE2-Luc POU recognition element reporter constructs were cloned in our laboratory by inserting a single copy of PRE1 and PRE2 into the pGL4.23 [*luc2*/minP] reporter construct (Promega). The Dreidel expression construct was created by subcloning the Dreidel sequence from a pHM6 vector, kindly provided by Professor K Avraham [40], into pSi (Promega).

For transient transfections, ND7 cells [41] were plated at approximately 2×10^5 cells/well density in six-well plates and cultured at 37°C with 5% carbon dioxide in L-glutamine-containing L-15 media supplemented with 10% heat-inactivated foetal bovine serum, 0.32% sodium bicarbonate, 0.25% glucose and 0.85% penicillin-streptomycin. The next morning, the culture medium was changed to DMEM with 10% heat-inactivated foetal bovine serum for at least one hour prior to transfection. Cells were transfected using a standard calcium precipitation method [42] with 100–200 ng reporter construct, 1–3 μg expression vector and 10 ng pRL-null vector (Promega). The total amount of DNA used per transfection was kept constant using pSi (Promega). Cells were then incubated for a minimum of six hours followed by washing and returning to normal culture media. Cells were harvested at least 24 hours later and assayed using the Dual-Luciferase kit (Promega).

Site-directed mutagenesis

The 4.2 kb-*Nr2f2*-Luc-Mut reporter construct was generated by overlap extension PCR [43] using the following mutagenic primers (mutated bases underlined): PRE1 forward mutagenic primer, CTTTTTAGCCGATTTGATCACTTTGATT; PRE1 reverse mutagenic primer, AATCAAAGTGATCAAATCGGC-TAAAAAG; PRE1 forward flanking primer, AAGCCTCCGGGTCGGGCCCGGAG; PRE1 reverse flanking primer, TCCGCGCTCCGGGGTCCAC; PRE2 forward mutagenic primer, GATAAAGTTGAGAGGAATTTATTT-TAATTGCAGGGTAACAATGAGGTGAAGTCTGGTGTT; PRE2 reverse mutagenic primer, AACACCAGACTTCACCT-CATTGTTACCCTGCAATTAAAATAAATTCCTCT-CAACTTTATC; PRE2 forward flanking primer, GCTTAAT-GAATTCCCATCACTTGC; PRE2 reverse flanking primer, GGAATTCTCACAATCAACTAGCGG. The purified mutated fragment of PRE2 was cloned into 4.2 kb-*Nr2f2*-Luc, replacing the wild type sequence. The purified mutated fragment of PRE1 was cloned into this plasmid to create a double-mutant 4.2 kb-*Nr2f2*-Luc-Mut. Further subcloning was carried out to correct presumed PCR errors though a c to t missense mutation remained 288 bp downstream from the intended PRE1 mutagenesis site. The double-mutant probe was used in luciferase assays as described above.

Transcription factor binding site evolutionary conservation analysis

The ECR Browser [44] was used to navigate to the region of the human genome (build "hg19") that corresponded to the mouse sequence for PRE1 (CTTTTTAGCATATTTGATCACTTT-GATT) or PRE2 (GGAATTTATTTTAATTGCATCATAA-CAATGAGGTGA). The view obtained was expanded to include approximately 70 bp of flanking sequence and adjusted to include only mouse, rat and chimpanzee alignments. This view was submitted to the Mulan software and the phylogenetic tree obtained was not adjusted. The results of the Mulan analysis were submitted to MultiTF using the 'vertebrates' and 'optimised for function' settings against the TRANSFAC Professional Version 10.2 library to identify evolutionarily conserved POU domain transcription factor binding sites. Positional weight matrices (PWMs) for POU-domain transcription factors were included in this analysis as POU family binding sites are clearly related [4] despite there being subtle sequence-specific differences in the recognition elements of different POU4 family members [45]:

OCT1_B, OCT1_Q5, OCT1_Q6, OCT4, OCT_C, OCT_Q6, POU1F1_Q6, POU3F2 and POU6F1.

Cryosectioning and immunocytochemistry

Adult (i.e. >P21) C57BL/6J mice, adult Sprague Dawley rats and P1 C57BL/6J mice were killed in accordance with Schedule I of the UK Animals (Scientific Procedures) Act (1986). This study was approved by the UCL Biological Services Ethical Review Committee. Inner ears were isolated and the apical bone of the cochlea was perforated to facilitate fixation in 4% paraformalde-hyde solution in PBS for one to two hours at room temperature or overnight at 4°C. Cryoprotection was carried out by incubation of samples in 30% sucrose in PBS at 4°C overnight. Tissue was mounted in 1% low gelling agarose (Sigma-Aldrich) in PBS which had been heated to 140°C and allowed to cool to 37°C. Once set, a block was cut, attached to a cryostat chuck with OCT compound and rapidly frozen by immersion in liquid nitrogen prior to cryosectioning at −20 to −25°C. Sections of 15 μm were cut using a CM1850 cryostat (Leica).

Prior to immunolabelling, sections were permeabilised and blocked for one hour at room temperature with blocking solution (10% goat serum in PBS) containing 0.1–0.5% Triton-X. Slides were then incubated in blocking solution with an antibody raised against NR2F2 (a kind gift from Dr M Studer) 1:1000 at 4°C overnight [26]. Following three five-minute PBS washes, 2 μg/ml Alexa 488-conjugated anti-rabbit secondary antibody, 1 Unit/ml Alexa 633-conjugated Phalloidin and 5 μM DAPI in PBS or blocking solution was added for two hours at room temperature. Following three further five-minute PBS washes, samples were mounted in Fluoromount G and visualised using a Zeiss LSM 510 Meta microscope. Relative expression comparisons were made between apical to basal cochlear turns within individual slides.

Results

Identification of *Nr2f2* as a putative target of POU4F3

In order to identify target genes of POU4F3 that may be involved in hair cell survival and maintenance, a subtractive hybridization was carried out in UB/OC-2 cells followed by serial virtual Northern analysis as described previously [15]. One putative upregulatory target of POU4F3 identified by this screen was Clone D8. This clone was sequenced and subjected to BLAST analysis against the ENSEMBL mouse cDNA database (based on NCBI m35 data) to identify known mouse cDNA sequences to which it corresponded. Clone D8 matched the 3′ region of the related *Nr2f1* and *Nr2f2* cDNA sequences which are highly homologous in this region. However, the match to *Nr2f2* (ENSMUST00000032768) was more extensive (343 bp, 100% sequence similarity: $p = 8.6 \times 10^{-273}$) than the match to *Nr2f1* (116 bp, 87% sequence similarity: $p = 3.3 \times 10^{-25}$). Therefore, this analysis identified *Nr2f2* as an upregulated transcript from the POU4F3 forward subtractive hybridization library.

As *Nr2f2* was identified as a putative upregulatory target of POU4F3 in the subtractive hybridization screen, its expression in UB/OC-2 cells was investigated as these cells constitutively express POU4F3 in their proliferating state and are derived from the mouse inner ear at E13, i.e. after the onset of NR2F2 expression in the developing ear [18,29]. *Nr2f2* mRNA and protein were found to be present in proliferating UB/OC-2 cells by reverse transcriptase PCR, western blot and immunohisto-chemistry (Figure S1). NR2F2 expression was localised to UB/OC-2 cell nuclei, consistent with its reported subcellular localisa-tion [26,29].

POU4F3 binds two sites in the *Nr2f2* 5′ flanking region identified by Genomatix software analysis

For POU4F3 to directly regulate a gene, it must bind to the target gene promoter via a sequence specific binding site. To establish whether *Nr2f2* is a direct or indirect target of POU4F3 we performed bioinformatic and functional analysis of the *Nr2f2* 5′ flanking region to identify any such binding sites. The Genomatix MatInspector and ModelInspector programmes were used to interrogate the most proximal 5 kb of the predicted *Nr2f2* promoter (GXP_158616, identified by the Gene2Promoter software) [35–38].

Using MatInspector software at the highest stringency (0.05) a number of putative POU4F3 binding sites were identified. Of these, the two most proximal to the *Nr2f2* transcriptional start site were selected for further analysis and designated POU regulation element (PRE) 1 and 2 (Figure 1a). PRE2 also overlapped with a SORY OCT1 module identified using the Genomatix ModelInspector software i.e. PRE2 is within an experimentally verified functional promoter subunit that contains predicted binding sites for both OCT1 and SORY.

The ability of POU4F3 to bind these putative recognition elements was tested by electrophoretic mobility-shift assay (EMSA) analysis. *In vitro* translated POU4F3 produced bandshifts for both PRE1 and PRE2 that were not reproduced by *in vitro* translated luciferase protein (Figure 1b), demonstrating POU4F3-specific binding to PRE1 and PRE2. In addition, endogenous UB/OC-2 cell POU4F3 was shown to bind to these two sequences in EMSA competition analysis (Figure 1c). Hence, UB/OC-2 cell nuclear protein extract binding was reduced with increasing amounts of unlabelled POU4F3 consensus sequence but not with an excess of an unrelated transcription factor binding site, demonstrating the presence of POU4F3 in the shifted complex (Figure 1c).

POU4F3 activates the 4.2 kb *Nr2f2* 5′ flanking region containing PRE1 and PRE2

Having demonstrated the ability of POU4F3 to bind to PRE1 and PRE2, we tested whether POU4F3 can regulate the *Nr2f2* 5′ flanking region that contains these sites. A luciferase reporter construct containing 4.2 kb of the *Nr2f2* 5′ flanking region (4.2 kb-*Nr2f2*-Luc, Figure 2a) was used in co-transfection studies in ND7 cells. This analysis demonstrated a dose-dependent increase in 4.2 kb-*Nr2f2*-Luc activity in response to increasing POU4F3 levels. Compared to basal activity, promoter activity was five times higher at the maximal amount of POU4F3 expression construct used (3 μg) and was not replicated with a non-DNA-binding POU4F3 mutant (*dreidel*), showing that this activation is dependent on POU4F3-specific DNA binding (Figure 2b).

These experiments were also conducted in UB/OC-2 cells but gave inconsistent results (data not shown). Experiments in this cell type may have been complicated by a number of factors. For example, proliferating UB/OC-2 cells endogenously express POU4F3 which may have been sufficiently expressed to mask the variation induced by the co-transfected POU4F3 expression construct. UB/OC-2 cells also vary their POU4F3 expression depending on differentiation status; it is, therefore, possible that POU4F3 expression varies more dynamically than is currently known. Given these possibilities, ND7 cells were used as POU4F3 activation of *Nr2f2* constructs was consistent in this assay.

To investigate whether POU4F3 activates 4.2 kb-*Nr2f2*-Luc via PRE1, PRE2 or a combination of both, each was cloned upstream of a minimal promoter in a luciferase vector to generate PRE1-Luc and PRE2-Luc (Figure 2a) and used in co-transfection studies with increasing amounts of POU4F3. In ND7 cells, co-transfection with a POU4F3 expression construct produced a 1.8-fold upregulation of PRE1-Luc (Figure 2c) and a 1.5-fold upregulation of PRE2-Luc (Figure 2d). In both cases, this activation required functional POU4F3 as *dreidel* mutant protein did not upregulate either binding site (Figure 2c & d).

Activation of 4.2 kb-*Nr2f2*-Luc is dependent on PRE1 and PRE2

To confirm that POU4F3-mediated regulation of the *Nr2f2* promoter is dependent on PRE1 and PRE2, both sites were mutated to attenuate POU4F3 binding. Mutant versions of PRE1 and PRE2 were synthesised and tested in EMSA analysis which showed effective attenuation of *in vitro* translated POU4F3 binding (Figure 3a). Both mutations were then introduced into the 4.2 kb-*Nr2f2*-Luc construct by overlap extension PCR to produce 4.2 kb-*Nr2f2*-Luc-Mut and the effect of these mutations on POU4F3-specific activation was investigated in co-transfection studies. Regulation of 4.2 kb-*Nr2f2*-Luc-Mut was attenuated to 1.3-fold compared to a two-fold activation of 4.2 kb-*Nr2f2*-Luc at the maximal amount of POU4F3 expression construct used (1 μg). This suggests that PRE1 and PRE2 are required for activation of the *Nr2f2* 5′ flanking region by POU4F3 (Figure 3b).

Evolutionary Conservation of PRE1 and PRE2

Similar to protein-coding regions of the genome, functionally important regulatory elements are conserved across species [46–48]. Therefore, the evolutionary conservation of the POU4F3 binding sites PRE1 and PRE2 within the *Nr2f2* 5′ flanking sequence was investigated to indicate the potential functional importance of these sites *in vivo*. The Mulan program performs an analysis of sequences from multiple species; constructs phylogenetic trees; creates alignments; and identifies evolutionarily conserved regions. The output of this analysis is used as input for the MultiTF program which identifies evolutionarily conserved transcription factor binding sites [48].

The MultiTF program relies on the TRANSFAC Professional Version 10.2 library [49], however, no PWM is available for POU4F3 in this database. Therefore, PWMs for closely related transcription factor binding site families (i.e. OCT and POU) were used as POU family binding sites are closely related [4]. This analysis revealed two evolutionarily conserved binding sites in the human region of *NR2F2* that correspond to the PRE1 locus in the mouse *Nr2f2* 5′ flanking region. The overlapping binding sites identified correspond to OCT1 and OCT_C PWMs (Figure 4a). No such evolutionarily conserved binding sites were identified for PRE2 which is less well-conserved than PRE1 across the species assessed (Figure 4b). This conservation suggests that PRE1 is more likely to play an important role in the regulation of *Nr2f2* expression *in vivo* over recent evolution.

Expression of NR2F2 in the embryonic rat inner ear

The inner ear's function is primarily achieved by mechanosensory hair cells which are surrounded by support cells and extracellular matrix that form a complicated microanatomy which is required to achieve the special senses of hearing and balance. Accordingly, the development of this organ requires a complex co-ordination of cellular proliferation, differentiation and morphogenetic movement [50].

POU4F3 and NR2F2 are both expressed in the embryonic mouse inner ear. In the developing cochlea, *Pou4f3* expression begins in the base and extends to include all hair cells of the inner ear by birth. This expression is unique to hair cells and correlates with the wave of hair cell differentiation [6–8,51]. NR2F2 is more

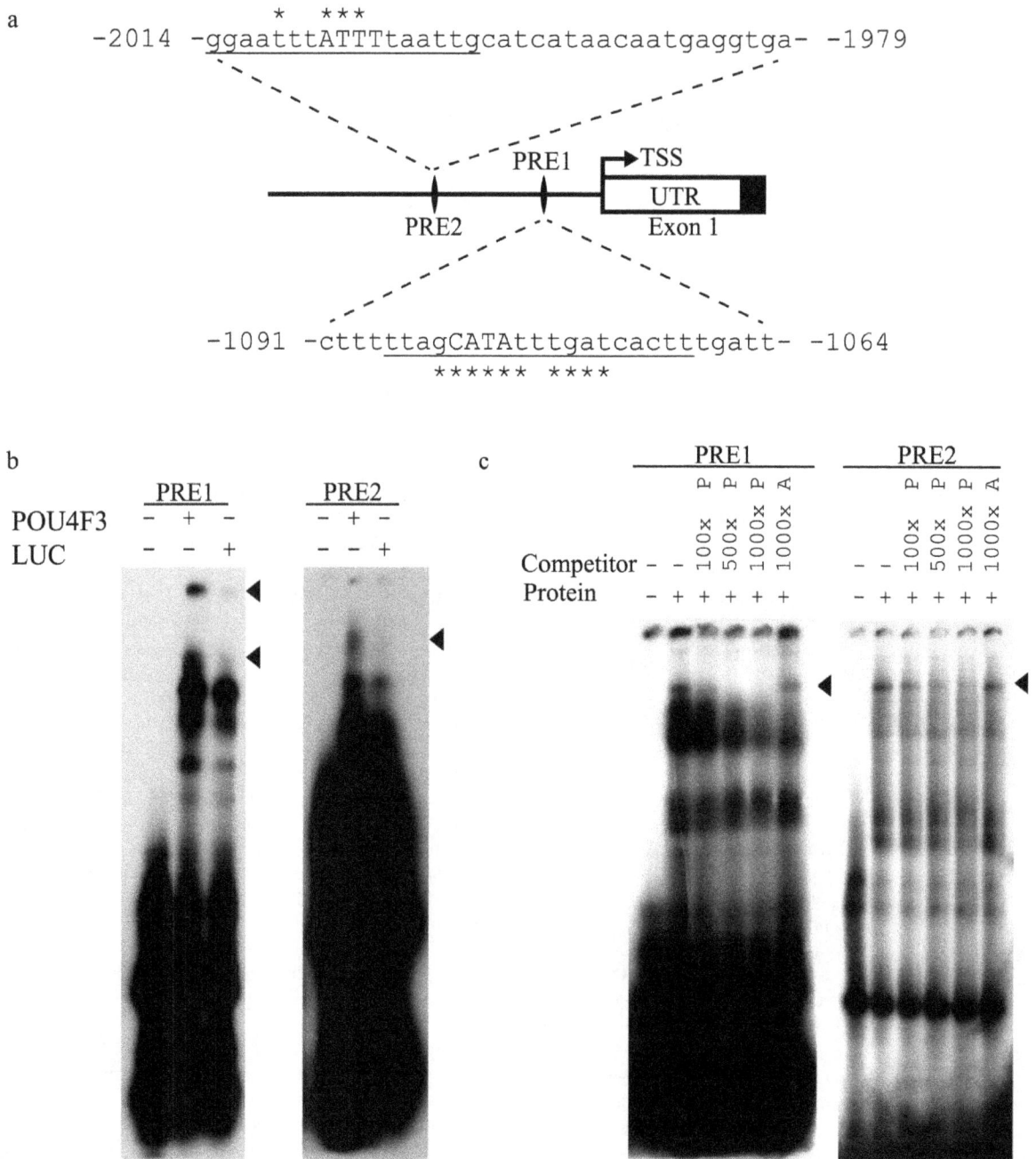

Figure 1. Identification and verification of POU4F3 recognition elements in the *Nr2f2* 5′ flanking region. *a.* Schematic diagram of the location of two putative POU4F3 recognition elements (PREs) identified in the *Nr2f2* 5′ flanking region and used in EMSA analysis. Underlined bases correspond to POU4F3 binding sites predicted by MatInspector software with capital letters denoting matches to the core sequence. The asterisks signify matrix position conservation >60/100 (*TSS*, transcriptional start site and *UTR*, un-translated region). *b.* Sequences shown in *a* were used as radiolabelled probes in EMSA analysis with *in vitro* translated POU4F3 or a Luciferase control. Bandshifts due to protein-specific binding by POU4F3 are indicated by arrowheads. *c.* The same probes were used in EMSA analysis with UB/OC-2 cell nuclear protein extract. Reactions were incubated either alone, with UB/OC-2 cell nuclear protein extract or with the nuclear protein extract and an excess of non-radiolabelled competitor oligonucleotide as indicated. The '*P*' suffix refers to non-radiolabelled POU4F3 binding sequence whereas the '*A*' suffix indicates a non-radiolabelled AP4 binding sequence. POU4F3-sequence-specific shifts are indicated by arrowheads.

widely expressed in the developing apical organ of Corti with its expression level gradually reducing to become more refined to the LER in the basal cochlea [29].

The putative interaction of POU4F3 and *Nr2f2* in the inner ear was investigated by further characterising the expression profile of NR2F2 in the E18 rat inner ear. NR2F2 expression was seen

throughout the developing cochlea at this age with strongest expression seen in the greater epithelial ridge, lesser epithelial ridge and supero-medial wall of the apical cochlear duct (Figure 5). Expression decreased in an apical-to-basal direction with NR2F2 expression being limited to the lesser epithelial ridge of the basal cochlear duct (Figure 5). This embryonic expression

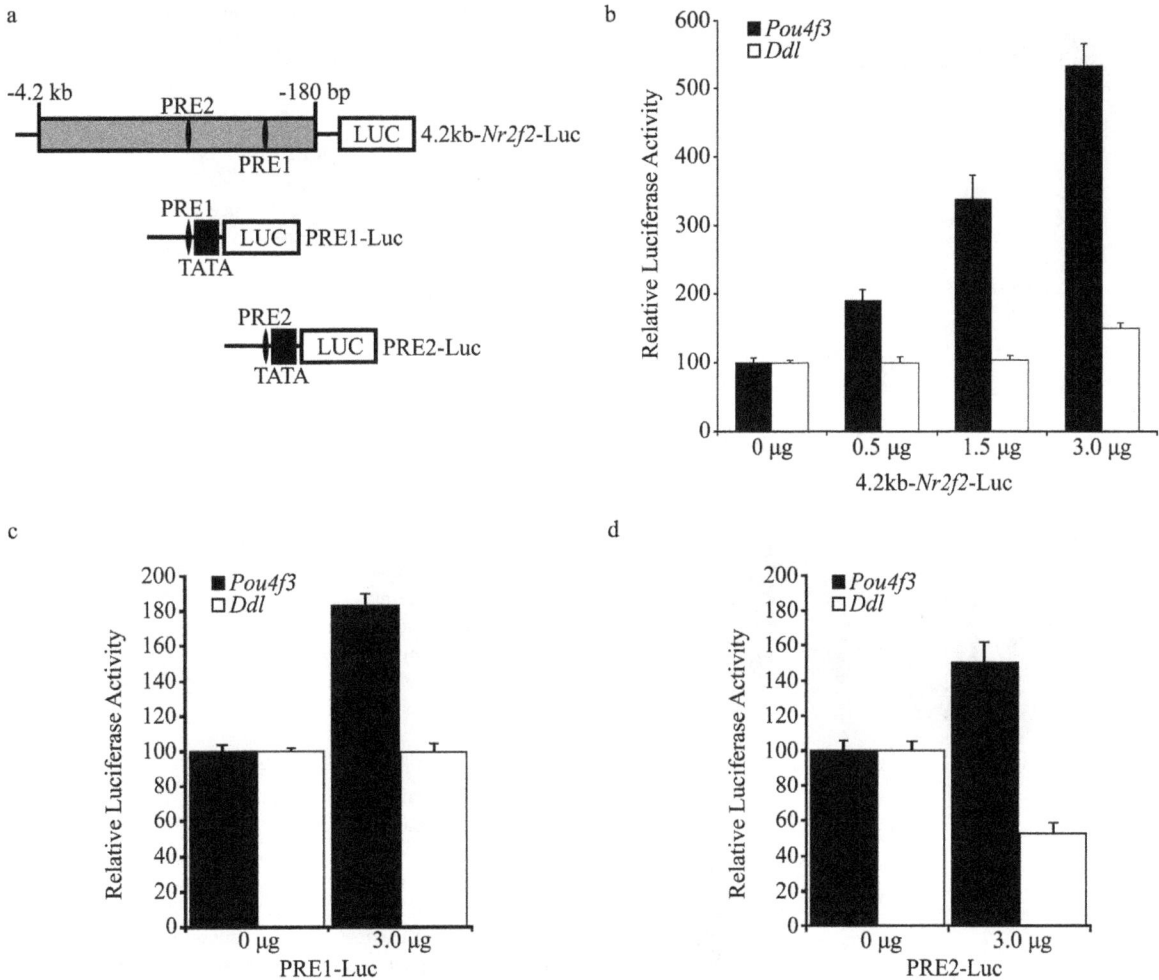

Figure 2. POU4F3-mediated activation of PRE1, PRE2 and a 4.2 kb *Nr2f2* **5' flanking sequence.** *a.* Schematic diagram of reporter constructs used in luciferase reporter assays in ND7 cells. POU4F3 recognition element (PRE) 1 and 2 are shown in an *Nr2f2* 5' flanking region fragment (*grey box*) cloned upstream of a luciferase reporter gene (*LUC*). The location of this fragment relative to the *Nr2f2* transcriptional start site is shown. *b.* Evaluation of the response of 4.2 kb-*Nr2f2*-Luc to increasing levels of POU4F3 or *dreidel* mutant POU4F3 (*Ddl*) expression construct. The luciferase activity of the reporter is normalised to its response to the empty expression vector and results are expressed relative to this. *c.* Response of the PRE1-Luc reporter construct in co-transfection experiments with POU4F3 and *dreidel* expression constructs. *d.* Evaluation of the response of PRE2-Luc in similar experiments to *c.* Error bars represent the s.e.m in *b, c* and *d* (n = 6 for each data point).

profile is in keeping with the previously reported expression profile of NR2F2 in the embryonic mouse inner ear [29].

NR2F2 expression in the postnatal mouse inner ear

The subtractive hybridization experiment that identified *Nr2f2* as a putative target of POU4F3 was conducted to elucidate the pathway through which POU4F3 affects hair cell differentiation and survival. In the murine model, hair cell differentiation requires POU4F3 in the embryonic and early postnatal period [8]. However POU4F3 is also required for hair cell survival and is expressed postnatally [10–12,17]. It must therefore influence its targets postnatally – though the targets required for hair cell maturation will not necessarily be identical to those required for maintenance. As NR2F2 is expressed postnatally in a number of organs, we investigated whether it is expressed in the postnatal inner ear to ascertain whether it is has the potential to play a role in hair cell maintenance as a POU4F3 target.

As in the embryonic cochlea, NR2F2 expression was widespread in the postnatal mouse cochlea both within and around the

sensory epithelium. Throughout the P1 cochlea a typical nuclear pattern of NR2F2 expression was seen in hair cells as well as a number of supporting cells of the sensory epithelium i.e. Hensen cells, Claudius cells and inner sulcus cells (Figure 6). Expression appeared stronger in the apical coil including the hair cells as compared to the basal coil and hair cells. This apical-to-basal expression gradient was maintained in the adult cochlea where NR2F2 expression in the basal cochlea further decreased so that NR2F2 expression was no longer observed in the nuclei of basal hair cells. As in the cochlea, NR2F2 expression is widespread though variable in the postnatal cristae where its expression is greatest within the sensory epithelium, including hair cells (Figure 6).

Discussion

POU4F3 expression is essential for hair cell differentiation and maintenance [6–8,10–12]. However, it's only known direct targets (BDNF, NT-3 and Caprin-1) and downstream targets (Gfi1 and Lhx3) do not sufficiently explain the link between POU4F3 and

a

b

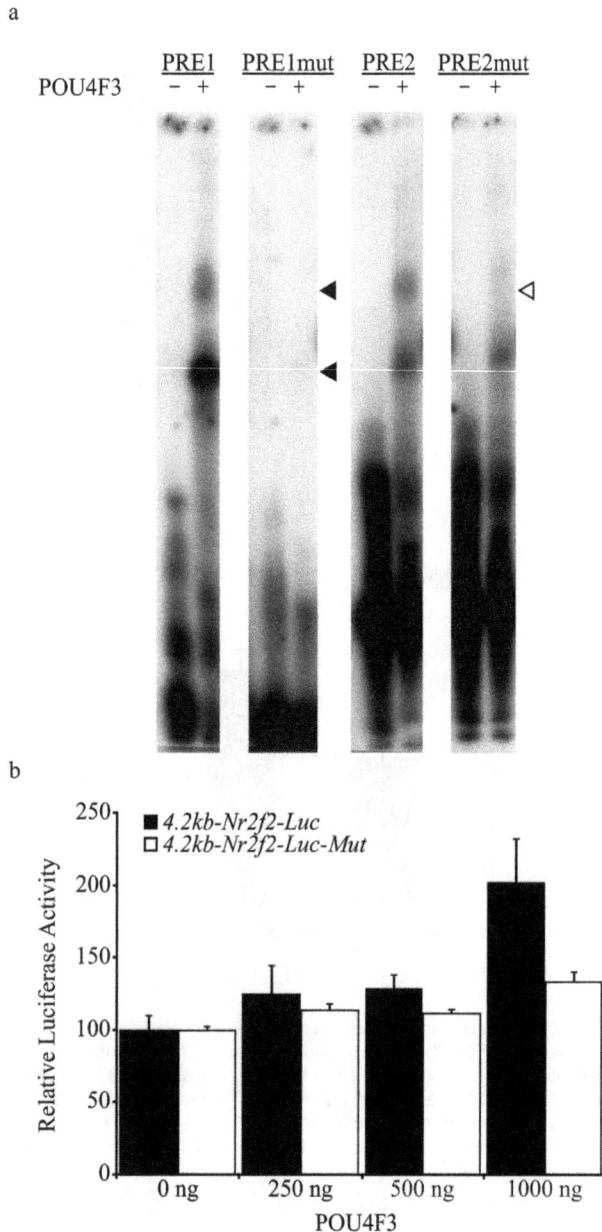

Figure 3. POU4F3 recognition element mutation attenuates POU4F3 binding and activation of the *Nr2f2* 5′ flanking region.
a. EMSA analysis of POU4F3 recognition element mutations on POU4F3 binding. Probes were either incubated alone or with *in vitro* translated POU4F3. *Closed arrowheads* indicate POU4F3-specific PRE1 bandshifts and the *open arrowhead* indicates the POU4F3-specific PRE2 bandshift. The mutation of each site severely compromises the ability of POU4F3 to bind to each sequence. *b*. Luciferase reporter assay upon co-transfection of POU4F3 with the 4.2 kb-*Nr2f2*-Luc-Mut reporter construct. Error bars represent the s.e.m (n = 6 for each data point).

the hair cell phenotype [14–16,40]. In this paper, we furthered the understanding of this pathway by identifying and verifying *Nr2f2* as a new regulatory target of POU4F3.

A combination of bioinformatic analysis, EMSA and reporter gene assays demonstrated that POU4F3 binds to two sites in the 4.2 kb *Nr2f2* 5′ flanking region and that the most proximal binding site (PRE1) is well-conserved through recent evolution. The sites identified were shown to be required for POU4F3-

mediated upregulation of this region as their mutation attenuates this effect and were sufficient to confer POU4F3 activation upon a heterologous promoter (Figure 2). These results suggest that *Nr2f2* is a direct regulatory target of POU4F3.

NR2F2 in the developing inner ear

Experiments in embryonic rats confirmed expression of NR2F2 in the developing cochlea (Figure 5) [29]. The widespread expression of NR2F2 indicates the presence of otic regulators of this gene. However, our data reveal that hair cell-selective *Nr2f2* expression could be regulated by POU4F3 given its hair cell-specific expression pattern in the inner ear [6,8].

As well as the POU4F3 binding sites described here, the *Nr2f2* 5′ flanking region is known to contain response elements for retinoic acid, sonic hedgehog, and Notch [52–54] which are all involved in inner ear development [50,55,56]. Furthermore, NR2F2 is known to mediate signalling between mesenchymal and endothelial compartments and it is thought that organs which require the differentiation of mesenchyme to epithelium display NR2F2 expression in the mesenchyme [57]. Epithelial-mesenchymal interactions give rise to most of the otic capsule and retinoic acid (which interacts with NR2F2) is thought to orchestrate mesenchyme-epithelial interactions in inner ear morphogenesis [54,57,58]. Though their relevance in the inner ear is not yet known, NR2F2 could act as an integrator of these important developmental pathways.

NR2F2 in the postnatal inner ear

The potential role of NR2F2 in hair cell survival and maintenance was investigated in the adult inner ear. In the postnatal mouse cochlea, NR2F2 expression is highest in the apex and falls to undetectable levels in basal hair cells. This pattern correlates with a previous report of higher apical POU4F3 expression in the postnatal rat cochlea [17]. The potentially matching expression pattern of these two transcription factors, if verified, supports the possibility that their interaction is active in the postnatal cochlea.

Apical hair cells are responsible for the detection of low frequency sounds and are the first hair cells of the cochlea to exit the cell cycle at approximately E13 in the developing mouse [59]. However, these hair cells are the last to complete their differentiation as the wave of hair cell differentiation proceeds from base to apex [30]. Though the gradient of NR2F2 expression seen in development mirrors hair cell differentiation this pattern persists despite full differentiation, i.e. after the onset of hearing. The nuclear expression pattern in these cells suggests that NR2F2 is acting as a transcription factor. However, as the postnatal role of NR2F2 is currently poorly understood, further work is required to determine its role in the hair cells of the inner ear.

The postnatal cochlear expression pattern of NR2F2 reported here is contrary to previously published data where it was not detected in the postnatal mouse organ of Corti [29]. This may be explained by several factors. For example, different antibodies were used in these experiments and tissue was obtained from different mouse strains [26,29,60]. As transcription factors typically have low expression levels, a small change in the sensitivity of the differing immunohistochemistry assays could account for this discrepancy.

In addition to its cochlear expression, NR2F2 expression was detected in the nuclei of hair cells and supporting cells of the postnatal mouse crista. The cristae form part of the vestibular apparatus and are required for the detection of rotational acceleration. A role for NR2F2 in this system would, therefore, imply a role in balance function. This is the first report of NR2F2

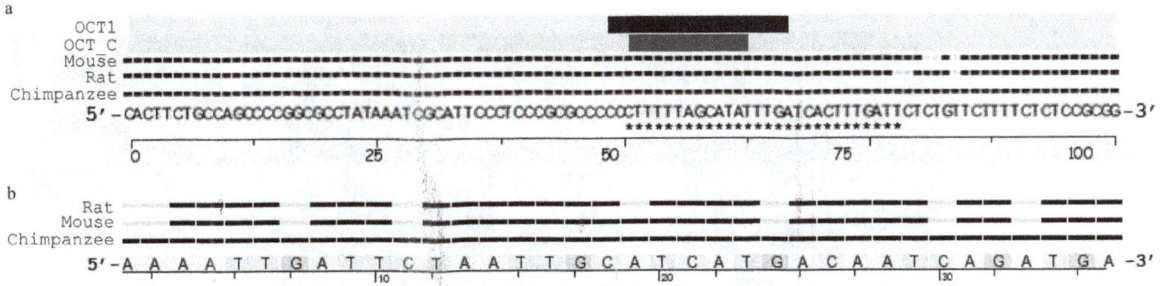

Figure 4. Evolutionary conservation of transcription factor binding sites in the *Nr2f2* 5′ flanking region. Schematic representation of the conservation of the region surrounding the human PRE1 sequence (*a*) and the human PRE2 sequence (*b*) compared to corresponding mouse, rat and chimpanzee sequences. *a*. The well-conserved PRE1 sequence is indicated by asterisks with evolutionarily conserved binding sites for the POU domain transcription factors OCT1 and OCT_C indicated by horizontal bars. Conserved bases are represented by black boxes with mismatches in grey. *b*. PRE2 is less well conserved with eight mismatched bases and no corresponding POU domain transcription factor binding sites in the TRANSFAC database. Conserved bases are again represented by black boxes with mismatches in grey.

expression in the vestibular system and further work to improve the understanding of its relative expression and function at this site, as compared to the cochlea, may assist in elucidation of the molecular basis for phenotypic differences between cochlear and vestibular hair cells.

NR2F2 function in the inner ear

Given the diversity of genes and pathways in which NR2F2 is involved, it is possible that NR2F2 could play a novel role in cochlear development and maintenance. However, despite knowledge of its regulation, the function of NR2F2 in the inner ear remains unknown. Furthermore, the complex regulation of *Nr2f2* and its various targets [61] make it difficult to differentiate any effects of POU4F3 from other activators of *Nr2f2* which are known to be important in inner ear differentiation e.g. sonic hedgehog [62]. The hair-cell-specific expression of POU4F3 suggests that its interaction with *Nr2f2* represents a regulatory mechanism that may be restricted to these sensory cells.

It is notable that a number of other reported regulatory targets of POU4F3 are transcription factors, including Lhx3 and Gfi-1 [16,40] suggesting that it lies at the head of a complex and diverse regulatory pathway and may act as master-regulator of hair cell function. This is certainly consistent with the evidence provided from both transgenic mice with targeted deletion of *Pou4f3* and humans with *POU4F3* mutations, which exhibit loss of hearing due to loss of hair cell function.

Postnatally, mitochondrial biogenesis has been implicated in hair cell survival following the administration of ototoxic medications and noise exposure [63,64]. NR2F2 has been shown to be involved in white adipose tissue mitochondrial biogenesis and may therefore influence this otoprotective function in postnatal hair cells [65]. Also, established roles of NR2F2 such as angiogenesis may be relevant to other cochlear cells that do not express POUF43 but do express NR2F2 [65].

Embryonic lethality at E10 makes it hard to assess the effect of *Nr2f2* knockout on the developing mouse inner ear and no relevant conditional knockout mouse exists [24]. In the adult

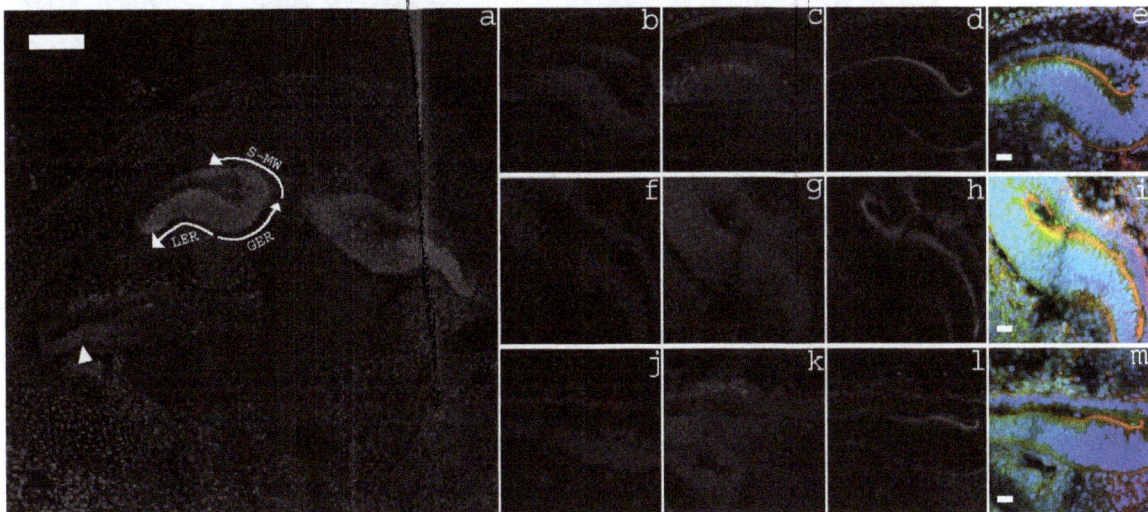

Figure 5. Expression of NR2F2 in the embryonic rat cochlea. *a*. Overview of NR2F2 expression throughout the cochlea at E18. NR2F2 is most strongly expressed in the cochlear duct. In the apical duct, strong expression is seen in the super-medial wall (*S-MW*), greater epithelial ridge (*GER*), and lesser epithelial ridge (*LER*). This expression is reduced to the lesser epithelial ridge in the basal duct (*arrowhead*). *b–e*. Apical cochlear duct; *f–i*. Middle cochlear duct; and *j–m*. Basal cochlear duct. *b, f & j*, cell nuclei stained with DAPI; *c, g & k*, NR2F2 labelling; *d, h & l*, filamentous actin labelled by Phalloidin; and *e, i & m*, Merge of DAPI (*blue*), NR2F2 (*green*) and Phalloidin (*red*), scale bars: 100 μm in *a* and 20 μm in *e, i & m*.

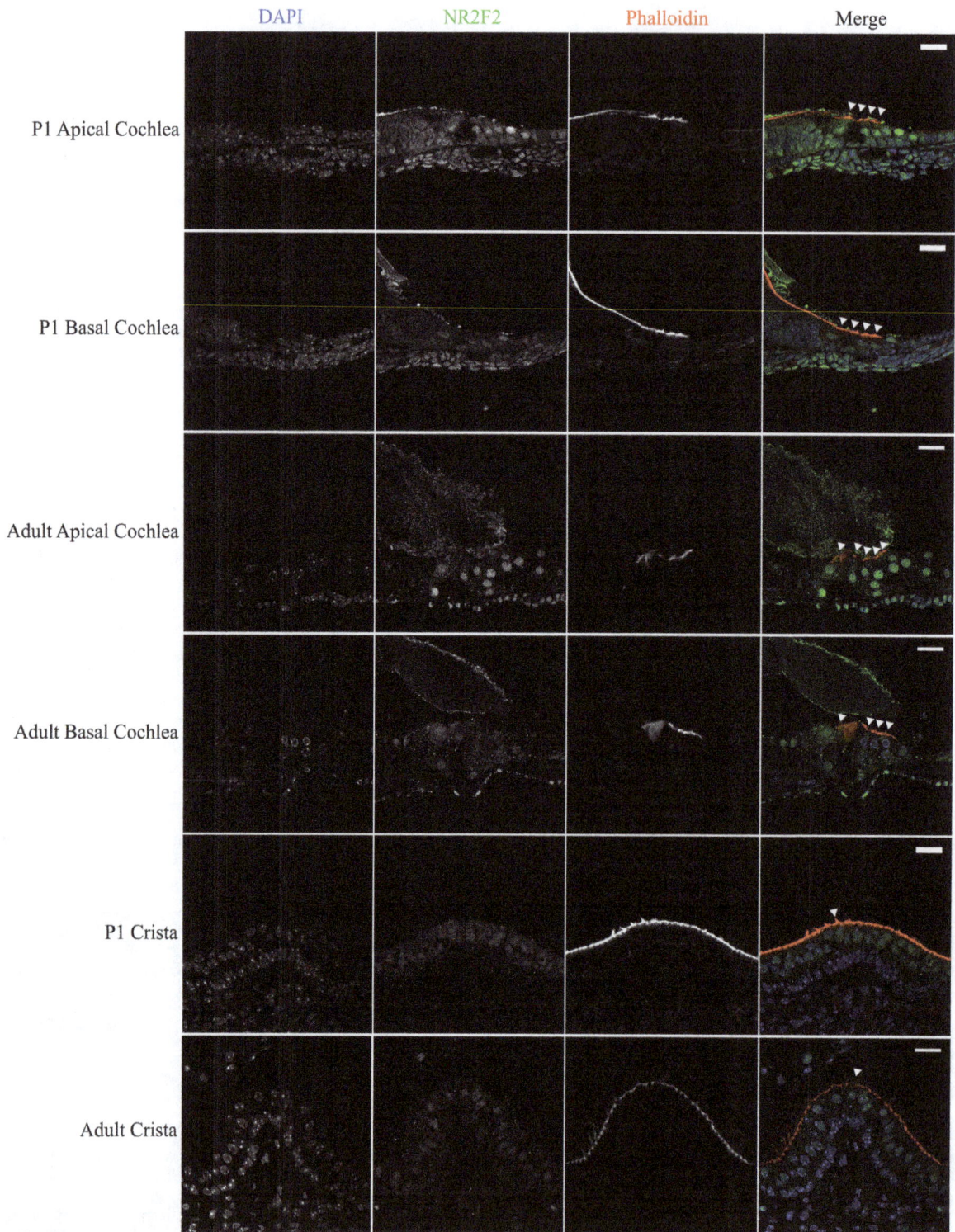

Figure 6. Expression of NR2F2 in the postnatal mouse inner ear. Cryotome sections of P1 and adult mouse inner ears were subjected to immunohistochemistry in order to characterise their postnatal NR2F2 expression pattern. NR2F2 labelling is seen throughout the sensory epithelia of the cochlea and cristae. In the cochlea, NR2F2 labelling is of greater intensity in the apex than the base. NR2F2 is expressed in the nuclei of apical hair cells at all ages examined. In P1 mice, NR2F2 expression is reduced in basal hair cells. In adult mice, the expression of NR2F2 in basal hair cell nuclei appears further reduced. NR2F2 is also expressed in the hair cells of the cristae of the semicircular canals and this expression is maintained into adulthood. In merged images DAPI is *blue*, NR2F2 is *green*, Phalloidin is *red*, *arrowheads* indicate hair cells and *scale bars*: 20 μm.

Nr2f2 knockout mouse, there is no overt phenotype, though hearing has yet to be assessed in these animals [66]. However, postnatal expression of NR2F2 in a number of organs – including the inner ear – suggests that it plays a functional role beyond development. Current evidence suggests that NR2F2 is most important in regeneration and de-differentiation in pathological conditions [65].

Therefore, beyond the verification of POU4F3 regulation of *Nr2f2 in vivo*, the characterisation of NR2F2 function in development and under pathological conditions in mature animals is essential to clarify the putative role of this versatile orphan nuclear receptor in the development and maintenance of hearing and balance.

Supporting Information

Figure S1   ***Nr2f2 expression in UB/OC-2 cells.*** *Nr2f2* expression was investigated in proliferating UB/OC-2 cells. *a*, cDNA was generated from proliferating UB/OC-2 cells and subjected to PCR which demonstrated amplification of the predicted *Nr2f2* fragment (213 bp). Integrity of the cDNA was demonstrated by amplification of *Gapdh* (450 bp) with sizes of amplified fragments measured against a 1 kb marker (*M*). Adjacent lanes have been removed for presentation purposes. *b*, The specificity of an anti-NR2F2 antibody was verified by western blot. A predominant 50 kDa band (*arrow*) with the expected

molecular weight of NR2F2 is detected. Immunohistochemistry was carried out with (*c*) and without (*d*, secondary antibody alone) an anti-NR2F2 antibody in UB/OC-2 cells. This analysis showed NR2F2 expression localised to OC-2 cell nuclei. *i*, nuclei were stained DAPI; *ii*, NR2F2 expression; *iii*, Phalloidin labelling; and *iv*, shows a merged image of *i*, *ii* and *iii* with DAPI in *blue*, NR2F2 in *green* and Phalloidin in *red*. *Scale bars*: 100 μm. (TIF)

Acknowledgments

We would like to thank Dr M Vasseur-Cognet, Dr M Studer and Professor K Avraham for their kind gifts of reagents. We would also like to thank Dr L Nolan for her general advice and technical assistance with the above experiments.

Author Contributions

Conceived and designed the experiments: CT ERT JEG SJD. Performed the experiments: CT JEG SJD. Analyzed the data: CT ERT JEG SJD. Contributed reagents/materials/analysis tools: CT ERT JEG SJD. Contributed to the writing of the manuscript: CT ERT JEG SJD.

References

1. Andersen B, Rosenfeld MG (2001) POU domain factors in the neuroendocrine system: lessons from developmental biology provide insights into human disease. Endocr Rev 22: 2–35.
2. Scott MP, Tamkun JW, Hartzell III GW (1989) The structure and function of the homeodomain. Biochim Biophys Acta - Rev Cancer 989: 25–48.
3. Aurora R, Herr W (1992) Segments of the POU domain influence one another's DNA-binding specificity. Mol Cell Biol 12: 455–467.
4. Verrijzer CP, Van der Vliet PC (1993) POU domain transcription factors. Biochim Biophys Acta 1173: 1–21.
5. Badea TC, Williams J, Smallwood P, Shi M, Motajo O, et al. (2012) Combinatorial expression of Brn3 transcription factors in somatosensory neurons: genetic and morphologic analysis. J Neurosci 32: 995–1007.
6. Erkman L, McEvilly RJ, Luo L, Ryan AK, Hooshmand F, et al. (1996) Role of transcription factors a Brn-3.1 and Brn-3.2 in auditory and visual system development. Nature 381: 603–606.
7. Xiang M, Gan L, Li D, Chen ZY, Zhou L, et al. (1997) Essential role of POU-domain factor Brn-3c in auditory and vestibular hair cell development. Proc Natl Acad Sci U S A 94: 9445–9450.
8. Xiang, Gao WQ, Hasson T, Shin JJ, Xiang M (1998) Requirement for Brn-3c in maturation and survival, but not in fate determination of inner ear hair cells. Development 125: 3935–3946.
9. Keithley EM, Erkman L, Bennett T, Lou L, Ryan AF (1999) Effects of a hair cell transcription factor, Brn-3.1, gene deletion on homozygous and heterozygous mouse cochleas in adulthood and aging. Hear Res 134: 71–76.
10. Collin RWJ, Chellappa R, Pauw R-J, Vriend G, Oostrik J, et al. (2008) Missense mutations in POU4F3 cause autosomal dominant hearing impairment DFNA15 and affect subcellular localization and DNA binding. Hum Mutat 29: 545–554.
11. Lee HK, Park H-J, Lee K-Y, Park R, Kim U-K (2010) A novel frameshift mutation of POU4F3 gene associated with autosomal dominant non-syndromic hearing loss. Biochem Biophys Res Commun 396: 626–630.
12. Vahava O, Morell R, Lynch ED, Weiss S, Kagan ME, et al. (1998) Mutation in Transcription Factor POU4F3 Associated with Inherited Progressive Hearing Loss in Humans. Science 279: 1950–1954.
13. Kim H-J, Won H-H, Park K-J, Hong SH, Ki C-S, et al. (2013) SNP linkage analysis and whole exome sequencing identify a novel POU4F3 mutation in autosomal dominant late-onset nonsyndromic hearing loss (DFNA15). PLoS One 8: e79063.
14. Clough RL, Sud R, Davis-Silberman N, Hertzano R, Avraham KB, et al. (2004) Brn-3c (POU4F3) regulates BDNF and NT-3 promoter activity. Biochem Biophys Res Commun 324: 372–381.
15. Towers ER, Kelly JJ, Sud R, Gale JE, Dawson SJ (2011) Caprin-1 is a target of the deafness gene Pou4f3 and is recruited to stress granules in cochlear hair cells in response to ototoxic damage. J Cell Sci 124: 1145–1155.
16. Hertzano R, Dror AA, Montcouquiol M, Ahmed ZM, Ellsworth B, et al. (2007) Lhx3, a LIM domain transcription factor, is regulated by Pou4f3 in the auditory but not in the vestibular system. Eur J Neurosci 25: 999–1005.
17. Gross J, Angerstein M, Fuchs J, Stute K, Mazurek B (2011) Expression analysis of prestin and selected transcription factors in newborn rats. Cell Mol Neurobiol 31: 1089–1101.
18. Rivolta MN, Grix N, Lawlor P, Ashmore JF, Jagger DJ, et al. (1998) Auditory hair cell precursors immortalized from the mammalian inner ear. Proc Biol Sci 265: 1595–1603.
19. Shiina N, Shinkura K, Tokunaga M (2005) A novel RNA-binding protein in neuronal RNA granules: regulatory machinery for local translation. J Neurosci 25: 4420–4434.
20. Shiina N, Yamaguchi K, Tokunaga M (2010) RNG105 deficiency impairs the dendritic localization of mRNAs for Na+/K+ ATPase subunit isoforms and leads to the degeneration of neuronal networks. J Neurosci 30: 12816–12830.
21. Solomon S, Xu Y, Wang B, David MD, Schubert P, et al. (2007) Distinct structural features of caprin-1 mediate its interaction with G3BP-1 and its induction of phosphorylation of eukaryotic translation initiation factor 2alpha, entry to cytoplasmic stress granules, and selective interaction with a subset of mRNAs. Mol Cell Biol 27: 2324–2342.
22. Tsai SY, Tsai MJ (1997) Chick ovalbumin upstream promoter-transcription factors (COUP-TFs): coming of age. Endocr Rev 18: 229–240.
23. Suzuki T, Moriya T, Darnel a D, Takeyama J, Sasano H (2000) Immunohistochemical distribution of chicken ovalbumin upstream promoter transcription factor II in human tissues. Mol Cell Endocrinol 164: 69–75.
24. Pereira FA, Qiu Y, Zhou G, Tsai M-J, Tsai SY (1999) The orphan nuclear receptor COUP-TFII is required for angiogenesis and heart development. Genes Dev 13: 1037–1049.
25. You L, Lin F, Lee C, DeMayo F, Tsai M (2005) Suppression of Notch signalling by the COUP-TFII transcription factor regulates vein identity. Nature 435: 98–104.
26. Tripodi M, Filosa A, Armentano M, Studer M (2004) The COUP-TF nuclear receptors regulate cell migration in the mammalian basal forebrain. Development 131: 6119–6129.
27. Takamoto N, You L-R, Moses K, Chiang C, Zimmer WE, et al. (2005) COUP-TFII is essential for radial and anteroposterior patterning of the stomach. Development 132: 2179–2189.
28. McCaffery P, Wagner E, O'Neil J, Petkovich M, Dräger UC (1999) Dorsal and ventral retinal territories defined by retinoic acid synthesis, break-down and nuclear receptor expression. Mech Dev 82: 119–130.
29. Tang LS, Alger HM, Lin F, Pereira FA (2005) Dynamic expression of COUP-TFI and COUP-TFII during development and functional maturation of the mouse inner ear. Gene Expr Patterns 5: 587–592.
30. Lim DJ, Anniko M (1985) Developmental morphology of the mouse inner ear. A scanning electron microscopic observation. Acta Otolaryngol Suppl 422: 1–69.
31. Tang LS, Alger HM, Pereira FA (2006) COUP-TFI controls Notch regulation of hair cell and support cell differentiation. Development 133: 3683–3693.
32. Ladias JAA, Karathanasis SK (1991) Regulation of the Apolipoprotein AI Gene by ARP-1, a Novel Member of the Steroid Receptor Superfamily. Science 251: 561–565.
33. Lopes da Silva S, Cox JJ, Jonk IJ, Kruijer W, Burbach JP (1995) Localization of transcripts of the related nuclear orphan receptors COUP-TF I and ARP-1 in the adult mouse brain. Brain Res Mol Brain Res 30: 131–136.
34. Hall TA (1999) BioEdit: a user-friendly biological sequence alignment editor and analysis program for Windows 95/98/NT. Nucleic Acids Symp Ser 41: 95–98.
35. Scherf M, Klingenhoff A, Werner T (2000) Highly specific localization of promoter regions in large genomic sequences by PromoterInspector: a novel context analysis approach. J Mol Biol 297: 599–606.
36. Quandt K, Frech K, Karas H, Wingender E, Werner T (1995) MatInd and MatInspector: new fast and versatile tools for detection of consensus matches in nucleotide sequence data. Nucleic Acids Res 23: 4878–4884.

37. Cartharius K, Frech K, Grote K, Klocke B, Haltmeier M, et al. (2005) MatInspector and beyond: promoter analysis based on transcription factor binding sites. Bioinformatics 21: 2933–2942.

38. Frech K, Werner T (1996) Specific Modelling of Regulatory Units in DNA Sequences. Nucleic Acids Res 24: 1212–1219.

39. Perilhou A, Tourrel-Cuzin C, Zhang P, Kharroubi I, Wang H, et al. (2008) The MODY1 gene for hepatocyte nuclear factor 4alpha and a feedback loop control COUP-TFII expression in pancreatic beta cells. Mol Cell Biol 28: 4588–4597.

40. Hertzano R, Montcouquiol M, Rashi-Elkeles S, Elkon R, Yücel R, et al. (2004) Transcription profiling of inner ears from Pou4f3(ddl/ddl) identifies Gfi1 as a target of the Pou4f3 deafness gene. Hum Mol Genet 13: 2143–2153.

41. Wood JN, Bevan SJ, Coote PR, Dunn PM, Harmar A, et al. (1990) Novel cell lines display properties of nociceptive sensory neurons. Proc Biol Sci 241: 187–194.

42. Kingston RE, Chen CA, Okayama H (2001) Calcium phosphate transfection. Curr. Protoc. Immunol. p. Unit 10.13.

43. Ling MM, Robinson BH (1997) Approaches to DNA mutagenesis: an overview. Anal Biochem 254: 157–178.

44. Ovcharenko I, Nobrega MA, Loots GG, Stubbs L (2004) ECR Browser: a tool for visualizing and accessing data from comparisons of multiple vertebrate genomes. Nucleic Acids Res 32: W280–6.

45. Gruber CA, Rhee JM, Gleiberman A, Turner EE (1997) POU domain factors of the Brn-3 class recognize functional DNA elements which are distinctive, symmetrical, and highly conserved in evolution. Mol Cell Biol 17: 2391–2400.

46. Loots GG, Locksley RM, Blankespoor CM, Wang ZE, Miller W, et al. (2000) Identification of a coordinate regulator of interleukins 4, 13, and 5 by cross-species sequence comparisons. Science 288: 136–140.

47. Elnitski L, Li J, Noguchi CT, Miller W, Hardison R (2001) A negative cis-element regulates the level of enhancement by hypersensitive site 2 of the beta-globin locus control region. J Biol Chem 276: 6289–6298.

48. Ovcharenko I, Loots GG, Giardine BM, Hou M, Ma J, et al. (2005) Mulan: multiple-sequence local alignment and visualization for studying function and evolution. Genome Res 15: 184–194.

49. Matys V, Kel-Margoulis O V, Fricke E, Liebich I, Land S, et al. (2006) TRANSFAC and its module TRANSCompel: transcriptional gene regulation in eukaryotes. Nucleic Acids Res 34: D108–10.

50. Bok J, Chang W, Wu DK (2007) Patterning and morphogenesis of the vertebrate inner ear. Int J Dev Biol 51: 521–533.

51. Raz Y, Kelley MW (1999) Retinoic acid signaling is necessary for the development of the organ of Corti. Dev Biol 213: 180–193.

52. Diez H, Fischer A, Winkler A, Hu C-J, Hatzopoulos AK, et al. (2007) Hypoxia-mediated activation of Dll4-Notch-Hey2 signaling in endothelial progenitor cells and adoption of arterial cell fate. Exp Cell Res 313: 1–9.

53. Krishnan V (1997) Mediation of Sonic Hedgehog-Induced Expression of COUP-TFII by a Protein Phosphatase. Science 278: 1947–1950.

54. Qiu Y, Krishnan V, Pereira FA, Tsai SY, Tsai MJ (1996) Chicken ovalbumin upstream promoter-transcription factors and their regulation. J Steroid Biochem Mol Biol 56: 81–85.

55. Romand R, Dolle P, Hashino E (2006) Retinoid Signaling in Inner Ear Development. J Neurobiol 66: 687–704.

56. Kelley MW (2006) Regulation of cell fate in the sensory epithelia of the inner ear. Nat Rev Neurosci 7: 837–849.

57. Pereira FA, Tsai MJ, Tsai SY (2000) COUP-TF orphan nuclear receptors in development and differentiation. Cell Mol Life Sci 57: 1388–1398.

58. Fekete DM (1999) Development of the vertebrate ear: insights from knockouts and mutants. Trends Neurosci 22: 263–269.

59. Ruben RJ (1967) Development of the inner ear of the mouse: a radioautographic study of terminal mitoses. Acta Otolaryngol: Suppl 220: 1–44.

60. Zhou C, Qiu Y, Pereira FA, Crair MC, Tsai SY, et al. (1999) The nuclear orphan receptor COUP-TFI is required for differentiation of subplate neurons and guidance of thalamocortical axons. Neuron 24: 847–859.

61. Park J-I, Tsai SY, Tsai M-J (2003) Molecular mechanism of chicken ovalbumin upstream promoter-transcription factor (COUP-TF) actions. Keio J Med 52: 174–181.

62. Krishnan V, Elberg G, Tsai MJ, Tsai SY (1997) Identification of a novel sonic hedgehog response element in the chicken ovalbumin upstream promoter-transcription factor II promoter. Mol Endocrinol 11: 1458–1466.

63. Hyde GE, Rubel EW (1995) Mitochondrial role in hair cell survival after injury. Otolaryngol Head Neck Surg 113: 530–540.

64. Choi C-H, Chen K, Vasquez-Weldon A, Jackson RL, Floyd RA, et al. (2008) Effectiveness of 4-hydroxy phenyl N-tert-butylnitrone (4-OHPBN) alone and in combination with other antioxidant drugs in the treatment of acute acoustic trauma in chinchilla. Free Radic Biol Med 44: 1772–1784.

65. Lin F-J, Qin J, Tang K, Tsai SY, Tsai M-J (2011) Coup d'Etat: an orphan takes control. Endocr Rev 32: 404–421.

66. Qin J, Tsai M-J, Tsai SY (2008) Essential roles of COUP-TFII in Leydig cell differentiation and male fertility. PLoS One 3: e3285.

4

Up-Regulation of Nerve Growth Factor in Cholestatic Livers and Its Hepatoprotective Role against Oxidative Stress

Ming-Shian Tsai[1,2], Yu-Chun Lin[3], Cheuk-Kwan Sun[4], Shih-Che Huang[3], Po-Huang Lee[1,5]*, Ying-Hsien Kao[3]*

1 Department of Surgery, E-DA Hospital, Kaohsiung, Taiwan, 2 The School of Medicine for Post-Baccalaureate, I-Shou University, Kaohsiung, Taiwan, 3 Department of Medical Research, E-DA Hospital, Kaohsiung, Taiwan, 4 Department of Medical Education, E-DA Hospital, Kaohsiung, Taiwan, 5 Department of Surgery, National Taiwan University Hospital, National Taiwan University College of Medicine, Taipei, Taiwan

Abstract

The role of nerve growth factor (NGF) in liver injury induced by bile duct ligation (BDL) remains elusive. This study aimed to investigate the relationship between inflammation and hepatic NGF expression, to explore the possible upstream molecules up-regulating NGF, and to determine whether NGF could protect hepatocytes from oxidative liver injury. Biochemical and molecular detection showed that NGF was up-regulated in cholestatic livers and plasma, and well correlated with systemic and hepatic inflammation. Conversely, systemic immunosuppression reduced serum NGF levels and resulted in higher mortality in BDL-treated mice. Immunohistochemistry showed that the up-regulated NGF was mainly localized in parenchymal hepatocytes. In vitro mechanistic study further demonstrated that TGF-β1 up-regulated NGF expression in clone-9 and primary rat hepatocytes. Exogenous NGF supplementation and endogenous NGF overexpression effectively protected hepatocytes against TGF-β1- and oxidative stress-induced cell death in vitro, along with reduced formation of oxidative adducted proteins modified by 4-HNE and 8-OHdG. TUNEL staining confirmed the involvement of anti-apoptosis in the NGF-exhibited hepatoprotection. Moreover, NGF potently induced Akt phosphorylation and increased Bcl-2 to Bax ratios, whereas these molecular alterations by NGF were only seen in the H_2O_2-, but not TGF-β1-treated hepatocytes. In conclusion, NGF exhibits anti-oxidative and hepatoprotective effects and is suggested to be therapeutically applicable in treating cholestatic liver diseases.

Editor: Carlos M. Rodriguez-Ortigosa, CIMA. University of Navarra, Spain

Funding: This study was supported in part by grants from EDA Hospital (EDAHP-100004 to MST and EDAHT-100021 to YHK). The funders had no role in study design, data collection and analysis, decision to publish, or preparation of the manuscript.

Competing Interests: The authors have declared that no competing interests exist.

* Email: pohuang1115@ntu.edu.tw (PHL); danyhkao@gmail.com (YHK)

Introduction

Cholestatic liver injury is not an uncommon clinical scenario, which can be caused by obstructed bile flow due to sclerosing cholangitis, periampullary tumor, cholelithiasis and prolonged parenteral nutrition use [1]. The pathological changes of liver associated with cholestasis include hepatocyte necrosis and apoptosis, neutrophil infiltration, bile duct epithelial proliferation, hepatic stellate cell activation and finally fibrosis. Production of reactive oxygen species (ROS) is among the key factors underlying liver injury [2,3].

Nerve growth factor (NGF) is vital for the differentiation, survival, and synaptic activity of the peripheral sympathetic and sensory nervous systems [4,5]. Moreover, NGF is up-regulated in various types of inflamed tissues [6] and shown to protect nerve cells against oxidative stress [7,8,9,10]. In our previous study, gastric perforation enhanced aortic as well as cardiac expression of both NGF mRNA and protein [11]. In liver, NGF has been demonstrated to play a role in regulating liver fibrosis [12,13,14], carcinogenesis [15,16], angiogenesis [15], and cholangiocyte proliferation [17]. In response to various chemical injuries, NGF

expression is up-regulated in the liver [18]. Although NGF has been reported to be up-regulated during experimental cholestatic injury [17], its role in hepatocytes following oxidative injury and its mechanism of regulation during cholestasis remain unclear. Moreover, little is known about the underlying mechanisms mediating NGF effects on hepatocytes.

In the present study, we hypothesized that cholestatic injury can up-regulate NGF expression in liver through an inflammatory signaling axis. We further investigated whether NGF is able to exert anti-apoptotic effects on hepatocytes and protect hepatocytes from various insults, including oxidative stress. We showed that NGF induced activation of PI3K/Akt and up-regulated the Bcl-2/Bax ratios in hepatocytes. Furthermore, NGF protects hepatocytes against TGF-β1 and hydrogen peroxide-induced oxidative damage. These data shed new light on the mechanism whereby NGF provides protection against oxidative injury and may be potentially relevant in the development of new therapeutic modalities for cholestatic liver injury.

Materials and Methods

Animals and ethics statement

Six to eight-week-old ICR male mice were raised *ad libitum* at 20–22°C with a 12 hr of light-dark cycle in the Animal Center of I-Shou University. All animal experimental procedures were approved by the Institute of Animal Care and Use Committee at E-DA Hospital (Affidavit of Approval of Animal Use Protocol No. IACUC-99018 and 100015) and performed in accordance with the Guide for the Care and Use of Laboratory Animals (NIH publication No. 85–23, National Academy Press, Washington, DC, USA, revised 1996). Mice were randomly divided into experimental groups. Cholestatic liver injury was induced by surgical procedures for common bile duct ligation (BDL) as previously described [19]. In brief, induction of anesthesia of mice was performed by inhalation of a gas mixture of 2.5% isoflurane and oxygen. After laparotomy under deep anesthesia, the common bile duct was doubly ligated and transected between the two ligatures and followed by abdominal closure with absorbable sutures. Postoperative analgesia was immediately performed by single subcutaneous injection with Ketoprofen at 5 mg/kg. For time-course observation, six mice were used for each time point. Specimens were collected at day zero for normal control and on 7 and 14 post operative days (POD) for BDL groups. For anti-inflammatory treatment, methylprednisolone sodium succinate (MP; Solu-Medrol, Pharmacia & Upjohn Company, New York, NY) or normal saline as solvent control was intraperitoneally administered under anesthesia at a dose of 5 mg/kg daily immediately after BDL surgery for 14 consecutive days. Six mice in each experimental group were used to observe survival and specimens were collected from survivors at end point.

Serum and liver tissue collection

At the time points indicated, 1.5 mL of whole blood was collected from the mice under deep anesthesia with inhalation of isoflurane, followed by direct percutaneous puncture of left ventricle. After centrifugation, sera were frozen at −80°C until analysis. Serum samples were used to determine biochemical parameter levels, including aspartate aminotransferase (AST), alanine aminotransferase (ALT) and total bilirubin, through a clinical automatic analyzer (Department of Laboratory Medicine, E-DA Hospital). Liver tissues were dissected and aliquoted into three parts for mRNA, protein and paraffin-embedded tissue sectioning.

ELISA

Serum cytokine levels were determined using commercially available ELISA kits (TNF-α and IL-6 from Biolegend, San Diego, CA; NGF from Millipore, Billerica, MA; TGF-β1 from R&D, Minneapolis, MN) according to manufacturer's instructions.

Reverse transcription (RT) and quantitative polymerase chain reaction (qPCR)

Total RNA was extracted from liver tissues or cultured cells using Trizol reagent (Invitrogen, Gaithersburg, MD). Two micrograms of total RNA was subject to RT-qPCR analysis as previously described [19]. In brief, an AMV reverse transcriptase system (Promega, Madison, WI) was used to generate complementary DNA. Real-time PCR amplification was performed on a thermal cycler (ABI 7500, Applied Biosystems, Foster City, CA) using the FastStart DNA Master^{PLUS} SYBR Green I kit (Roche, Castle Hill, Australia) under the following cycling conditions: one cycle of 95°C for 10 min, 45 cycles of 95°C for 15 s, 60°C for 5 s, and 72°C for 20 s. An extra melting curve protocol was used at the final step to validate specificity of PCR reaction. The primer sequences were: *β-actin*, 5'-TCC TGT GGC ATC CAC GAA ACT-3' (forward) and 5'-GAA GCA TTT GCG GTG GAC GAT-3' (reverse); *NGF*, 5'- ACG CAG CTT TCT ATC CT-3' (forward) and 5'- TTT AGT CCA GTG GGC TTC-3' (reverse); *TGF-β1*, 5'-CGT CAG ACA TTC GGG AAG C-3' (forward) and 3'-CAG CCA CTC AGG CGT ATC A-3' (reverse); *TNF-α*, 5'- TGA ACT TCG GGG TGA TCG GTC -3' (forward) and 5'-AGC CTT GTC CCT TGA AGA GAA -3' (reverse); *IL-6*, 5'-ATG AAC AAC GAT GAT GCA CTT G -3' (forward) and 5'-TAA GTC AGA TAC CTG ACA ACA G -3' (reverse). Parallel amplification of *β-actin* was used as the internal control. Fold change of each gene was calculated by the comparative Ct method.

Western blotting analysis

Liver tissues and cellular total proteins were extracted with an ice-cold RIPA lysis buffer containing protease inhibitor cocktail (Roche, Indianapolis, IN) and phosphatase inhibitors (1 mM sodium fluoride and 1 mM sodium orthovanadate), followed by protein measurement using a Coomassie protein assay kit (Pierce Biotechnology, Rockford, IL). SDS-PAGE, electrotransfer, and immunodetection were performed as previously described [20]. For detection, antibodies against NGF, 4-hydroxynonenal (4-HNE) and 8-hydroxydeoxyguanosine (8-OHdG) were from Millipore (Temecula, CA), TGF-β1, Bcl-2, Bax, phosphor- and total Akt were from Cell Signaling (Danvers, MA), and β-actin from Santa Cruz (Santa Cruz, CA); Bcl-2 and Bax from Trevigen (Gaithersburg, MD). For the purpose of semi-quantitative analysis, images of enhanced chemiluminescent signal were digitally documented on an imaging system (BioSpectrum, UVP, Upland, CA) and densitometrically analyzed using ImageJ software (NIH, USA). Relative protein levels were expressed as induction folds by calculating the density ratios between interest proteins and internal control and normalizing to negative control. For determination of cellular levels of oxidative adducts, densities of the major immunoreactive signals ranging from 40 to 90 kDa and those from 50 to 70 kDa were summed up and considered as formation of 4-HNE and 8-OHdG-related protein adducts, respectively. Fold changes were calculated by normalization with respective internal controls and expressed as folds of negative control.

Immunohistochemistry (IHC)

Formalin-fixed and paraffin-embedded mouse liver sections were used for IHC staining as previously described [20]. Briefly, the deparaffinized and rehydrated sections were treated for antigen retrieval and incubated with anti-NGF polyclonal antibodies (1:200 dilution) at 4°C overnight. The antigenicity in tissue sections was visualized with an HRP-linked polymer Envision detection system (DAKO, Glostrup, Denmark) followed by hematoxylin counterstaining. Normal liver sections treated with normal rabbit IgG at equimolar concentration were used as negative controls.

Cell culture and viability assay

For cytokine stimulation experiments, primary hepatocytes were isolated from male Fisher 344 rats (220-260 g) using a two-step collagenase perfusion method as previously described [21]. Primary hepatocytes were plated onto the plates pre-coated with type I collagen at a density of 5×10^5 cells/well. For NGF gene transfection, clone-9 hepatocytes (BCRC no. 60201) were purchased from Bioresource Collection and Research Center (Hsin-Chu, Taiwan) and maintained in F-12K medium (Sigma) with 10% heat-inactivated fetal calf serum (Invitrogen, Logan, UT) and

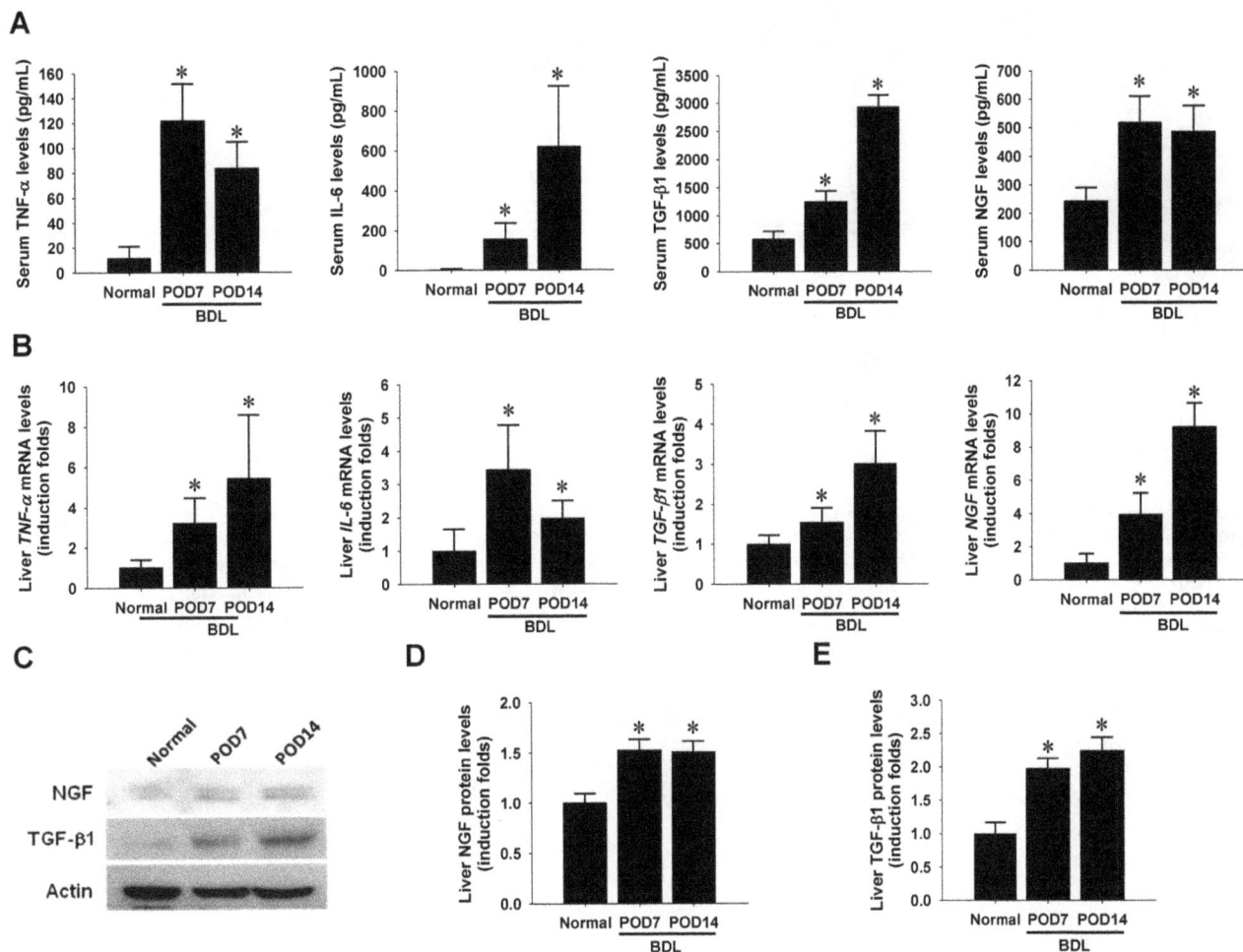

Figure 1. Increased serum NGF levels and intrahepatic NGF up-regulation in mice receiving bile duct ligation (BDL) surgery. Normal mice (n = 6) and those receiving BDL surgery were sacrificed at 7 or 14 post-operative days (POD7, n = 4) or (POD14, n = 4). (A) Serum levels of NGF and inflammatory cytokines, including TNF-α, IL-6 and TGF-β1 were determined by ELISA detection. (B) Liver tissue extracts were collected for mRNA isolation and subjected to RT-qPCR analysis. (C) Contents of NGF and TGF-β1 proteins in pooled liver extracts in experimental groups were measured using Western blot detection. Subsequent densitometrical analysis showed that cholestatic injury increased protein abundance of NGF (D) and TGF-β1 (E). All data are shown in mean±SEM. * indicates P <0.05 as compared to normal controls.

regular antibiotics [19]. For cell viability determination, an MTT-based cellular assay was performed as previously described [22].

NGF gene cloning, plasmid construction, and gene transfection

Full length NGF cDNA was cloned from a human fetal brain cDNA library (Stratagene, La Jolla, CA) by PCR reaction. The PCR primers used to clone the human full-length NGF cDNA (1052 bp) were designed based on the NGF sequence in the GenBank database (accession number, NM_002506; using 5′-CCG CTC GAG AGA GAG CGC TGG GAG C -3′ as forward primer, and 5′- TCC CCC GGG TTT ATG CTT CCA AAA -3′ as reverse primer). The PCR-amplified NGF cDNA was cloned into the pCR-Blunt II-TOPO vector and transformed into E. coli competent cells provided in the Zero Blunt TOPO PCR cloning kit (Invitrogen, Carlsbad, CA). After DNA sequencing analysis, the full length NGF cDNA fragment was subcloned into the XhoI and SmaI sites of pCMS-EGFP mammalian expression vector with enhanced green fluorescent protein (pCMS-EGFP, Invitrogen), thus giving rise to a recombinant plasmid containing NGF cDNA

(pCMS-NGF). Afterwards, the transformed cells were grown at 37°C until log phase (OD$_{600\ nm}$~0.5–0.9). The plasmid DNA was amplified and prepared for in vitro gene transfection using a liposome-based gene delivery system (Lipofectamine 2000, Invitrogen) according to manufacturer's instructions. Briefly, 1~4 µg plasmid DNA and Lipofectamine solution were mixed under serum-reduced condition and added into cultured cells at 70% confluency. After 24 hrs of incubation, the medium was replaced with fresh medium containing G418. The overexpressed GFP was observed under fluorescent microscopy, whereas the NGF gene transfection efficiency was assessed at both transcriptional and translational levels by using RT-qPCR and ELISA assays, respectively.

Apoptotic detection by TUNEL staining

Primary rat hepatocytes grown on chamber slides were fixed with ice-cold paraformaldehyde after receiving treatment, and subjected to a TUNEL-based in situ cell death detection assay. TUNEL signals were detected and visualized with DAB color formation using a standard protocol provided by manufacturer

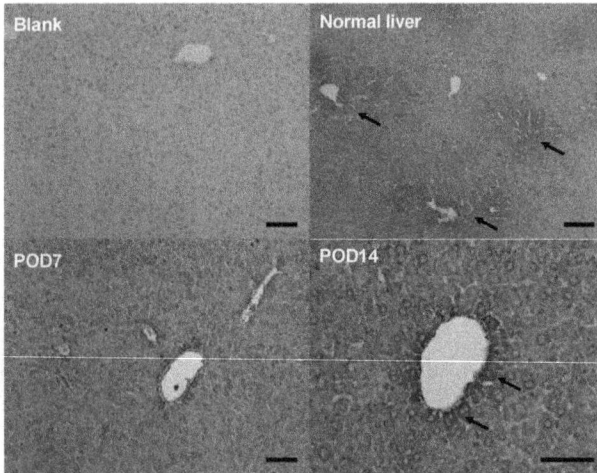

Figure 2. Parenchymal localization of NGF peptide in hepatocytes with cholestatic injury. The mice receiving bile duct ligation surgery were sacrificed at POD7 and POD14 and the liver tissues were fixed with formalin. Paraffin-embedded tissue sections were subjected to immunohistochemisty NGF staining. Arrows indicate the parenchymal localization of NGF antigenicity in zone 3. Bars, 50 μm.

(Roche, Mannheim, Germany). Slides were counterstained with hematoxylin. Quantification of the nuclear positive signals in each group was performed by counting at least 20 randomly selected images at high-power fields under microscopy and the positivity was shown in percentage of total cells.

Statistics

In vivo data were presented as mean±standard error of mean (SEM), while in vitro as mean±standard deviation (SD). Signif-

icance among groups was determined by one-way analysis of variance (ANOVA) followed by the Bonferroni post hoc test. A p value less than 0.05 was considered statistically significant.

Results

Up-regulated NGF expression in parenchymal hepatocytes of cholestatic livers

To elucidate the role played by NGF in cholestatic liver injuries, mice serum and liver tissues were collected at 7 days and 14 days post BDL operation. Elevated plasma levels of AST, ALT, and total bilirubin confirmed the effectiveness of surgery-induced cholestatic injury in mouse livers (**Figure S1**). ELISA detection showed that the pro-inflammatory cytokine levels including TNF-α, IL-6, and TGF-β1 in those mice with liver injury were significantly elevated. In parallel, the plasma NGF levels also increased along with the progression of liver fibrosis (**Figure 1A**). To investigate whether the cholestatic insult triggers de novo synthesis of NGF in livers, total RNA and protein extracts were used for further molecular measurements. RT-qPCR analysis indicated that the intrahepatic transcript contents of *TNF-α, IL-6, TGF-β1*, and *NGF* genes were remarkably up-regulated at POD14 of BDL surgery (**Figure 1B**), while gene expression patterns correlated with those of serum peptides. Western blotting demonstrated a similar increasing trend between intrahepatic NGF and TGF-β1 peptides (**Figure 1C, 1D, 1E**).

Localization of up-regulated NGF expression in liver parenchymal hepatocytes

To better characterize NGF localization in normal and injured livers, formalin-fixed and paraffin-embedded liver sections were applied to histopathological examinations. IHC staining results for NGF peptides showed that, in normal liver, NGF was expressed at constitutively lower levels and its antigenic signal was not homogeneously distributed in all liver lobules but only seen in

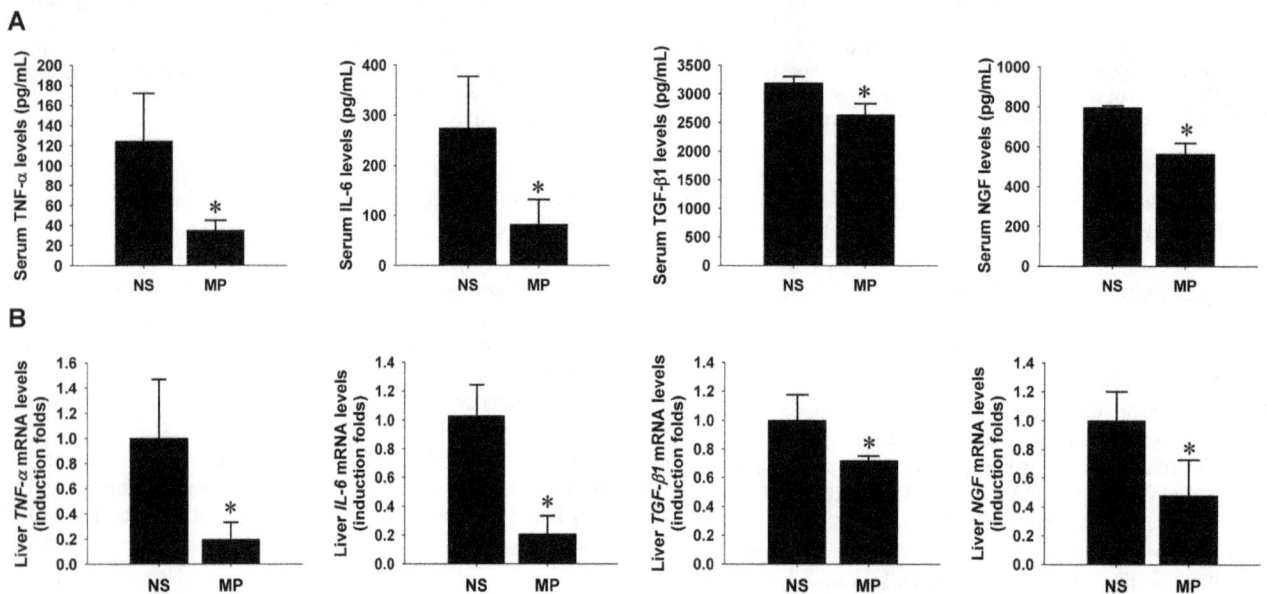

Figure 3. Amelioration of systemic inflammation and NGF up-regulation in cholestatic liver-bearing mice by methylprednisolone (MP). The mice receiving bile duct ligation surgery immediately underwent intraperitoneal administration with either normal saline (NS, n = 5) or MP (n = 3) at 1 mg/kg/day for 14 consecutive days. Mice plasma and liver extracts were collected and subjected to ELISA (A) and RT-qPCR analyses (B), respectively. Inflammatory cytokines measured included TNF-α, IL-6, NGF, and TGF-β1. Data are shown in mean±SEM. * indicates P<0.05 compared to NS group.

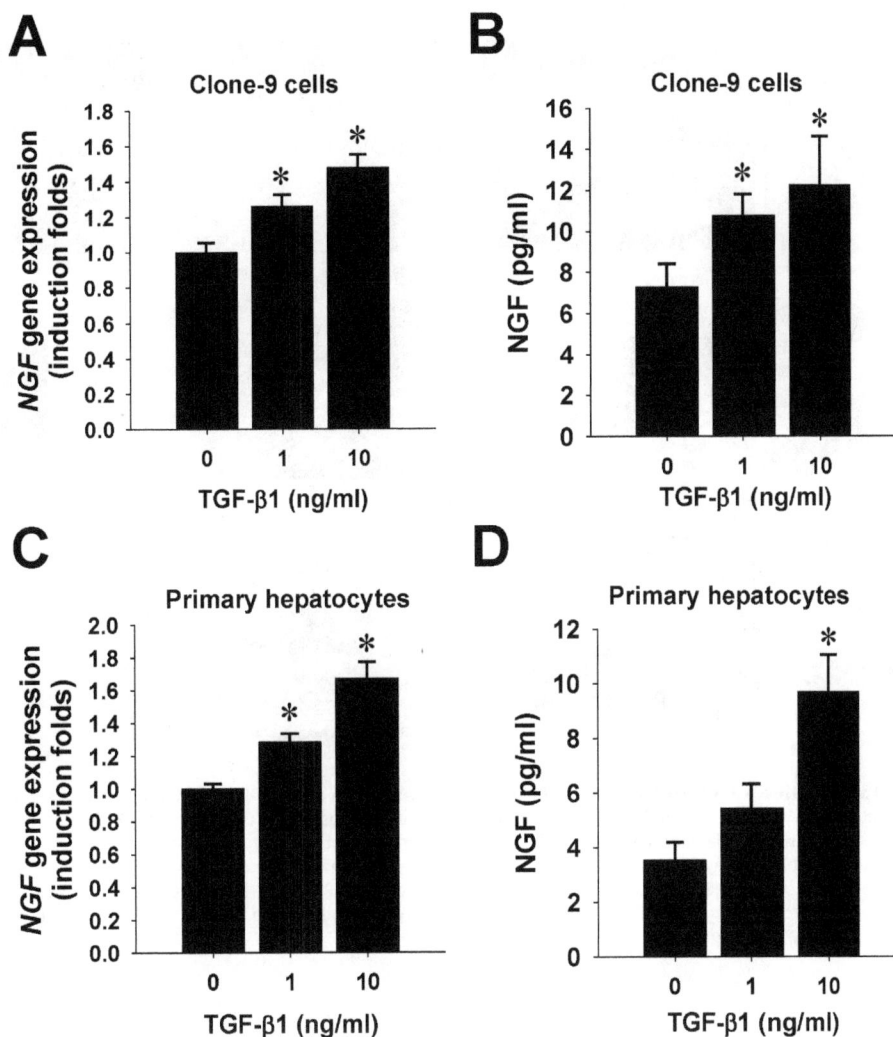

Figure 4. Up-regulation of NGF expression in rodent hepatocytes by TGF-β1. Both clone-9 and primary rat hepatocytes were used for NGF induction experiments. Clone-9 cells (A, B) and primary hepatocytes isolated from rats (C, D) were grown on collagen I-coated dishes and treated with TGF-β1 at the indicated doses (ng/mL) for 6 h. Total RNA was extracted and subjected to qPCR analysis for *NGF* mRNA levels (A, C). Besides, conditioned media for 24 hrs of treatment were collected for NGF ELISA detection (B, D). Note that TGF-β1 at 10 ng/mL remarkably increased de novo synthesis of NGF in both cultured hepatocytes. Data are representative results from three independent experiments and shown in mean±SD. * indicates $P<0.05$ compared to negative control group.

limited areas around the central vein zone 3 (**Figure 2**). After cholestatic injury, the constitutive NGF antigenicity level was apparently up-regulated and homogeneously distributed throughout the hepatic lobules at POD7. Again, stronger NGF antigenicity was seen mainly localized to the cytoplasms of parenchymal hepatocytes in injured livers at POD14. The above findings highlight the significance of the NGF up-regulation in liver fibrosis and suggest that it may play a pathophysiological role therein.

Amelioration of systemic and intrahepatic inflammation and suppression of NGF up-regulation by MP treatment

Since NGF has been previously reported to be up-regulated by inflammatory cytokines in diseased bladder [6], aorta [23], heart [11], and livers [17,18,24,25], we next to determine whether anti-inflammatory treatment could ameliorate the cholestasis-associated NGF up-regulation in injured livers. A synthetic glucocorticoid drug (MP) was used to suppress systemic immunactivity and thereby clarify the causal relationship between inflammation and systemic

and/or hepatic NGF up-regulation. Biochemistry data showed that MP treatment effectively suppressed AST (**Figure S2A**), but not the elevated plasma levels of ALT (**Figure S2B**) and total bilirubin (**Figure S2C**). Although Western blotting revealed that the MP treatment did not affect NGF and TGF-β1 peptide contents in cholestatic livers (**Figure S2D**), ELISA data demonstrated that MP prominently reduced serum levels of TNF-α, IL-6, TGF-β1, and NGF peptides in the surviving mice with cholestatic liver injury (**Figure 3A**). Similarly, RT-qPCR data also showed that MP treatment remarkably lowered the transcript contents of *TNF-α, IL-6, TGF-β1,* and *NGF* genes in injured livers (**Figure 3B**). These findings strongly suggested that MP administration not only systemically suppressed host immunity but also locally lowered NGF de novo synthesis in livers, which might, at least in part, underlie the reduction of NGF contents in plasma pools. More intriguingly, systemic immunosuppression by MP administration immediately after BDL surgery caused a higher mortality (3 out of 6) compared to that in normal saline controls (1 out of 6) (**Figure S3**), implicating that NGF may play a hepatoprotective role in livers and

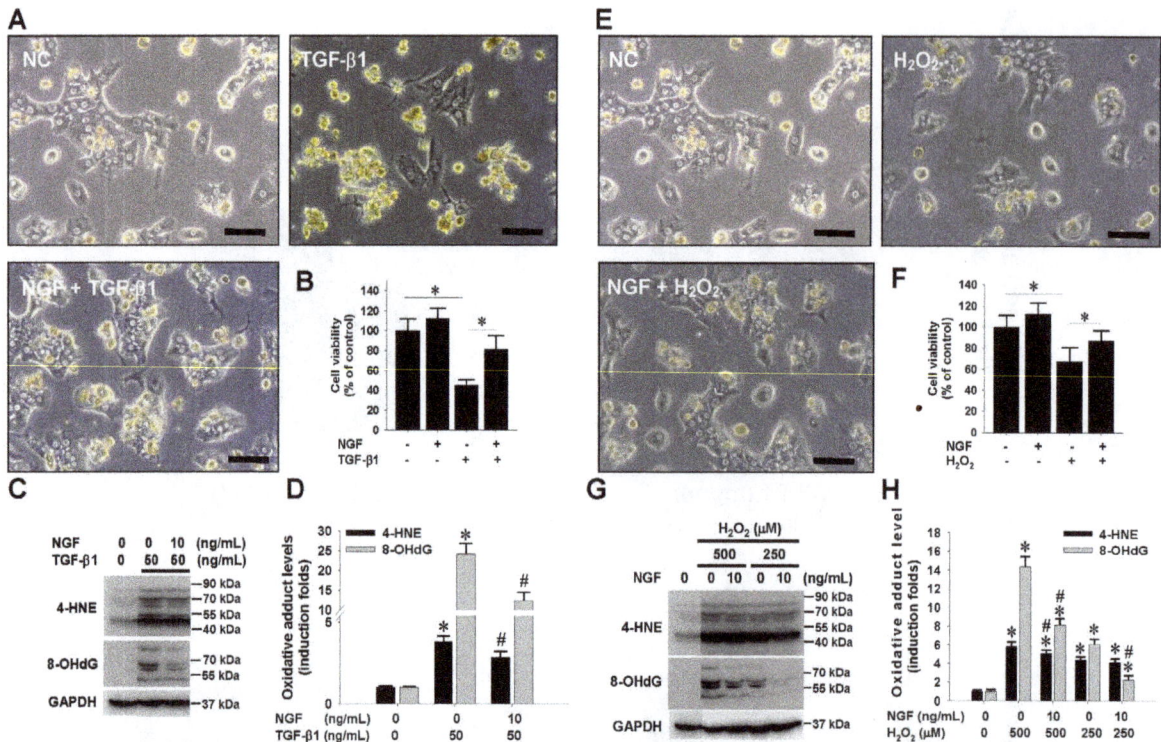

Figure 5. In vitro hepatoprotective effects of exogenous NGF supplementation on TGF-β1-induced and oxidative cell death. Primary rat hepatocytes were treated with recombinant NGF at 20 ng/mL for 24 hrs and exposed to TGF-β1 and H₂O₂ for another 24 hrs, followed by morphological observation (A, E) and cell viability assay (B, F). Representative microphotographs were shown (*Bar* = 50 μm). Cellular viability was determined by the MTT-based viability assay. Western blots (C, G) and subsequent densitometrical analyses (D, H) show that NGF pretreatment attenuated the elevation of 4-HNE and 8-OHdG modified protein levels induced by TGF-β1 and H₂O₂. Data are representative results from three independent experiments and expressed as mean±SD. * indicates $P<0.05$, as compared between groups or with negative control. # indicates $P<0.05$ compared with corresponding NGF-negative groups.

that suppression of systemic NGF levels at an acute stage may aggravate cholestatic injury and be lethal.

Up-regulation of NGF in rodent hepatocytes by TGF-β1

Since pro-inflammatory cytokines such as TGF-β1 have been demonstrated to up-regulate NGF expression in pancreatic stellate cells [26], we next sought to answer whether TGF-β1 or other pro-inflammatory cytokines are responsible for the NGF up-regulation in parenchymal hepatocytes of cholestatic livers. The result of a pilot cytokine screening showed that TNF-α, IL-1β, and IL-6 did not stimulate NGF gene transcription in primary rat hepatocytes (**Figure S4**). To further determine the regulatory role of TGF-β1, a line of clone-9 hepatocytes and primary rat hepatocytes were treated with TGF-β1 and the NGF expression was quantified at both transcription and translation levels. The RT-qPCR data revealed that exogenous TGF-β1 significantly increased NGF gene transcripts in both clone-9 cells (**Figure 4A**) and primary hepatocytes (**Figure 4C**). Similarly, ELISA for conditioned media also showed that exogenous TGF-β1 significantly increased soluble NGF peptide release from both clone-9 cells (**Figure 4B**) and primary hepatocytes (**Figure 4D**), supporting that NGF expression in parenchymal hepatocytes was up-regulated by TGF-β1.

Exogenous NGF treatment attenuated TGF-β1- and H₂O₂-induced hepatotoxicity and oxidative stress

To better understand the hepatoprotective effect of NGF, primary rat hepatocytes were pretreated with recombinant NGF

for 24 hrs, followed by treatment with either TGF-β1 or H₂O₂. Morphological observation (**Figure 5A, 5E**) and simultaneous cell viability evaluation demonstrated that pretreatment with exogenous NGF significantly rescued both the TGF-β1- (**Figure 5B**) and H₂O₂-elicited hepatotoxicity (**Figure 5F**). Since NGF was previously reported to play an anti-oxidative role in the nervous system [10,27,28], we next determined whether NGF protects hepatocytes through ameliorating oxidative stresses, NGF-pretreated hepatocytes were exposed to TGF-β1 and H₂O₂ insults and the intracellular levels of the proteins injured by 4-HNE and 8-OHdG adducts were determined. Western blotting and subsequent densitometrical results clearly showed that NGF pretreatment prominently suppressed the elevation of 4-HNE and 8-OHdG modified protein levels in hepatocytes induced by TGF-β1 (**Figure 5C, 5D**) and H₂O₂ (**Figure 5G, 5H**). Parallel Western detection confirmed the existence of two NGF receptors, TrkA and p75 NTR, in primary hepatocytes (**Figure S5**), supporting the integrity of NGF signaling machinery wherein.

Endogenous NGF overexpression protected cultured hepatocytes against TGF-β1- and H₂O₂-induced cell death and oxidative stresses

To mimic NGF overexpression in hepatocytes, clone-9 hepatocytes were transfected with pCMS plasmids carrying either EGFP or full-length NGF cDNA. ELISA confirmed that NGF gene delivery after 48 hrs of transfection drove clone-9 cells to significantly produce soluble NGF peptides (**Figure 6D**). We next

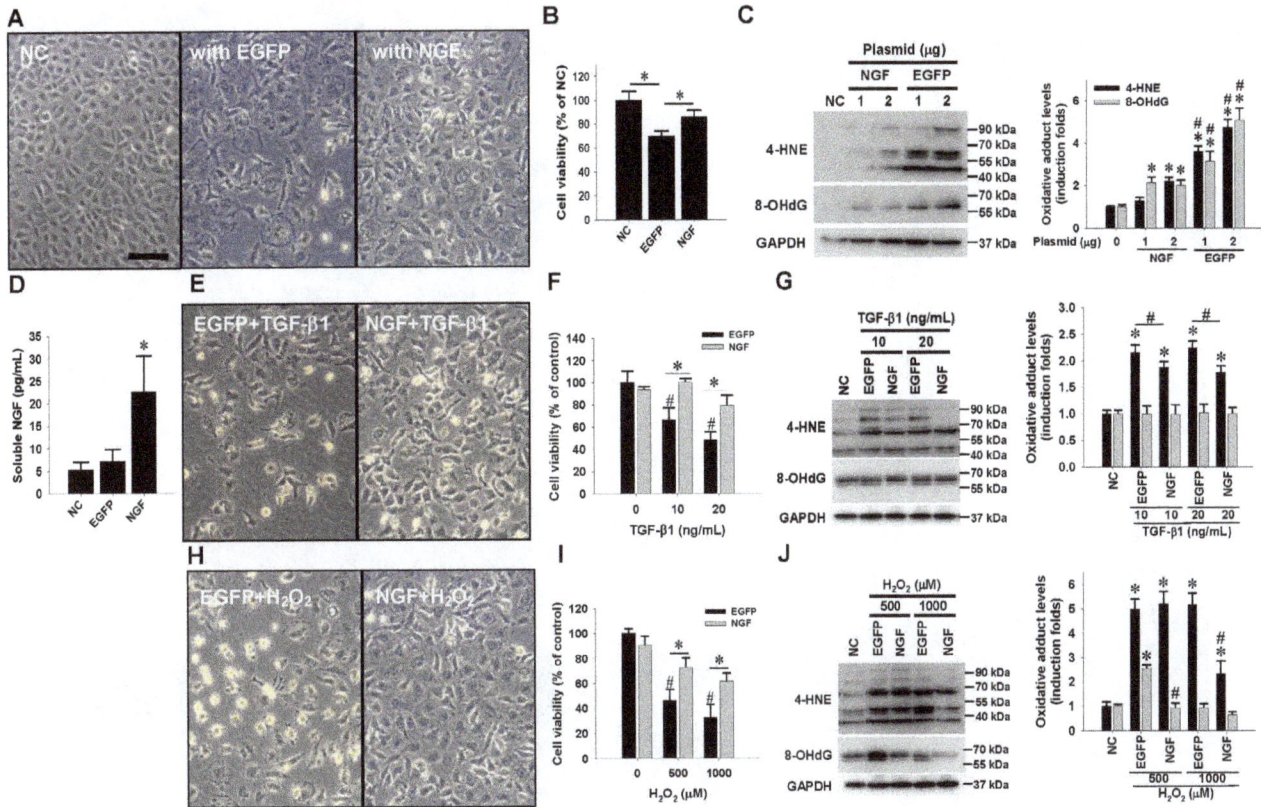

Figure 6. In vitro hepatoprotective effects of endogenous NGF overexpression on TGF-β1-induced and oxidative cell death. Clone-9 hepatocytes were transfected with either pCMS plasmid encoding EGFP (EGFP) or full-length NGF cDNA (NGF) using Lipofectamine reagent for 48 hrs, followed by morphological documentation (A, E, H) and viability determination (B, F, I). *Bar* = 50 μm. ELISA showed that NGF gene transfection for 48 hrs significantly induced soluble NGF production in conditioned medium (D). The clone-9 hepatocytes transfected with plasmids were exposed to either TGF-β1 or H_2O_2 for 24 hrs. The MTT cell viability assay showed that NGF overexpression not only prevented transfection-induced cytotoxicity (B) but also reduced TGF-β1 (F) and H_2O_2 (I) cytotoxicity. # and * indicate $P<0.05$ compared with negative control (NC) and between groups, respectively. Western blotting results showed that NGF overexpression attenuated the elevation of cellular oxidative adduct formation, including 4-HNE and 8-OHdG modified proteins, induced by plasmid transfection (C), TGF-β1 (G), and H_2O_2 insults (J). Data are representative results from three independent experiments, and normalized to NC. Density data are expressed as mean±SD. * indicates $P<0.05$ compared with NC; # indicates $P<0.05$ compared with corresponding EGFP- or NGF-transfected group.

tested the ability of NGF to affect viability in response to hepatotoxic insults. Transfection with EGFP plasmid alone reduced hepatocyte viability, while NGF overexpression prevented transfection-induced cytotoxicity as revealed by cell morphological observation (**Figure 6A**) and a cell viability assay (**Figure 6B**). Not surprisingly, transfection with plasmids expressing EGFP gave rise to higher cellular levels of both 4-HNE and 8-OHdG adduct-modified proteins than those in the cells overexpressing NGF (**Figure 6C**). To determine whether NGF overexpression ameliorated the pro-apoptotic and oxidative stimuli, NGF-overexpressing clone-9 cells were further treated with either TGF-β1 or H_2O_2 treatment for 24 hrs. Morphological observation clearly showed that NGF overexpression reduced the TGF-β1- and H_2O_2-induced cytotoxicity (**Figure 6E, 6H**). The cell viability assay consistently confirmed the NGF-driven hepatoprotective effects against both insults (**Figure 6F, 6I**). Again, NGF overexpression effectively reduced the increased formation of both oxidative adducts caused by TGF-β1 and H_2O_2 treatment (**Figure 6G, 6J**).

Anti-apoptosis was involved in NGF-mediated hepatoprotection against oxidative stress

To confirm the involvement of anti-apoptogenesis in the NGF-exhibited hepatoprotection against oxidative stress, in situ TUNEL detection was used to quantify the cellular apoptotic events under in vitro hepatotoxic injury. The TUNEL staining (**Figure 7A**) and quantitative results (**Figure 7B**) clearly indicated that both TGF-β1 and H_2O_2 insults significantly increased nuclear apoptotic signals in treated primary rat hepatocytes, while NGF pretreatment effectively prevented the increased hepatocytic apoptosis induced by both agents, supporting that NGF may functionally protect hepatocytes against TGF-β1- and oxidation-induced hepatocellular apoptosis.

Involvement of NGF-induced Akt phosphorylation and increase of Bcl-2/Bax ratios in hepatocytes under oxidative stress

Since the NGF-up-regulated Bcl-2 expression is responsible for its anti-oxidative ability in the nervous system [29], we next examined whether NGF modulates anti- and/or pro-apoptotic machineries in hepatocytes in vitro. Western blotting results showed that NGF at a dose of 10 ng/mL or higher for 24 hrs significantly induced phosphorylation of Akt (**Figure 8A, 8B**), the upstream mediator of Bcl-2. Concomitantly, NGF enhanced expression of anti-apoptotic protein Bcl-2, but suppressed that of pro-apoptotic protein Bax. Taken together, the Bcl-2-to-Bax ratios

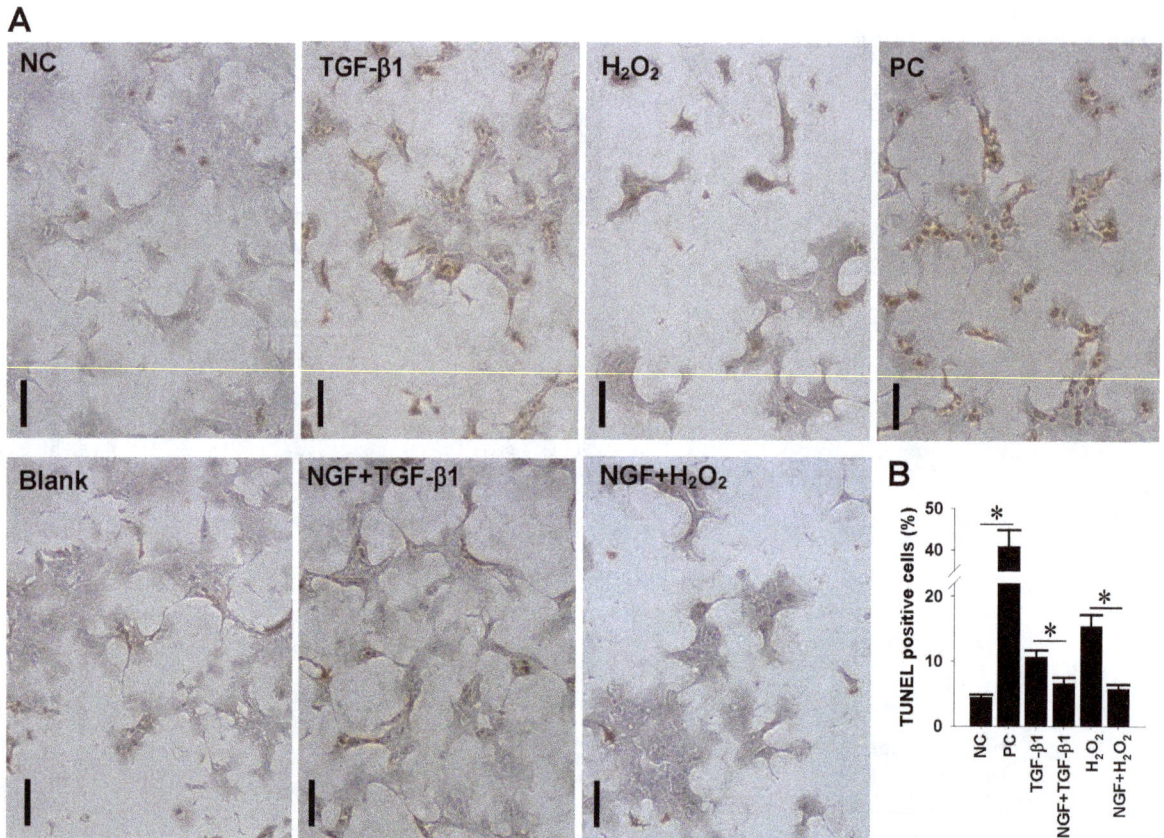

Figure 7. Anti-apoptotic effect of exogenous NGF pretreatment on TGF-β1-induced and oxidative cell death of primary hepatocytes. Primary rat hepatocytes grown on collagen-coated chamber slides were treated with recombinant NGF at 20 ng/mL for 24 hrs and exposed to either TGF-β1 at 50 ng/mL or H_2O_2 at 500 μM for another 24 hrs. After treatments, cells were fixed with paraformaldehyde and subjected to in situ TUNEL staining for cellular apoptotic events. (A) Representative TUNEL staining images are shown (Bars = 10 μm). Positive control (PC) was performed by treating cells with DNase I. (B) Nuclear TUNEL-positive signals were quantified by measuring 20 high-power fields per group and the positivity are shown in mean±SEM. * indicates $P<0.05$ as compared with negative control (NC) or between groups.

were significantly increased (**Figure 8C**). To further determine whether NGF supplementation ameliorates the disruption of anti-apoptotic machinery induced by TGF-β1 and H_2O_2, primary hepatocytes with or without NGF pretreatment were under exposure to either insult. Western blotting indicated that the NGF-up-regulated Akt phosphorylation (**Figure 8D, 8E**) and the increased Bcl-2-to-Bax ratios (**Figure 8D, 8F**) were only seen in the cells with H_2O_2 oxidative insult, but not in those with TGF-β1 treatment. These findings support that the up-regulated NGF in injured livers possesses anti-apoptotic benefit for hepatocyte survival through activating Akt signaling and restoring the equilibrium between Bcl-2 and Bax.

Discussion

Using a cell line and primary culture hepatocytes, this study is the first to show the hepatoprotective ability of NGF against oxidative stress and TGF-β1, both being key mediators in cholestatic liver injury [1,19]. Moreover, we also demonstrated that cholestasis-related inflammatory signaling such as TGF-β1 was able to induce NGF expression in cultured hepatocytes. The findings in the present study improved our understanding about the pathobiological role of NGF and the molecular mechanisms in the pathogenesis of cholestatic liver injury.

Enhanced hepatic expression of NGF during cholestasis

Local or systemic inflammation has been shown to induce NGF production in various tissue types, including the bladder [6], aorta [23] and heart [11]. NGF up-regulation in livers has been previously demonstrated under a wide range of pathological scenarios, including hepatotoxin-induced fibrosis [24], ischemia-reperfusion injury [25], oxidative injury [18], and cholestatic injury [17]. Consistent to the findings of previous studies, the present study demonstrated that BDL resulted in enhanced hepatic expression of both NGF mRNA and protein, which is temporarily related to the elevation of inflammatory cytokines in both liver extracts and plasma, including IL-6, TNF-α, and TGF-β1 (**Figure 1**). This observation implicates that inflammatory signaling may serve as an upstream player to modulate hepatic NGF expression. Although systemic anti-inflammation by MP did not effectively reduce hepatic content of NGF peptides (**Figure S2**), we still noted that it prominently suppressed the up-regulated NGF transcription in livers and the elevated NGF peptides in plasma of the BDL animals (**Figure 3**). Together with the fact that soluble NGF could be released from the NGF-overexpressing hepatocytes (**Figure 6D**), all the evidence supports the concept that hepatic NGF production is induced by inflammation signaling and eventually contributes to the systemic pool in this rodent BDL model. Moreover, it is worth to emphasize that the systemic immunosuppression by MP treatment immediately after BDL

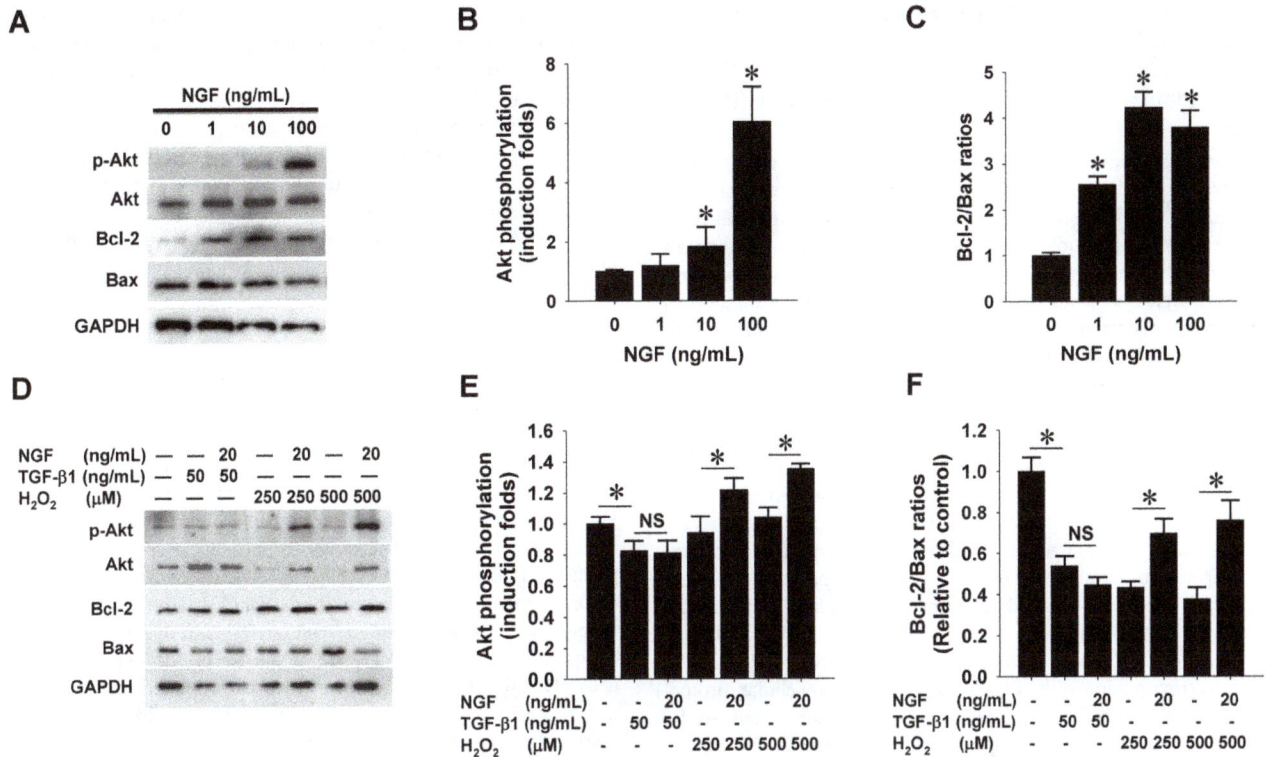

Figure 8. In vitro biomodulatory effect of NGF in cultured hepatocytes. (A) Clone-9 hepatocytes were treated with recombinant NGF for 24 hrs at indicated doses. Protein lysates were collected and subjected to Western blot detection for phosphor-Akt, total Akt, Bcl-2 and Bax expression levels. The relative Akt phosphorylation (B) and the ratios of Bcl-2 to Bax protein levels (C) were densitometrically measured. (D) Primary rat hepatocytes were exposed to TGF-β1 with or without 24 hrs of NGF pretreatment and the lysates were subjected to Western blotting. Subsequent densitometry showed that NGF pretreatment significantly increased Akt phosphorylation in the cells with H_2O_2, but not TGF-β1 insult (E). Similarly, NGF prominently ameliorated the down-regulated ratio of Bcl-2 and Bax proteins only in the cells with H_2O_2 exposure (F). Data are representative results from three independent experiments and shown as mean±SD. * indicates $P<0.05$ compared with negative control or between groups. NS, not significant.

surgery resulted in not only a lower plasma NGF levels but also a higher mortality (**Figure S3**). This result may, at least in part, reflect the hepatoprotective effect of systemic NGF peptides and highlight again the biological significance of the NGF up-regulation during the acute stage of liver injury.

Our in vitro mechanistic study further demonstrated that TGF-β1 but not IL-6 or TNF-α, stimulated NGF production by hepatocytes in a dose-dependent manner (**Figure 4**). Consistently,

other lines of evidence also showed that TGF-β1 up-regulates NGF in pancreatic stellate cells [26] and dental pulp cells[30], while the signaling pathways involved include activin-like kinase-5 [26] and mitogen-activated protein kinase [30]. Moreover, we also observed that NGF could induce the expression of TGF-β1 in hepatocytes (data not shown). In fact, a mutual regulation between TGF-β1 and NGF has been previously noted in the nervous system [31,32]. Further studies are warranted to explore the

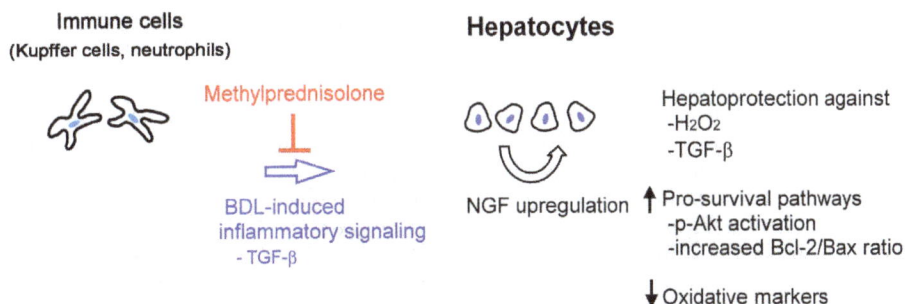

Figure 9. Hypothetical scheme showing the regulatory mechanisms and hepatoprotective roles of NGF. When the liver encounters BDL-induced cholestatic injury, inflammatory signals will induce NGF up-regulation, which can be blocked by systemic MP administration. In vitro study shows that TGF-β1 may be one of the upstream molecules that induce NGF expression in parenchymal hepatocytes. NGF is able to (1) ameliorate hepatocyte cell death caused by exogenous hydrogen peroxide and TGF-β1, (2) enhance pro-survival pathways, including p-Akt and Bcl-2/Bax ratio, (3) decrease intracellular oxidative adduct formation.

biological significance and the mechanisms underlying the mutual regulation of TGF-β1 and NGF in parenchymal hepatocytes.

Anti-oxidative and hepatoprotective effects of NGF on hepatocytes

Oxidative stress plays an important role in the cellular interactions, and is crucial during the pathogenesis of cholestatic liver injury and fibrosis [33,34]. ROS production can result from activation of resident macrophages (Kupffer cells) and recruitment of neutrophils and monocytes into the liver [2,35]. Moreover, cytokines released from inflammatory cells and accumulation of bile acids per se can increase oxidative stress and/or suppress anti-oxidative machinery within hepatocytes [36,37,38]. Upon exposure to oxidative stress, hepatocytes may develop several mechanisms, including impaired mitochondrial function, activation of the Akt pathway and disequilibrium of Bcl-2/Bax ratio [3]. Despite these intrinsic defense mechanisms, persistent accumulation of ROS within the liver will inevitably cause hepatocyte cell death and eventually promote fibrosis [38]. In this study, we demonstrated that exogenous NGF supplement and endogenous overexpression protected cultured rodent hepatocytes against hydrogen peroxide-induced hepatocellular death (**Figure 5**). Meanwhile, the oxidative markers, including 4-HNE and 8-OHdG-modified protein levels, were significantly reduced in the NGF-overexpressing cells (**Figure 6**). The hepatocellular protection of exogenous NGF supplement was demonstrated to be mediated through its anti-apoptotic effect (**Figure 7**). Mechanistically, the NGF-enhanced Akt phosphorylation and recovered Bcl-2/Bax ratios were involved therein (**Figure 8**). We thus propose that NGF is able to activate various cellular mechanisms of not only anti-oxidative stress but also cell survival (**Figure 9**). As for the contradictory result that NGF supplement did not change status of Akt phosphorylation and Bcl-2/Bax ratios (**Figure 8D**) but effectively ameliorated TGF-β1-induced cytotoxicity (**Figure 5**), it is most likely that different sets of apoptotic regulators may participate in different apoptogenic scenarios. In fact, not only Bcl but also inhibitor of apoptosis protein (IAP) family members could confer resistance to the induced hepatocyte apoptosis [36,39,40]. This issue awaits further elucidation.

In the context of NGF-elicited anti-oxidative effect, NGF deprivation has long been known to induce oxidative stress in the nervous system [27]. Moreover, NGF also inhibited the oxidative stress-induced apoptosis of PC12 cells [10,28]. The anti-oxidative effect of NGF is related to increased expression of Bcl-2 protein family members [29], activation of mammalian target of rapamycin (mTOR) signaling [7], phospho-Akt pathway [41] and up-regulation of free radical scavenging enzymes [28]. In line with our findings, oxidative stress itself was more recently reported to up-regulate NGF mRNA expression in livers, while functional blockade of NGF with a neutralizing antibody increased hepatic oxidative stress and decreased glutathione production [18], supporting an intimately regulatory relationship between NGF and oxidative stress. Moreover, exogenous NGF treatment was shown to increase hepatocyte intracellular glutathione levels through the TrkA signaling pathway [42]. It will be interesting to further elucidate the exact mechanisms involved in the hepatoprotective effects of NGF.

All the different liver component cells under cholestatic insult must adjust their cell behaviors and modulate interactions with one another accordingly. In addition to the anti-oxidative and hepatoprotective effects, NGF also plays multifunctional roles in an autocrine or paracrine manner and modulates fibrogenesis and liver regeneration during liver injuries. For instance, NGF, also reportedly produced by cholangiocytes, was demonstrated to promote cholangiocyte proliferation [17], which is one of the hallmarks of cholestasis. Another well-studied effect of NGF is to induce apoptosis of hepatic stellate cells and consequently resolve liver fibrosis [14,43,44]. These studies, along with our findings, highlight the pathobiological significance of NGF among different types of cells within normal and diseased livers. Although the present study demonstrated the molecular mechanisms of NGF regulation and its protective effects in hepatocytes, more studies are needed to determine whether the pathological changes of cholestasis could be ameliorated through manipulation of NGF receptor signal axis in vivo.

Study limitations

A few limitations of this study should be addressed. First, although we showed TGF-β1, but not IL-6 or TNF-α, enhanced NGF expression in vitro, it does not mean that TGF-β1 is the only upstream molecule regulating NGF or it is sufficient alone to regulate NGF in vivo. Experimental animal studies using TGF-β1 knock-out or knock-in animals may be helpful to clarify the causal relationship. Second, the downstream effects of NGF depend greatly on the type of its receptors. TrkA signaling tends to be cytoprotective, while p75 usually mediates proapoptotic effects. It is generally believed that hepatocytes mainly express TrkA receptors [41,42], whereas p75 is a marker for hepatic stellate cells [12,43]. Although we confirmed the existence of TrkA and p75 receptors in both normal liver parenchyma (**Figure S6**) and cultured primary hepatocytes (**Figure S5**), to date it is still unclear whether and how the expression of NGF receptors changes during liver injury. Therefore, to elucidate which receptor is responsible for the NGF-exhibited hepatoprotection is imperative to further understand the role of NGF in various liver diseases.

In conclusion, the present study demonstrates that NGF up-regulation is related to the inflammatory process during experimental cholestatic liver injury. We also found, for the first time, NGF is able to protect hepatocytes against oxidation-induced hepatocellular death, along with amelioration of cellular oxidative stress. Therefore, NGF supplementation is suggested to be therapeutically applicable in cholestatic liver injury.

Supporting Information

Figure S1 Serum biochemistry data from mice after cholestatic injury. Normal mice (n = 6) and those receiving BDL surgery were sacrificed at 7 or 14 post-operative days (POD7, n = 4) or (POD14, n = 4). Collected mouse sera were subjected to biochemical measurements, including AST (A), ALT (B), and total bilirubin (C). Gray boxes represent quartile deviation of groups. Data are shown in mean±SEM. * indicates $P<0.05$ as compared to the normal control.
(DOC)

Figure S2 Intrahepatic NGF protein expression was not changed by methylprednisolone (MP) treatment. The mice receiving bile duct ligation surgery underwent intraperitoneal administration with either normal saline (NS, n = 5) or MP (n = 3) at 5 mg/kg/day for 14 days. Mice sera were collected and subjected to biochemical analyses, including AST (A), ALT (B), and total bilirubin (C). The liver tissue were collected for protein isolation and subsequently subjected to Western blotting detection (D). Note that MP treatment only suppressed serum AST levels. Although MP treatment significantly reduced plasma NGF levels, it did not prevent the cholestasis-induced NGF up-regulation in livers. Data are shown in mean±SEM. * indicates $P<0.05$ compared to NS group.
(DOC)

Figure S3 Effect of methylprednisolone (MP) treatment on survival of mice with experimental cholestatic liver injury. Normal saline (NS) was used as solvent control group. (DOC)

Figure S4 TNF-α, IL-1β and IL-6 did not induce NGF expression. Primary hepatocytes were isolated from rats and grown on collagen I-coated dishes. Cells were treated with TNF-α, IL-1β, and IL-6 at the indicated doses (ng/mL) for 6 hrs. Total RNA was extracted and subjected to qPCR analysis for *NGF* mRNA levels. Expression levels were normalized to internal control actin gene. Note that none of cytokines increased transcription of NGF gene in cultured primary hepatocytes. Data are shown in mean±SD. * indicates $P<0.05$ compared to the negative controls (NC). (DOC)

Figure S5 Expression levels of TrkA and p75NTR in primary rat hepatocytes. Primary hepatocytes isolated from rat livers were treated with either recombinant TGF-β1 at 50 ng/mL or H_2O_2 at 500 μM for 24 hrs. Protein lysates were collected and subjected to Western blot detection for TrkA and p75NTR expression. (DOC)

Figure S6 Immunohistochemistry showing parenchymal localization of TrkA and p75NTR peptides in normal mouse livers. Formalin-fixed and paraffin-embedded mouse liver tissues were subjected to immunohistochemical staining and counterstaining with hematoxylin. Note that homogenous pattern and sparsely spotted distribution of TrkA and p75 NTR, two NGF cognate receptors, were seen in parenchyma of normal mouse livers. Images at right panel are the magnified rectangular area indicated by dashed lines in left images. Bars = 100 μm. (DOC)

Acknowledgments

Authors thank Ms. Shang-Chieh Lu, Chia-Wei Lin, and Mr. Po-Han Chen for their excellent technical assistance. We also acknowledge Dr. Randall Widelitz for editing the manuscript.

Author Contributions

Conceived and designed the experiments: MST PHL YHK. Performed the experiments: YCL CKS. Analyzed the data: YCL CKS SCH. Contributed reagents/materials/analysis tools: CKS SCH. Wrote the paper: MST PHL YHK.

References

1. Hofmann AF (2002) Cholestatic liver disease: pathophysiology and therapeutic options. Liver 22 Suppl 2: 14–19.
2. Jaeschke H (2011) Reactive oxygen and mechanisms of inflammatory liver injury: Present concepts. J Gastroenterol Hepatol 26 Suppl 1: 173–179.
3. Marin JJ, Hernandez A, Revuelta IE, Gonzalez-Sanchez E, Gonzalez-Buitrago JM, et al. (2013) Mitochondrial genome depletion in human liver cells abolishes bile acid-induced apoptosis: Role of the Akt/mTOR survival pathway and Bcl-2 family proteins. Free Radic Biol Med 61C: 218–228.
4. Snider WD (1994) Functions of the neurotrophins during nervous system development: what the knockouts are teaching us. Cell 77: 627–638.
5. Lockhart ST, Turrigiano GG, Birren SJ (1997) Nerve growth factor modulates synaptic transmission between sympathetic neurons and cardiac myocytes. J Neurosci 17: 9573–9582.
6. Guerios SD, Wang ZY, Boldon K, Bushman W, Bjorling DE (2008) Blockade of NGF and trk receptors inhibits increased peripheral mechanical sensitivity accompanying cystitis in rats. Am J Physiol Regul Integr Comp Physiol 295: R111–122.
7. Cao GF, Liu Y, Yang W, Wan J, Yao J, et al. (2011) Rapamycin sensitive mTOR activation mediates nerve growth factor (NGF) induced cell migration and pro-survival effects against hydrogen peroxide in retinal pigment epithelial cells. Biochem Biophys Res Commun 414: 499–505.
8. Cao Y, Liu JW, Yu YJ, Zheng PY, Zhang XD, et al. (2007) Synergistic protective effect of picroside II and NGF on PC12 cells against oxidative stress induced by H2O2. Pharmacol Rep 59: 573–579.
9. Kirschner PB, Jenkins BG, Schulz JB, Finkelstein SP, Matthews RT, et al. (1996) NGF, BDNF and NT-5, but not NT-3 protect against MPP+ toxicity and oxidative stress in neonatal animals. Brain Res 713: 178–185.
10. Satoh T, Sakai N, Enokido Y, Uchiyama Y, Hatanaka H (1996) Free radical-independent protection by nerve growth factor and Bcl-2 of PC12 cells from hydrogen peroxide-triggered apoptosis. J Biochem 120: 540–546.
11. Tsai MS, Chung SD, Liang JT, Ko YH, Hsu WM, et al. (2010) Enhanced expression of cardiac nerve growth factor and nerve sprouting markers in rats following gastric perforation: the association with cardiac sympathovagal balance. Shock 33: 170–178.
12. Kendall TJ, Hennedige S, Aucott RL, Hartland SN, Vernon MA, et al. (2009) p75 Neurotrophin receptor signaling regulates hepatic myofibroblast proliferation and apoptosis in recovery from rodent liver fibrosis. Hepatology 49: 901–910.
13. Lin N, Hu K, Chen S, Xie S, Tang Z, et al. (2009) Nerve growth factor-mediated paracrine regulation of hepatic stellate cells by multipotent mesenchymal stromal cells. Life Sci 85: 291–295.
14. Trim N, Morgan S, Evans M, Issa R, Fine D, et al. (2000) Hepatic stellate cells express the low affinity nerve growth factor receptor p75 and undergo apoptosis in response to nerve growth factor stimulation. Am J Pathol 156: 1235–1243.
15. Kishibe K, Yamada Y, Ogawa K (2002) Production of nerve growth factor by mouse hepatocellular carcinoma cells and expression of TrkA in tumor-associated arteries in mice. Gastroenterology 122: 1978–1986.
16. Xu LB, Liu C, Gao GQ, Yu XH, Zhang R, et al. (2010) Nerve growth factor-beta expression is associated with lymph node metastasis and nerve infiltration in human hilar cholangiocarcinoma. World J Surg 34: 1039–1045.
17. Gigliozzi A, Alpini G, Baroni GS, Marucci L, Metalli VD, et al. (2004) Nerve growth factor modulates the proliferative capacity of the intrahepatic biliary epithelium in experimental cholestasis. Gastroenterology 127: 1198–1209.
18. Valdovinos-Flores C, Gonsebatt ME (2013) Nerve growth factor exhibits an antioxidant and an autocrine activity in mouse liver that is modulated by buthionine sulfoximine, arsenic, and acetaminophen. Free Radic Res 47: 404–412.
19. Kao YH, Chen CL, Jawan B, Chung YH, Sun CK, et al. (2010) Upregulation of hepatoma-derived growth factor is involved in murine hepatic fibrogenesis. J Hepatol 52: 96–105.
20. Hu TH, Huang CC, Liu LF, Lin PR, Liu SY, et al. (2003) Expression of hepatoma-derived growth factor in hepatocellular carcinoma. Cancer 98: 1444–1456.
21. Pinkse GG, Voorhoeve MP, Noteborn M, Terpstra OT, Bruijn JA, et al. (2004) Hepatocyte survival depends on beta1-integrin-mediated attachment of hepatocytes to hepatic extracellular matrix. Liver Int 24: 218–226.
22. Chiba T, Yokosuka O, Fukai K, Kojima H, Tada M, et al. (2004) Cell growth inhibition and gene expression induced by the histone deacetylase inhibitor, trichostatin A, on human hepatoma cells. Oncology 66: 481–491.
23. Tsai MS, Ko YH, Hsu WM, Liang JT, Lai HS, et al. (2011) Enhanced aortic nerve growth factor expression and nerve sprouting in rats following gastric perforation. J Surg Res 171: 205–211.
24. Oakley F, Trim N, Constandinou CM, Ye W, Gray AM, et al. (2003) Hepatocytes express nerve growth factor during liver injury: evidence for paracrine regulation of hepatic stellate cell apoptosis. Am J Pathol 163: 1849–1858.
25. Ohkubo T, Sugawara Y, Sasaki K, Maruyama K, Ohkura N, et al. (2002) Early induction of nerve growth factor-induced genes after liver resection-reperfusion injury. J Hepatol 36: 210–217.
26. Haas SL, Fitzner B, Jaster R, Wiercinska E, Gaitantzi H, et al. (2009) Transforming growth factor-beta induces nerve growth factor expression in pancreatic stellate cells by activation of the ALK-5 pathway. Growth Factors 27: 289–299.
27. Nair P, Tammariello SP, Estus S (2000) Ceramide selectively inhibits apoptosis-associated events in NGF-deprived sympathetic neurons. Cell Death Differ 7: 207–214.
28. Sampath D, Jackson GR, Werrbach-Perez K, Perez-Polo JR (1994) Effects of nerve growth factor on glutathione peroxidase and catalase in PC12 cells. J Neurochem 62: 2476–2479.
29. Maroto R, Perez-Polo JR (1997) BCL-2-related protein expression in apoptosis: oxidative stress versus serum deprivation in PC12 cells. J Neurochem 69: 514–523.
30. Yongchaitrakul T, Pavasant P (2007) Transforming growth factor-beta1 up-regulates the expression of nerve growth factor through mitogen-activated protein kinase signaling pathways in dental pulp cells. Eur J Oral Sci 115: 57–63.
31. Lindholm D, Hengerer B, Zafra F, Thoenen H (1990) Transforming growth factor-beta 1 stimulates expression of nerve growth factor in the rat CNS. Neuroreport 1: 9–12.

32. Cosgaya JM, Aranda A (1995) Nerve growth factor regulates transforming growth factor-beta 1 gene expression by both transcriptional and posttranscriptional mechanisms in PC12 cells. J Neurochem 65: 2484–2490.

33. Ljubuncic P, Tanne Z, Bomzon A (2000) Evidence of a systemic phenomenon for oxidative stress in cholestatic liver disease. Gut 47: 710–716.

34. Assimakopoulos SF, Mavrakis AG, Grintzalis K, Papapostolou I, Zervoudakis G, et al. (2008) Superoxide radical formation in diverse organs of rats with experimentally induced obstructive jaundice. Redox Rep 13: 179–184.

35. Gujral JS, Farhood A, Bajt ML, Jaeschke H (2003) Neutrophils aggravate acute liver injury during obstructive cholestasis in bile duct-ligated mice. Hepatology 38: 355–363.

36. Herrera B, Alvarez AM, Beltran J, Valdes F, Fabregat I, et al. (2004) Resistance to TGF-beta-induced apoptosis in regenerating hepatocytes. J Cell Physiol 201: 385–392.

37. Franklin CC, Rosenfeld-Franklin ME, White C, Kavanagh TJ, Fausto N (2003) TGFbeta1-induced suppression of glutathione antioxidant defenses in hepatocytes: caspase-dependent post-translational and caspase-independent transcriptional regulatory mechanisms. FASEB J 17: 1535–1537.

38. Czaja MJ (2002) Induction and regulation of hepatocyte apoptosis by oxidative stress. Antioxid Redox Signal 4: 759–767.

39. Schoemaker MH, Ros JE, Homan M, Trautwein C, Liston P, et al. (2002) Cytokine regulation of pro- and anti-apoptotic genes in rat hepatocytes: NF-kappaB-regulated inhibitor of apoptosis protein 2 (cIAP2) prevents apoptosis. J Hepatol 36: 742–750.

40. Brenner C, Galluzzi L, Kepp O, Kroemer G (2013) Decoding cell death signals in liver inflammation. J Hepatol 59: 583–594.

41. Lu J, Wu DM, Hu B, Zheng YL, Zhang ZF, et al. (2010) NGF-Dependent activation of TrkA pathway: A mechanism for the neuroprotective effect of troxerutin in D-galactose-treated mice. Brain Pathol 20: 952–965.

42. Li JF, Shu JC, Tang SH, Deng YM, Fu MY, et al. (2013) beta-Nerve growth factor attenuates hepatocyte injury induced by D-galactosamine in vitro via TrkA NGFR. Mol Med Rep 8: 813–817.

43. Suzuki K, Tanaka M, Watanabe N, Saito S, Nonaka H, et al. (2008) p75 Neurotrophin receptor is a marker for precursors of stellate cells and portal fibroblasts in mouse fetal liver. Gastroenterology 135: 270–281 e273.

44. Passino MA, Adams RA, Sikorski SL, Akassoglou K (2007) Regulation of hepatic stellate cell differentiation by the neurotrophin receptor p75NTR. Science 315: 1853–1856.

5

Reprogramming Suppresses Premature Senescence Phenotypes of Werner Syndrome Cells and Maintains Chromosomal Stability over Long-Term Culture

Akira Shimamoto[1]*, **Harunobu Kagawa**[1], **Kazumasa Zensho**[1], **Yukihiro Sera**[1], **Yasuhiro Kazuki**[2], **Mitsuhiko Osaki**[2,3], **Mitsuo Oshimura**[2], **Yasuhito Ishigaki**[4], **Kanya Hamasaki**[5], **Yoshiaki Kodama**[5], **Shinsuke Yuasa**[6], **Keiichi Fukuda**[6], **Kyotaro Hirashima**[7], **Hiroyuki Seimiya**[7], **Hirofumi Koyama**[8], **Takahiko Shimizu**[8], **Minoru Takemoto**[9], **Koutaro Yokote**[9], **Makoto Goto**[10], **Hidetoshi Tahara**[1]*

1 Department of Cellular and Molecular Biology, Graduate School of Biomedical & Health Sciences, Hiroshima University, Hiroshima, Japan, 2 Department of Biomedical Science, Institute of Regenerative Medicine and Biofunction, Graduate School of Medical Science, Tottori University, Yonago, Japan, 3 Division of Pathological Biochemistry, Faculty of Medicine, Tottori University, Yonago, Japan, 4 Medical Research Institute, Kanazawa Medical University, Kahoku, Ishikawa, Japan, 5 Department of Genetics, Radiation Effects Research Foundation, Hiroshima, Japan, 6 Department of Cardiology, Keio University School of Medicine, Tokyo, Japan, 7 Division of Molecular Biotherapy, The Cancer Chemotherapy Center, Japanese Foundation For Cancer Research, Tokyo, Japan, 8 Department of Advanced Aging Medicine, Chiba University Graduate School of Medicine, Chiba, Japan, 9 Department of Clinical Cell Biology and Medicine, Chiba University Graduate School of Medicine, Chiba, Japan, 10 Division of Orthopedic Surgery & Rheumatology, Tokyo Women's Medical University Medical Center East, Tokyo, Japan

Abstract

Werner syndrome (WS) is a premature aging disorder characterized by chromosomal instability and cancer predisposition. Mutations in *WRN* are responsible for the disease and cause telomere dysfunction, resulting in accelerated aging. Recent studies have revealed that cells from WS patients can be successfully reprogrammed into induced pluripotent stem cells (iPSCs). In the present study, we describe the effects of long-term culture on WS iPSCs, which acquired and maintained infinite proliferative potential for self-renewal over 2 years. After long-term cultures, WS iPSCs exhibited stable undifferentiated states and differentiation capacity, and premature upregulation of senescence-associated genes in WS cells was completely suppressed in WS iPSCs despite *WRN* deficiency. WS iPSCs also showed recapitulation of the phenotypes during differentiation. Furthermore, karyotype analysis indicated that WS iPSCs were stable, and half of the descendant clones had chromosomal profiles that were similar to those of parental cells. These unexpected properties might be achieved by induced expression of endogenous telomerase gene during reprogramming, which trigger telomerase reactivation leading to suppression of both replicative senescence and telomere dysfunction in WS cells. These findings demonstrated that reprogramming suppressed premature senescence phenotypes in WS cells and WS iPSCs could lead to chromosomal stability over the long term. WS iPSCs will provide opportunities to identify affected lineages in WS and to develop a new strategy for the treatment of WS.

Editor: Zhongjun Zhou, The University of Hong Kong, Hong Kong

Funding: This work was supported by a Grant-in-Aid for Challenging Exploratory Research No. 25670030 (to A.S) and for Scientific Research No. 20014015 (to H.T) and No. 24590902 (to M.G) from the Ministry of Education, Culture, Sports, Science and Technology of Japan. This work was also supported by a Health and Labor Sciences Research Grant from the Ministry of Health Labor and Welfare of Japan (to A.S). The funders had no role in study design, data collection and analysis, decision to publish, or preparation of the manuscript.

Competing Interests: The authors have declared that no competing interests exist.

* Email: shim@hiroshima-u.ac.jp (AS); toshi@hiroshima-u.ac.jp (HT)

Introduction

Werner syndrome (WS) is a rare human autosomal recessive disorder characterized by early onset of aging-associated diseases, chromosomal instability, and cancer predisposition [1,2]. Fibroblasts from WS patients exhibit premature replicative senescence [3], and *WRN*, a gene responsible for the disease, encodes a RecQ-type DNA helicase [4–7], that is involved in maintenance of chromosome integrity during DNA replication, repair, and recombination [8,9]. WRN helicase is known to interact with a variety of proteins associated with DNA metabolism including

proteins of replication fork progression, base excision repair, and telomere maintenance [8,9]. The dysfunction of WRN helicase causes defects in telomeric lagging-strand synthesis and telomere loss during DNA replication [10]. Further, it is also reported that telomere loss caused by a defect in WRN helicase involves chromosome end fusions that are suppressed by telomerase [11]. These observations suggest that premature senescence in WS cells reflects defects in telomeric lagging-strand synthesis followed by accelerated telomere loss during DNA replication.

Somatic cell reprogramming follows the introduction of several pluripotency genes including Oct3/4, Sox2, Klf4, c-myc, Nanog and Lin-28 into differentiated cells such as dermal fibroblasts, blood cells, and other cell types [12–17]. During reprogramming, somatic cell-specific genes are suppressed, and embryonic stem cell (ESC)-specific pluripotency genes are induced, leading to the generation of iPSCs with undifferentiated states and pluripotency [18]. In addition, ESC-like infinite proliferative potential is directed by induction of the endogenous telomere reverse-transcriptase catalytic subunit (hTERT) gene and the reactivation of telomerase activity during reprogramming [13,18].

Recently, Cheung et al. demonstrated that cells from WS patients were successfully reprogrammed into iPSCs with restored telomere function, suggesting that the induction of hTERT during reprogramming suppresses telomere dysfunction in WS cells lacking WRN [19]. However, the effects of long-term culture on the undifferentiated states, self-renewal abilities, and differentiation potentials of WS iPSCs remain unknown. In a previous study, progressive telomere shortening and loss of self-renewal ability were observed in iPSCs from dyskeratosis congenital patient cells in a long-term culture [20], warranting the evaluation of the properties of patient cell-derived iPSCs with telomere dysfunctions over the long term.

In this study, we cultured WS iPSCs with self-renewal capacity and infinite proliferative potential for over 2 years and reported similar properties to those of normal iPSCs including undifferentiated states and differentiation ability. Notably, WS iPSCs maintained stable karyotypes and their potential to recapitulate premature senescence phenotypes during differentiation over the long term. The present data demonstrate that reprogramming suppresses premature senescence phenotypes in WS cells by reversing the aging process and restoring telomere maintenance over the long term.

Materials and Methods

Cell lines

WS patients were diagnosed on the basis of clinical symptoms and WRN gene mutations. A0031 WS patient fibroblasts from a 37-year-old male were obtained from Goto Collection of RIKEN Bioresource Center (https://www.brc.riken.jp/lab/cell/english/index_gmc.shtml) [21], and WSCU01 patient fibroblasts were isolated from a 63-year-old Japanese male who was diagnosed at Chiba University. Both fibroblast isolates had type 4/6 heterozygous mutations. TIG-3 human fetal lung-derived fibroblast cells and WS patient-derived fibroblasts were used to generate iPSC lines. PLAT-A cells (kindly provided from Dr. Toshio Kitamura) were used to produce retroviruses [22]. SNL 76/7 (SNL) cells (DS pharma biomedical) were used as feeder layers for reprogramming of fibroblasts and maintenance of iPSCs. The human fibroblast-derived iPSC line iPS-TIG114-4f1 was obtained from the National Institute of Biomedical Innovation [23].

PLAT-A cells, TIG-3 fibroblasts, TIG-114 fibroblasts from the 36-year-old male, and SNL cells were grown in the Dulbecco's modified Eagle's medium (DMEM; Sigma) supplemented with 10% fetal bovine serum (FBS; Hyclone) and antibiotics (Invitrogen). WS fibroblasts were maintained on collagen-coated dishes (Nitta Gelatin), SNL cells were maintained on gelatin-coated dishes (Nitta Gelatin), and iPSCs were maintained in the ES medium comprising Knockout DMEM (Invitrogen) supplemented with 20% Knockout Serum Replacement (Invitrogen), glutamine, non-essential amino acids, β-mercaptoethanol and 4-ng/ml basic FGF. All cells were maintained at 37°C under 5% CO_2 atmosphere.

Generation of iPSCs

The generation of iPSCs was performed as described previously [13]. Briefly, 2×10^6 PLAT-A cells were plated in T25 flasks (Biocoat, BD Falcon), and were transfected with 4 µg pMXs-OCT3/4, SOX2, KLF-4, and c-myc (Addgene) 1 day later. Twenty-four hours after transfection, the culture medium was replaced with a fresh medium and cells were incubated for 24 h prior to harvest of viral supernatants. Viral supernatants containing Yamanaka factors were combined in even ratios.

For reprogramming experiments, 3×10^5 fibroblasts were seeded on 60-mm dishes and were infected with viral supernatants containing Yamanaka factors in the presence of 8 µg/ml polybrene 1 day later. Four days after infection, fibroblasts were harvested, and 1×10^5 cells were reseeded onto mitomycin C-inactivated SNL feeder layers on 100-mm dishes. Twenty-four hours after reseeding, the medium was replaced with the ES medium, and cultures were maintained by replacing the medium every other day. Approximately 30 days after retroviral transduction, emerging iPSC colonies with ESC colony-like flat and round shapes were picked up by mechanical dissection and were plated onto fresh feeder layers on 4-well plates (Thermo Scientific Nunc). Subsequently, iPSC lines were established by successive passages onto fresh feeder layers with split ratios between 1:3 and 1:5 using dispase (Roche Applied Science).

Alkaline phosphatase activity

Undifferentiated states of emerging colonies were examined using alkaline phosphatase staining. After formalin fixation, colonies were stained with reaction buffer containing 100 mM Tris-Cl (pH 8.5), 0.25 mg/ml Naphthol AS-BI phosphate (Sigma) and 0.25 mg/ml fast red violet LB salt (Sigma).

Embryoid body formation and in vitro differentiation

Clumps of iPSCs were transferred to non-adherent polystyrene dishes containing the ES medium without basic FGF to form embryoid bodies (EBs). The medium was replaced every other day. After 8 days of floating culture, EBs were transferred onto gelatin-coated plates and were maintained in DMEM supplemented with 10% FBS, β-mercaptoethanol, and antibiotics for another 8 days. For detection of senescence phenotypes during differentiation, Y-27632-treated iPSCs were dissociated into single cell suspensions with Accutase (Innovative Cell Technologies) and 1×10^4 cells were transferred into 96-well V-shaped bottom plates (Greiner Bio-One) to form evenly sized EBs. After 12 days of EB formation in the ES medium without basic FGF, EBs were cultured in DMEM supplemented with 10% FBS, β-mercaptoethanol, and antibiotics.

Teratoma formation

After harvest, 1×10^6 iPSCs were injected into the testes of a severe combined immunodeficient (SCID) mice (CREA, Japan). Three months after injection, tumors were dissected and were fixed using 4% paraformaldehyde. Subsequently, dissected tumor tissues were embedded in paraffin and were sliced and stained with hematoxylin and eosin.

Western blot

Whole cell lysates were prepared in SDS sample buffer and subjected to electrophoresis on 8% SDS-polyacrylamide gels, and separated proteins were transferred onto PVDF membranes (FluoroTrans W, Pall Corporation). Membranes were blocked with TBS-T containing 5% skim milk and were then incubated with anti-WRN (1:500, 4H12, Abcom) or anti-β-actin (1:30000,

Ac-15, Sigma) monoclonal antibodies for 3 h at room temperature. Membranes were then washed with TBS-T and were incubated with horseradish peroxidase-conjugated anti-mouse IgG (1:5000, NA931V, GE) for 1 h at room temperature. Chemiluminescence reactions were performed using Western Lightning Plus-ECL (PerkinElmer) and were detected using exposure of x-ray films.

Mutation analysis

The DNA fragments mut.4 (c.3139-1G>C) and mut.6 (c.1105C>T) were amplified with the primer pairs WS_mut4_U, GGTAAACGGTGTAGGAGTCTGC and WS_mut4_L, CTTGTGAGAGGCCTATAAACTGG, and WS_mut6_U, TGAAGATTCAACTACTGGGGGAGTAC and WS_ mut6_L, ACGGGAATAAAGTCTGCCAGAACC, respectively, using genomic DNA as a template. Mutations were analyzed by direct sequencing using these PCR primers.

Short tandem repeat (STR) analysis

Genomic DNAs were purified from WS fibroblasts and their derivative iPSC clones using phenol/chloroform extraction and were then used for analysis using a Cell ID System (Promega). PCR products were analyzed using an Applied Biosystems 3130xl Genetic Analyzer and GeneMapper software.

Gene expression profiling

Cy3-labeled total RNAs were hybridized onto Human Genome U133 Plus 2.0 Arrays (GeneChip, Affymetrix). Arrays were then scanned using the GeneChip Scanner 3000 7G (Affymetrix), and the obtained data were analyzed by Affymetrix Expression Console Software. The microarray dataset has been deposited in the NCBI Gene Expression Omnibus database under Series Accession GSE62114.

Measurement of telomere length

Genomic DNAs were digested using *Hinf*I restriction enzyme (TakaraBio), and were subjected to electrophoresis on 1% agarose gels. Size-fractionated DNAs were transferred onto Hybond-N+ membranes (GE). Membranes were hybridized with a digoxigenin-labeled $(CCCTAA)_4$ probe, and TRFs were detected using TeloTAGGG Telomere Length Assays (Roche Applied Science) according to the manufacturer's instructions.

RT-PCR and real-time qRT-PCR analysis of mRNA expression

Total RNA was prepared using RNeasy spin columns (Qiagen) according to the manufacturer's instructions. RT-PCR was performed with 0.1 μg of total RNA using SuperScript One-Step RT-PCR (Invitrogen). Semi-quantitative analysis was performed after converting total RNA into cDNA using a High Capacity RNA-to-cDNA kit (Life Technologies), and real-time PCR was performed using a Rotor-Gene SYBR Green PCR kit (Qiagen). Relative gene expression levels were analyzed according to the ΔΔCt method using Ct values of GAPDH mRNA as an internal control. Primer sequences are listed in Tables S1 and S2.

Immunofluorescence cytochemistry

Following fixation of iPSCs and differentiated cells with 4% paraformaldehyde for 15 min at 4°C, cells were permeabilized with 0.1% Triton X-100, washed with PBS containing 2% BSA, and incubated with primary antibodies diluted in PBS containing 2% BSA.

Primary antibodies against Nanog (1:200, Cell Signaling, D73G4), SSEA-4 (1:200, Cell Signaling, MC813), Tra-1-60 (1:200, Cell Signaling, #4746), Tra-1-81 (1:200, Cell Signaling, #4745), βIII-tubulin (1:200, Millipore, TU-20), desmin (1:200, Neomarkers, RB-9014-P0), vimentin (1:200, Santa Cruz, V9), and α-fetoprotein (1:500, Sigma, HPA010607) were detected using the secondary antibodies Alexa 488-conjugated anti-goat IgG (1:500, Invitrogen, A11055), Alexa 488-conjugated anti-mouse IgG (1:500, Invitrogen, A11001), Alexa 488-conjugated anti-mouse IgM (1:500, Invitrogen, A21042), and Alexa 488-conjugated anti-rabbit IgG (1:500, Invitrogen, A11013). Cell nuclei were stained with 1- μg/ml 4',6-diamidino-2-phenylindole (DAPI).

Karyotype analysis

After culturing iPSCs in the ES medium containing 100-ng/ml colcemid for 5 h at 37°C, cells were harvested using trypsin and were treated with 0.075 M KCl for 15 min at 37°C. Cells were then fixed in Carnoy's fluid, and chromosome slides were prepared. G-banding analysis was conducted using a previously described method [24].

M-FISH was performed with the Multi-color probe kit "24XCyte" (MetaSystems, Altlussheim, Germany) according to the manufacturer's protocol with slight modifications. Briefly, probes were denatured at 75°C for 5 min and were hybridized to metaphase spreads, which were denatured in 0.07 N NaOH at room temperature for 1 min. Slides were then incubated at 37°C for 2 nights and were then washed in 0.4× SSC at 72°C for 2 min, in 2× SSC containing 0.05% Tween 20 at room temperature for 30 s, and in 2× SSC at room temperature for 1 min, and the mounting medium (DAPI, 125 ng/ml) and a cover slip were applied. Acquisition and analysis of M-FISH images were performed using a CytoVision ChromoFluor System (Applied Imaging, Newcastle upon Tyne, UK).

Transduction of hTERT gene

PT67 retrovirus packaging cells (Takara Bio USA, Madison, WI, USA) were transfected with pMSCV-hTERT-puro using GenePorter II according to the manufacturer's protocol. After 24 h, the culture medium was replaced, cells were incubated for a further 24 h period, and viral supernatants were harvested, A0031 and WSCU01 WS fibroblasts were infected with viral supernatant in the presence of 8 μg/ml polybrene. Confluent infected cells were then split into 2 new dishes, and puromycin selection of infected cells was initiated at the following passage. Confluent infected cells were then passaged in 4-fold dilutions, leading to an increase in 2 population doubling levels for each passage.

SA-β-gal assay

SA-β-gal staining was performed as described by Debacq-Chainiaux et al. [25].

Ethical statement

This study was approved by the Ethics Review Board of the Graduate School of Medicine, Chiba University and was conducted in accordance with the Declaration of Helsinki. Written informed consents were obtained from patients prior to tissue harvesting and iPSC generation, and patients were entitled to the protection of confidential information. Genome/gene analyses performed in this study were approved by the Ethics Committee for Human Genome/Gene Analysis Research at Hiroshima University. All animal experiments were performed in strict compliance with the protocol approved by the Institutional Animal Care and Use Committee of Tottori University (13-Y-

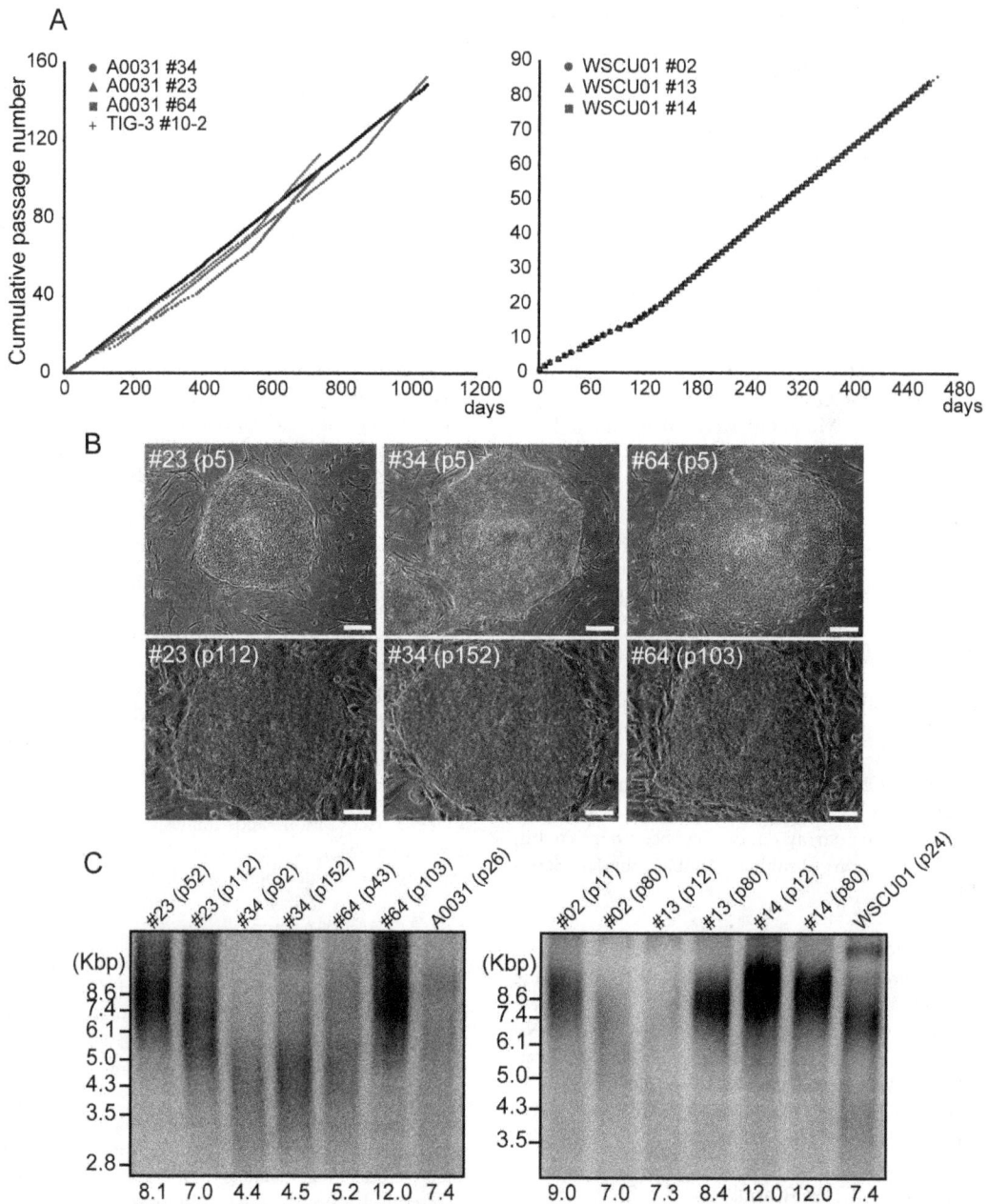

Figure 1. Infinite Proliferation of WS iPSCs after Long-Term Culture. (A) Cumulative passage number for WS iPSCs. (B) Colony morphologies of A0031-derived WS iPSC clones in early and late passages. Bars = 100 μm. (C) TRF length analysis of WS iPSC clones in early and late passages.

18), and the Animal Care and Use Committee of Chiba University (25–131). All recombinant DNA experiments were performed in strict conformance with the guidelines of the Institutional Recombinant DNA Experiment Safety Committee at Hiroshima University.

Results

Infinite proliferative potential of WS iPSCs after long-term culture

To determine whether reprogramming provides WS cells with infinite proliferative potential, we generated iPSCs from WS patient fibroblasts. Morphologically distinct colonies from parental cells emerged after transduction of Yamanaka factors using retroviruses and showed elevated alkaline phosphatase activity (Figures S1A and S1B). Colonies were picked up, and 6 WS iPSC lines were established using fibroblasts from 2 independent WS patients after several passages. In western blotting analysis using an anti-WRN antibody, WRN protein was not detected in WS iPSCs but was expressed in both normal fibroblasts and iPSCs (Figure S2A). Direct sequencing analysis of WS iPSCs identified compound heterozygous Mut4/Mut6 mutations in the *WRN* gene similar to those observed in parental cells, and the derivation of WS iPSCs from parental cells was confirmed by STR analysis (Figures S2B and S2C). Finally, the 6 WS iPSC lines #23, #34, and #64 from A0031 and #02, #13, and #14 from WSCU01 were successfully established.

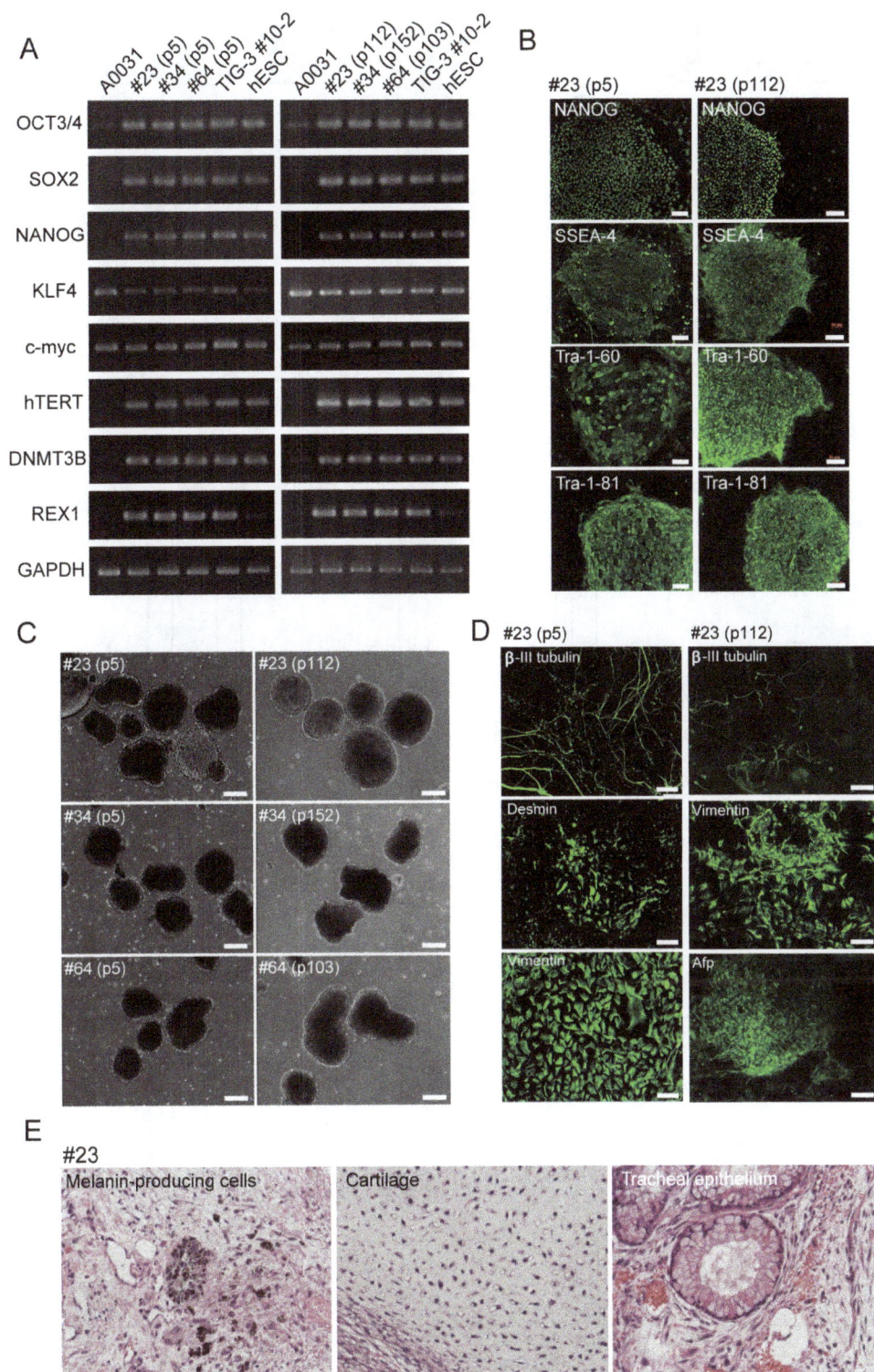

Figure 2. Sustained ESC-like characteristics of WS iPSCs after Long-Term Culture. (A) Expression of pluripotency genes in A0031-derived WS iPSC clones in early and late passages. (B) Expression of hESC markers in A0031-derived WS iPSC clone #23 in early and late passages. Bars = 100 μm. (C) EB formation in A0031-derived WS iPSC clones from early and late passages. Bars = 100 μm. (D) Immunocytochemical analysis of differentiation of EBs into 3 germ layers for A0031-derived iPSC clone #23 in early and late passages. β-III tubulin (ectoderm), desmin (mesoderm), vimentin (mesoderm and parietal endoderm), and α-fetoprotein (Afp, endoderm). Bars = 100 μm. (E) Hematoxylin and eosin histology of teratomas from A0031-derived iPSC clone #23. Formation of all 3 germ layers is shown including melanin-producing cells (ectoderm), cartilage (mesoderm), and tracheal epithelium (endoderm).

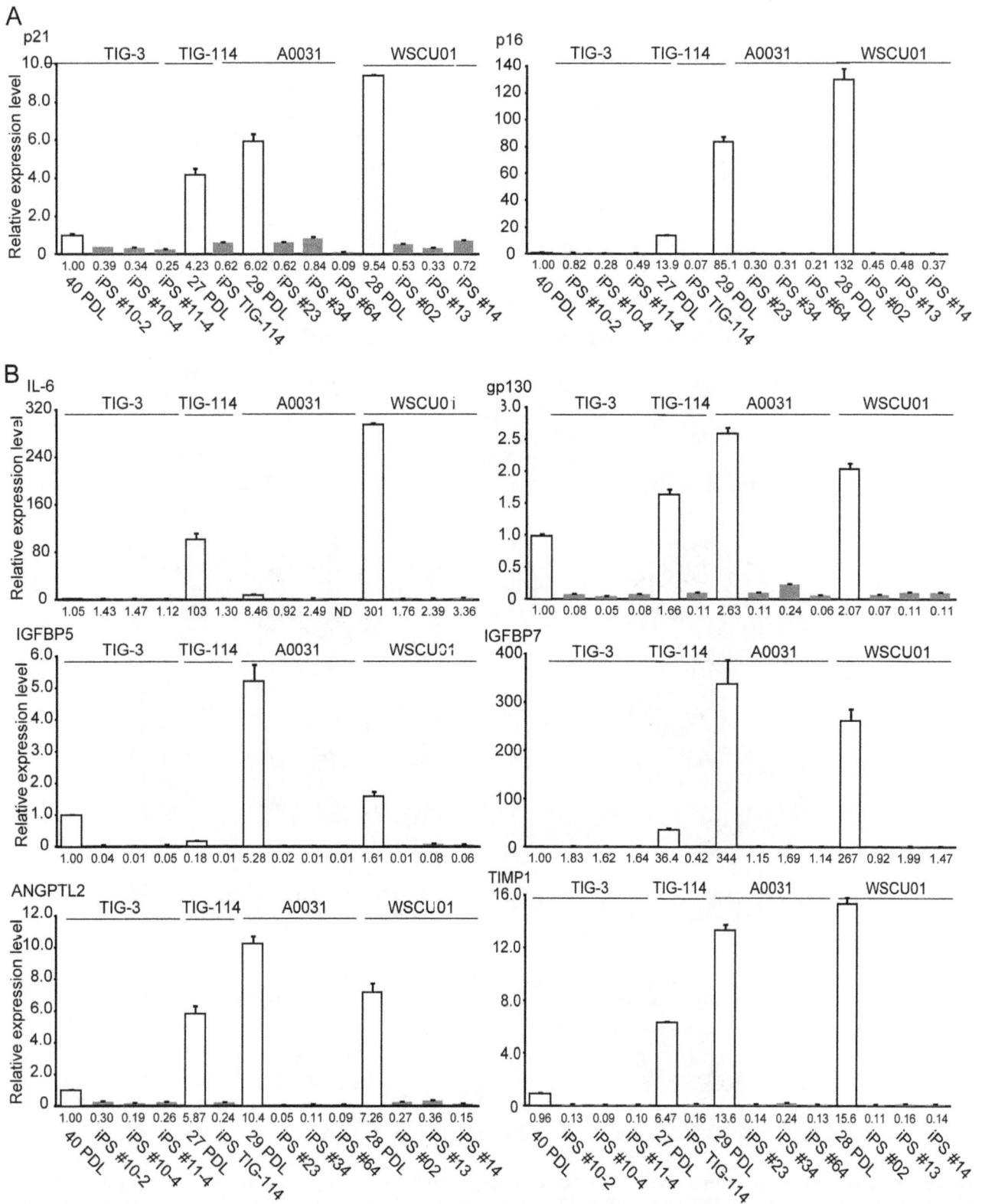

Figure 3. Suppression of Senescence-Associated Gene Expression in Reprogrammed WS iPSCs. (A) Expression of CDKI genes in parental fibroblasts and iPSCs. White columns show relative expression levels in the parental fibroblasts TIG-3, TIG-114, A0031, and WSCU01, and gray columns show those of their derived iPSC clones. Numbers under the horizontal axis in each graph show relative values in mRNA expression compared with that in TIG-3 fibroblasts. Values represent means of three technical replicates ± SD. (B) Expression of SASP genes in parental fibroblasts and iPSCs. Each graph is shown as in (A).

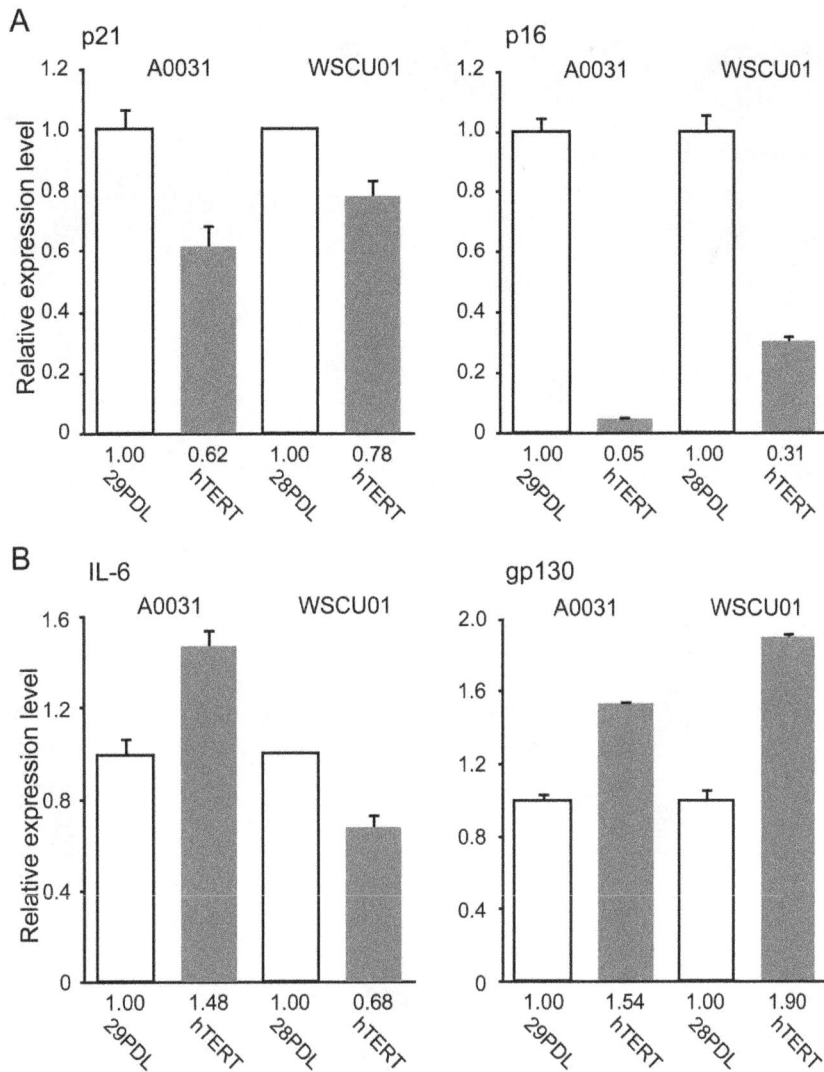

Figure 4. Reprogramming of the SASP gene loci is mediated by factors other than activated telomerase. (A) Expression of CDKI genes in WS fibroblasts and their hTERT-transduced derivatives. White columns show relative expression levels in A0031 and WSCU01 fibroblasts, and gray columns show those of their hTERT-transduced derivatives. Numbers under the horizontal axis in each graph show relative values in mRNA expression compared with that in parental fibroblasts. Values represent means of three technical replicates ± SD. (B) Expression levels of SASP genes in WS fibroblasts and their hTERT-transduced derivatives. Each graph is shown as in (C).

WS iPSC lines from A0031 were cultured for 120 continuous passages over 2 years without morphological changes or loss of growth capacity (Figures 1A and 1B). Moreover, iPSC lines from WSCU01 proliferated for a year (Figures 1A and S1C). Average terminal restriction fragment (TRF) lengths in clones #23, #34, and #64 (A0031) were decreased, invariable, and increased during long-term culture, respectively, and similar telomere dynamics were observed in WSCU01-derived iPSC clones (Figure 1C).

Sustained ESC-like characters of WS iPSCs after long-term culture

To determine the persistence of ESC-like characteristics in WS iPSCs, we compared undifferentiated states and differentiation potentials between WS iPSCs from early and late passages. WS iPSC lines expressed pluripotency genes and hESC-specific surface markers during early passages (around p10), and during late passages (around p100; Figures 2A, 2B, S3 and S4). These iPSC

lines also showed sustained formation of embryoid bodies and differentiation into 3 germ layers (Figures 2C, 2D, and S5). Furthermore, at around p50, WS iPSC lines generated teratomas that contained tissue structures of all 3 germ layers. These were consistent with those shown in normal iPSC lines after transplantation into the testes of SCID mice (Figures 2E and S6). Thus, reprogrammed WS fibroblasts acquired infinite proliferative potential, and the ESC-like characteristics of the resulting iPSCs were maintained for more than 2 years.

Suppression of senescence-associated gene expression in WS iPSCs after long-term culture

Global gene expression analysis using DNA chips showed pronounced similarities among pluripotent stem cells including WS iPSCs. However, marked differences between WS iPSC and WS fibroblasts were observed (Figure S7). Heat map analysis also showed a high analogy of global gene expression profiles in these

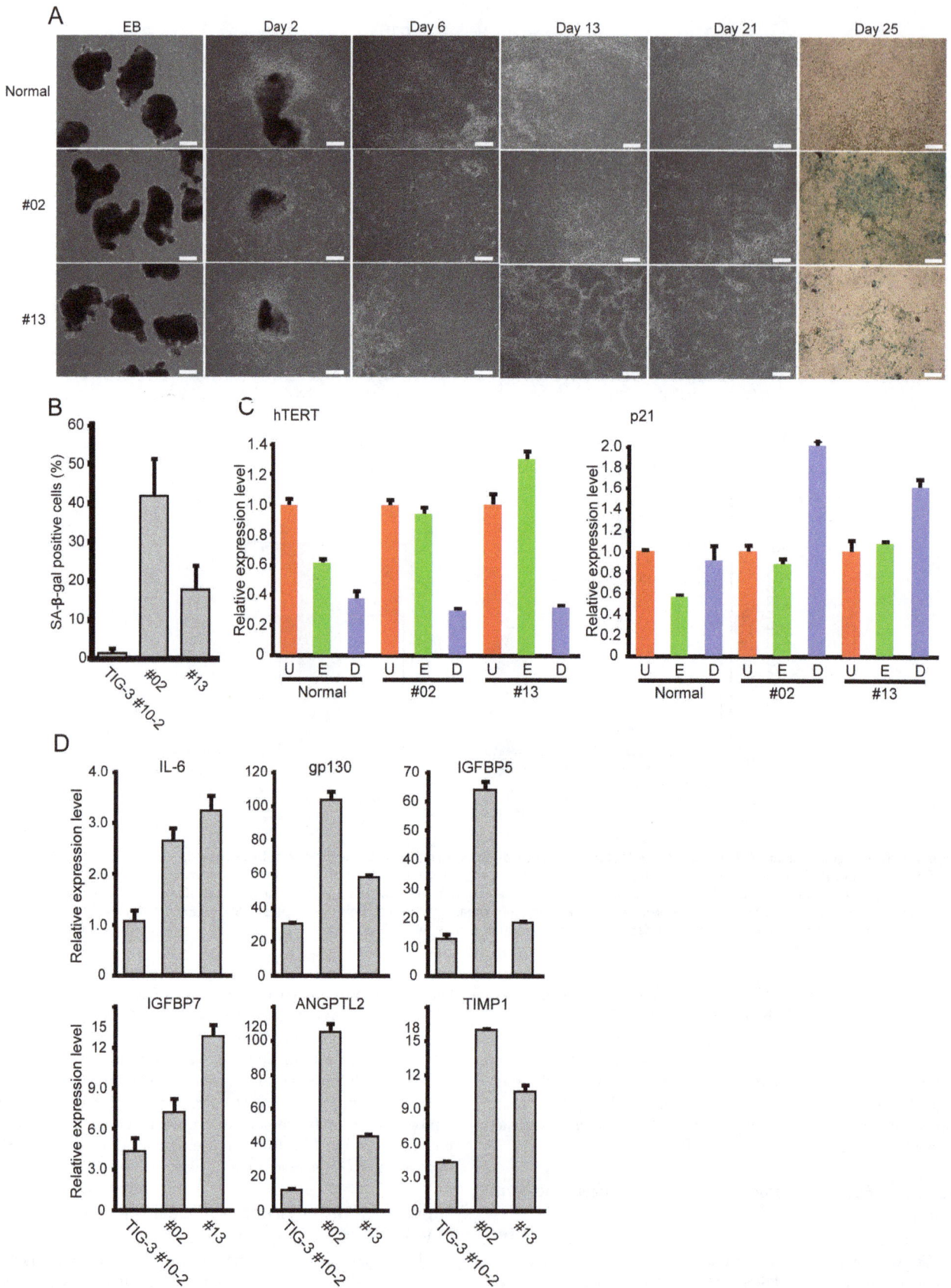

Figure 5. Recapitulation of Premature Senescence Phenotypes in Differentiated Cells from WS iPSCs. (A) Differentiation of EBs from normal (TIG-3) and WS (WSCU01 #02 and #13) iPSCs. Differentiated cells from WS iPSCs showed premature senescence. SA-β-gal staining was performed on day 25 of differentiation. Bars = 100 μm. (B) Percentage of senescent cells after 25 days of differentiation. SA-β-gal-positive cells were

counted in three randomly selected fields with 40× magnification. Values represent means of the three fields ± SD. (C) Expression of hTERT and p21 mRNAs in undifferentiated iPSCs ("U," red columns), EBs after 12 days of formation ("E," green columns), and differentiated cells after 25 days of differentiation ("D," blue columns). Values represent means of three technical replicates ± SD. (D) Expression of SASP genes in differentiated cells from normal (TIG-3) and WS (WSCU01 #02 and #13) iPSCs after 25 days of differentiation. Graphs shows fold changes relative to undifferentiated iPSCs. Values represent means of three technical replicates ± SD.

pluripotent stem cell lines, but distinctly different profiles from those of WS fibroblasts (Figures S8A). Recent studies of aging have identified senescence-induced inflammatory and secretory factors that are collectively referred to as the senescence-associated secretory phenotype (SASP) and are the hallmarks of aging. It is widely accepted that age-associated inflammatory responses contribute to human aging mechanisms [26]. Accordingly, we observed downregulation of SASP secretory factors, including inflammatory cytokines, growth factors and MMPs, in both normal and WS iPSCs compared with WS fibroblasts (Figures S8B). Subsequently, we performed real-time qRT-PCR analysis using PDL-matched normal and patient fibroblasts, and their iPSC derivatives which were maintained in long-term culture. Although relative expression levels of the senescence-associated cyclin-dependent kinase inhibitor (CDKI) genes *p21Waf1/Cip1* and *p16INK4a* in normal fibroblasts correlated with the donor age, the expression levels of these genes were higher in WS fibroblasts than in normal fibroblasts, indicating that replicative senescence was prematurely induced in WS cells (Figure 3A). However, expression levels of these genes were significantly reduced in all iPSC clones from normal and WS cells (Figure 3A), suggesting that these gene loci are reprogrammed to the same degree in normal and WS iPSCs. Thus, we examined the expression of the typical SASP genes *IL-6* and *gp130* [27] and found higher expression levels in WS fibroblasts than in normal fibroblasts (Figure 3B). Moreover, expression levels of these genes drastically decreased in both normal and WS iPSCs compared with parental fibroblasts. Similarly, expression levels of the SASP genes *IGFBP5, IGFBP7, ANGPTL2,* and *TIMP1* ([28–31] were significantly decreased in both normal and WS iPSCs compared with parental fibroblasts (Figures 3B).

Reprogramming of the SASP gene loci is mediated by factors other than activated telomerase

WS fibroblasts were previously shown to bypass premature senescence following introduction of the telomerase gene *hTERT*

[32], Similarly, the present WS cells bypassed premature replicative senescence, and hTERT allowed cell division for over 150 PDL in A0031 cells, and 40 PDL in WSCU01 cells compared with parental cells that became senescent at less than 30 PDL (Figures S9A and S9B). TRF length analysis showed that hTERT-expressing WS cells acquired longer telomeres during passages than parental cells (Figures S9C). To examine whether the expression of hTERT was sufficient to suppress the upregulation of aging-associated genes in WS cells, we compared expression levels of CDKI and SASP genes between WS fibroblasts and their hTERT-expressing derivatives. Whereas a decline in p21wafl/cip1 and p16INK4a mRNA expression was observed in hTERT-expressing cells (Figure 4A), IL-6 and gp130 expression was not suppressed following the introduction of hTERT, suggesting that reprogramming of the SASP gene loci is mediated by factors other than activated telomerase (Figure 4B). The present data show complete suppression of premature senescence phenotypes in WS cells using transcription factor-induced reprogramming and suggest that persistence of the undifferentiated state and pluripotency are crucial for reversing the aging process.

Recapitulation of premature senescence phenotypes in differentiated cells from WS iPSCs

To establish cell lineages that prematurely senesced, EBs consisting of equal numbers of iPSCs maintained in long-term culture were differentiated in serum-containing medium. Differentiated cells from WS iPSC-derived EBs were outgrown less rapidly than those from normal iPSC-derived EBs (Figure 5A, Day 2). These cells exhibited flat and enlarged morphology (Figure 5A, Day 6, 13, and 21) and became positive for SA-β-gal staining (Figure 5A, Day 25, and Figure 5B). Whereas expression levels of hTERT were downregulated equally in differentiated cells from normal and WS iPSCs, p21 mRNA was more highly induced in differentiated cells from WS iPSCs than those from normal iPSCs (Figure 5C). Expression levels of the SASP genes were also significantly increased in differentiated cells from WS iPSCs

Table 1. Results of chromosome analysis of WS iPSC clones and their parenral cells.

Cell lines	Numbers of cells analyzed by G-banding	Numbers of cells analyzed by M-FISH	Karyotypes
A0031	20 (13/7)	ND	46,XY,del(8)(q22q24)/46,XY,t(1;14)(p34.1;q13),t(4;7)(p15.2;q22),del(8)(q22q24)
iPS#23	20	10	46,XY,t(1;14)(p34.1;q13),t(4;7)(p15.2;q22),del(8)(q22q24),der(21)t(17;21)(?;q22.3)
iPS#34	20	10	46,XY,t(1;14)(p34.1;q13),t(4;7)(p15.2;q22),del(8)(q22q24)
iPS#64	20	10	46,XY,t(1;14)(p34.1;q13),t(4;7)(p15.2;q22),del(8)(q22q24),der(19)t(2;19)(?;p13.3)
WSCU01	20	ND	46,XY,normal
iPS#02	20	10	47,XY,+del(20)(p?)
iPS#13	20	10	46,XY,normal
iPS#14	20	10	46,XY,normal

Abbreviations: t, translocation; del, deletion; der, derivative chromosome; p, short arm; q, long arm.

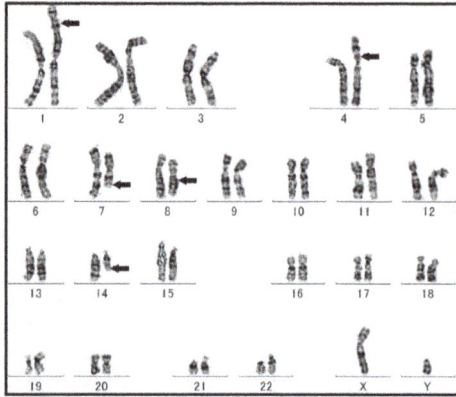

A A0031

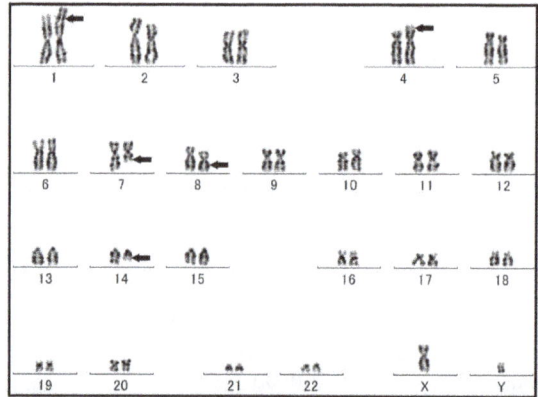

B #34

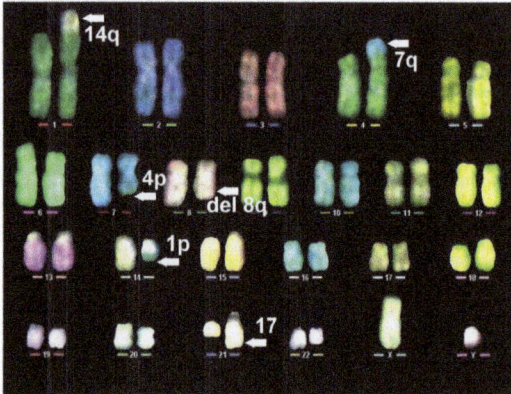

C #23

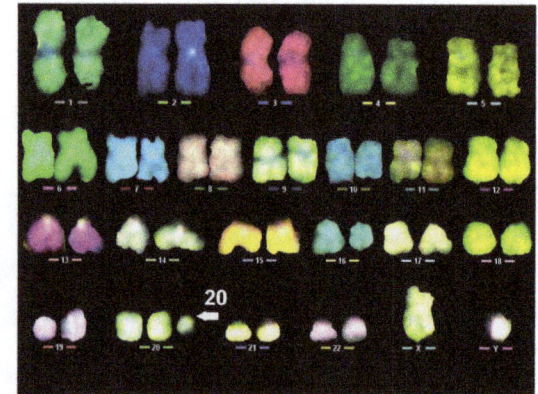

F #02

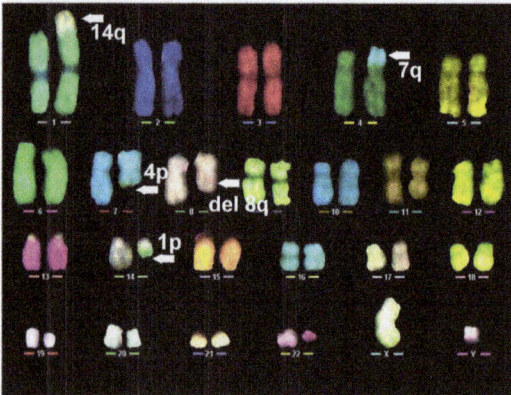

D #34

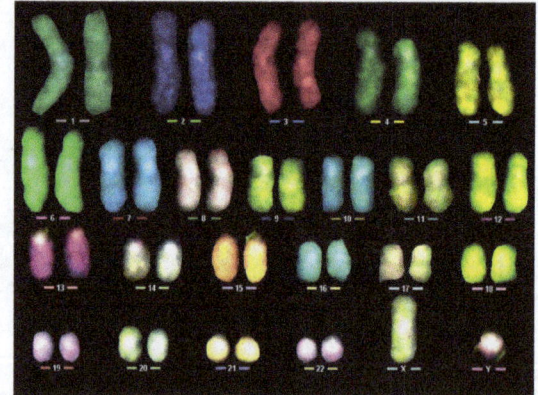

G #13

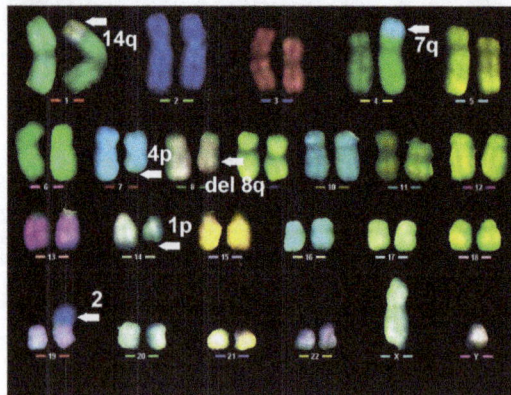

E #64

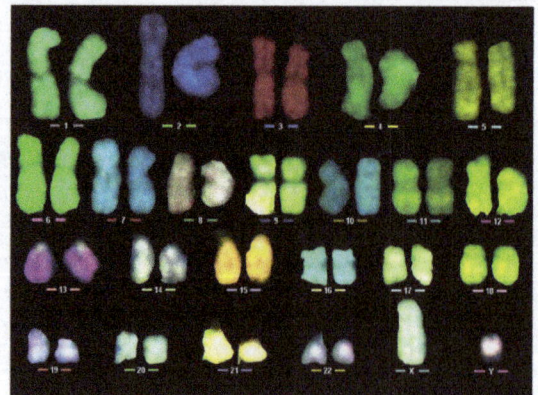

H #14

Figure 6. Karyotype Analysis of WS iPSCs. Chromosomal profiles of G-band analysis. (A) Parental A0031 fibroblast and (B) A0031-derived iPSC clone #34. Arrows indicate translocation breakpoints. Chromosomal profiles of M-FISH analysis. A0031-derived iPSC clones (C) #23, (D) #34, and (E) #64 and WSCU01-derived iPSC clones (F) #02, (G) #13, and (H) #14. Arrows indicate translocation breakpoints or an extra chromosome.

compared with those from normal iPSCs (Figure 5D). These results demonstrated recapitulation of premature senescence phenotypes with downregulation of hTERT in differentiated cells from WS iPSCs.

Karyotype analysis of WS iPSCs

WS is characterized by genomic instability, and gene translocation events have been observed during culture of patient-derived cells [33]. Because reprogramming of somatic cells and subsequent maintenance of iPSCs involves extensive cell division, WS iPSCs may acquire additional chromosomal abnormalities. Thus, we compared chromosomal profiles of long-term cultured WS iPSC clones with those of parental WS fibroblasts by karyotype analysis. The subsequent G-banding stain and multicolor fluorescence *in situ* hybridization (M-FISH) analysis are summarized in Table 1.

Chromosomal profiles of parental A0031 WS fibroblasts showed mosaicism with the following abnormal karyotypes: 46, XY with a deletion in 8q and 46, XY with a deletion in 8q along with reciprocal translocations between 1p and 14q, and 4p and 7q (Figure 6A). These karyotypes support previous observations of chromosomal instability in WS cells [33]. Whereas, 1 of the derived iPSC clones (#34) had the same chromosomal profile as its parent cells (Figures 6B and 6D), the other 2 A0031-derived iPSC clones (#23 and #64) had the translocations 21q and 19p, respectively, in addition to those of the parental karyotype (Table 1, Figures 6C and 6E). Moreover, whereas parental WSCU01 fibroblasts and 2 of their derived iPSC clones (#13 and #14) had normal karyotypes (Table 1, Figures 6G and 6H), the remaining iPSC clone #02 carried the abnormal karyotype 47 (XY with an additional aberrant chromosome derived from chromosome 20; Table 1, Figure 6F).

The observation that 3 of 6 WS iPSC clones had the same karyotypes as their parental cells after approximately 100 passages suggests that karyotypes of WS cells are stabilized following reprogramming.

Discussion

In this study, we demonstrated that WS fibroblasts could be reprogrammed into iPSCs using Yamanaka factors, and the resulting iPSCs showed unlimited proliferative capacity that was sufficient for self-renewal over a period of 2 years. WS iPSCs also exhibited undifferentiated states and differentiation potential after long-term culture. Subsequently, we showed that WS iPSCs maintain immortality and ESC-like characteristics that indicate corrected telomere dysfunction following reprogramming of WS cells. Although WRN was not essential for generation of iPSCs, WRN helicase may protect genome integrity by mechanism other than the maintenance of telomere in iPSCs.

TRF length analysis indicated that WS iPSC lines maintained telomere with size variation in each clone. It is known that human iPSCs derived from normal somatic cells showed varied telomere length, and variation of telomere length among human iPSC clones is thought to partly depend on acquired telomerase activity associated with their reprogrammed states [34,35]. Therefore, variation of telomere length observed among WS iPSC clones would be due to clonal variation in telomerase activity rather than telomere dysfunction associated with *WRN* deficiency.

Normal human iPSCs are known to acquire genomic instability with a high incidence of additions, deletions and translocations [36,37]. In contrast, chromosomal aberrations are frequently caused by telomere dysfunctions in WS fibroblasts following the induction of cell cycle progression [11]. Nonetheless, the present data show unexpected maintenance of chromosomal profiles in WS iPSC clones during long-term culture for more than 100 passages although half of these clones acquired additional chromosomal abnormalities. Previously, the introduction of hTERT reduced the chromosomal aberrations in cells from WS patients [11]. In agreement, the present data indicate endogenous hTERT expression in WS iPSCs, but not in parental fibroblasts, suggesting that reprogramming suppresses chromosomal instability in WS cells by reactivating telomerase.

Previous studies show that WS fibroblasts express inflammatory cytokines [38] and WS is associated with inflammatory conditions such as atherosclerosis, diabetes and osteoporosis [39–43]. The present data indicate that both CDKI and SASP genes are prematurely induced in WS fibroblasts compared with PDL-matched normal fibroblasts. However, expression levels of these genes were completely suppressed in WS iPSCs to the same degree observed in normal iPSCs. In contrast, hTERT did not suppress SASP genes in WS fibroblasts, as shown by previous study [44], although a decline in p21waf1/cip1 and p16INK4a mRNAs was observed. Taken together, these observations suggest that pluripotency-associated transcription factor-induced reprogramming reverses the aging process in both normal and WS cells. Furthermore, differentiated cells from EBs of long-term cultured WS iPSCs showed premature senescence phenotypes, thus demonstrating that WS iPSCs stably maintained their potential to recapitulate premature senescence phenotypes during differentiation over the long term. In addition, embryoid body-mediated iPSC differentiation recapitulated premature senescence phenotypes in WS iPSCs, suggesting that it would provide a simple and rapid way to identify cell lineages affected in WS.

In the present study, we demonstrated the potential of WS iPSCs to proliferate infinitely and differentiate into various cell types, which could be used to provide patient cells in large quantities over the long term. Because WS-specific iPSCs may be differentiated into multiple cell types, their experimental use may resolve the major pathogenic processes of WS for which cell types available from patients are usually limited to lymphocytes and/or fibroblasts. The present technologies may also be used to develop cell transplantation therapies for WS patients using gene-corrected patient cells. The present observations indicate that WS iPSCs may be a powerful tool for understanding normal aging and the pathogenesis of WS.

Supporting Information

Figure S1 Generation of WS iPSCs. (A) Generation of iPSCs. Normal (TIG-3) and Werner syndrome (A0031 and WSCU01) fibroblasts are shown in the left panels, and emergence of morphologically distinct ESC-like colonies from parental cells is shown in the right panels. (B) Alkaline phosphatase activity of ESC-like colonies derived from TIG-3 and A0031 fibroblasts. (C) Colony morphologies of WSCU01-derived WS iPSC clones in early and late passages. Bars = 100 μm.
(EPS)

Figure S2 Evidences that WS iPSCs were derived from patients. (A) Western blot analysis of WRN helicase protein in WS iPSCs. (B) Direct sequencing analysis identified compound heterozygous mut.4/mut.6 mutations in WS iPSCs. Mut.4 is a C to G substitution at the splice-donor site bordered by exon 26, as shown by an arrow in the illustration of the double-strand base sequence. Obtained pherograms show antisense peak shapes. A peak corresponding to mut.4 in normal TIG-3 fibroblast shows a single "C," whereas the WS iPSC clone #34 from A0031 fibroblasts gave double peaks showing "G" in addition to "C." Mut.6 is a T to C substitution in exon 9. A peak corresponding to mut.6 in normal cells showed a single "C," whereas WS iPSC gave double peaks showing "T" in addition to "C." C, blue; G, black; T, red; A, green. (C) STR analysis of A0031-derived iPSC clone #34, showing that iPSC clone #34 was derived from the parental A0031 fibroblasts.
(EPS)

Figure S3 Expression of pluripotency genes in WSCU01-derived WS iPSC clones in early and late passages.
(EPS)

Figure S4 Expression of hESC markers in WS iPSCs in early and late passages. A0031-derived clones #34, and #64, and WSCU01-derived clones #02, #13, and #14 are shown. Bars = 100 μm.
(EPS)

Figure S5 Immunocytochemistry for differentiation of embryoid bodies into 3 germ layers for WS iPSCs in early and late passages. A0031-derived clones #34, and #64, and WSCU01-derived clones #02, #13, and #14 are shown. Bars = 100 μm.
(EPS)

Figure S6 Hematoxylin and eosin histology of teratomas derived from iPSCs. Hematoxylin and eosin histology of teratomas derived from iPSCs. The normal TIG-3 fibroblast-derived clone #10-2, A0031-derived clones #34, and #64, and the WSCU01-derived clone #02 are shown. Formation of all 3 germ layers is shown with melanin-producing cells and glial tissue (ectoderm), cartilage (mesoderm) and intestinal epithelia. Glands are lined by columnar epithelia and tracheal epithelium (endoderm).
(EPS)

Figure S7 Figure Scatter plots comparing gene expression profiles.
(EPS)

Figure S8 Analysis of senescence-associated gene expression in iPSCs. (A) Heat map analysis of WS iPSC #34 and parental WS A0031 fibroblasts, normal TIG-3 fibroblast-derived iPSCs, and hESC; 3277 probes with >5-fold differences in expression between A0031 fibroblast and WS iPSC were included in the heat map. (B) Heat map analysis of the gene profiles of secreted protein probes with >2-fold differences in expression between A0031 fibroblasts and the 3 pluripotent stem cell lines WS iPSC, TIG-3 iPSC, and hESC.
(EPS)

Figure S9 hTERT bypassed premature replicative senescence of WS fibroblasts. (A) Morphologies of growing normal TIG-3 fibroblasts, and A0031 and WSCU01 WS fibroblasts. WS fibroblasts showed premature senescence. SA-β-gal staining was performed for WSCU01 (lower). Bars = 100 μm. (B) Cumulative population doubling levels for hTERT-expressing WS cells. (C) TRF lengths of A0031 fibroblasts and their TERT-transduced derivatives.
(EPS)

Table S1
(EPS)

Table S2
(EPS)

Acknowledgments

We are grateful to Miho Kusuda-Furue (National Institute of Biomedical Innovation), Hidenori Akutsu (National Center for Child Health and Development) and Haruhiko Koseki (RCAI RIKEN) for their help, encouragement and suggestions. We also thank M. K. F and Bunsyo Shiotani for the critical review of draft manuscripts, and the Analysis Center of Life Science, Natural Science Center for Basic Research and Development of Hiroshima University for processing microarray data.

Author Contributions

Conceived and designed the experiments: AS. Performed the experiments: AS HK KZ YS YK MO MO HK TS KH HS YI KH YK. Analyzed the data: AS YK MO MO TS KH HS YI KH YK. Contributed reagents/materials/analysis tools: MG MT KY SY KF HT. Wrote the paper: AS MG. Final approval of the version to be published: AS HT.

References

1. Goto M, Miller RW, Ishikawa Y, Sugano H (1996) Excess of rare cancers in Werner syndrome (adult progeria). Cancer Epidemiol Biomarkers Prev 5: 239–246.
2. Goto M (2000) Werner's syndrome: from clinics to genetics. Clin Exp Rheumatol 18: 760–766.
3. Salk D, Au K, Hoehn H, Martin GM (1981) Effects of radical-scavenging enzymes and reduced oxygen exposure on growth and chromosome abnormalities of Werner syndrome cultured skin fibroblasts. Hum Genet 57: 269–275.
4. Yu CE, Oshima J, Fu YH, Wijsman EM, Hisama F, et al. (1996) Positional cloning of the Werner's syndrome gene. Science 272: 258–262.
5. Oshima J, Yu CE, Piussan C, Klein G, Jabkowski J, et al. (1996) Homozygous and compound heterozygous mutations at the Werner syndrome locus. Hum Mol Genet 5: 1909–1913.
6. Goto M, Imamura O, Kuromitsu J, Matsumoto T, Yamabe Y, et al. (1997) Analysis of helicase gene mutations in Japanese Werner's syndrome patients. Hum Genet 99: 191–193.
7. Matsumoto T, Imamura O, Yamabe Y, Kuromitsu J, Tokutake Y, et al. (1997) Mutation and haplotype analyses of the Werner's syndrome gene based on its genomic structure: genetic epidemiology in the Japanese population. Hum Genet 100: 123–130.
8. Shimamoto A, Sugimoto M, Furuichi Y (2004) Molecular biology of Werner syndrome. Int J Clin Oncol 9: 288–298.
9. Rossi ML, Ghosh AK, Bohr VA (2010) Roles of Werner syndrome protein in protection of genome integrity. DNA Repair (Amst) 9: 331–344.
10. Crabbe L, Verdun RE, Haggblom CI, Karlseder J (2004) Defective telomere lagging strand synthesis in cells lacking WRN helicase activity. Science 306: 1951–1953.
11. Crabbe L, Jauch A, Naeger CM, Holtgreve-Grez H, Karlseder J (2007) Telomere dysfunction as a cause of genomic instability in Werner syndrome. Proc Natl Acad Sci U S A 104: 2205–2210.
12. Takahashi K, Yamanaka S (2006) Induction of pluripotent stem cells from mouse embryonic and adult fibroblast cultures by defined factors. Cell 126: 663–676.
13. Takahashi K, Tanabe K, Ohnuki M, Narita M, Ichisaka T, et al. (2007) Induction of pluripotent stem cells from adult human fibroblasts by defined factors. Cell 131: 861–872.
14. Yu J, Vodyanik MA, Smuga-Otto K, Antosiewicz-Bourget J, Frane JL, et al. (2007) Induced pluripotent stem cell lines derived from human somatic cells. Science 318: 1917–1920.
15. Aoi T, Yae K, Nakagawa M, Ichisaka T, Okita K, et al. (2008) Generation of pluripotent stem cells from adult mouse liver and stomach cells. Science 321: 699–702.
16. Stadtfeld M, Hochedlinger K (2010) Induced pluripotency: history, mechanisms, and applications. Genes Dev 24: 2239–2263.

17. Okita K, Yamanaka S (2011) Induced pluripotent stem cells: opportunities and challenges. Philos Trans R Soc Lond B Biol Sci 366: 2198–2207.
18. Stadtfeld M, Maherali N, Breault DT, Hochedlinger K (2008) Defining molecular cornerstones during fibroblast to iPS cell reprogramming in mouse. Cell Stem Cell 2: 230–240.
19. Cheung HH, Liu X, Canterel-Thouennon L, Li L, Edmonson C, et al. (2014) Telomerase protects werner syndrome lineage-specific stem cells from premature aging. Stem Cell Reports 2: 534–546.
20. Batista LF, Pech MF, Zhong FL, Nguyen HN, Xie KT, et al. (2011) Telomere shortening and loss of self-renewal in dyskeratosis congenita induced pluripotent stem cells. Nature 474: 399–402.
21. Goto M, Ishikawa Y, Sugimoto M, Furuichi Y (2013) Werner syndrome: A changing pattern of clinical manifestations in Japan (1917~2008). Biosci Trends 7: 13–22.
22. Morita S, Kojima T, Kitamura T (2000) Plat-E: an efficient and stable system for transient packaging of retroviruses. Gene Ther 7: 1063–1066.
23. Amps K, Andrews PW, Anyfantis G, Armstrong L, Avery S, et al. (2011) Screening ethnically diverse human embryonic stem cells identifies a chromosome 20 minimal amplicon conferring growth advantage. Nat Biotechnol 29: 1132–1144.
24. Ohtaki K, Sposto R, Kodama Y, Nakano M, Awa AA (1994) Aneuploidy in somatic cells of in utero exposed A-bomb survivors in Hiroshima. Mutat Res 316: 49–58.
25. Debacq-Chainiaux F, Erusalimsky JD, Campisi J, Toussaint O (2009) Protocols to detect senescence-associated beta-galactosidase (SA-betagal) activity, a biomarker of senescent cells in culture and in vivo. Nat Protoc 4: 1798–1806.
26. Goto M (2008) Inflammaging (inflammation + aging): A driving force for human aging based on an evolutionarily antagonistic pleiotropy theory? Biosci Trends 2: 218–230.
27. Salama R, Sadaie M, Hoare M, Narita M (2014) Cellular senescence and its effector programs. Genes Dev 28: 99–114.
28. Kojima H, Kunimoto H, Inoue T, Nakajima K (2012) The STAT3-IGFBP5 axis is critical for IL-6/gp130-induced premature senescence in human fibroblasts. Cell Cycle 11:
29. Wajapeyee N, Serra RW, Zhu X, Mahalingam M, Green MR (2008) Oncogenic BRAF induces senescence and apoptosis through pathways mediated by the secreted protein IGFBP7. Cell 132: 363–374.
30. Tabata M, Kadomatsu T, Fukuhara S, Miyata K, Ito Y, et al. (2009) Angiopoietin-like protein 2 promotes chronic adipose tissue inflammation and obesity-related systemic insulin resistance. Cell Metab 10: 178–188.
31. Gilbert LA, Hemann MT (2010) DNA damage-mediated induction of a chemoresistant niche. Cell 143: 355–366.
32. Wyllie FS, Jones CJ, Skinner JW, Haughton MF, Wallis C, et al. (2000) Telomerase prevents the accelerated cell ageing of Werner syndrome fibroblasts. Nat Genet 24: 16–17.
33. Salk D, Au K, Hoehn H, Stenchever MR, Martin GM (1981) Evidence of clonal attenuation, clonal succession, and clonal expansion in mass cultures of aging Werner's syndrome skin fibroblasts. Cytogenet Cell Genet 30: 108–117.
34. Mathew R, Jia W, Sharma A, Zhao Y, Clarke LE, et al. (2010) Robust activation of the human but not mouse telomerase gene during the induction of pluripotency. FASEB J 24: 2702–2715.
35. Vaziri H, Chapman K, Guigova A, Teichroeb J, Lacher M, et al. (2010) Spontaneous reversal of the developmental aging of normal human cells following transcriptional reprogramming. Regen Med 5: 345–363.
36. Taapken SM, Nisler BS, Newton MA, Sampsell-Barron TL, Leonhard KA, et al. (2011) Karotypic abnormalities in human induced pluripotent stem cells and embryonic stem cells. Nat Biotechnol 29: 313–314.
37. Martins-Taylor K, Nisler BS, Taapken SM, Compton T, Crandall L, et al. (2011) Recurrent copy number variations in human induced pluripotent stem cells. Nat Biotechnol 29: 488–491.
38. Kumar S, Vinci JM, Millis AJ, Baglioni C (1993) Expression of interleukin-1 alpha and beta in early passage fibroblasts from aging individuals. Exp Gerontol 28: 505–513.
39. Murano S, Nakazawa A, Saito I, Masuda M, Morisaki N, et al. (1997) Increased blood plasminogen activator inhibitor-1 and intercellular adhesion molecule-1 as possible risk factors of atherosclerosis in Werner syndrome. Gerontology 43 Suppl 1: 43–52.
40. Yokote K, Hara K, Mori S, Kadowaki T, Saito Y, et al. (2004) Dysadipocytokinemia in werner syndrome and its recovery by treatment with pioglitazone. Diabetes Care 27: 2562–2563.
41. Rubin CD, Zerwekh JE, Reed-Gitomer BY, Pak CY (1992) Characterization of osteoporosis in a patient with Werner's syndrome. J Am Geriatr Soc 40: 1161–1163.
42. Davis T, Kipling D (2006) Werner Syndrome as an example of inflamm-aging: possible therapeutic opportunities for a progeroid syndrome? Rejuvenation Res 9: 402–407.
43. Goto M, Sugimoto K, Hayashi S, Ogino T, Sugimoto M, et al. (2012) Aging-associated inflammation in healthy Japanese individuals and patients with Werner syndrome. Exp Gerontol 47: 936–939.
44. Choi D, Whittier PS, Oshima J, Funk WD (2001) Telomerase expression prevents replicative senescence but does not fully reset mRNA expression patterns in Werner syndrome cell strains. FASEB J 15: 1014–1020.

6

Genomic Analysis of *Sleeping Beauty* Transposon Integration in Human Somatic Cells

Giandomenico Turchiano[1], Maria Carmela Latella[1], Andreas Gogol-Döring[2,3], Claudia Cattoglio[4], Fulvio Mavilio[1,5], Zsuzsanna Izsvák[6], Zoltán Ivics[7], Alessandra Recchia[1]*

1 Center for Regenerative Medicine, Department of Life Sciences, University of Modena and Reggio Emilia, Modena, Italy, 2 German Centre for Integrative Biodiversity Research (iDiv) Halle-Jena-Leipzig, Leipzig, Germany, 3 Institute of Computer Science, Martin Luther University Halle-Wittenberg, Halle, Germany, 4 Howard Hughes Medical Institute, Department of Molecular and Cell Biology, University of California, Berkeley, Berkeley, California, United States of America, 5 Genethon, Evry, France, 6 Max Delbruck Center for Molecular Medicine, Berlin, Germany, 7 Division of Medical Biotechnology, Paul Ehrlich Institute, Langen, Germany

Abstract

The *Sleeping Beauty* (SB) transposon is a non-viral integrating vector system with proven efficacy for gene transfer and functional genomics. However, integration efficiency is negatively affected by the length of the transposon. To optimize the SB transposon machinery, the inverted repeats and the transposase gene underwent several modifications, resulting in the generation of the hyperactive SB100X transposase and of the high-capacity "sandwich" (SA) transposon. In this study, we report a side-by-side comparison of the SA and the widely used T2 arrangement of transposon vectors carrying increasing DNA cargoes, up to 18 kb. Clonal analysis of SA integrants in human epithelial cells and in immortalized keratinocytes demonstrates stability and integrity of the transposon independently from the cargo size and copy number-dependent expression of the cargo cassette. A genome-wide analysis of unambiguously mapped SA integrations in keratinocytes showed an almost random distribution, with an overrepresentation in repetitive elements (satellite, LINE and small RNAs) compared to a library representing insertions of the first-generation transposon vector and to gammaretroviral and lentiviral libraries. The SA transposon/SB100X integrating system therefore shows important features as a system for delivering large gene constructs for gene therapy applications.

Editor: Sebastian D. Fugmann, Chang Gung University, Taiwan

Funding: Funding was received for this study from Italian Ministry of University and Research-FIRB 2008 (AR), DEBRA international (AR) and the European Research Council (GT-SKIN) (FM). The funders had no role in study design, data collection and analysis, decision to publish, or preparation of the manuscript.

Competing Interests: The authors have declared that no competing interests exist.

* Email: alessandra.recchia@unimore.it

Introduction

The *Sleeping Beauty* (SB) transposon is a member of the Tc1/*mariner* transposon superfamily. Tc1/*mariner* elements are generally 1,300–2,400 bp in length and contain a single gene coding for the transposase that is flanked by terminal inverted repeats (IR). The IRs of SB host a pair of binding sites containing short, 15–20 bp direct repeats (DRs). Both the outer and the inner pairs of transposase-binding sites are required for transposition. The SB transposase binds the IRs in a sequence-specific manner, and mediates precise cut-and-paste transposition in a wide variety of vertebrate cells including human cells [1–3]. For this reason, the SB-based integration system is a valuable tool for functional genomics in several model organisms and represents a promising vector for human gene therapy [4,5]. However, a major bottleneck of any transposon-based application is the low transposition efficiency. Therefore, considerable effort was dedicated to improve the SB integration machinery by modifying its IRs and systematically mutating the transposase gene. In 2002, Cui et al. carefully explored the structure and functions of the IRs. They modified the outer and inner DR sites of both IRs and the spacer sequence between the DRs generating a new version of transposon IR,

called T2, with fourfold increased transposition efficiency [6]. However, the transpositional activity of this system (and that of the first-generation transposon [7]) is negatively affected by the size of transposon, resulting in an exponential drop for every kb introduced between the two IR.

In 2004, Zayed et al. constructed the "sandwich" (SA) version of the transposon vector [8]. The SA IR consists of two complete transposon elements in a head to head orientation, flanking a DNA expression cassette, thereby forming a sandwich-like arrangement. Mutation of the 5′ terminal CA nucleotides of the right IR abolishes cleavage at the innermost transposon ends; therefore, only the four terminal DRs represent the catalytic substrate for the "cut and paste" transposition. The SA transposon showed a 3.7-fold enhanced activity over first generation transposon to integrate ~7.5 kb-DNA sequence upon SB10 transposase delivery. Five years later, a transposase 100-fold more active than SB10, named SB100X, was developed by a high-throughput, PCR-based DNA shuffling strategy [1]. The improved integration efficiency associated with SB transposition opened new avenues for its application. The hyperactive SB100X transposase was employed to obtain highly efficient germline transgenesis in pigs [9,10] rabbits [11] and rodents [12,13], stable transfer of therapeutic genes in clinical relevant cells [1,14–18], and reprogramming of mouse embryonic and human foreskin fibroblasts into iPS cells [19].

In this study, we investigated the integration efficiency of large expression cassettes mediated by the optimized SB elements: the SA transposon and the SB100X transposase. We report a side-by-side comparison between the SA and the T2 transposons carrying DNA cargo of increasing length. We performed a deep molecular characterization of SA-mediated integrants in epithelial cell lines and in primary immortalized keratinocytes stressing the SB system with cargos up to 18 kb. These data provide evidence for stability of SB-mediated integration and the reproducibility of the cut-and-paste mechanism even with large transposons embedded between two double IRs. Moreover, clonal analysis reveals a linear correlation between transposon copies harboured into the genomic DNA and their expression, an important characteristic for gene therapy application. Finally, high-resolution, genome-wide mapping of SA integrations in human keratinocytes revealed a close-to-random integration pattern with respect to genes and chromosomes, highlighting a relative low risk of genotoxicity as previously reported for SB transposition in cell lines [20–23]. Interestingly, the high-throughput analysis of SA integration sites showed an overrepresentation of integration events into repetitive elements (RE) of the human genome, in particular satellite, small RNA and LINE elements.

Materials and Methods

Cell culture

HeLa cells were cultured using DMEM medium (Lonza) added with 10% Fetal Bovine Serum (FBS), 1% L-Glutamine (L-Gln) and 1% Penicillin-Streptomycin (Pen/Strep). For each experiment, an aliquot of cryo-preserved HeLa cells was thawed and plated on 8 cm dishes. Upon reaching 80–90% of confluency, cells were re-plated on 6-wells culture plates at a concentration of 2×10^5 cells/well. After 24 h, cultures in each well were at 70–80% confluency, ready to be transfected.

Mouse NIH3T3 fibroblast cell line was maintained in Dulbecco's Modified Eagle's medium (Euroclone), supplemented with 10% bovine serum.

We have used SV40 immortalized keratinocytes derived from a patient affected by generalized atrophic benign epidermolysis bullosa (GABEB) produced by Borradori et al. [24] and kindly provided by J.W. Bauer. GABEB cells were cultivated in EpiLife medium supplemented with human keratinocyte growth supplement (HKGS) (Invitrogen, US). EpiLife is a serum-free keratinocyte culture medium with a low calcium (0.06 mM) concentration supplemented with HKGS which results in a final concentration of 0.2% (v/v) BPE, 5 lg/mL bovine insulin, 0.18 *lg/mL* hydrocortisone, 5 lg/mL bovine transferrin and 0.2 ng/mL human EGF. Upon reaching 80–90% of confluency, cells were re-plated on 6-wells culture plates at a concentration of 2.3×10^5 cells/well. After 24 h, cultures in each well were at 70–80% confluency, ready to be transfected.

Plasmid constructs

The plasmid carrying the T2 IRs including a Venus reporter gene driven by the chicken β actin promoter fused to CMV early enhancer element (CAGGS) and the construct coding for the SB100X were described in Mates et al. [1]; the SA transposon IRs were described in Zayed et al. [8]. The CAGGS Venus expression cassette was *Dra III* excised from pT2 3.2 and introduced into *EcoRV* digested pSA to obtain pSA 5.7. pT2 3.2 and pD28 [25]

were digested with *XbaI* to clone a non coding DNA of 2.7 kb from pD28 into the transposon.

Two fragments of the first intron of the HPRT gene were PCR amplified and cloned into the pCR 2.1 (TOPO cloning kit, Invitrogen) plasmid. The pT2 10 plasmid was cloned ligating the pT2 CAGGS Venus *SpeI* with *NheI* fragment of the amplified HPRT intron 1. The pT2 14 plasmid derives from pT2 10 digested with *ClaI* ligated to the *NotI* fragment of the amplified HPRT intron 1. Finally, pT2 18 was obtained by ligating a third sequence amplified from the HPRT intron 1 with pT2 14 through *EcoRI* restricted ends. The pSA 5.7 plasmid was digested with *NheI* and ligated to the *NheI* non coding fragment of the HPRT gene to obtain the pSA 9.7. Then the pSA 9.7 was digested with *PmeI* enzyme and ligated with a *PvuII* fragment of the HPRT intron 1 to obtain the pSA 14. To enlarge the pSA14, a sequence amplified from the intron 3 of the Lamb3 gene was introduced by *EcorV* compatible ends to obtain pSA 18.

Transfection-based transposition and calculation of transposition efficiency

HeLa and GABEB cells were both transfected with FugeneHD transfection reagent (Roche). For each sample 2 µg of DNA were added to 100 µl of either DMEM (for HeLa) or EpiLife (for GABEB). The media used for this transfection reaction mix were not added with FBS, L-Gln or Pen/Strep.

The transposon/transposase amounts of plasmid DNA were calculated to respect the stoichiometric ratio of 1:1 or, for transposon >10 kb, 2:1, in a total quantity of 2 µg. 2 µg of transposon-only plasmid were used for non-transposed control.

Each transfection reaction mix was complexed with 6 µl of FugeneHD (10 µl with SA and T2 18 in GABEB cells) and subsequently mixed by pulse-vortexing for a few seconds. The mixes were thereafter left at room temperature for 10′ in order to allow the formation of lipoplexes. After the 10′ had expired, each mix was added drop-by-drop to a cell culture sample, which was subsequently incubated at 37°C.

HeLa cells were transfected with Calcium Phosphate method using 15 µg of 14- or 18 kb transposons mixed with the plasmid carrying the transposase expression cassette.

The percentage of Venus+ cells was determined 2 and 20–30 days post-transfection via flow cytometry and the transposition efficiency was calculated as: Venus+ cells at 20–30 days post transfection/Venus+ cells at Day 2×100. Cells that were only transfected with the transposon plasmid represented the control for background integration events.

Transposed clones were analysed via flow cytometry to determine the presence of doublets and the Venus mean fluorescence intensity (MFI).

Isolation of single cell clones

GABEB cells were limiting diluted to obtain a concentration of 0.5 cell/well, plated onto lethally irradiated NIH3T3 cells and cultured in keratinocyte growth medium, a DMEM and Ham's F12 media mixture (2:1) containing FCS (10%), penicillin-streptomycin (1%), glutamine (2%), insulin (5 µg/ml), adenine (0.18 mM), hydrocortisone (0.4 µg/ml), cholera toxin (0.1 nM), and triiodothyronine (2 nM). After 1 week, the medium was replaced by EpiLife medium supplemented with HKGS. After 2 weeks GABEB cells were trypsinised at subconfluence and re-plated without the NIH 3T3 feeder-layer in EpiLife HKGS medium.

HeLa cells were seeded to obtain a concentration of 0.3 cells/well in a 96 well plate in DMEM medium complemented with 10% FBS.

Southern blot analysis

Ten μg of genomic DNA, extracted from $1-5\times10^6$ cells by a QIAmp DNA Mini kit (Qiagen), were digested overnight with *NheI* (SA 9.7-derived clones) and *AflII* (T2 10-derived clones) to verify the copy number of the transposed cassette, or with *NcoI* (SA 9.7-derived clones) and *MfeI* plus *NdeI* (T2 10-derived clones) to verify the integrity of the transposed cassette. Digested gDNA was run on a 0,8% agarose gel, transferred to a nylon membrane (Duralon, Stratagene) by Southern capillary transfer and probed with 2×10^7 cpm ^{32}P-labeled Venus probe according to standard techniques [26].

PCR screening for episomal SB vectors

About 100 ng of template gDNA were used in a PCR reaction. Primers capable to amplify the Amp resistance gene or the SB100X transposase (**Table S1**) were used to detect genomic integrations of SA 9.7 backbone and SB100X, respectively. PCR conditions were as follows: 30″ at 94°C, 30″ at 58°C and 30″ at 72°C for 30 cycles.

LM-PCR and bioinformatic analysis

Integration sites were amplified by Linker Mediated PCR (LM-PCR), as described [27]. Briefly, genomic DNA was extracted from $0.5-5\times10^6$ transposed cells and digested with *MseI* and *XhoI* enzyme to prevent amplification from internal mutated IR fragments. An *MseI* double-stranded linker was then ligated and LM-PCR performed with nested primers specific for the linker and SA IR/DR (**Table S1**).

LM-PCR derived amplicons were run on a Roche/454 GS FLX using titanium chemistries by GATC Biotech AG Next Gen Lab. A valid integration contained: the TAGpSAIR nested primer and the entire SA IR/DR sequence up to a TA dinucleotide.

Alignment pipeline. 31,603 sequencing reads were tested for the presence of the SA IR sequence and TA dinucleotide. The SA IR and any primer sequences were trimmed, and the remaining reads starting with TA dinucleotides were mapped to the human genome (hg19) using NCBI BLAST (blastn with default parameters). We kept only reads which were mapped to a single genomic site with at least 90% sequence identity and an E-value of at most 0.05. Only reads which could be mapped from their 5′ end onwards were considered for further analysis. Redundant reads mapping to identical genomic positions were collapsed. This way we got 2019 unique SA integration sites.

For the statistical analysis we generated 10,000 control sites in-silico taking into account the bias introduced by LM-PCR techniques. We first generated artificial reads starting with TA dinucleotide of the human genome in a way that the control sequences had both the length and the frequency of *MseI* restriction sites (TTAA) as observed in real sequencing reads. The artificial reads were then processed by the same mapping criteria used for the SA sites.

RM blast analysis. Analyses of repetitive element were performed with RepeatMasker Blast (http://repeatmasker.org) [28]. To achieve reliable and comparable results we processed the raw sequences trimming out the primer sequences used in LM-PCR, the IR/LTR/linker specific sequences following the primers. Resulting reads were further trimmed till the 40th nucleotide discarding every sequence with less than 40 nucleotides. Finally, we collapsed the reads that were either identical or with one mismatch. A two-sample test for proportions was used for pairwise comparison of the RE within the different datasets.

For statistical analysis we created control sets as follows. We first randomly sampled 1 Million sequences 49 bp in length from the human reference genome (hg19). Then we discarded all sequences

not starting with TA. The resulting set of 65,826 TA-weighted sequences was used as a background for T*neo* SB integrations. For a second random control set we first randomly sampled 10 Million sequences of length 120 bp from the genome. Then we discarded all sequences not starting with TA, or either not containing the *MseI* restriction motif TTAA or having a TTAA within the first 39 bp of the sequence. After removing the part of the sequences following the first occurrence of TTAA, we received 292,917 sequences of lengths between 40 bp and 120 bp, which were weighted for TA and *MseI* and could be used as a background for SA integrations. We passed the generated sequences through the same filtering/trimming pipeline as the actual integration reads.

A third random control set of 45,235 genomic sequences weighted for *MseI* was adapted from Cattoglio et al. [29] and used as a background for MLV and HIV integrations.

Bidirectional PCR mapping on GABEB clones

Transposon integrations in GABEB clones were amplified by LM-PCR as described. PCR products were shotgun-cloned (TOPO TA cloning kit, Invitrogen) and then sequenced. Sequences between the TA and the linker primers were mapped onto the human genome by the BLAT genome browser (UCSC Human Genome hg19). Sequences featuring a unique best hit with ≥90% identity to the human genome were considered genuine integration sites. To confirm the genuine integration in both directions we design primers on the genomic region hit and performed a direct PCR in conjunction with the pSAIR specific primer for the SA IR sequence (**Table S1**). The derived amplicons were loaded on agorose gel and checked for the expected length.

Results

Efficiency of T2 and SA transposons

The sandwich (SA) transposon vector has superior ability to transpose >10 kb transgenes with respect to the first-generation transposon when SB10 transposase was provided [8]. Nevertheless, the T2 transposon, resulting from site-specific mutations in the IR sequences and insertion of double TA flanking each IR, has been demonstrated to have a four-fold enhanced activity over the first-generation transposon construct [6]. A side-by-side comparison of SA and T2 transposon was needed to address the transposition efficiency of increasing DNA cargoes and to verify their molecular behaviours once integrated into the human genome.

We generated SA- and T2-based plasmids (SA 5.7 and T2 3.2 **Figure 1**) keeping the Venus reporter gene as standard expression cassette. Increasing sizes of a non-coding human stuffer DNA (4-, 8.3- and 12.3 kb in the SA plasmid; 6.8-, 10.8- and 14.8 kb in the T2 plasmid) were introduced between the two IR/DR to produce transposons of comparable length. For the sake of simplicity, we named these plasmids with the transposon construct type and the size of the transposable cassette expressed in kilobases (**Figure 1**).

Transposition experiments were performed in HeLa cells and in immortalized primary keratinocytes derived from patients affected by Generalized Atrophic Benign Epidermolysis Bullosa (GABEB), an inherited skin adhesion defect. All the experiments aimed at the identification of the integration efficiency of the IR-flanked transgene were measured by long-term Venus fluorescence in the absence of selective pressure. We co-transfected the SB100X transposase-expressing plasmid together with transposon plasmids in two different molar ratios (1:1 or 1:2) depending on the transposon length. Larger cargos required more transposon DNA to reach good transfection efficiency.

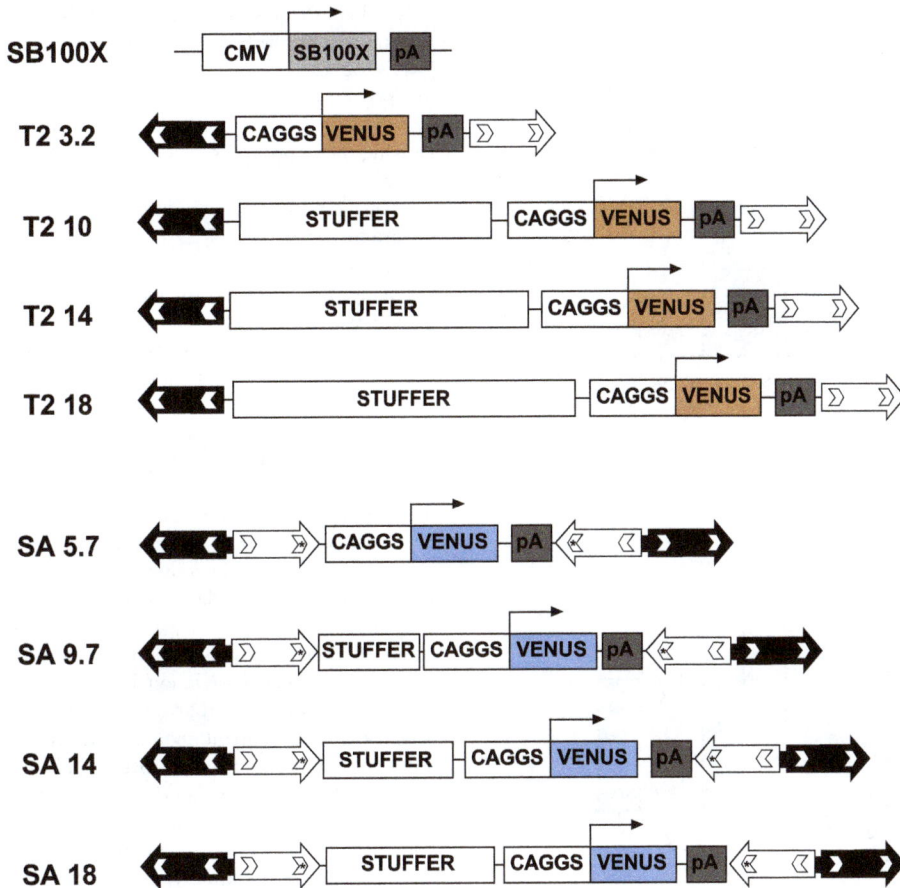

Figure 1. Transposon fleet. Schematic representation of the generated plasmids. SB100X carries the Hyperactive *Sleeping Beauty* transposase coding sequence placed under the control of the CMV promoter and followed by an SV40 poly-Adenylation (pA) signal. The transposons T2 and SA possess the expression cassette consisting of the CAGGS promoter, VENUS reporter gene and SV40 pA signal. The stuffer DNA represented has variable increasing size. The arrows represent the IR/DR ends recognised by the transposase. SA constructs are characterized by the presence of two complete IR/DR at each ends (white and black arrows) and the asterisks underline the IR mutated site not recognized as a catalytic substrate by the transposase. Numbers following T2 or SA abbreviation indicate the size in kilobases of the transposed cassette.

At least three independent experiments for each cell type and transposon were performed in order to reduce variability due to the transfection procedure. Mock-transfected HeLa and GABEB cells, and cells transfected with the T2 or SA Venus constructs alone were used as controls (in the absence of transposase, no transposition event should occur and residual reporter gene expression after long periods would only be attributable to noise or to rare random plasmid integration events). Transgene expression all along the culture period (up to 31 days) was measured via flow cytometry to follow the trend of the signal that persists in presence of SB100X and drops without the transposase (**Figure S1**).

The transposition efficiency was normalized by transfection efficiency (numbers of cells that received the plasmids after transfection) and calculated as the ratios between the percentage of Venus+ cells at the endpoint (20–31 days) and the percentage of transfected cells 2–3 days after DNA delivery to the cells. The endpoint of each experiment is achieved when the percentage of Venus+ cells in the sample transfected with the transposon alone stabilized to less than ~0.5%.

Figure 2A and 2B show the transposition rate obtained in HeLa and GABEB cells. As previously reported [1,8], the transposition efficiency was inversely proportional to the transposon size. In HeLa cells, the transposition efficiency dropped 7.8

fold (from 58.5% to 7.5%) when increasing the cargo payload from 3.2 kb to 18 kb, independently of the transposon structure (T2 or SA). Interestingly, this size-dependent effect was less pronounced in GABEB cells. In this cell type the decrease was of 1.8 fold (from 44% to 24%) for T2 and SA and the transposition rate for 18 kb transposons remained approximately 24% compared to the 7.5% in HeLa cells.

Clonal molecular analysis

Although we performed a molecular characterization of almost all T2 and SA vectors in HeLa or GABEB cells (**Table S2.**), we focused our genomic analysis on a relatively large T2 and SA transposons cassette (10 kb) and on GABEB keratinocytes. Bulk populations of transposed cells were sorted for Venus expression 20–35 days post transfection and cloned by limiting dilution. Genomic DNA extracted from each clone was first investigated by PCR for the presence of the transposon backbone and SB100X expressing plasmid. Notably, we scored 14.8% of clones (8 out of 54) positive for the Ampicillin sequence present within the transposon backbone about 60 days post transfection, while few (2 out 54) of the analysed clones were positive for the SB100X sequence (**Table S2**).

A

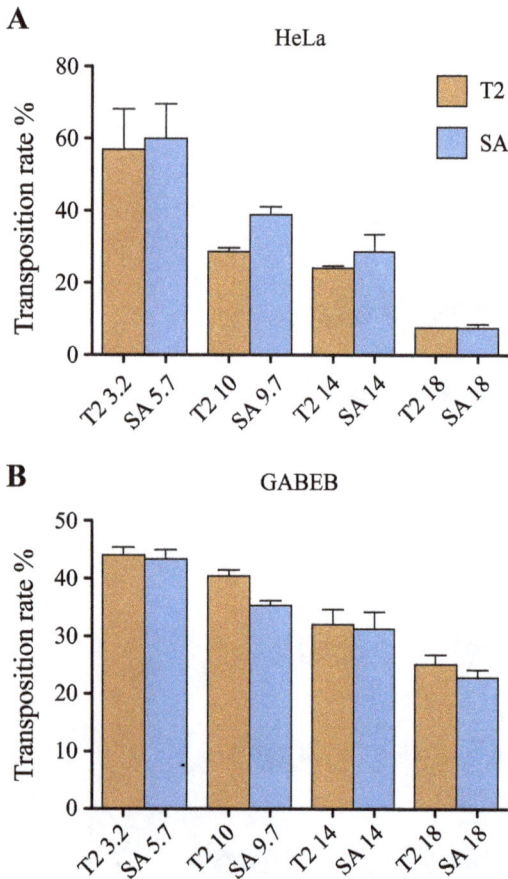

HeLa

(Chart with Transposition rate % on Y axis, legend T2 (tan) and SA (blue), X axis categories: T2 3.2, SA 5.7, T2 10, SA 9.7, T2 14, SA 14, T2 18, SA 18)

B

GABEB

(Chart with Transposition rate % on Y axis, X axis categories: T2 3.2, SA 5.7, T2 10, SA 9.7, T2 14, SA 14, T2 18, SA 18)

Figure 2. Transposition efficiency. HeLa (A) and GABEB (B) cells were co-transfected with the T2 and SA transposons- and transposase-carrying plasmids. The transposition rate, on the Y axis, is derived by the ratio between the percentage of Venus+ cells at about 20 and 2 days post transfection. Data are representative of three independent experiments (mean ± SEM; $n = 3$).

We next performed Southern blotting on the genomic DNA of 16 clones for each transposon type to determine the transgene copies harboured in the genome and their integrity. To this end, we digested the genomic DNA with *AflII* (T2 clones) or *NheI* (SA clones) that release fragments longer than 3.4 and 4.2 kb. Hybridization with a Venus-specific probe showed that most of the SA treated samples (13 out of 16) carry a single integrated transposon, only 1 clone (#26) had 3 copies, and 2 out of 16 clones contained 2 copies (#8, #13) resulting in an average copy number of 1.3. Surprisingly, 16 GABEB clones obtained with T2, harbour 1 to 7 copies with an average of 3 integrated transposons per clone (**Figure 3A**). **In general we observed that the mean copy number is more affected by the transfection efficiency (Table S2) respect to the size and type of transposons.**

Further restriction analysis performed with *MfeI* and *NdeI* on 9 T2 clones and with *NcoI* on 8 SA clones showed that all clones harbour the full-length transposon cassette (**Figure 3B**). Among the 21 integrated transposons in the 9 T2 clones, only one, belonging to clone #3, is shorter than expected. None of the 13 integrated transposons in the 8 SA clones was rearranged.

To unequivocally prove that all the integration events mediated by SA transposition resulted from a genuine "cut and paste" mechanism, we mapped the insertion site at both transposon ends using an adapted version of Linker-Mediated PCR (LM-PCR)

[27]. Ten Venus-expressing GABEB clones, derived from transposition of the SA 5.7 plasmid, were examined. Six integrants (#1, 4, 7, 13, 14, 16) belonging to 5 clones were bi-directionally mapped by LM-PCR. Additional 21 integrants were revealed by LM-PCR and confirmed by specific PCR on the genomic region flanking the opposite IR (**Figure 3C**). Importantly, almost all the integration events occurred without genomic rearrangements, deletions or insertions, in the target sites. Only 2 out of 27 integrations (#26 and #27 belonging to clone 34) could not bi-directionally confirmed.

Finally, we correlated the expression level of the reporter gene with the copy number of the transposon. The positional effect variegation primarily observed with retroviral and lentiviral vectors [30] could lead to the silencing of the therapeutic gene delivered by the vector. We asked weather the SB integrations would be affected by this phenomenon. We correlated the expression of Venus protein, measured by Mean Fluorescence Intensity (M.F.I.), with copy number of either the SA and T2 transposon, as determined by Southern blot or q-PCR analyses of 62 GABEB clones. For comparison, we analysed the M.F.I of a GFP reporter gene, driven by the human Keratin 14 promoter, in 70 HaCaT clones isolated upon LV transduction. A linear correlation curve was traced to retrieve the R^2 coefficient of determination. Transposon samples show an $R^2 = 0.759$ with a statistically defined correlation between two variables ($P_N = 0.6$). LV samples display an $R^2 = 0.001$ with a null defined correlation (**Figure 4**). Independent analysis of transposed clones obtained in different cells (HaCaT and GABEB) and carrying a reporter gene driven by PGK or CAGGS promoter showed comparable results indicating common directly correlation between MFI and copy number (data not shown). We conclude that SB integrants tend to express their cargo faithfully, and multi-copy integrants express in a copy-number dependent manner, consistent with earlier observations [31].

Integration pattern analysis

In the last few years, several papers described the integration profile of the SB, *piggyBac* (PB), and Tol2 transposons [20–23,32–36]. Here we report the integration profile and preference of the sandwich compared with the first-generation SB transposon [20] in human epithelial cells. To generate a library of SA integration events, we transfected 20 million GABEB cells with SA transposon- and SB100X-carrying plasmids. The 20% of Venus-positive cells were sorted three days after transfection to enrich the population expressing the reporter gene. A 90%-pure sorted population was kept in culture for 3 weeks to dilute the un-integrated SA vector reaching a stable 78% Venus+ bulk population. We used LM-PCR and pyrosequencing to generate 6,084 non-redundant SA-linked genomic sequences in human immortalized GABEB keratinocytes. The Blast alignment retrieved 2,019 unambiguously mapped integration sites. As a control, 10,000 random unique sequences were generated in silico balancing the biases introduced by the LM-PCR (amplicon lenght and *MseI* proximity) and the availability of the TA dinucleotides in the genome. In the analysis we also annotated a large dataset (59,169 hits) generated in HeLa cells transposed with the first-generation T*neo* transposon and selected for 2 weeks with neomycin [20]. The integration sites and control sites were annotated as transcriptional start site (TSS)-proximal when mapping in the ±2.5 kb window around a TSS, intragenic when mapping within a transcription unit, and intergenic in all other cases. Among SA integrations, 58.6% were in an intergenic position, 38.9% were within the transcribed portion of at least 1 gene, and 2.5% was within a 5 kb window encompassing the TSS

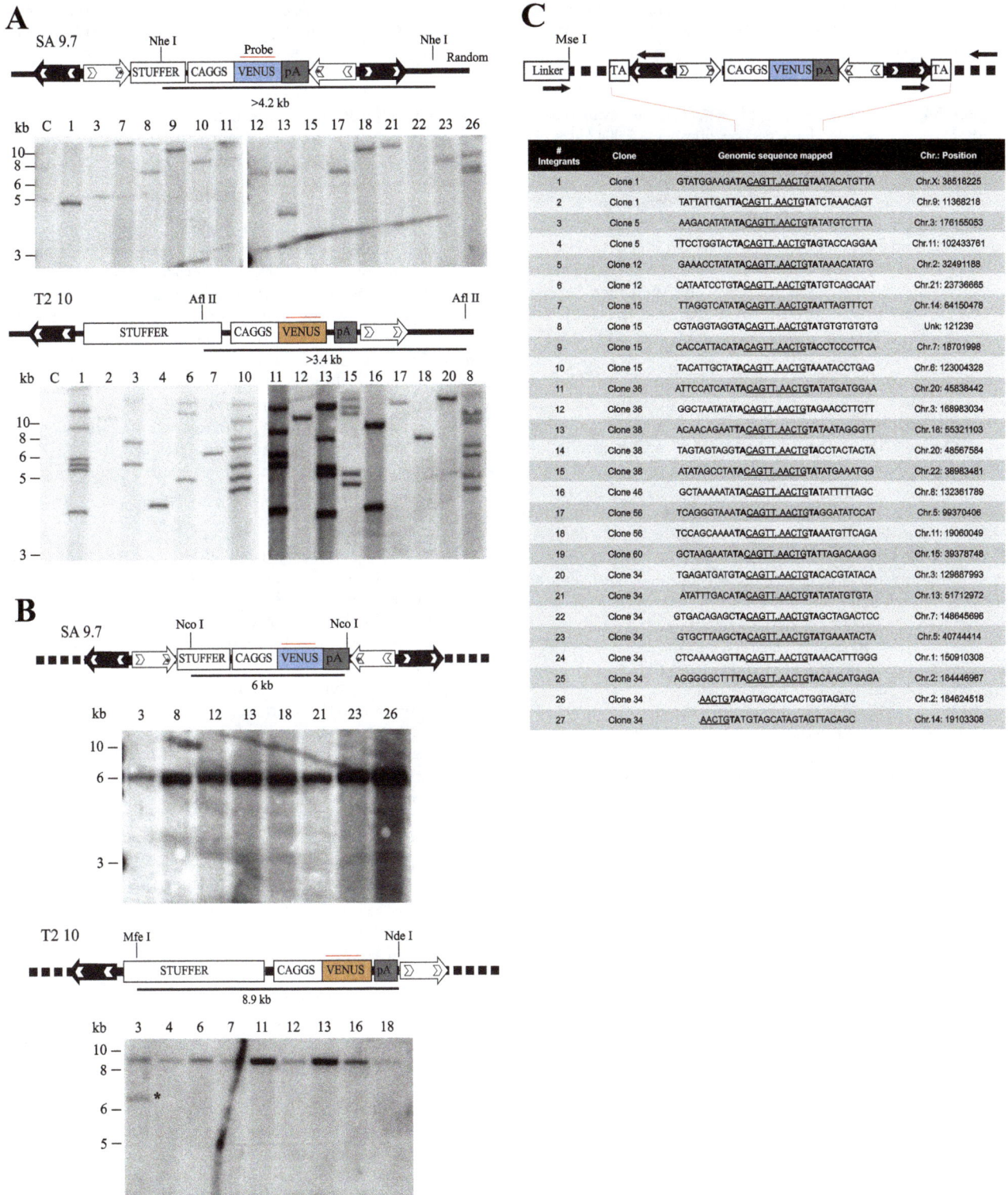

# Integrants	Clone	Genomic sequence mapped	Chr.: Position
1	Clone 1	GTATGGAAGATACAGTT..AACTGTAATACATGTTA	Chr.X: 38518225
2	Clone 1	TATTATTGATTACAGTT..AACTGTATCTAAACAGT	Chr.9: 11368218
3	Clone 5	AAGACATATATACAGTT..AACTGTATATGTCTTTA	Chr.3: 176155053
4	Clone 5	TTCCTGGTACTACAGTT..AACTGTAGTACCAGGAA	Chr.11: 102433781
5	Clone 12	GAAACCTATATACAGTT..AACTGTATAAACATATG	Chr.2: 32491188
6	Clone 12	CATAATCCTGTACAGTT..AACTGTATGTCAGCAAT	Chr.21: 23736665
7	Clone 15	TTAGGTCATATACAGTT..AACTGTAATTAGTTTCT	Chr.14: 64150478
8	Clone 15	CGTAGGTAGGTACAGTT..AACTGTATGTGTGTGTG	Unk: 121239
9	Clone 15	CACCATTACATACAGTT..AACTGTACCTCCCTTCA	Chr.7: 18701998
10	Clone 15	TACATTGCTATACAGTT..AACTGTAAATACCTGAG	Chr.6: 123004328
11	Clone 36	ATTCCATCATATACAGTT..AACTGTATATGATGGAA	Chr.20: 45838442
12	Clone 36	GGCTAATATATACAGTT..AACTGTAGAACCTTCTT	Chr.3: 188983034
13	Clone 38	ACAACAGAATTACAGTT..AACTGTATAATAGGGTT	Chr.18: 55321103
14	Clone 38	TAGTAGTAGGTACAGTT..AACTGTACCTACTACTA	Chr.20: 48567584
15	Clone 38	ATATAGCCTATACAGTT..AACTGTATATGAAATGG	Chr.22: 38983481
16	Clone 46	GCTAAAAATATACAGTT..AACTGTATATTTTTAGC	Chr.8: 132361789
17	Clone 56	TCAGGGTAAATACAGTT..AACTGTAGGATATCCAT	Chr.5: 99370406
18	Clone 56	TCCAGCAAAATACAGTT..AACTGTAAATGTTCAGA	Chr.11: 19060049
19	Clone 60	GCTAAGAATATACAGTT..AACTGTATTAGACAAGG	Chr.15: 39378748
20	Clone 34	TGAGATGATGTACAGTT..AACTGTACACGTATACA	Chr.3: 129887993
21	Clone 34	ATATTTGACATACAGTT..AACTGTATATATGTGTA	Chr.13: 51712972
22	Clone 34	GTGACAGAGCTACAGTT..AACTGTAGCTAGACTCC	Chr.7: 148645696
23	Clone 34	GTGCTTAAGCTACAGTT..AACTGTATGAAATACTA	Chr.5: 40744414
24	Clone 34	CTCAAAAGGTTACAGTT..AACTGTAAACATTTGGG	Chr.1: 150910308
25	Clone 34	AGGGGGCTTTTACAGTT..AACTGTACAACATGAGA	Chr.2: 184446967
26	Clone 34	AACTGTAAGTAGCATCACTGGTAGATC	Chr.2: 184624518
27	Clone 34	AACTGTATGTAGCATAGTAGTTACAGC	Chr.14: 19103308

Figure 3. Molecular characterization of the *Sleeping Beauty*-mediated integration events in GABEB cell clones. (A) Southern blot analysis of genomic DNA from GABEB cell clones digested with *Nhel* (SA clones) or *AflII* (T2 clones), single cutter in the transposon cassette, and hybridized to a Venus probe. A single band higher than 4.2 kb (SA clones) and 3.4 kb (T2 clones) indicates integration of one copy of the transposon into the genome. Multiple Venus-specific bands correspond to repeated integration events. (B) Southern Blot analysis of genomic DNA from 8 (SA) and 9 (T2) clones digested with *Ncol* (SA clones) or *Mfel* and *Ndel* (T2 clones). The expected Venus-specific band corresponding to 6 kb for SA and 8.9 kb for T2 transposon indicates the correct integration of the transposons into the genome. C, mock-transfected cells; red bars, Venus-specific probe. Clone showing rearrangement of the transposon cassette is highlighted by black asterisk. (C) Bi-directional mapping of the junctions between transposon and genomic DNA. The table summarizes 27 integrations belonging to 10 single clones. For each integrant, the underlined sequence

represents a portion of the transposon IRs, left (CAGTT) and right (AACTG) separated by dots; TA dinucleotide (in bold) is the target site correctly duplicated after transposition. Hit chromosomes and positions are reported. UnK, unknown region of the human genome based on UCSC hg19 assembly.

(**Figure 5A**; the complete list of sequences is available in GenBank database with the accession number SRP047118). In general, the distribution of the SB integrants in both datasets is fairly random and resembles the composition of the human genome showing no statistical differences compared to their relative controls, i.e. all p-values (both two-sample tests for proportions and Fisher's Exact Tests) were $>10^{-2}$.

We then analysed the frequency of human repetitive elements in the transposon libraries, SA and T*neo* [20], and their relative weighted controls availing of the RM Blast browser [28]. For comparison we also analysed two viral-derived integration datasets (MLV and HIV) generated in human CD34+ multipotent hematopoietic progenitor cells (HPCs) [29] and their control library weighted for *MseI* restriction site distribution. The raw data generated by deep sequencing of the LM-PCR (applied to SA, MLV and HIV treated cells) and LAM-PCR (applied in [20]) products were filtered and trimmed in order to rescue the genuine integration events (see materials and methods). After filtering and trimming we retrieved 6,084 and 165,887 unique sequences in SA and first-generation vector libraries, respectively, and 37,873 and 31,204 unique sequences from MLV and HIV datasets, respectively. We generated large control datasets taking into account the bias introduced by the respective technique. In particular from the hg19 genome database we retrieved 45,235 control reads weighted for *MseI*, 65,570 sequences weighted for the presence of TA dinucleotide hit by the SB transposons, and 209,913 sequences *MseI*- and TA-double weighted.

The RM Blast analysis revealed an overrepresentation of REs in the SA integrations (34%) with respect to the TA and *MseI*-weighted control (14%) and to all the other datasets analysed (**Figure 5B**). In particular, Satellite, small RNA and LINE elements were enriched in the SA library (24-, 7.6- and 3.5-fold increase over the background, respectively) whereas in the first-

generation vector library only a slight increase in the satellite and simple repeats elements was measurable (3.5- and 2.6-fold over the background, respectively); comparable LINE frequency was detected.

Besides the higher frequency in the satellite elements, the two SB transposon datasets share a slight under-representation of SINE, LTR and DNA transposable elements in comparison with their random control libraries. We introduced MLV and HIV libraries to compare the frequency of integration into RE generated by a retroviral integrase-mediated integration mechanism. The RM Blast analysis pointed out that viral vectors disfavour integration in RE (14–16% vs 24%), and, in particular, satellite, LTR and LINE elements are underrepresented. These data clearly confirm a difference in the integration site selection between viral vectors and SB transposons and identify new signatures in the SA integrome that should be taken into consideration when using them as tools for genetic manipulation.

Discussion

The SB transposon IRs were mutated to improve their capacity to be mobilized and, to date, there is not a direct comparison that define genetic characteristics of the T2 and SA IRs [8]. In this study, HeLa cells and GABEB keratinocytes [24] were transfected with a panel of T2 or SA transposons carrying size-increasing Venus expression cassette in combination with SB100X plasmid (Figure 1). Transfection rate was higher in HeLa than in GABEB cells (**Figure S1**) and the transposition efficiency was inversely proportional to the transposon size (**Figure 2**). Interestingly, HeLa cells were severely affected by the transposon size compared to primary immortalized cells. These results suggest that the transposase activity could be favoured by some cellular factor differentially expressed in GABEB and HeLa cells. Nonetheless, T2 and SA constructs carrying cargos of comparable size showed similar transposition efficiency in both cell lines. From these data we can conclude that the T2 IR construct is interchangeable with the SA construct with some advantages: T2 has shorter IRs thereby it could accept a larger cargo cassette.

Transposed GABEB and HeLa populations were subjected to limiting dilution to obtain a single cell derived expansion. The derived clones were employed to characterize several molecular parameters: transposon-independent insertion, copy number, genomic stability, faithful transposition activity, correlation between copy number and expression of the integrated cassette. The SB100X sequence was retrieved in 6 out of 211 analysed clones while almost 14% of the clones (30 clones) were found positive for the transposon backbone sequences (**Table S2**). We hypothesize that the plasmid backbone carrying the transposon could have some advantages to remain episomal or to integrate in the genome. The transposon excision step from the plasmid leaves the backbone with a double strand break that induce recruitment of the endogenous repair machinery and integration into the cell genome. We also analysed the copy number of the clones. **Figure 3A** shows an average of 1.3 SA copies/clone while T2 copy number spans from 1 to 7 transposons with an average of 3 copies. This difference mostly depends on the transfection efficiency as confirmed by the analysis of the other transposed cell populations generated in this study. Therefore, it is possible to fine tune this parameter by adjusting the ratios of the two SB

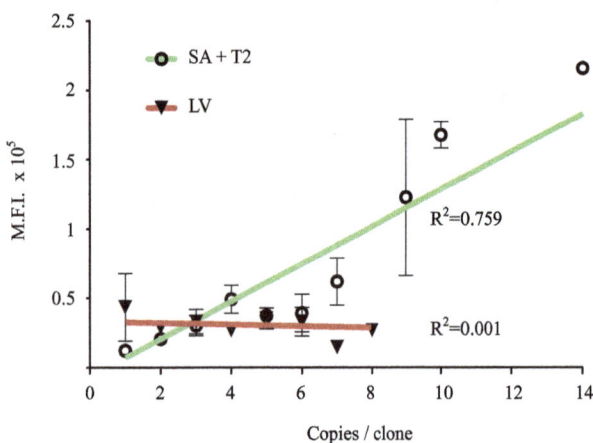

Figure 4. Correlation between copy number and expression of the integrated cassette. Mean fluorescence intensity (M.F.I.) of 62 GABEB clones positive for Venus expressing SB transposons are represented as circles; triangles indicate the M.F.I. of 70 GFP+ HaCaT clones transduced with a lentiviral vector (LV). Standard deviation bars are present for those clones carrying the same copy number. R^2 coefficients of determination were extracted from the linear regression plot, green line for SA and T2 transposons and red line for LV.

A

	SA	Random TA MseI		Tneo	Random TA		MLV	HIV	Random MseI
	n=6,084	n=209,913		n=165,887	n=65,570		n=37,873	n=31,204	n=45,235
Low complexity	0.01	0.06		* 0.04	0.07		0.07	0.08	0.08
Simple repeats	1.10	1.17		** 3.99	1.55		1.03	1.16	1.14
Satellite	** 2.37	0.10		** 0.53	0.15		** 0.01	* 0.08	0.16
Small RNA	** 0.23	0.03		0.03	0.02		* 0.10	** 0.15	0.05
DNA elements	0.76	1.00		* 0.87	1.00		** 0.60	0.83	0.90
LTR elements	* 1.99	2.67		** 2.12	3.00		** 1.72	** 1.29	3.49
LINEs	** 26.86	7.68		* 7.66	7.99		** 2.18	** 3.59	8.41
SINEs	** 0.54	1.52		** 1.56	3.87		** 8.23	9.00	9.46

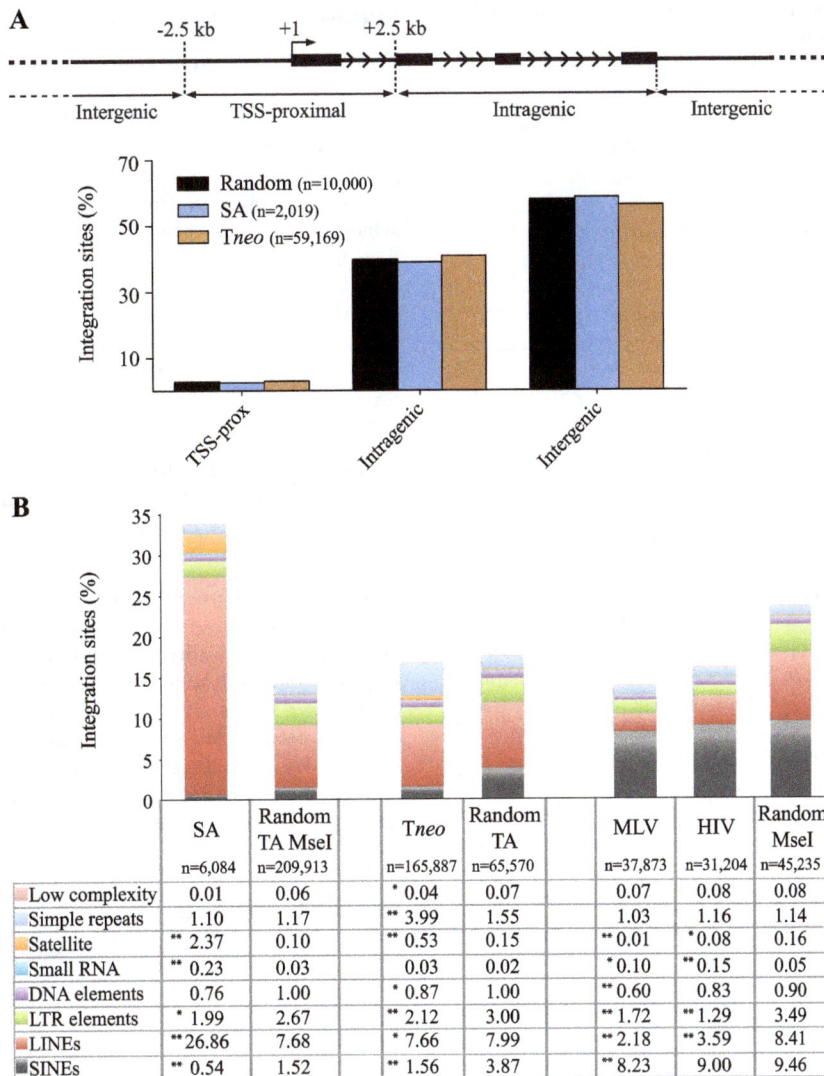

Figure 5. Integration pattern analysis. (A) Integration sites were annotated as "TSS-proximal" when occurring within a distance of ± 2.5 kb from the gene's TSS, as "Intragenic" when occurring in a gene body and as "Intergenic" in all other cases. Black bars represent exons of a schematic gene, arrowhead indicates the direction of transcription. Distribution of SA, Tneo and random integration sites in the genome is plotted accordingly to defined annotations. (B) Distribution of Repetitive Elements in SB SA and Tneo libraries, in MLV and HIV libraries. Relative weighted random libraries were reported: TA and *MseI*-weighted for SA, TA-weighted for Tneo and *MseI*-weighted for MLV and HIV libraries. **$p \leq 10^{-3}$, *$p \leq 10^{-2}$.

components used for transfection or, as previously reported, bypass the transfection procedure through the viral delivery of transposase and transposon by adenoviral vector [37], integration defective lentiviral vector [38,39], retroviral particle [40] and adeno-associated vectors [41].

We were able to associate copy number of the transposon with the expression level of the Venus fluorescence gene. Mean Fluorescence Intensity does follow a direct proportion with the copies harboured (**Figure 4**). In contrast, expression of the reporter gene in lentiviral-mediated integrants does not correlate with copy number and is more subjected to the activity of surrounding genomic sequences [42,43].

Next, the integrated transposons in these clones were also analysed for their integrity via Southern Blot. Retroviral and lentiviral vectors can rearrange during the reverse transcription step resulting in partially-deleted integrated proviruses, a frequent occurrence in transgene hosting repetitive sequences [44,45]. The

SB mediated integration, by contrast, does not require reverse transcription and thus is expected to preserve the integrity of the transgene. Ninety-eight percent of the integrants, resulting from T2 and SA transposition, have a correct size (**Figure 3 B**).

The sandwich transposon has a doubled IR/DR structure at both ends with 8 transposase binding sites in total. In principle, every transposase unit, bound to one DR site, could interact with the others to create different chiasm geometries (also described in [6]); some of these conformations could modify the integration activity resulting in chromosomal aberrations. To investigate the fidelity of the transposition process 10 GABEB clones were mapped bi-directionally by LM-PCR and transposon-genome junction was amplified by site-specific PCR. Twenty-five integrations, out of 27 (92.6%), were validated for a canonical transposition event with the TA target site duplication signature at both ends (**Figure 3C**). Two integrations mapped by LM-PCR were not confirmed in the opposite transposon end suggesting rearrangements probably caused by the repair mechanism occurred in the transposition break.

LM-PCR was also employed to derive a high-definition map of SA/SB100X integration sites in the genome of a transposed GABEB bulk population. This analysis is commonly applied to integrating vectors (i.e. retroviral and lentiviral vectors) because it allows to evaluate genotoxicity [46,47] and to understand molecular mechanism driving the integration towards specific regions of the genome [29,48–50]. The technique returned 2,019 SA unambiguously mappable integration sites randomly distributed throughout the human genome, in accordance with previously published data on first-generation transposon [20,23] (**Figure 5A**). For gene therapy purposes, the SB system results in a safer integration profile compared to other integrating vector such as Tol2, PB transposon and retroviral vectors [20–23,32,33], which favor TSS-proximal regions or gene body sequences.

Although the integration site distribution in relation to genes was found close to random, the RM Blast analysis shows a significant bias distribution of SA integrations in repetitive elements (RE), particularly in satellite, LINE and small RNA genes. It could be that these genomic regions are favourable for integration due to their base composition (TA-richness) or there might be molecular mechanisms that actively recruit the transposon/transposase complex at specific RE sites [51–55].

Curiously, the frequency of RE elements in the first-generation transposon library and its weighted control were comparable. Differently from the SA (obtained in 80% Venus expressing immortalized keratinocytes without selective pressure), the first-generation transposon library derives from transposed HeLa cells [56] selected for two weeks by antibiotic resistance. This culture condition could negatively select those integrations landing into poorly expressed genomic loci or into heterochromatin regions [35]. Nevertheless, the first-generation transposon integrations were slightly increased into satellite regions and SINE, whereas LTR and DNA elements were underrepresented compared to the background.

These data identify some common features in SB datasets. Conversely, MLV and HIV-derived viral vectors disfavour integration in RE (satellite, LTR and LINE accordingly also to [57]) suggesting an active role of viral integrase in the selection of integration sites that could better support the expression, replication and survival of the viral progeny. The genomic features newly identified in the SA integrome raise an interesting matter that needs to be deeply investigated for future application.

Supporting Information

Figure S1 Transposition trend. Expression of Venus fluorescence protein was detected by cytofluorimetric analyses at different time points in HeLa cells (A) and in GABEB cells (B). The days post transfection (p.t.) are plotted on the X axis, while the percentage of Venus+ cells are represented on the Y axis. The maximum expression from a transfected reporter gene was achieved 2 days p.t (black vertical dotted line). The plot shows the samples co-transfected with SB100X and transposon plasmids: T2 3.2 or the SA 5.7 (blue), T2 10 or SA 9.7 (green), T2 14 or SA 14 (purple), and the 18 kb transposons (orange). SA constructs represented with dashed lines, T2 with continuous lines. In gold are represented the negative controls transfected with T2 3.2 or SA 5.7 alone, without the SB100X plasmid.
(EPS)

Table S1   List of primers used for plasmid episomial amplification, LM-PCR, and site-specific amplification of the SA-genome junctions.
(DOC)

Table S2   Transposed clones were analysed to show the following parameters: number of retrieved Venus⁺ clones for each bulk; percentage of Venus⁺ cells in bulk populations 48 hours p.t.; percentage of stable Venus expressing cells in bulk populations; percentage of clones positive for the Ampicillin or SB100X sequence carried by transfected plasmids; mean copy number retrieved by Southern blot analysis; recombinant events detected in transposed clones by Southern blot analysis.
(DOC)

Acknowledgments

We thank Davide Pietrobon for compiling the scripts used for filtering and trimming the raw sequences from SB and viral libraries.

Author Contributions

Conceived and designed the experiments: GT MCL. Performed the experiments: GT MCL. Analyzed the data: GT MCL CC AGD FM Z. Izsvák Z. Ivics AR. Wrote the paper: GT AR.

References

1. Mates L, Chuah M, Belay E, Jerchow B, Manoj N, et al. (2009) Molecular evolution of a novel hyperactive Sleeping Beauty transposase enables robust stable gene transfer in vertebrates. Nature genetics 41: 753–761.

2. Ivics Z, Izsvak Z, Minter A, Hackett PB (1996) Identification of functional domains and evolution of Tc1-like transposable elements. Proc Natl Acad Sci U S A 93: 5008–5013.

3. Ivics Z, Hackett P, Plasterk R, Izsvak Z (1997) Molecular reconstruction of Sleeping Beauty, a Tc1-like transposon from fish, and its transposition in human cells. Cell 91: 501–510.

4. Hackett PB, Largaspada DA, Cooper LJ (2010) A transposon and transposase system for human application. Mol Ther 18: 674–683.

5. Ivics Z, Kaufman C, Zayed H, Miskey C, Walisko O, et al. (2004) The Sleeping Beauty transposable element: evolution, regulation and genetic applications. Current issues in molecular biology 6: 43–55.

6. Cui Z, Geurts A, Liu G, Kaufman C, Hackett P (2002) Structure-function analysis of the inverted terminal repeats of the sleeping beauty transposon. Journal of molecular biology 318: 1221–1235.

7. Izsvak Z, Ivics Z, Plasterk R (2000) Sleeping Beauty, a wide host-range transposon vector for genetic transformation in vertebrates. Journal of molecular biology 302: 93–102.

8. Zayed H, Izsvak Z, Walisko O, Ivics Z (2004) Development of hyperactive sleeping beauty transposon vectors by mutational analysis. Mol Ther 9: 292–304.

9. Ivics Z, Garrels W, Mates L, Yau TY, Bashir S, et al. (2014) Germline transgenesis in pigs by cytoplasmic microinjection of Sleeping Beauty transposons. Nat Protoc 9: 810–827.

10. Garrels W, Holler S, Taylor U, Herrmann D, Niemann H, et al. (2014) Assessment of fetal cell chimerism in transgenic pig lines generated by sleeping beauty transposition. PLoS One 9: e96673.

11. Ivics Z, Hiripi L, Hoffmann OI, Mates L, Yau TY, et al. (2014) Germline transgenesis in rabbits by pronuclear microinjection of Sleeping Beauty transposons. Nat Protoc 9: 794–809.

12. Katter K, Geurts AM, Hoffmann O, Mates L, Landa V, et al. (2013) Transposon-mediated transgenesis, transgenic rescue, and tissue-specific gene expression in rodents and rabbits. FASEB J 27: 930–941.

13. Ivics Z, Mates L, Yau TY, Landa V, Zidek V, et al. (2014) Germline transgenesis in rodents by pronuclear microinjection of Sleeping Beauty transposons. Nat Protoc 9: 773–793.

14. Jin Z, Maiti S, Huls H, Singh H, Olivares S, et al. (2011) The hyperactive Sleeping Beauty transposase SB100X improves the genetic modification of T cells to express a chimeric antigen receptor. Gene Ther 18: 849–856.

15. Liu L, Sanz S, Heggestad AD, Antharam V, Notterpek L, et al. (2004) Endothelial targeting of the Sleeping Beauty transposon within lung. Mol Ther 10: 97–105.

16. Belur LR, Frandsen JL, Dupuy AJ, Ingbar DH, Largaespada DA, et al. (2003) Gene insertion and long-term expression in lung mediated by the Sleeping Beauty transposon system. Mol Ther 8: 501–507.

17. Zhu J, Kren B, Park C, Bilgim R, Wong P, et al. (2007) Erythroid-specific expression of beta-globin by the sleeping beauty transposon for Sickle cell disease. Biochemistry 46: 6844–6858.

18. Wilber A, Linehan JL, Tian X, Woll PS, Morris JK, et al. (2007) Efficient and stable transgene expression in human embryonic stem cells using transposon-mediated gene transfer. Stem Cells 25: 2919–2927.

19. Grabundzija I, Wang J, Sebe A, Erdei Z, Kajdi R, et al. (2013) Sleeping Beauty transposon-based system for cellular reprogramming and targeted gene insertion in induced pluripotent stem cells. Nucleic Acids Res 41: 1829–1847.

20. Ammar I, Gogol-Doring A, Miskey C, Chen W, Cathomen T, et al. (2012) Retargeting transposon insertions by the adeno-associated virus Rep protein. Nucleic acids research 40: 6693–6712.

21. Huang X, Guo H, Tammana S, Jung Y-C, Mellgren E, et al. (2010) Gene transfer efficiency and genome-wide integration profiling of Sleeping Beauty, Tol2, and piggyBac transposons in human primary T cells. Mol Ther 18: 1803–1813.

22. Huang X, Wilber A, Bao L, Tuong D, Tolar J, et al. (2006) Stable gene transfer and expression in human primary T cells by the Sleeping Beauty transposon system. Blood 107: 483–491.

23. Voigt K, Gogol-Doring A, Miskey C, Chen W, Cathomen T, et al. (2012) Retargeting sleeping beauty transposon insertions by engineered zinc finger DNA-binding domains. Mol Ther 20: 1852–1862.

24. Borradori L, Chavanas S, Schaapveld R, Gagnoux-Palacios L, Calafat J, et al. (1998) Role of the bullous pemphigoid antigen 180 (BP180) in the assembly of hemidesmosomes and cell adhesion–reexpression of BP180 in generalized atrophic benign epidermolysis bullosa keratinocytes. Experimental cell research 239: 463–476.

25. McCormack W, Seiler M, Bertin T, Ubhayakar K, Palmer D, et al. (2006) Helper-dependent adenoviral gene therapy mediates long-term correction of the clotting defect in the canine hemophilia A model. Journal of thrombosis and haemostasis 4: 1218–1225.

26. Sambrook J, Russell DW (2001) Molecular cloning: a laboratory manual. CSHL press 1.

27. Schmidt M, Hoffmann G, Wissler M, Lemke N, Mussig A, et al. (2001) Detection and direct genomic sequencing of multiple rare unknown flanking DNA in highly complex samples. Hum Gene Ther 12: 743–749.

28. Smit AFA, Hubley R, Green P (1996-2010) RepeatMasker Open-3.0. Available: http://repeatmasker.org. Accessed 2014 Oct 23.

29. Cattoglio C, Facchini G, Sartori D, Antonelli A, Miccio A, et al. (2007) Hot spots of retroviral integration in human CD34+ hematopoietic cells. Blood 110: 1770–1778.

30. Cavazza A, Cocchiarella F, Bartholomae C, Schmidt M, Pincelli C, et al. (2013) Self-inactivating MLV vectors have a reduced genotoxic profile in human epidermal keratinocytes. Gene Ther 20: 949–957.

31. Garrels W, Mates L, Holler S, Dalda A, Taylor U, et al. (2011) Germline transgenic pigs by Sleeping Beauty transposition in porcine zygotes and targeted integration in the pig genome. PLoS One 6: e23573.

32. Hackett P, Largaespada D, Switzer K, Cooper L (2013) Evaluating risks of insertional mutagenesis by DNA transposons in gene therapy. Translational research.

33. Yant S, Wu X, Huang Y, Garrison B, Burgess S, et al. (2005) High-resolution genome-wide mapping of transposon integration in mammals. Molecular and cellular biology 25: 2085–2094.

34. Zhang W, Muck-Hausl M, Wang J, Sun C, Gebbing M, et al. (2013) Integration profile and safety of an adenovirus hybrid-vector utilizing hyperactive sleeping beauty transposase for somatic integration. PLoS One 8: e75344.

35. de Jong J, Akhtar W, Badhai J, Rust AG, Rad R, et al. (2014) Chromatin landscapes of retroviral and transposon integration profiles. PLoS Genet 10: e1004250.

36. Wang Y, Wang J, Devaraj A, Singh M, Jimenez Orgaz A, et al. (2014) Suicidal autointegration of sleeping beauty and piggyBac transposons in eukaryotic cells. PLoS Genet 10: e1004103.

37. Yant S, Ehrhardt A, Mikkelsen J, Meuse L, Pham T, et al. (2002) Transposition from a gutless adeno-transposon vector stabilizes transgene expression in vivo. Nature biotechnology 20: 999–1005.

38. Moldt B, Miskey C, Staunstrup N, Gogol-Doring A, Bak R, et al. (2011) Comparative genomic integration profiling of Sleeping Beauty transposons mobilized with high efficacy from integrase-defective lentiviral vectors in primary human cells. Mol Ther 19: 1499–1510.

39. Field AC, Vink C, Gabriel R, Al-Subki R, Schmidt M, et al. (2013) Comparison of lentiviral and sleeping beauty mediated alphabeta T cell receptor gene transfer. PLoS One 8: e68201.

40. Galla M, Schambach A, Falk C, Maetzig T, Kuehle J, et al. (2011) Avoiding cytotoxicity of transposases by dose-controlled mRNA delivery. Nucleic acids research 39: 7147–7160.

41. Zhang W, Solanki M, Muther N, Ebel M, Wang J, et al. (2013) Hybrid adeno-associated viral vectors utilizing transposase-mediated somatic integration for stable transgene expression in human cells. PloS one 8.

42. Moiani A, Paleari Y, Sartori D, Mezzadra R, Miccio A, et al. (2012) Lentiviral vector integration in the human genome induces alternative splicing and generates aberrant transcripts. The Journal of clinical investigation 122: 1653–1666.

43. Cesana D, Sgualdino J, Rudilosso L, Merella S, Naldini L, et al. (2012) Whole transcriptome characterization of aberrant splicing events induced by lentiviral vector integrations. The Journal of clinical investigation 122: 1667–1676.

44. Titeux M, Pendaries V, Zanta-Boussif MA, Decha A, Pironon N, et al. (2010) SIN retroviral vectors expressing COL7A1 under human promoters for ex vivo gene therapy of recessive dystrophic epidermolysis bullosa. Mol Ther 18: 1509–1518.

45. Holkers M, Maggio I, Liu J, Janssen JM, Miselli F, et al. (2013) Differential integrity of TALE nuclease genes following adenoviral and lentiviral vector gene transfer into human cells. Nucleic Acids Res 41: e63.

46. Aiuti A, Cassani B, Andolfi G, Mirolo M, Biasco L, et al. (2007) Multilineage hematopoietic reconstitution without clonal selection in ADA-SCID patients treated with stem cell gene therapy. The Journal of clinical investigation 117: 2233–2240.

47. Biffi A, Bartolomae C, Cesana D, Cartier N, Aubourg P, et al. (2011) Lentiviral vector common integration sites in preclinical models and a clinical trial reflect a benign integration bias and not oncogenic selection. Blood 117: 5332–5339.

48. Bushman F, Lewinski M, Ciuffi A, Barr S, Leipzig J, et al. (2005) Genome-wide analysis of retroviral DNA integration. Nature reviews Microbiology 3: 848–858.

49. Bushman F (2007) Retroviral integration and human gene therapy. The Journal of clinical investigation 117: 2083–2086.

50. Montini E, Cesana D, Schmidt M, Sanvito F, Bartholomae C, et al. (2009) The genotoxic potential of retroviral vectors is strongly modulated by vector design and integration site selection in a mouse model of HSC gene therapy. The Journal of clinical investigation 119: 964–975.

51. Liu G, Geurts AM, Yae K, Srinivasan AR, Fahrenkrug SC, et al. (2005) Target-site preferences of Sleeping Beauty transposons. J Mol Biol 346: 161–173.

52. Vigdal T, Kaufman C, Izsvak Z, Voytas D, Ivics Z (2002) Common physical properties of DNA affecting target site selection of sleeping beauty and other Tc1/mariner transposable elements. Journal of molecular biology 323: 441–452.

53. Olson W, Zhurkin V (2011) Working the kinks out of nucleosomal DNA. Current opinion in structural biology 21: 348–357.

54. Foltz D, Jansen L, Black B, Bailey A, Yates J, et al. (2006) The human CENP-A centromeric nucleosome-associated complex. Nature cell biology 8: 458–469.

55. Masumoto H, Nakano M, Ohzeki J-I (2004) The role of CENP-B and alpha-satellite DNA: de novo assembly and epigenetic maintenance of human centromeres. Chromosome research 12: 543–556.

56. De Luca M, Pellegrini G, Mavilio F (2009) Gene therapy of inherited skin adhesion disorders: a critical overview. The British journal of dermatology 161: 19–24.

57. Carteau S, Hoffmann C, Bushman F (1998) Chromosome structure and human immunodeficiency virus type 1 cDNA integration: centromeric alphoid repeats are a disfavored target. J Virol 72: 4005–4014.

7

The pCri System: A Vector Collection for Recombinant Protein Expression and Purification

Theodoros Goulas[1]*, Anna Cuppari[1¤a], Raquel Garcia-Castellanos[1¤b], Scott Snipas[2], Rudi Glockshuber[3], Joan L. Arolas[1], F. Xavier Gomis-Rüth[1]*

1 Proteolysis Lab, Molecular Biology Institute of Barcelona, CSIC, Barcelona Science Park, Helix Building, Barcelona, Spain, 2 Sanford-Burnham Medical Research Institute, La Jolla, California, United States of America, 3 Institute of Molecular Biology and Biophysics, Department of Biology, Zurich, Switzerland

Abstract

A major bottleneck in structural, biochemical and biophysical studies of proteins is the need for large amounts of pure homogenous material, which is generally obtained by recombinant overexpression. Here we introduce a vector collection, the pCri System, for cytoplasmic and periplasmic/extracellular expression of heterologous proteins that allows the simultaneous assessment of prokaryotic and eukaryotic host cells (*Escherichia coli*, *Bacillus subtilis*, and *Pichia pastoris*). By using a single polymerase chain reaction product, genes of interest can be directionally cloned in all vectors within four different rare restriction sites at the 5′end and multiple cloning sites at the 3′end. In this way, a number of different fusion tags but also signal peptides can be incorporated at the N- and C-terminus of proteins, facilitating their expression, solubility and subsequent detection and purification. Fusion tags can be efficiently removed by treatment with site-specific peptidases, such as tobacco etch virus proteinase, thrombin, or sentrin specific peptidase 1, which leave only a few extra residues at the N-terminus of the protein. The combination of different expression systems in concert with the cloning approach in vectors that can fuse various tags makes the pCri System a valuable tool for high throughput studies.

Editor: Dipankar Chatterji, Indian Institute of Science, India

Funding: This study was supported in part by grants from European, Spanish, and Catalan agencies (FP7-HEALTH-2010-261460 "Gums&Joints"; FP7-PEOPLE-2011-ITN-290246 "RAPID"; FP7-HEALTH-2012-306029-2 "TRIGGER"; BFU2012-32862; CSD2006-00015; and 2014SGR9). The funders had no role in study design, data collection and analysis, decision to publish, or preparation of the manuscript.

Competing Interests: The authors have declared that no competing interests exist.

* Email: fxgr@ibmb.csic.es (FXGR); thgcri@ibmb.csic.es (TG)

¤a Current address: Structural MitoLab, Molecular Biology Institute of Barcelona, CSIC, Barcelona Science Park, Barcelona, Spain
¤b Current address: Protein Expression Core Facility, Institute for Research in Biomedicine (IRB Barcelona), Barcelona Science Park, Barcelona, Spain

Introduction

Researchers performing biochemical, biophysical and biological studies on proteins commonly require large amounts of pure homogeneous material, which cannot usually be purified from natural sources. Alternatively, proteins are over-expressed heterologously in various systems incorporating host cells of bacterial, yeast, insect, or mammalian origin [1–3]. A critical step in protein production, after target selection, is to examine as many parameters as possible and to identify the most promising strategy for protein expression and purification with a minimum of resources and time.

Prior information on the protein of interest is crucial. An extensive search in databases such as NCBI (http://www.ncbi.nlm.nih.gov), UniProt (http://www.uniprot.org) and PDB (http://www.pdb.org) for known homologous proteins may identify possible problems and appropriate solutions for subsequent experiments. In addition, it is advisable to test protein orthologs of different origin, including distantly related or unrelated species (bacteria, archaea, and eukaryotes). At this point, analysis of the primary and secondary structure of both the encoding mRNA and the translated polypeptide may anticipate downstream problems.

There is a plethora of freely available software and databases for identifying protein families and sequence conservation patterns (PFAM) [4], putative signal peptides (SPs; SignalIP) [5], lipoboxes (DOLOP) [6], glycosylation, phosphorylation and other posttranslational modifications [7], transmembrane domains (TOPCONS, TMHMM, BOCTOPUS) [8–10], and unfolded/disordered regions (DisEMBL, PONDR, PSIPRED Protein Sequence Analysis Workbench) [11–13]. Protein location within the cell, i.e. cytoplasmic, periplasmic, or extracellular (PSORT, http://psort.hgc.jp), provides an indication of the requirements of the protein for proper folding, including disulfide bond formation and the need for special chaperons in each cellular compartment [14–16]. Further prediction of the secondary structure content (JPRED, LOMETS) [17,18] can give clues about possible protein domains and motifs, a characterisation which may prove useful for chopping full-length multi-domain proteins into globular moieties. In general, successful recombinant protein expression depends on the removal of wild-type SP, lipoboxes, posttranslational signals, low-complexity regions, hydrophobic residues at the protein termini and membrane spanning regions, while conserving the boundaries of globular domains [19].

In parallel, cDNA characterisation is important in designing the cloning strategy and identifying potential problems at the transcriptional and translational levels. Although these processes are affected by a number of exo- and endo-nucleases, the stability of the resulting mRNA is critical in protein expression experiments [20]. mRNA can be protected by introducing sequences at the 5′ untranslated regions (UTRs) and stem loop structures at the 3′ UTRs [21]. The GC base content (>70%) may affect levels of expression and can be easily determined by sequence analysis software. Rare codons (GCUA 2.0) [22], especially consecutive ones, are frequently found in heterologous genes and may lead to translational errors due to ribosomal stalling [23,24]. Such codon bias can be remedied by replacing selected codons or, if necessary, by overall gene optimisation using appropriate software (OPTI-MIZER) [25]. Once the above requirements are fulfilled, the gene can be inserted into the vector by directional cloning using restriction enzymes that do not cut within the gene sequence (NEBcutter) [26]. Efficiency of translation termination can be increased by introducing strong stop codons (UAA, especially in context when followed by a U base, or consecutive ones) at the end of the translated gene [27]. Although present in many expression vectors, transcription terminators can be included downstream of the transcribed gene if instability is predicted [28]. Finally, sources of cDNA can be found in the Mammalian Gene Collection (http://mgc.nci.nih.gov/) and at the home page of Culture Collection of the World (http://www.ecotao.com/holism/agric/hpcc.html).

No expression system is generic for all target proteins, so both bacterial and eukaryotic systems need to be explored. *Escherichia coli* provides the cheapest expression host, and it is the most widely used but its machinery is not as sophisticated as that of eukaryotic hosts, and it cannot always express well folded proteins of variable origin [15]. Other alternatives often need to be tested, including bacterial systems such as *Bacillus subtilis* [29] and more advanced eukaryotic systems such as the yeasts *Pichia pastoris* [1] and *Saccharomyces cerevisiae* [30], the baculovirus expression system in insect cells [3], mammalian cells [31], or cell-free systems using prokaryotic extracts [32], which have highly variable cost-efficiency ratios.

With *E. coli* alone, many variables can be tested in order to improve expression levels and achieve proper protein folding [2,33]. A number of specialised strains carrying mutations [34,35] or plasmids that co-express proteins favouring expression at the transcriptional or translational level (e.g. pRARE or pLysE/pLysS) are available [24,36]. Coupled expression of exogenous chaperones can assist in proper folding and prevent protein aggregation [37,38]. Expression can also be influenced by other parameters, such as the culture method (e.g. batch fermentation, fed batch and dialysis fermentation) [39], cell growth media composition (lysogeny broth (LB), the enriched terrific broth (TB), two times yeast and tryptone broth (2×YT), and auto-induction media) [40], and culture conditions like temperature (18–37°C), shaking, aeration and other physical variables. All these factors can affect production levels, secretion, protein folding, solubility and host proteolytic activity [41,42].

The many systems for introducing fusion tags currently available were originally developed to facilitate the detection and purification of recombinant proteins. Tags such as polyhistidine (His$_6$-tag) and streptavidin-binding peptide (Strep-tag) allow purification by affinity chromatography and protein detection by Western blotting [43,44], and others such as C-terminally fused green fluorescent protein (GFP) are an indispensable tool for membrane protein biochemists [45]. Finally, several studies have shown that the introduction of tags at the N- or C-terminus of proteins can improve expression levels by providing an optimized environment for translation initiation and mRNA protection, protein solubility [46–48], and carrier-driven crystallisation experiments [49].

Here we present a collection of vectors with which various expression systems and fusion tags can be evaluated simply and effectively. We examine the applicability of this system and provide several test cases, which support its robustness and versatility. This vector collection, which has been extensively tested and modified, is freely available to the scientific community under Addgene (https://www.addgene.org).

Materials and Methods

Genetic manipulations and vector preparation

Three series of vectors were generated on the basis of vectors available from the European Molecular Biology Laboratory (pETMBP-1a, pETTRX-1a, and pETGST), Novagen (pET-26b, and pET-28a), MoBiTec (pHT-01, and pHT-43), Invitrogen (pPICZA and pPICZαA), and from the Glockshuber laboratory (pRBI-DsbC) [50]. The inserted sequences for pCri-11, 13, and 14 were amplified from pET-15b-SUMO1 [51], pMIS3.0E [52], and pKLSLt [53], respectively. All vectors were prepared for directional cloning in *Nco*I or *Nde*I restriction sites at the 5′end and in *Xho*I at the 3′end. The gene coding for GFP (UniProt code: B6UPG7; 729 bp), including a multiple cloning site (MCS; from pETMBP-1a; 52 bp), was introduced into all vectors. The insert was cloned between the *Nco*I or *Nde*I and *Xho*I restriction sites and was modified to contain an *Msc*I or *Nhe*I restriction site immediately after the *Nco*I and *Nde*I sites, respectively. Standard cloning techniques were used throughout [54]. Polymerase chain reaction (PCR) primers and DNA modifying enzymes were purchased from Sigma-Aldrich and Thermo-Scientific, respectively. PCR was performed using Phusion high-fidelity DNA polymerase (Thermo-Scientific) according to the manufacturer's instructions and following a standard optimisation step of a thermal gradient in each reaction. For vector preparation, a number of insertions and mutations introduced or eliminated nucleotide sequences. We followed a PCR-based strategy described elsewhere [55], including a *Dpn*I digestion step to remove parental DNA. Digestion with restriction enzymes was carried out according to standard protocols. When necessary, a second round of digestion was performed before the final DNA purification step. DNA was purified from PCR reactions, enzymatic reactions, agarose gel band extractions, and vector extractions using OMEGA-Biotek purification kits. Chemically competent *E. coli* DH5α, BL21 (DE3), and Origami 2 (DE3) cells (Novagen) were prepared and transformed following Hanahan method [56]. Competent cells of *P. pastoris* KM71H (Invitrogen) and *B. subtilis* WB800N (MoBiTec) were prepared according to the manufacturer's instructions.

Protein expression and purification

For expression trials, *mecR1* (UniProt code: P0A0B0; an integral-membrane metallopeptidase) was cloned into vector pCri-8a and 13a; the gene coding for fragilysin (UniProt code: O86049; Ala212-Asp397; a soluble metalloendopeptidase) into pCri-1a, 4a, 6a and 8a; *gfp* into pCri-1a, 4a, 6a, 8a, 11a, and 14a; the gene coding for carboxypeptidase A2 (CPA2; UniProt code: P48052; Leu19-Tyr419; a soluble metalloexopeptidase) into pCri-8a, 9a, 16a, and 18a; and the gene coding for peptide-N-glycosidase F (PNGase F; UniProt code: P21163; Ala41-Asp354; a soluble glycosidase) into pCri-4a and 8a. The constructs were transformed in *E. coli* BL21 (DE3), Origami 2 (DE3), or *B. subtilis* cells and plated on LB plates supplemented with antibiotics (30 µg/mL kanamycin or 5 µg/mL chloramphenicol). A single colony was inoculated in 5 mL LB broth and incubated overnight at 30°C with stirring at 250 rpm. 1 mL of the pre-inoculum was used to inoculate 100 mL of LB broth and cells were left to grow at 37°C until $OD_{600 \text{ nm}} \approx 0.7$–0.8. Subsequently, cells were incubated with 0.4–1 mM isopropyl-β-D-1-thiogalactopyranoside (IPTG) to induce protein expression and kept for 5 h at 37°C or overnight at 20°C.

For expression trials in *P. pastoris* cells, vectors were linearized with *PmeI* restriction enzyme and transformed using the *Pichia* EasyComp transformation kit (Invitrogen). Cells were inoculated in low salt yeast peptone dextrose (YPD) plates supplemented with 100 µg/mL zeocin and incubated for 3–4 days at 28°C. Colonies were selected and grown in 100 mL buffered complex glycerol medium (BMGY) at 28°C until an $OD_{600 \text{ nm}} \approx 2$. Cells were then harvested, resuspended in buffered complex methanol medium (BMMY), and protein expression was induced with 0.5% methanol.

Cells were separated from the growth media by centrifugation at 8,000×g for 30 min at 4°C. Secreted proteins were collected from the growth media and dialysed in buffer A (50 mM Tris-HCl, 250 mM NaCl, pH 7.5), and cytoplasmic proteins were extracted from the cells in the same buffer. For lysis, cells were sonicated with 3 pulses of 5 min each at 40% amplitude (Branson digital sonifier). Samples were collected before and after centrifugation (30,000×g for 30 min at 4°C) representing total and soluble protein fractions, respectively.

Selected samples were further purified by affinity chromatography using either nickel-nitrilotriacetic acid- (Ni-NTA), maltose binding protein- (MBP) or glutathione S-transferase- (GST) HiTrap columns, or a Sepharose 4B matrix column (GE Healthcare Life Sciences). 10 mL of crude protein extract was applied to the columns, followed by three washes with buffer A. Proteins were eluted with buffer A supplemented with either 300 mM imidazole (Ni-NTA-affinity), 10 mM maltose (MBP-affinity), 10 mM reduced glutathione (GST-affinity) or 20 mM lactose (Sepharose-affinity). Finally, samples were buffer-exchanged to buffer B (20 mM Tris-HCl, 150 mM NaCl, pH 7.4) using a PD-10 desalting column (GE Healthcare Life Sciences). Samples were kept at 4°C at all times.

For expression and purification of MecR1, the cultures were scaled up to 6L, the collected cells were broken with a cell disrupter (Constant Cell Disruption Systems) at 2.4kBar and non-disrupted cells and cell debris were removed by centrifugation at 20,000×g for 45 min in a Sorvall centrifuge. Membranes were collected by ultracentrifugation at 150,000×g for 2 h at 4°C in a Beckman Optima L-90K using a 50.2 Ti rotor (Beckman) and 26.3-ml polycarbonate bottles with cap assembly (Beckman). Collected membranes were homogenized using a glass Potter and solubilized under gentle stirring by overnight incubation at 4°C in buffer C (50 mM Tris-HCl, 300 mM NaCl, 10 mM imidazole, 1 mM 1,4-dithio-D-threitol, pH 8.0) containing 100 mM lauryl-dimethylamine N-oxide (LDAO; Sigma) and EDTA-free proteinase inhibitor cocktail tablets (Roche). Non-solubilized proteins were removed by ultracentrifugation as described above. The sample was incubated overnight at 4°C with Ni-NTA resin (Invitrogen). The bound protein was batch purified in an open column (Bio-Rad), washed extensively, and the tagged protein eluted with buffer C plus 300 mM imidazole. The sample was desalted using a PD-10 column in buffer C containing 5 mM LDAO.

Fusion-tag removal by proteinase cleavage

Tobacco etch virus (TEV) proteinase and sentrin specific proteinase 1 (SENP1) were over-expressed in *E. coli* BL21 (DE3) pLysE cells using pET28-based vectors, which attach an N-terminal His$_6$-tag. Cultures (typically 4L) were grown in LB broth at 37°C until an $OD_{600 \text{ nm}} \approx 0.7$–0.8, induced with 0.5 mM IPTG, and incubated either overnight at 20°C or for 5 h at 30°C for TEV proteinase or SENP1 expression, respectively. Subsequently, cells were collected by centrifugation at 5,000×g for 30 min at 4°C and partially purified by Ni-NTA affinity chromatography as previously described [57,58]. Proteinases were stored at −80°C in buffer D (20 mM Tris-HCl, 50 mM NaCl, pH 7.5, 30% glycerol). Proteinase cleavage trials of tagged-proteins were performed overnight at 4°C in buffer B using various protein:proteinase ratios. For trials with thrombin (GE Healthcare Life Sciences), 2 units of proteinase were used to process 25 µg of protein in 100 µL of buffer C at room temperature and aliquots were taken at various time points.

Enzymatic assays

For hydrolytic activity measurements, PNGase F and fragilysin were partially purified by Ni-NTA-affinity chromatography as described above. Glycosidase activity of PNGase F was tested against the glycoprotein ribonuclease B (RNase B; New England Biolabs) at a w/w ratio of 1:5 PNGase F/RNase B and a final protein concentration of 0.5 mg/mL. Reactions were incubated overnight at 4°C and analysed by SDS-PAGE. Peptidase activity of fragilysin was tested against BODIPY FL-casein (Invitrogen) as previously described [59]. Crude protein extracts of CPA2 were used for assays after an initial activation with partial tryptic digestion in a w/w ratio of 1/100 of CPA2/trypsin at room temperature for 1 h. The activated protein was incubated with furyl-acryloyl-L-phenylalanine-L-phenylalanine (0.05 mM; Sigma) in buffer B and the activity was monitored by measurement of the absorbance change at 330 nm.

Western-blot analysis

Protein samples were analyzed by Tricine-SDS-PAGE, transferred to Hybond ECL membranes (GE Healthcare Life Sciences), and finally blocked overnight at room temperature with 20 mL of blocking solution (137 mM NaCl, 2.7 mM KCl, 4.3 mM Na_2HPO_4, 1.47 mM NaH_2PO_4, 0.05% Tween 20) containing 1.5% bovine serum albumin. MecR1 was detected by immunoblot analysis using custom polyclonal antibodies (Eurogentec) at dilution 1:1,000 and a secondary antibody (goat anti-rabbit IgG (HL) peroxidase-conjugated antibody; Pierce) at dilution 1:5,000 (both in blocking solution). The immune complexes were detected using an enhanced chemiluminescence system (Super Signal West Pico Chemiluminescent; Pierce) according to the manufacturer's instructions. Membranes were exposed to hyperfilm ECL films (GE Healthcare Life Sciences).

Miscellaneous

Denatured protein samples were analyzed by 10%–15% Tricine-SDS-PAGE [60] and stained with Coomassie-brilliant blue. Protein concentrations were routinely determined by absorbance at 280 nm, and, wherever necessary, corrected by the BCA protein assay method (Thermo Scientific) using bovine serum albumin as a standard. Protein identification by peptide mass fingerprinting was performed at the Protein Chemistry Facility of Centro de Investigaciones Biológicas (Madrid, Spain). Figures of vector maps were prepared with GENEIOUS (Biomatters).

Results and Discussion

Description of the pCri System

We generated a collection of vectors for recombinant protein overexpression in two bacterial (*E. coli* and *B. subtilis*) and one eukaryotic (*P. pastoris*) host strains. Vectors, available from commercial sources or laboratories, were initially modified by inserting new nucleotide sequences or point mutations, and finally evaluated for functionality. Most of the *E. coli* vectors are pET based [61] with the exception of pCri-12, which is based on pTrc99a [50]. The bacillus and yeast vectors are based on pHT [62] and pPICZ series [63], respectively, and can be stably propagated in *E. coli* cells when antibiotic resistance is conferred (Tables 1–3). In all vectors, protein expression is achieved by IPTG induction, except for the yeast vectors, for which methanol is required.

The collection consists of 29 vectors grouped into three main categories (Tables 1–3). Based on the available 5′end restriction sites for target gene cloning, the vectors are sorted into pCri-a and pCri-b series using either *Nco*I and *Msc*I or *Nde*I and *Nhe*I sites, respectively (Fig. 1 and Fig. S1). The pCri-a series is further separated into pCri-a and pCri-a-Strep based on the fusion tag that can be attached at the C-terminus of the target protein. Within each category, the vectors allow obtaining constructs with different fusion tags or expression in a particular host organism. Usage of the aforementioned 5′end restriction sites incorporates a methionine start codon, thus obviating the need to introduce it into the target gene during PCR amplification. An MCS universal for all vectors has been placed at the 3′end, which encodes seven rare restriction sites not found in most of the vectors (see vector maps for more details; Fig. S1). For convenience and tracking during vector preparation, a GFP insert is cloned within all vectors. The inserted genes can be sequenced from either terminus with specific primers as detailed in Table 4.

Preparation is greatly simplified, as only two restriction sites are used for directional cloning of a target gene into a large series of vectors. Although newer cloning techniques are now available (e.g. ligation independent cloning system [64]), this method was satisfactory. Cloning of target genes of variable size spanning from 150 to 7,000 base pairs was routinely performed with a success rate of more than seven out of ten positive clones when genes were cloned between an *Nco*I or *Nde*I and a *Xho*I site. To achieve reproducible results, it was essential to repeat double digestions of the vectors with all the restriction enzyme combinations.

Applications and main considerations of the pCri System

The choice and use of a suitable vector should be based on the properties of the target protein and the needs of the experiment in question. Here, in an effort to evaluate the functionality of the collection and to provide a rationale for the use of the vectors, we cloned and expressed several proteins of different origin and function:

Fusion tags assisting in protein purification. The pCri System allows the fusion of a His_{6-8}-tag at the N-terminus of the target protein, which can be in tandem with larger tags such MBP [43], GST [65], small ubiquitin-like modifier (SUMO) [66], and the β-trefoil lectin module of protein LSL$_{150}$ from the mushroom *Laetiporus sulphureus* (LSL) [53] (Tables 1–3). The C-terminus of the target protein can likewise be furnished with a His_6-tag or a Strep-tag if the stop codon of the amplified gene is omitted. These tags add a functionality to the target protein, which is commonly used as a first purification step through affinity chromatography [43,53,65]. On this basis, we cloned and expressed GFP in pCri-

1a, 4a, 6a, 8a, 11a, and 14a. The proteins were purified by Ni-NTA affinity chromatography except for MBP, GST, and LSL fusion products, which were purified by their specific affinity resins (Fig. 2A). Nickel or cobalt affinity chromatography of His_6-tagged proteins are among the most commonly used methods for purification, but others using the affinity properties of MBP or GST, and the recently reported LSL$_{150}$, can provide better purification results under mild elution conditions. This choice among alternative affinity purification systems allows the best purification method to be used for each target protein. Moreover, many of those tags can be used to track poorly expressed proteins by Western-blot analysis, as they are otherwise undetectable by Coomassie-stained SDS-PAGE.

Fusion tags assisting in protein solubility. In addition, several studies showed that tags such as N-utilisation substrate A (NUSA), MBP, or the smaller GST and SUMO have positive effects on the cargo protein due to their solubility-enhancing or chaperoning properties [2,66,67]. Nevertheless, their working mechanism is still controversial, with several studies suggesting a more passive role due to their excellent solubility properties rather than a direct influence on the folding of their partner [47]. For example, fragilysin (Ala212-Asp397) [59], a bacterial enterotoxin metallopeptidase, was expressed in high amounts in fusion with MBP, TRX, GST, and His_6-tag, both at 37°C or 20°C (Fig. 2B). However, only MBP rendered the protein soluble during low temperature expression trials, whereas other fusions or expression at higher temperatures produced protein prone to aggregation. The protein remained in solution even after MBP removal (Fig. 2C) but catalytically inactive against fluorescent-labelled casein, indicating at least partial misfolding. Similar results were obtained when fragilysin was expressed with the smaller Z-tag (≈10 kDa) [59], indicating that fusion proteins may have a positive effect on target solubility without necessarily implying that it will be well folded and active. Nevertheless, these fusion tags can have an application in the expression of proteins with known solubility problems that need to be temporally stabilised until an adequate condition/solution is found [67].

Expression of proteins requiring disulfide bonds and other posttranslational modifications. Correct folding and stabilization requires the formation of disulfide bonds in many proteins. These can be formed in oxidising environments as found in the periplasmic and extracellular environment of bacteria, or in specialised organelles of eukaryotes. *B. subtilis* has a large secretory capacity, whereas *E. coli* secretion is mainly limited to the periplasm [68,69]. In *P. pastoris*, proteins are first driven to the endoplasmic reticulum and, after folding, they are secreted to the extracellular medium [70]. The pCri System includes vectors that fuse SP specialised for protein translocation to these cellular compartments. pCri-9 and 12 can be used with *E. coli* cells, whereas pCri-16 and 18 are suitable for expression in *P. pastoris* and *B. subtilis*, respectively. In the case of pCri-12, a disulfide-bond isomerase C (DsbC) is coexpressed with the target protein and provides additional support in the correct pairing of disulfide bonds in the periplasm [50].

As a test protein, we used human CPA2, which is commonly expressed in *P. pastoris* cells [71]. Unexpectedly, expression trials indicated that the protein is produced not only in the extracellular environment of *P. pastoris* but also in the cytoplasm and periplasm of *E. coli* cells (Fig. 2D). In contrast, *B. subtilis* did not express the protein either extracellularly or intracellularly. In all cases, the protein was soluble and correctly processed after limited tryptic digestion, showing activity against small substrates. However, this is not always the case. Besides the oxidising conditions other proteins may often participate in correct folding, including

Table 1. pCri-a vectors.

Name	Vector size (~kbp)[a]	Fusion tag[b]	Extra residues after fusion removal	Fusion molecular mass (~kDa)[c]	Resistance marker[d]	Based on vector	Cells
pCri-1a	7.2	**His₆-MBP-TEV-insert-His₆-tag**	Gly-Ala	43.1	Kan^r	pETMBP-1a	*E. coli*
pCri-4a	6.4	**His₆-TRX-TEV**-insert-**His₆-tag**	Gly-Ala	14.3	Kan^r	pETTrx-1a	*E. coli*
pCri-6a	6.7	**His₆-GST-TEV**-insert-**His₆-tag**	Gly-Ala	28.9	Kan^r	pETGST-1a	*E. coli*
pCri-7a	6.0	insert-**His₆-tag**	-	-	Kan^r	pET-28a	*E. coli*
pCri-8a	6.1	**His₆-TEV**-insert-**His₆-tag**	Gly-Ala	2.4	Kan^r	pET-28a	*E. coli*
pCri-9a	6.1	**SP** (pelB)-insert-**His₆-tag**	-	-	Kan^r	pET-26b	*E. coli*
pCri-11a	6.3	**His₆-SUMO**-insert-**His₆-tag**	Ala	12.6	Kan^r	pET-26b	*E. coli*
pCri-12a	5.2	**SP** (ompA)-insert-**His₆-tag** and DsbC coexpression	-	-	Amp^r	pRBI-DsbC	*E. coli*
pCri-13a	6.4	**His₈-MISTIC-THR**[e]-insert-**His₆-tag**	Gly-Ser-Gly₃-Ala	16.2	Kan^r	pET-28a	*E. coli*
pCri-14a	6.5	**His₆-LSL-TEV**-insert-**His₆-tag**	Gly-Ala	19.6	Kan^r	pET-28a	*E. coli*
pCri-15a	4.0	**His₆-TEV**-insert	Gly-Ala	2.4	Zeo^r	pPICZA	*P. pastoris*
pCri-16a	4.4	**SP** (α-factor)-insert-**His₆-tag**	-	-	Zeo^r	pPICZαA	*P. pastoris*
pCri-17a[f]	8.6	**His₆-TEV**-insert-**His₆-tag**	Gly-Pro	2.4	Amp^r, Cm^r	pHT-01	*B. subtilis*
pCri-18a[f]	8.7	**SP** (samyQ)-insert-**His₆-tag**	-	-	Amp^r, Cm^r	pHT-43	*B. subtilis*

[a]Calculated vector sizes based on the electrophoretic mobility with a Lambda DNA marker as standard.
[b]For introducing a His₆-tag at the C-terminus of the target protein use a reverse primer without a stop codon.
[c]Molecular mass of C-terminal His₆-tag ~1 kDa.
[d]Kan^r, 30–50 μg/mL; Amp^r, 50–100 μg/mL; Cm^r, 5 μg/mL; Zeo^r, 25–100 μg/mL.
[e]Thrombin cleavage site (THR).
[f]Ampicillin resistance in *E. coli* and chloramphenicol resistance in *B. subtilis*.

Table 2. pCri-b vectors.

Name	Vector size (~kbp)[a]	Fusion tag[b]	Extra residues after fusion removal	Fusion molecular mass (~kDa)[c]	Resistance marker[d]	Based on vector	Cells
pCri-1b	7.2	His$_6$-**MBP-TEV**-insert-**His$_6$-tag**	Gly-His	43	Kan^r	pCri-1a	E. coli
pCri-4b	6.4	His$_6$-**TRX-TEV**- insert-**His$_6$-tag**	Gly-Ala-His	14.3	Kan^r	pCri-4a	E. coli
pCri-6b	6.7	His$_6$-**GST-TEV**-insert-**His$_6$-tag**	Gly-His	28.9	Kan^r	pCri-6a	E. coli
pCri-7b	6.0	insert-**His$_6$-tag**	-	-	Kan^r	pCri-7a	E. coli
pCri-8b	6.1	His$_6$-**tag-TEV**-insert-**His$_6$-tag**	Gly-His	2.4	Kan^r	pCri-8a	E. coli
pCri-9b	6.1	**SP** (pelB)-insert-**His$_6$-tag**	-	-	Kan^r	pCri-9a	E. coli
pCri-11b	6.3	His$_6$-**SUMO**-insert-**His$_6$-tag**	His	12.6	Kan^r	pCri-11a	E. coli
pCri-12b	5.2	**SP** (ompA)-insert-**His$_6$-tag** and DsbC coexpression	-	-	Amp^r	pCri-12a	E. coli
pCri-14b	6.5	His$_6$-**LSL-TEV**-insert-**His$_6$-tag**	Gly-His	19.6	Kan^r	pCri-14a	E. coli
pCri-15b	4.0	His$_6$-**TEV**-insert-**His$_6$-tag**	Gly-His	2.4	Zeo^r	pCri-15a	P. pastoris

[a]Calculated vector sizes based on the electrophoretic mobility with a Lambda DNA marker as standard.
[b]For introducing a His$_6$-tag at the C-terminus of the target protein use a reverse primer without a stop codon.
[c]Molecular mass of C-terminal His$_6$-tag ~1 kDa.
[d]Kan^r, 30–50 µg/mL; Amp^r, 50–100 µg/mL; Cm^r, 5 µg/mL; Zeo^r, 25–100 µg/mL.

Table 3. pCri-a-Strep vectors.

Name	Vector size (~kbp)[a]	Fusion tag[b]	Extra residues after fusion removal	Fusion molecular weight (~kDa)[c]	Resistance marker[d]	Based on vector	Cells
pCri-1a-Strep	7.2	His$_6$-**MBP-TEV**-insert-**Strep-tag**	Gly-Ala	43.1	Kan^r	pCri-1a	E. coli
pCri-4a-Strep	6.4	His$_6$-**TRX-TEV**-insert-**Strep-tag**	Gly-Ala	14.3	Kan^r	pCri-4a	E. coli
pCri-7a-Strep	6.0	insert-**Strep-tag**	-	-	Kan^r	pCri-7a	E. coli
pCri-8a-Strep	6.1	His$_6$-**TEV**-insert-**Strep-tag**	Gly-Ala	2.4	Kan^r	pCri-8a	E. coli
pCri-9a-Strep	6.1	**SP** (pelB)-insert-**Strep-tag**	-	-	Kan^r	pCri-9a	E. coli

[a]Calculated vector sizes based on the electrophoretic mobility with a Lambda DNA marker as standard.
[b]For introducing a Strep-tag at the C-terminus of the target protein use a reverse primer without a stop codon.
[c]Molecular mass of C-terminal Strep-tag ~1.3 kDa.
[d]Kan^r, 30–50 µg/mL.

A

B

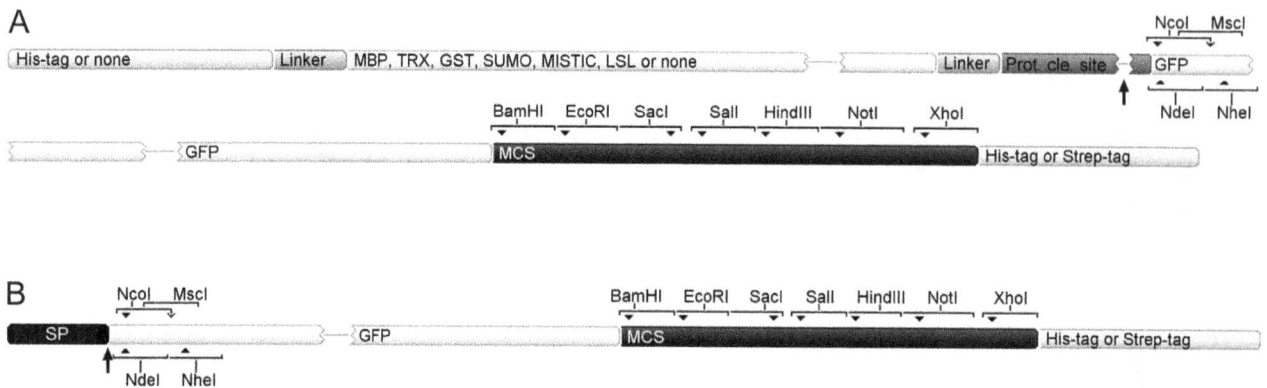

Figure 1. Vector overview of the pCri System. (A) Vectors for cytoplasmic protein expression. **(B)** Vectors for periplasmic and extracellular protein expression. An N-terminal His$_6$-tag can be fused in all vectors for intracellular expression except of pCri-7. Other tags can also be fused including MBP, TRX, GST, SUMO, MISTIC, and LSL (Table 1–3). In all vectors, a C-terminal His$_6$-tag or Strep-tag is attached if a stop codon is omitted within the target gene. Black arrows indicate the proteinase (i.e. TEV, SENP1 or thrombin) and signal peptide (SP) cleavage sites. Restriction sites allowing directional cloning are also shown. For more details regarding each vector, refer to Fig. S1.

oxidases, foldases, isomerases and specialised chaperones [69]. Moreover, disulfide bond formation is not the only factor in proper protein folding and stability, and further posttranslational modifications (e.g. glycosylation) may be required, which can be provided by *P. pastoris* [70].

Another approach for disulfide bond formation exploits the oxidising cytoplasm of thioredoxin reductase B (*trx*B$^-$) and glutathione reductase (*gor*$^-$) mutant *E. coli* cells (Origami 2) [34]. In contrast to the commonly used BL21 cells, Origami 2 efficiently expressed PNGase F, either with pCri-4a or 8a, soluble and catalytically active against RNase B (Fig. 2E and 2F). The protein contains disulfide bonds that require an oxidising environment, which is adequately formed in the cytoplasm of mutant cells. In addition, the combined use of thioredoxin A (TRX) as fusion protein in pCri-4a and expression in Origami 2 can lead to the overexpression of small multi-disulfide proteins, among others [69,72]. This system takes advantage of TRX,

which acts as an oxidant when it operates in an oxidized milieu found in mutant cells [34], thus providing an additional mechanism for disulfide bond formation within the cytoplasm. TRX is subsequently removed by TEV proteinase cleavage in the presence of selected amounts of redox agents to assist in correct disulfide bond pairing [72,73].

Expression of membrane proteins. Membrane proteins are among the targets most requested and at the same time difficult to express and purify. To address this issue, a vector was prepared, which fuses a small protein from *B. subtilis* with target proteins (pCri-13a). This protein, known as the membrane-integrating sequence for translation of integral-membrane protein constructs (MISTIC), folds autonomously into membranes, simultaneously dragging the tagged-protein to the cell membrane [74]. Moreover, this vector contains a longer His$_8$-tag in tandem with MISTIC in order to provide higher affinity for Ni-NTA affinity purification.

Table 4. Sequencing primers for the pCri System.

Primer Name	Vector (pCri)	Sequence site	Primer sequence
T7 promoter	1–14a,b except 12a–b*	5'end	Universal
T7 terminator	1–14a,b except 12a–b*	3'end	Universal
seq-pCri-1	1a and b internal*	5'end	GAAATCATGCCGAACATCCC
seq-pCri-4	4a and b internal*	5'end	GCGGCAACCAAAGTGGGTGCAC
seq-pCri-6	6a and b internal*	5'end	GACCATCCTCCAACTAGTG
seq-pCri-11	11a and b internal*	5'end	CAAAAGAACTGGGAATG
5'seq-pCri-12	12a** and b**	5'end	GATAACGAGGGCAAAAAATG
3'seq-pCri-12	12a* and b*	3'end	CAAAGTAAACAACATAAAAC
seq-pCri-13	13a internal*	5'end	CAGATTTTATCCATCTC
seq-pCri-14	14a and b internal*	5'end	CTTCTGGAATCACCCTC
5' AOX1	15a,b* and 16a**	5'end	Universal
3' AOX1	15a,b* and 16a*	3'end	Universal
5'seq-pCri-17	17a* and 18a**	5'end	CTTATCACTTGAAATTG
3'seq-pCri-17	17a* and 18a*	3'end	GATTTTATTAGTACAGGGAC

*Hybridises before *Nco*I, *Nde*I or *Xho*I restriction sites.
**Hybridises before the SP.

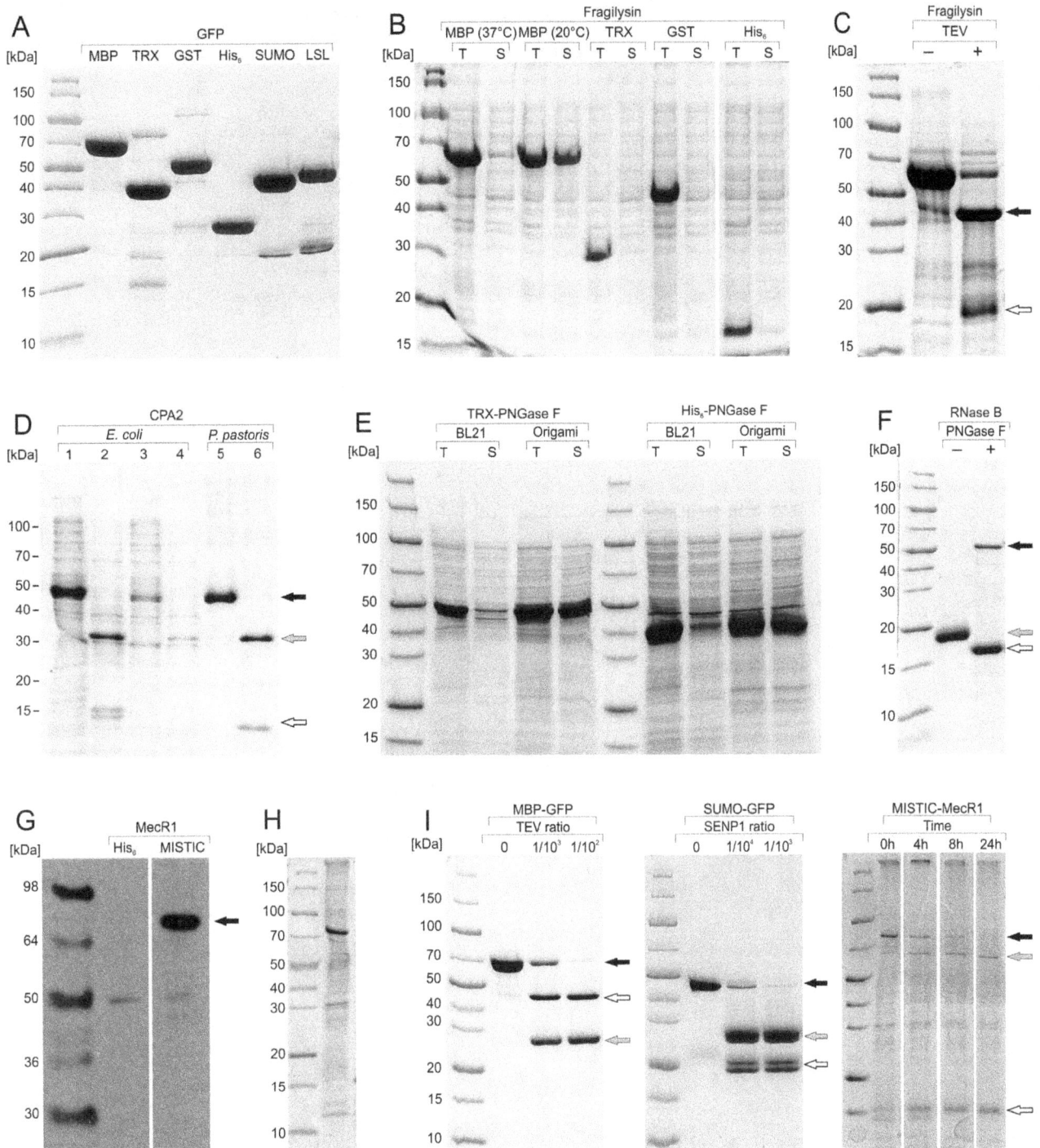

Figure 2. Protein expression and purification trials using the pCri System. (**A**) The GFP gene was cloned into pCri-1a, 4a, 6a, 8a, 11a, and 14a, the proteins expressed in *E. coli* BL21 cells, and subsequently purified by Ni-NTA-affinity chromatography except for MBP, GST, and LSL fusion products, which were purified by their respective specific affinity resins. (**B**) The gene coding for fragilysin was cloned into pCri-1a, 4a, 6a and 8a, and expressed in *E. coli* Origami 2 cells. Total (T) and soluble (S) fractions of crude protein extracts were further analysed by SDS-PAGE. All expression trials were performed at 20°C except for pCri-1a, which was also performed at 37°C. (**C**) Partially purified MBP-fragilysin before (−) and after (+) TEV proteinase cleavage. Arrows indicate the soluble fraction of fragilysin (white) and the MBP (black) after TEV proteinase cleavage. (**D**) Expression of CPA2 intracellularly (lanes 1 and 2) or periplasmatically (lanes 3 and 4) in *E. coli* cells, and extracellularly (lanes 5 and 6) in *P. pastoris* cells. Lanes indicate samples before (1, 3 and 5) and after (2, 4 and 6) tryptic digestion. Arrows indicate the pro-CPA2 (black), the mature form (grey) and the pro-peptide (white) after tryptic cleavage. (**E**) The PNGase F gene was cloned into pCri-4a and 8a and expressed overnight at 20°C in *E. coli* BL21 and Origami 2 cells. Total (T) and soluble (S) fractions of crude protein extracts were further analysed by SDS-PAGE. (**F**) Activity of affinity-purified TRX-PNGase F against glycosylated RNase B. (+) and (−) indicate presence and absence of PNGase F. Arrows indicate the PNGase F (black), native RNase B (grey) and deglycosylated RNase B (white). (**G**) MecR1 was expressed in *E. coli* BL21 using pCri-8a or 13a, and soluble fractions were analysed by Western blotting with specific antibodies as detailed in "Materials and Methods". A black arrow indicates the detected MecR1. (**H**) Partially purified MISTIC-MecR1 after Ni-NTA-affinity purification. (**I**) Partially purified MBP-GFP, SUMO-GFP and MISTIC-MecR1 were digested with TEV proteinase,

SENP1 or thrombin, respectively. For TEV proteinase and SENP1 digestions various ratios of proteinase:tagged-protein were tested in overnight incubations at 4°C, whereas for thrombin digestions 2 units of proteinase were used to digest 25 μg of protein for various times at room temperature. Arrows indicate tagged-protein (black), target protein (grey) and fused-tag (white) after proteinase cleavage.

For evaluation purposes, we cloned and expressed MecR1, a membrane metallopeptidase from *Staphylococcus aureus* implicated in methicillin resistance [75]. Detectable levels of expression were only achieved when the protein was fused with MISTIC, whereas mere fusion with an N-terminal His$_6$-tag was unsuccessful (Fig. 2G). Moreover, expression yields of the protein were sufficient (0.4 mg of affinity purified protein per litre of culture) to enable partial purification by Ni-NTA affinity chromatography after solubilisation of the membranes with the zwitterionic detergent LDAO (Fig. 2H). Although further studies are required to assess the folding state of this protein, fusion with MISTIC allowed us to express it in milligram amounts. Many other membrane proteins were also expressed in fusion with various MISTIC constructs, indicating that this system could be an alternative approach for membrane proteins that are difficult to express [52].

Removal of fusion tags. In most cases, release of the target protein from any fused tag is desirable. In the pCri System, a TEV proteinase cleavage site is introduced immediately after the tag in all vectors except for pCri-11 and pCri-13, in which a SENP1 or thrombin site is found, respectively (Fig. S1). TEV proteinase is a highly specific enzyme that recognises an hexapeptide sequence [76], whereas SENP1 further offers robustness and high proteolytic activity in addition to high specificity, usually requiring only minute amounts for tag removal [77]. Moreover, use of a thrombin cleavage site in pCri-13a was necessary due to the low efficiency of TEV proteinase in the presence of detergents which are required during membrane-protein solubilisation [78]. In addition, linker sequences (Gly-Ser)$_5$ and Gly$_3$-Ala were introduced before and after the thrombin recognition site, respectively, to improve access for proteinase cleavage (Fig. S1) [79].

Tag removal was achieved with variable amounts of endopeptidases, different incubation times and temperatures (Fig. 2I). These studies indicated that optimisation trials are needed in each case to identify the best conditions for complete digestion (e.g. buffer, temperature, proteinase:substrate ratio). Proteinase cleavage and tag removal result in the incorporation of one or two extra residues at the N-terminus of the expressed protein except for pCri-4b and pCri-13a, which attach three and six residues, respectively (Tables 1–3).

Conclusions

Here we introduce a vector collection designed for large-scale recombinant protein overexpression, and demonstrate its suitability in a series of test proteins. The choice of a suitable expression vector should be based on target and tag properties. The availability of a range of fusion tags allows the choice between different affinity purification methods. Moreover, some tags were included for specific use, such as MISTIC and TRX, which are intended for expression of membrane and disulfide rich proteins, respectively. In general, our common strategy first explores the effects of the presence or absence of N- or C- terminal tags (e.g. His$_6$-tag or Strep-tag) on each construct under different host cell growth conditions. Omission of the tag or alternation of the position can drastically influence the expression and solubility of the protein. If this approach is ineffective the chances of optimising the expression by testing other fusion combinations are reduced. Several reports showed the beneficial effects of the fusions on target solubility [2,67]. However, this is not always the case: the protein is often dragged into solution, rather than acting as a chaperone for the proper folding of its fusion partner. Removal of the fusion tag can revert the positive effect and cause precipitation [47,48]. If this occurs, then modified constructs, other homologous targets or even other expression systems need to be explored, including bacterial and eukaryotic cells that can be easily tested using the vector collection of the pCri System.

Supporting Information

Figure S1 Partial nucleotide sequence and translation of the pCri System vectors.
(DOC)

Acknowledgments

We thank Prof. José M. Mancheño (Madrid, Spain) for providing pKLSLt DNA, and Robin Rycroft for critical reading of the manuscript.

Author Contributions

Conceived and designed the experiments: TG FXGR. Performed the experiments: TG AC RGC. Analyzed the data: TG AC RGC FXGR. Contributed reagents/materials/analysis tools: SS RG FXGR. Wrote the paper: TG SS RGC RG JLA FXGR.

References

1. Cregg JM, Cereghino JL, Shi J, Higgins DR (2000) Recombinant protein expression in Pichia pastoris. Mol Biotechnol 16: 23–52.
2. Sørensen HP, Mortensen KK (2005) Advanced genetic strategies for recombinant protein expression in Escherichia coli. J Biotechnol 115: 113–128.
3. Kost T, Condreay JP, Jarvis DL (2005) Baculovirus as versatile vectors for protein expression in insect and mammalian cells. Nat Biotechnol 23: 567–575.
4. Punta M, Coggill PC, Eberhardt RY, Mistry J, Tate J, et al. (2012) The Pfam protein families database. Nucleic Acids Res 40: D290–D301.
5. Petersen TN, Brunak S, von Heijne G, Nielsen H (2011) SignalP 4.0: Discriminating signal peptides from transmembrane regions. Nat Methods 8: 785–786.
6. Madan Babu M, Sankaran K (2002) DOLOP-Database of bacterial lipoproteins. Bioinformatics 18: 641–643.
7. Zhou F, Xue Y, Yao X, Xu Y (2006) A general user interface for prediction servers of proteins' post-translational modification sites. Nat Protoc 1: 1318–1321.
8. Krogh A, Larsson B, von Heijne G, Sonnhammer EL (2001) Predicting transmembrane protein topology with a hidden Markov model: Application to complete genomes. J Mol Biol 305: 567–580.

9. Hayat S, Elofsson A (2012) BOCTOPUS: Improved topology prediction of transmembrane β barrel proteins. Bioinformatics 28: 516–522.
10. Bernsel A, Viklund H, Hennerdal A, Elofsson A (2009) TOPCONS: consensus prediction of membrane protein topology. Nucleic Acids Res 37: W465–468.
11. Linding R, Jensen LJ, Diella F, Bork P, Gibson TJ, et al. (2003) Protein disorder prediction: Implications for structural proteomics. Structure 11: 1453–1459.
12. Li X, Romero P, Rani M, Dunker AK, Obradovic Z (1999) Predicting protein disorder for N-, C-, and internal regions. Genome Inf 10: 30–40.
13. McGuffin LJ, Bryson K, Jones DT (2000) The PSIPRED protein structure prediction server. Bioinformatics 16: 404–405.
14. Lund PA (2001) Microbial molecular chaperones. Adv Microb Physiol 44: 93–140.
15. Baneyx F, Mujacic M (2004) Recombinant protein folding and misfolding in Escherichia coli. Nat Biotechnol 22: 1399–1408.
16. Driessen AJM, Nouwen N (2008) Protein translocation across the bacterial cytoplasmic membrane. Annu Rev Biochem 77: 643–667.
17. Cole C, Barber JD, Barton GJ (2008) The Jpred-3 secondary structure prediction server. Nucleic Acids Res 36: W197–W201.

18. Wu S, Zhang Y (2007) LOMETS: A local meta-threading-server for protein structure prediction. Nucleic Acids Res 35: 3375–3382.
19. Dong A, Xu X, Edwards AM (2007) In situ proteolysis for protein crystallization and structure determination. Nat Methods 4: 1019–1021.
20. Hall MN, Gabay J, De'barbouille M, Schwartz M (1982) A role for mRNA secondary structure in the control of translation initiation. Nature 295: 616–618.
21. Massé E, Escorcia FE, Gottesman S (2003) Coupled degradation of a small regulatory RNA and its mRNA targets in Escherichia coli. Genes Dev 17: 2374–2383.
22. Fuhrmann M, Hausherr A, Ferbitz L, Schödl T, Heitzer M, et al. (2004) Monitoring dynamic expression of nuclear genes in Chlamydomonas reinhardtii by using a synthetic luciferase reporter gene. Plant Mol Biol 55: 869–881.
23. McNulty DE, Claffee BA, Huddleston MJ, Kane JF (2003) Mistranslational errors associated with the rare arginine codon CGG in Escherichia coli. Protein Expr Purif 27: 365–374.
24. Gustafsson C, Govindarajan S, Minshull J (2004) Codon bias and heterologous protein expression. Trends Biotechnol 22: 346–353.
25. Puigbò P, Guzmán E, Romeu A, Garcia-Vallvé S (2007) OPTIMIZER: A web server for optimizing the codon usage of DNA sequences. Nucleic Acids Res 35: W126–W131.
26. Vincze T, Posfai J, Roberts RJ (2003) NEBcutter: A program to cleave DNA with restriction enzymes. Nucleic Acids Res 31: 3688–3691.
27. Poole ES, Brown CM, Tate WP (1995) The identity of the base following the stop codon determines the efficiency of in vivo translational termination in Escherichia coli. EMBO J 14: 151–158.
28. Newbury SF, Smith NH, Robinson EC, Hiles ID, Higgins CF (1987) Stabilization of translationally active mRNA by prokaryotic REP sequences. Cell 48: 297–310.
29. Westers L, Westers H, Quax WJ (2004) Bacillus subtilis as cell factory for pharmaceutical proteins: A biotechnological approach to optimize the host organism. Biochim Biophys Acta 1694: 299–310.
30. Porro D, Sauer M, Branduardi P, Mattanovich D (2005) Recombinant protein production in yeasts. Mol Biotechnol 31: 245–259.
31. Aricescu AR, Lu W, Jones EY (2006) A time- and cost-efficient system for high level protein production in mammalian cells. Acta Crystallographica D62: 1243–1250.
32. Schwarz D, Junge F, Durst F, Frölich N, Schneider B, et al. (2007) Preparative scale expression of membrane proteins in Escherichia coli-based continuous exchange cell-free systems. Nat Protoc 2: 2945–2957.
33. Makrides SC (1996) Strategies for achieving high-level expression of genes in Escherichia coli. Microbiol Mol Biol Rev 60: 512–538.
34. Stewart EJ, Aslund F, Beckwith J (1998) Disulfide bond formation in the Escherichia coli cytoplasm: An in vivo role reversal for the thioredoxins. EMBO J 17: 5543–5550.
35. Miroux B, Walker JE (1996) Over-production of proteins in Escherichia coli: Mutant hosts that allow synthesis of some membrane proteins and globular proteins at high levels. J Mol Biol 260: 289–298.
36. Studier FW (1991) Use of bacteriophage T7 lysozyme to improve an inducible T7 expression system. J Mol Biol 219: 37–44.
37. Hartl FU, Hayer-Hartl M (2002) Molecular chaperones in the cytosol: From nascent chain to folded protein. Science 295: 1852–1858.
38. De Marco A (2007) Protocol for preparing proteins with improved solubility by co-expressing with molecular chaperones in Escherichia coli. Nat Protoc 2: 2632–2639.
39. Shiloach J, Fass R (2005) Growing E. coli to high cell density-a historical perspective on method development. Biotechnol Adv 23: 345–357.
40. Studier FW (2005) Protein production by auto-induction in high density shaking cultures. Protein Expr Purif 41: 207–234.
41. Baneyx F, Georgiou G (1992) Degradation of secreted proteins in Escherichia coli. Ann NY Acad Sci 665: 301–308.
42. Yee L, Blanch HW (1992) Recombinant protein expression in high cell density fed-batch cultures of Escherichia coli. Biotechnol 10: 1550–1556.
43. Nallamsetty S, Waugh DS (2007) A generic protocol for the expression and purification of recombinant proteins in Escherichia coli using a combinatorial His6-maltose binding protein fusion tag. Nat Protoc 2: 383–391.
44. Schmidt TG, Skerra A (2007) The Strep-tag system for one-step purification and high-affinity detection or capturing of proteins. Nat Protoc 2: 1528–1535.
45. Drew D, Newstead S, Sonoda Y, Kim H, von Heijne G, et al. (2008) GFP-based optimization scheme for the overexpression and purification of eukaryotic membrane proteins in Saccharomyces cerevisiae. Nat Protoc 3: 784–798.
46. Waugh DS (2005) Making the most of affinity tags. Trends Biotechnol 23: 316–320.
47. Nallamsetty S, Waugh D (2006) Solubility-enhancing proteins MBP and NusA play a passive role in the folding of their fusion partners. Protein Expr Purif 45: 175–182.
48. Esposito D, Chatterjee DK (2006) Enhancement of soluble protein expression through the use of fusion tags. Curr Opin Biotechnol 17: 353–358.
49. Smyth DR, Mrozkiewicz MK, McGrath WJ, Listwan P, Kobe B (2003) Crystal structures of fusion proteins with large-affinity tags. Protein Sci 12: 1313–1322.
50. Maskos K, Huber-Wunderlich M, Glockshuber R (2003) DsbA and DsbC-catalyzed oxidative folding of proteins with complex disulfide bridge patterns in vitro and in vivo. J Mol Biol 325: 495–513.
51. Drag M, Salvesen GS (2008) DeSUMOylating enzymes-SENPs. IUBMB Life 60: 734–742.
52. Kefala G, Kwiatkowski W, Esquivies L, Maslennikov I, Choe S (2007) Application of Mistic to improving the expression and membrane integration of histidine kinase receptors from Escherichia coli. J Struct Funct Genomics 8: 167–172.
53. Angulo I, Acebrón I, de Las Rivas B, Muñoz R, Rodríguez-Crespo I, et al. (2011) High-resolution structural insights on the sugar-recognition and fusion tag properties of a versatile beta-trefoil lectin domain from the mushroom Laetiporus sulphureus. Glycobiology 21: 1349–1361.
54. Sambrook J, Russell WD (2001) Molecular cloning: A laboratory manual. Third Edit. Cold Spring Harbor (NY).
55. Hemsley A, Arnheim N, Toney MD, Cortopassi G, Galas DJ (1989) A simple method for site-directed mutagenesis using the polymerase chain reaction. Nucleic Acids Res 17: 6545–6551.
56. Hanahan D (1983) Studies on transformation of Escherichia coli with plasmids. J Mol Biol 166: 557–580.
57. Kapust RB, Tözsér J, Fox JD, Anderson DE, Cherry S, et al. (2001) Tobacco etch virus protease: Mechanism of autolysis and rational design of stable mutants with wild-type catalytic proficiency. Protein Eng 14: 993–1000.
58. Reverter D, Lima CD (2009) Preparation of SUMO proteases and kinetic analysis using endogenous substrates. Methods Mol Biol 497: 225–239.
59. Goulas T, Arolas JL, Gomis-Rüth FX (2011) Structure, function and latency regulation of a bacterial enterotoxin potentially derived from a mammalian adamalysin/ADAM xenolog. Proc Natl Acad Sci 108: 1856–1861.
60. Schägger H (2006) Tricine-SDS-PAGE. Nat Protoc 1: 16–22.
61. Studier WF, Rosenberg AH, Dunn JJ, Dubendorf JW (1990) Use of T7 RNA polymerase to direct expression of cloned genes. Methods Enzymol 185: 60–89.
62. Phan TTP, Nguyen HD, Schumann W (2012) Development of a strong intracellular expression system for Bacillus subtilis by optimizing promoter elements. J Biotechnol 157: 167–172.
63. Daly R, Hearn MTW (2005) Expression of heterologous proteins in Pichia pastoris: a useful experimental tool in protein engineering and production. J Mol Recognit 18: 119–138.
64. Aslanidis C, de Jong PJ (1990) Ligation-independent cloning of PCR products (LIC-PCR). Nucleic Acids Res 18: 6069–6074.
65. Harper S, Speicher DW (2011) Purification of proteins fused to glutathione S-transferase. Methods Mol Biol 681: 259–280.
66. Peroutka Iii RJ, Orcutt SJ, Strickler JE BT (2011) SUMO fusion technology for enhanced protein expression and purification in prokaryotes and eukaryotes. Methods Mol Biol 705: 15–30.
67. Kapust RB, Waugh DS (1999) Escherichia coli maltose-binding protein is uncommonly effective at promoting the solubility of polypeptides to which it is fused. Protein Sci 8: 1668–1674.
68. Simonen M, Palva I (1993) Protein secretion in Bacillus species. Microbiol Rev 57: 109–137.
69. De Marco A (2009) Strategies for successful recombinant expression of disulfide bond-dependent proteins in Escherichia coli. Microb Cell Fact 8: 26.
70. Macauley-Patrick S, Fazenda ML, McNeil B, Harvey LM (2005) Heterologous protein production using the Pichia pastoris expression system. Yeast 22: 249–270.
71. Reverter D, Garcia-Saez I, Catasus L, Vendrell J, Coll M, et al. (1997) Characterisation and preliminary X-ray diffraction analysis of human pancreatic procarboxypeptidase A2. FEBS Lett 420: 7–10.
72. Arolas JL, Botelho TO, Vilcinskas A, Gomis-Rüth FX (2011) Structural evidence for standard-mechanism inhibition in metallopeptidases from a complex poised to resynthesize a peptide bond. Angew Chemie 50: 10357–10360.
73. Sanglas L, Aviles FX, Huber R, Gomis-Rüth FX, Arolas JL (2009) Mammalian metallopeptidase inhibition at the defense barrier of Ascaris parasite. Proc Natl Acad Sci U S A 106: 1743–1747.
74. Roosild TP, Greenwald J, Vega M, Castronovo S, Riek R, et al. (2005) NMR structure of Mistic, a membrane-integrating protein for membrane protein expression. Science 307: 1317–1321.
75. Marrero A, Mallorqui-Fernández G, Guevara T, Garcia-Castellanos R, Gomis-Rüth FX (2006) Unbound and acylated structures of the MecR1 extracellular antibiotic-sensor domain provide insights into the signal-transduction system that triggers methicillin resistance. J Mol Biol 361: 506–521.
76. Kapust RB, Tözsér J, Copeland TD, Waugh DS (2002) The P1′ specificity of tobacco etch virus protease. Biochem Biophys Res Commun 294: 949–955.
77. Butt TR, Edavettal SC, Hall JP, Mattern MR (2005) SUMO fusion technology for difficult-to-express proteins. Protein Expr Purif 43: 1–9.
78. Vergis JM, Wiener MC (2011) The variable detergent sensitivity of proteases that are utilized for recombinant protein affinity tag removal. Protein Expr Purif 78: 139–142.
79. Dvir H, Choe S (2009) Bacterial expression of a eukaryotic membrane protein in fusion to various Mistic orthologs. Protein Expr Purif 68: 28–33.

8

Novel Biogenic Aggregation of Moss Gemmae on a Disappearing African Glacier

Jun Uetake[1,2]*, **Sota Tanaka**[3], **Kosuke Hara**[4], **Yukiko Tanabe**[5], **Denis Samyn**[6], **Hideaki Motoyama**[3], **Satoshi Imura**[3], **Shiro Kohshima**[7]

1 Transdisciplinary Research Integration Center, Minato-ku, Tokyo, Japan, 2 National Institute of Polar Research, Tachikawa, Tokyo, Japan, 3 Faculty of Science, Chiba University, Chiba, Chiba, Japan, 4 Graduate School of Science, Kyoto University, Kyoto, Japan, 5 Institute for Advanced Study, Waseda University, Shinjuku-ku, Tokyo, Japan, 6 Department of Mechanical Engineering, Nagaoka University of Technology, Nagaoka, Niigata, Japan, 7 Wildlife Research Center, Kyoto University, Kyoto, Kyoto, Japan

Abstract

Tropical regions are not well represented in glacier biology, yet many tropical glaciers are under threat of disappearance due to climate change. Here we report a novel biogenic aggregation at the terminus of a glacier in the Rwenzori Mountains, Uganda. The material was formed by uniseriate protonemal moss gemmae and protonema. Molecular analysis of five genetic markers determined the taxon as *Ceratodon purpureus*, a cosmopolitan species that is widespread in tropical to polar region. Given optimal growing temperatures of isolate is 20–30°C, the cold glacier surface might seem unsuitable for this species. However, the cluster of protonema growth reached approximately 10°C in daytime, suggesting that diurnal increase in temperature may contribute to the moss's ability to inhabit the glacier surface. The aggregation is also a habitat for microorganisms, and the disappearance of this glacier will lead to the loss of this unique ecosystem.

Editor: Lucas C.R. Silva, University of California Davis, United States of America

Funding: This study was partially supported by Ministry of Education, Culture, Sports, Science and Technology Grant-in-Aid for Scientific Research (A) No. 22241005 and by Proposal for Seeds of Transdisciplinary Research from the Transdisciplinary Research Integration Center, and by an NIPR publication subsidy. The funders had no role in study design, data collection and analysis, decision to publish, or preparation of the manuscript.

Competing Interests: The authors have declared that no competing interests exist.

* Email: juetake@nipr.ac.jp

Introduction

Many psychrophilic and psychrotolerant microorganisms inhabit supraglacial environments, which have recently been recognized as an important biome [1]. Cryonite granules, dark spherical aggregates typically 1 mm in diameter have been frequently observed on ablation zones of glaciers in many parts of the world [2]. These cryoconites consist of mineral particles, organic matter, and microorganisms, which are mainly formed by the aggregation of filamentous cyanobacteria [3]. These cryoconite harbor a diverse range of microorganisms, and studies of their molecular diversity have revealed microbial communities of bacteria [4] and archaea [5,6], as well as algae, fungi, amoebas, and invertebrates such as tardigrades [5]. These microbial communities play an important role in the carbon and nitrogen cycles on the glacier [1,7].

Other types of biological aggregations have been reported from supraglacier ecosystems in Iceland and Alaska, namely, globular moss aggregations known as 'glacier mice' [8,9] or 'moss polster' [10]. These are lenticular moss cushions (0.02 to 0.1 m in diameter) and are composed of a moss envelope covering an internal clast formed from glacial sediment and airborne particles [11]. These moss cushions are expected to impact the ecology and nutrient cycle of the supraglacial ecosystem [11], and also provide a favorable habitat for a variety of invertebrates, including Collembola, Tardigrades, and Nematoda [12].

Previous biological studies have frequently examined mid-latitude and polar glaciers, however, the tropical glaciers are have been studied rarely, except for New Guinea [13]. In equatorial Africa, glaciers persist in three major mountain regions (Mt. Kilimanjaro in Tanzania, Mt. Kenya in Kenya, and the Rwenzori Mountains in Uganda), which have not been previously been targeted in surveys of glacier biology. The Rwenzori glaciers are shrinking rapidly and are expected to disappear by 2020 due to climatic warming [14] and/or lowered humidity and lowered cloudiness [15], as measured by aerial photography and satellite imagery [14].

During a biological field survey on a glacier near the summit of Mt. Stanley, the highest peak in Uganda and in the Rwenzoris, we found a large, black bioaggregation (average long and short axes: 18.1 mm and 12.7 mm) in the supraglacial environment, greater than cryoconites. Examination revealed that these granular structures were formed by filamentous moss gemmae and protonema, not cyanobacteria. This is the first report to describe this habitat for such a structure, which we classified as a "glacial moss gemmae aggregation" (GMGA). In order to identify the material we measured the structure (size and mass) and isolated the dominant moss species using both culture and molecular techniques. Furthermore, we examined the photosynthetic activity of isolates under various temperature and radiation conditions.

Materials and Methods

Glacier characteristics and sampling

The Rwenzori Mountains (5109 m above sea level) contain the third-highest mountain in Africa, and are straddling the equator along the border of Uganda and the Democratic Republic of the Congo (Fig. 1). Since the LIA (Lac Gris Stage: 19th centry or just before), glaciers in the Rwenzoris have been shrinking; in 1906 the Elena glacier was estimated to cover 6.5 km^2 [16] and by 2003 it had decreased to approximately 1 km^2 [14]. During this period, glaciers on Mts. Emin, Gessi, and Luigi disappeared completely, leaving only glaciers remaining on three major peaks: Mts. Speke, Baker, and Stanley. In this study, we surveyed Stanley Plateau, the largest glacier on Mt. Rwenzori. Stanley Plateau is a flat sloped glacier that flows from Mt. Stanley's Alexandra Peak, and is around 1 km long and 0.1–0.3 km wide (Fig. 2a).

In February 2012 and 2013, we collected surface ice samples, including biological debris, at three sites: ST1 (N00°22′31.3″, E29°52′40.26″), ST2 (N00°22′34.74″, E29°52′37.2″) and ST3 (N00°22′52.32″, E29°52′24.6″). At each site, 5 samples were collected from different 0.1×0.1 m areas and stored in 50 ml plastic bottles for cell counts, and 5 samples from different areas (samples size not measured) and placed in 8 ml plastic bottles for DNA analysis and isolation. In the glacier foreland, located about 10 m from ST1, we collected shoots of bryophyte on dried GMGAs and placed them in 8 ml plastic bottles with RNAlater (Life Technologies, Carlsbad, USA). All samples were collected using pre-cleaned stainless steel scoops and spoons.

Samples for cell counts were fixed with 3% formaldehyde and stored at room temperature. All other samples were kept cold around 0°C in large stainless steel vacuum flasks with glacial ice samples until transport to Kasese, Uganda, the closest city to Rwenzori Mountains National Park. There, samples for molecular analysis were kept frozen around −20°C and samples for isolation were kept cold around 0°C, until they could be transported to the lab for analysis at the National Institute of Polar Research (Tokyo, Japan).

In the field, the internal temperature of the GMGAs at ST1 was measured using a waterproof temperature logger (R-52i; T&D, Matsumoto, Japan between 9–13 Feb. 2013) with 0.6 m sensor probe (TR-5106; T&D, Matsumoto, Japan). Probes were inserted into the center of two GMGAs and were monitored by camera (Optio WG-2: Ricoh, Tokyo, Japan) at intervals to ensure that the measuring apparatus was not disturbed.

Microscopic observation of biological materials on ice surface

The samples were cold-preserved prior to isolation and identification of species. After formaldehyde fixation, 0.1–0.4 ml of 12–60 fold diluted samples were filtered through a hydrophilic polytetrafluoroethylene membrane (Omnipore JGWP01300; Merck Millipore, Billerica, USA) with diameter 13 mm and pore size 0.2 μm. We observed and counted cell concentrations from one-quarter of the membrane, using a fluorescent microscope (IX71 and 81; Olympus, Tokyo, Japan).

18S r RNA gene molecular cloning

DNA of approximately 0.3 g was extracted from samples ST1, ST2, and ST3 using the Fast DNA SPIN Kit for soil (MP Biomedicals, Santa Ana, USA) according to the manufacturer's instructions. Extracted DNA was diluted to 1.62 ng/μl with water (Ambion Nuclease-Free Water; Life Technologies, Carlsbad, USA). Five aliquots from each site were combined for DNA amplification. Thermal cycling was performed with an initial denaturation step at 98°C for 3 min, followed by 25 cycles of denaturation at 98°C for 10 s, annealing at 55°C for 30 s, and elongation at 72°C for 1.5 min, using Ex Taq HS DNA polymerase (Takara, Shiga, Japan) and the primer pair of Euk A (5′-ACCTGGTTGATCCTGCCAGT-3′) and EukB (5′-GATCCTTCTGCAGGTTCACCTAC′). Cycling was completed by a final elongation step at 72°C for 3 min. The PCR-amplified DNA fragments were cloned into the pCR4 vector of the TOPO TA cloning kit (Invitrogen, Carlsbad, USA). Clones obtained from the libraries were sequenced using the 3130×l Genetic Analyzer (Life technologies, Carlsbad, USA) at the National Institute of Polar Research. All sequences were assembled using CodonCode Aligner (CodonCode Corporation, Centerville, USA) and assembled full-length sequences of 18S rRNA were aligned with the eukaryotic Silva database [17] using mothur ver. 1.27.0 [18]. Tentative chimeric sequences were removed using both the reference and *de novo* modes of Uchime [19] implemented in mothur software package. All good-quality sequences with more than 97% similarity were clustered into operational taxonomic units (OTU).

Isolation of moss and molecular identification

Fragments of cold-preserved GMGA samples were inoculated in liquid Bold's basal medium (BBM) [20] in a laminar flow bench and incubated at 4°C for 1 month. Protonemata that grew directly from observed gemmae (Fig. 2 c,d) were transplanted to fresh BBM liquid medium and a 1-month-incubation was repeated. Isolated protonemata were kept in BBM and 1.5% agar medium before extraction and analysis. DNA of a single cluster of protonema in liquid medium was extracted with the Fast DNA SPIN Kit for soil, and 4 different regions (18S rRNA; chloroplast genes, *trn*L, *rps*4 and *atpB-rbc*L intergenic spacer; and mitochondria gene, *nad*5) were amplified by using Ex Taq HS DNA polymerase (Takara, Shiga, Japan). Thermal cycling for 18S rRNA was carried out following Remias *et al.* [21], with 35 cycles of denaturation at 98°C for 10 s, annealing at 54°C for 30 s, and elongation at 72°C for 1 min 45 s, using the primer pair NS1 (5′-GTAGTCATATGCTTGTCTC-3′) and 18L(5′-CACCTACG-GAAACCTTGTTACGACTT-3′). For chloroplast *trn*L, thermal cycling was performed with 35 cycles of denaturation at 98°C for 10 s, annealing at 60°C for 30 s, and elongation at 72°C for 1 min 45 s and primer pair trnC (5′- CGAAATCGGTAGACGC-TACG-3′) and trnF (5′-ATTTGAACTGGTGACACGAG-3′), following Taberlet *et al.* [22]. For chloroplast *rps*4, thermal cycling was performed according to Nadot *et al.* [23] and Souza-Chies *et al.* [24], with 35 cycles of denaturation at 98°C for 10 s, annealing at 55°C for 30 s, and elongation at 72°C for 1 min 45 s using primer pair rps5 (5′-ATGTCCCGTTATCGAGGACCT-3′) and trnS (5′-TACCGAGGGTTCGAATC-3′). For chloroplast *atpB-rbc*L intergenic spacer, thermal cycling was performed according to Chiang *et al.* [25], with 35 cycles of denaturation at 98°C for 10 s, annealing at 49°C for 30 s, and elongation at 72°C for 30 s using primer pair *atpB*-1 (5′-ACATCKAR-TACKGGACC-3′) and *rbc*L-1 (5′-AACACCAGCTTTRAATC-CAA-3′). Lastly, for mitochondria nad5, thermal cycling was performed according to Shaw *et al.* [26], with 35 cycles of denaturation at 98°C for 10 s, annealing at 52°C for 30 s, and elongation at 72°C for 1 min 45 s using primer set *nad*5F4 (5′-GAAGGAGTAGGTCTCGCTTCA-3′) and nad5R3 (5′-AAAACGCCTGCTGTTACCAT-3′). Some of shoots of Bryophyta on dried GMGAs were picked up by tweezers and DNA was analyzed by same method as isolated protonema.

Figure 1. Location map of Rwenzori Mountains in Republic of Uganda.

Photosynthetic rate of GMGA and isolate

Photosynthetic rate at 7 different incubation temperatures (5, 10, 15, 20, 25, 30, and 40°C) was measured using a pulse amplitude modulation (PAM) fluorometer (Water-PAM, Waltz, Effeltrich, Germany) with Win-control software for control and analysis following Tanabe *et al.* [27]. PAM fluorometer is useful to measure the electron transport rate (ETR) of isolate underdifferent incubation factor [27]. For incubation temperatures of 5, 10, and 15°C, photosynthetically active radiation (PAR) intensities were 3, 64, 94, 144, 215, 305, 422, 687, and 1000 mmol photons/m^2/s. After these measurments, we had changed to another PAM device, because this device is obviously unstable only under 40°C due to mechanical trouble. Then, we used another device and measured again from 20°C. Results of 20, 25, 30°C are almost same as previous analysis, and measurement was stable at 40°C in next time. For incubation temperatures of 20, 25, 30, and 40°C, PAR

intensities were 8, 62, 92, 140, 209, 297, 412, 674, and 986 mmol photons/m^2/s. After a 30 s exposure, a saturating pulse of > 2000 mmol photons/m^2/s was applied for 0.4 sat 5°C in a temperature-controlled incubator. The gain value of the photoelectric multiplier (PM-Gain) was set to 3 for all measurements. After incubation of each sample at 5°C for 60 min in dark conditions, a tissue sample of GMGA and isolated protonemata were transferred to the quartz cuvette of the fluorometer. After measurement, incubation temperature was raised by 5°C to 10°C and incubated for 1 h, after which the temperature was raised by 5°C again and incubated for 1 h repeatedly until incubation temperature was 40°C. Light curves were obtained by running a rapid light curve protocol in Win-control software. The photosynthetic rate expressed as relative electron transport rate *rETR* [28] was as follows:

Figure 2. Research site and glacier moss gemmae aggregation (GMGA). a) Stanley Plateau glacier and Margarita Peak from Mt. Baker, b) Glacier ice surface covered by GMGAs, c) GMGAs (grid cells beneath GMGA are 1×1 mm), d) Cross section of GMGA (scale bar: 2 mm).

$$rETR = (Fm - Fm')/Fm' \times PAR. \qquad (1)$$

Here, F and Fm' are the transient and maximum fluorescence levels at certain actinic light intensities at a given time and (Fm'–F)/Fm' indicates Photosystem II (PSII) yield. Non-photochemical quenching (NPQ) was calculating by the following equation:

$$NPQ = (Fm - Fm')/Fm' \times PAR, \qquad (2)$$

where Fm is the maximum fluorescence level of non-illuminated samples.

Ethics Statement

Uganda Wildlife Authority and Uganda National Council for Science and Technology authorized all field researchs in Rwenzori Mountains National Park.

Results

Morphological features of GMGA

We found ellipsoidal blackish bioaggregations covering the glacier surface at ST1 (Fig. 2b). We collected and sampled 96 of these bioaggregations, as well as measured their long and short axes, thickness, and mass (Fig. 2c,d). The average long axis, short axis, thickness, and mass were 18.7 mm, 12.7 mm, 8.3 mm, and 1.6 g, respectively (Fig. 3), and these aggregations were clearly larger than cryoconites (average diameter: 1.1 mm; 3). Short axis length was well correlated with long axis length ($R^2 = 0.705$), but not as well correlated with thickness and mass ($R^2 = 0.543$ and 0.616, respectively). This means that the structure of this

bioaggregation is not spherical but is instead flattened. The bioaggregation was composed of many gemmae and protonema. The gemmae were germinating and developing filamentous protonema, and gemmae were formed repeatedly on protonema. Many moss gemmae were observed especially on the surface of the bioaggregation (Fig. 4a,b), from the top 1–2 mm of the cross section (Fig. 2d). The main framework of these structures was formed by moss gemmae, so we named this structure as "glacial moss gemmae aggregation (GMGA)".

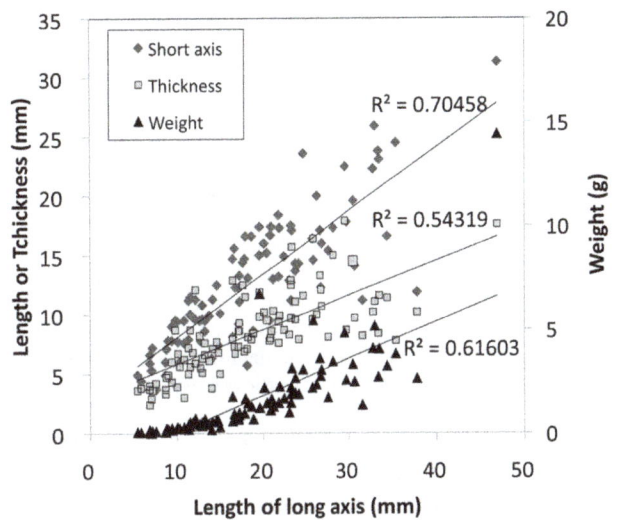

Figure 3. Size (long and short axes, thickness) and mass distribution of GMGAs.

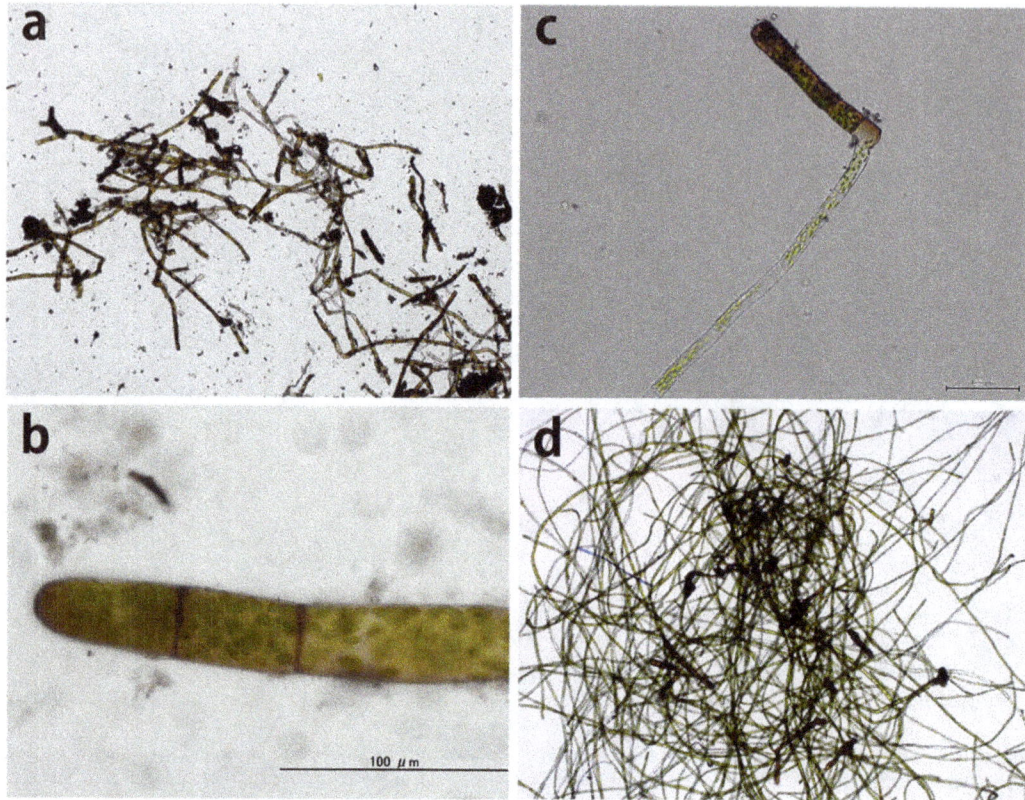

Figure 4. Cells of moss gemmae and protonemata. a,b) Moss protonemal cells formed the main frame of the GMGA (scale bar: 100 μm), c) Moss protonema grew from gemmae below 4°C (scale bar: 100 μm), d) protonemal cells for molecular identification after incubation below 4°C for 1 month in liquid Bold's basal medium.

The gemmae are filamentous, composed of 1–2 rows of 2–20 cells with slightly thickened brownish cell walls, 100–200 um long in maximum. These morpholocial characteristics were well agree with rhizoidal gemmae of cosmopolitan moss, *Ceratodon purpureus* (Hedw.) Brid. described by Imura and Kanda (1986) based on Antarctic specimens [29].

Molecular identification of moss species

We obtained a total of 81 clones from the GMGAs by 18S rRNA gene PCR-cloning. Sixty-three clones (77.8%) were clustered into the same OTU (AB858433: Table 1). The remaining 18 clones were of cercozoa, green algae, and fungi.

The 18S rRNA, *rps*4, *trn*L, *atp*B-*rbc*L intergenic spacer and *nad*5 gene sequence of the isolated protonemata (Fig. 4c,d) that grew from the observed gemmae and shoots on dried GMGA were summarized in Table 1. These high-percentage matches (more than 99.9% similarity) from five different regions of the protonemata show that the isolated moss is indeed *C. purpureus*. Also results from four different regions of shoots on dried GMGA show that this specie belonging to genus: *Bryum*, however we could not identify species level from these regions.

Optimum temperature and PAR of GMGA and isolated protonemata of *C. purpureus*

Internal (center) temperature changes was measured during the 2013 field season of two *in situ* GMGAs in ST1 and one dried GMGA found on a rock in glacier foreland (Fig. 5). Temperature change data show clear diurnal cycles with daily exposure to below-freezing temperatures daily. Maximum temperatures reached 8–10°C for *in situ* GMGAs and above 20°C for dried GMGA. The daytime increase in temperature was due to absorption of thermal radiation, but was variable due to decreases in radiation from frequent cloud cover and cooling by glacier ice.

A photosynthetic light curve was measured using a PAM fluorometer under different temperatures of GMGA and the two isolates (Fig. 6). The ETR of GMGA and isolates is high between 20–30°C, and highest at 25°C for GMGA. Electron transport was detected in all three samples even at temperatures as low as 5°C, but ETR was zero or extremely low at 40°C. These results indicate that the optimum temperatures of GMGA and isolated *C. purpureus* are around 25°C. The ETR of GMGA and the isolates was high at low PAR levels (305 and 422 μmol/m²/s) at 5–15°C; however, ETR was high at a medium PAR level (687 μmol/m²/s) at 20–30°C (Fig. 5). Therefore, the optimal PAR value for *in situ* temperatures (0–10°C) is likely between 305–422 μmol/m²/s. The highest ETR value in GMGAs was 674 μmol/m²/s, which was approximately twice that of the other two isolates (Fig. 6).

Distribution of GMGAs on the glacier

GMGAs of *Ceratodon purpureus* were observed only at site ST1, the glacier terminus (Fig. 7). The organic carbon mass (62.72±19.39 g/m²) was highest at the terminus. This record is higher than current highest glacial organic carbon mass (38.5±12.4 g/m²) from Qiyi Glacier, China [30]. Moreover, organic carbon mass at our sites without observed GMGAs (ST2:20.73±8.91 g/m², ST3:23.55±6.89 g/m²) were roughly equal to the average high organic carbon mass at Qiyi Glacier (mean: 25.4±16.5 g/m²) [26].

Table 1. Sequences list of five different genetic regions from three different sample types (1: GMGA_cloning, 2: isolate protonema and 3: dried GMGA_cloning) and their closest relatives.

Gene type	Genetic region	Sample type	Accession number	Length (bp)	Closest relative specie	Accession number of relative	Identity (%)	sequence match (bp)
Ribosomal RNA gene	18S rRNA	1: GMGA_cloning	AB858433	1819	Ceratodon sp. AM2008N12	KC291530	99.8	1721/1724
		2: isolate protonema			Ceratodon purpureus	Y08989	99.7	1751/1757
					Ceratodon purpureus	KC291530	99.9	1677/1678
					Ceratodon purpureus	Y08989	99.7	1673/1679
					GMGA_cloning (this study)	AB858433	100	1679/1679
		3: dried GMGA_cloning	AB872997	1697	Bryum caespiticium	AF023703	100	1697/1697
Chloroplast gene	rps4	2: isolate protonema	AB848717	674	Ceratodon purpureus	FJ572605	100	623/623
					Ceratodon purpureus	FJ572589	100	625/625
					Ceratodon purpureus	AF435271	100	561/561
					Ceratodon purpureus	AY908122	100	652/652
					Trichodon cylindricus*	AY908125	94.1	622/661
		3: dried GMGA_cloning	AB872999	684	Bryum cyathiphyllum	AF521683	99.9	667/668
	trnL	2: isolate protonema	AB848718	482	Ceratodon purpureus	FJ572485	100	482/482
					Ceratodon purpureus	AF435310	100	482/482
					Glyphomitrium humillimum*	EU246911	94.2	438/465
		3: dried GMGA_cloning	AB873000	520	Bryum cyathiphyllum	AY150351	100	492/492
	atpB-rbcL intergenic spacer	2: isolate protonema	AB980065	637	Ceratodon purpureus	AY881031	100	621/621
					Ceratodon purpureus	AY881034	100	621/621
					Ceratodon purpureus	AY881052	100	598/598
					Cheilothela chloropus*	AY881063	89.9	571/635
Mitocondrial gene	nad5	2: isolate protonema	AB848719	1112	Ceratodon purpureus	AY908859	99.9	1093/1094
					Ceratodon purpureus	AY908862	99.9	1090/1091
					Trichodon cylindricus*	AY908863	99.5	1089/1095
		3: dried GMGA_cloning	AB872998	1107	Bryum argenteum	AY908945	100	1082/1082

*show second highly related species.

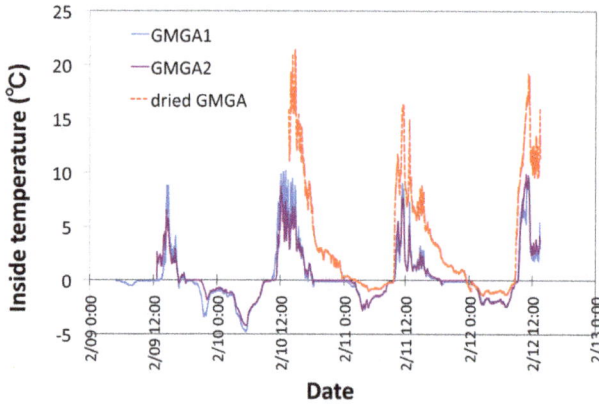

Figure 5. Internal changes in temperature of 2 GMGA and 1 dried GMGA left on a rock during the 2013 research period (February 9–12, 2013).

Discussion

Glacial Moss Gemmae Aggregation (GMGA) is a novel moss aggregation

Mosses, in the form of "glacier mice", have been previously recorded from supraglacial habitats [11,12]; however, the structure of GMGA is completely different from that of "glacier mice". Whereas "glacier mice" are formed by the moss shoots, that level of cellular differentiation was not detected in GMGAs. These findings report first description of developing moss gemmae and protonema in the supraglacial environments. *Ceratodon purpureus*, which formed the GMGA observed in this study, is a cosmopolitan moss species widely distributed throughout entire continents [31] and is known to grow in extreme environments (i.e. polluted sites including highway shoulders and on coal and heavy metal mine tailings) [32]. *Ceratodon purpureus* also occurs in the cryosphere in high alpine areas, Antarctica [33–35]. In the Rwenzori Mountains, unfortunately inhabitation of *C. purpureus* around glacier had not directly observed by authors, however, *C. purpureus* has been detected at elevations from 2800 m to 3700 m [36] and *C. purpureus* specimen (PC0106302) taken at just below the Speke Glacier (4480 m a.s.l.) are stored in Muséum National d'Histoire Naturelle, Paris, France. These evidences would show that possibility of dispersal of spore or gemma from near glacier and deposition on the glacier surface by local wind circulation.

Adaptation of GMGA isolate to warmer temperature

The optimum temperature of polar mosses are widly distributed from 2°C to 35°C according to species [37]. The optimum temperature for the *C. purpureus* isolates (25°C; Fig. 6) is normal value even in polar region, but this was higher than that for *C. purpureus* in Antarctica as previously reported, which was 15°C in the liquid and agar cultures [33]. Moreover, another study showed that the optimum temperature for photosynthesis in *C. purpureus* is around 15°C, but significant carbon fixation occurs at 5°C [35]. Although the measurement of optimum temperature by measuring fluorescence of chlorophyll used in the present study is an indirect measurement of growth (e.g., [33]), these values reflect photosynthetic ability at each temperature (e.g., [35]). Therefore, the populations of *C. purpureus* from this Ugandan glacier have likely adapted to a higher optimum temperature than Antarctic populations.

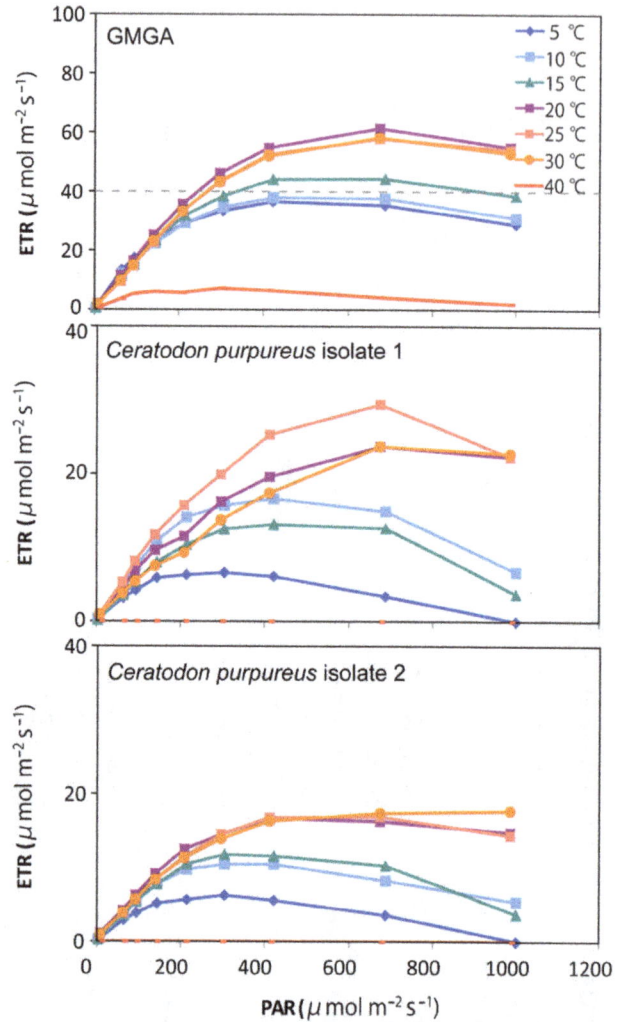

Figure 6. Photosynthetic light curves of GMGA and the two isolates under different incubation temperatures (from 5°C to 40°C) using a pulse amplitude modulation fluorometer.

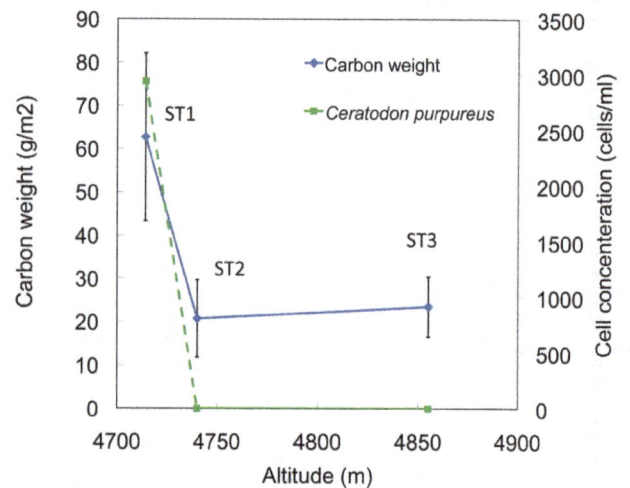

Figure 7. Distribution of carbon mass and *Ceratodon purpureus* cell concentrations on Stanley Plateau Glacier.

Cold and light stress in isolated *C. purpureus*

Polar mosses tend to adapt broad range of favorable temperature. For example, relationship between net assimilation rate (NAR) and temperature of *Drepanocladus uncinatus* in Signy Island show that optimum temperature is 20°C and NAR are more than 40% from 0°C to 30°C (5°C interval) by using cold incubation sample (5°C in light/−5°C in dark) [37]. Relatively higher temperature optimum and broad range of favorable temperature is similar to our result of GMGA and *C. purpureus* isolates. Optimum temperature shift with environmental temperature change were reported from some of experiments [38,39], but *Drepanocladus uncinatus* in Signy Island [37] did not correspond to these studies. Also in our study, optimum temperature did not shift to lower temperature inspite of preincubation temperature at 4°C. The internal temperature of the GMGAs (Fig. 5) was below the optimum (Fig. 6), therfore, the daytime internal temperature of GMGA (10°C) is not optimum but favorable for growth of Ugandan glacial *C. purpureus*. In this temperature range of internal GMGAs (−5°C to 10°C), the optimum PAR is below the warmer temperature (Fig. 6). This phenomenon may be able to explain photoinhibition from cold stress. In low-temperature conditions, PSII is inhibited due to decreased rate of repair of damaged D1 protein and increased excitation pressure [40].

Polar mosses are able to survive short-term freezing and thawing cycles in summer and prolonged freezing in winter. *Bryum argenteum* taken from tropical to polar origin showed no damage after 10 days with a temperature regime of 5°C in light/−5°C in dark, and grew slowly under these conditions [37]. In Rwenzori, *C. purpureus* live under similar dirnal cycle (from −5°C to 10°C) through year and GMGAs structure would be formed slowly. After being deposited and frozen at below −20°C for 2.5 years, regeneration of the *C. purpureus* from east Antarctica is very active [33]. *Ceratodon purpureus* protonemata from GMGA can be isolated and grown at 4°C after half a year of cryopreservation at −80°C. Therefore, isolate of *C. purpureus* in this study have potential to survive both short-term and prolonged freezing stress.

PAR in this natural environment is higher than optimum PAR at 5–15°C. Although we did not measure PAR directly, an automatic weather station with a radiation meter was installed beside the Stanley Plateau (N0°22'34.55'', E 29°52'43.24''; 4750 m above sea level) by the Stations at High Altitude for Research on the Environment project [41]. According to their data, and assuming no significant divergence with today's conditions, the maximum diurnal seasonal shortwave radiation (c.a. 400 W/m^2) occurs around 2:00 PM. PAR (400–700 nm) generally comprises 50% of total solar radiation reaching the Earth's surface. Assuming a conversion rate of 1 W/m^2 to 4.57 μmol/m^2/s, our estimated maximum PAR is 914 μmol/m^2/s when maximum shortwave radiation is 400 W/m^2. This value is higher than the optimum PAR in any temperature and this may cause photoinhibition in low temperature.

These results indicate that both low temperature and high radiation on the glacier are stress factors for *C. purpureus*. In acidic rivers in Japan, *Dicranella heteromalla* (Hedw.) Schimp. remains in a prolonged protonema stage for several growing seasons without producing shoots or sporophytes [42], which researchers concluded was due to the water's extraordinarily low pH (1.9–2.1). A similar prolonged protonema phase of the moss *Scopelophila cataractae* (Mitt.) Broth was reported in copper-rich sites as well in Japan [43]. Therefore, the low temperature and high radiation stress on the glacier, may keep *C. purpureus* in the gemmae and protonemal stage instead of developing into shoots.

Higher photosynthesis activity of GMGA than of isolates

The value of ETR from GMGAs was twofold that of isolates (Fig. 6), possibly due to differences in nutrient condition and the effects of other photosynthetic microorganisms. Growth conditions of moss may be more suitable in GMGA than in the artificial medium (liquid BBM) used in this study, because GMGA contains sufficient nutrients for effective growth. Yet, GMGA is not a simple aggregation of only moss, but also contains many other microorganisms. For example, we observed *Cylindrocystis brebissonii* cells and red snow algae [44], which is related to the green algae commonly found in supraglacial environments, in the GMGAs. Both moss and green algae affect the total photosynthetic activity of GMGAs.

Possible process of GMGA formation

If biological material in sites without GMGAs form a thick deposition layer (more than a few millimeters), the temperature below the surface would increase to above 0°C, the same as in GMGAs. If this is so, then the invasion of gemmae of *C. purpureus* adapted to warmer temperatures on the cold glacier surface can be attributed to this increased subsurface temperature. Similar temperature increases occur in other glaciers (e.g., Qiyi Glacier); however, GMGA-like structures and growth of moss gemmae have not been found on any other glacier, despite studies of glacier biology being conducted around the world [45]. This may relate to possible inhabitation of *C. purpureus* near the glacier and unique features of the Rwenzori; namely, the lack of a clear seasonal temperature cycle. On Stanley Plateau, diurnal temperature change in all seasons is in a range of approximately 0°C to 5°C [41]. Consequently, long periods of freezing do not exist, permitting microorganisms to grow throughout the year. Therefore, we suppose that at least these two factors (the internal temperature rise and the long growth season) may contribute to the formation of GMGAs.

However, GMGAs are disproportionally dominant near the glacier terminus. The high number of GMGAs observed at ST1 must be supported by factors specific to that site. Although our data does not let us reach firm conclusions about the distribution of GMGAs, downward transpotation by surface melting water and availability of sunlight is a likely candidate factor.

Glacier surface melt of Stanley Plateau homogenously spread over the ice area and remarkable water channels on surface are few. Takeuchi [46] speculated cryoconite granules (around 1 mm diameter) are more stable from meltwater than unicellular microorganisms due to larger size character. GMGAs are much larger diameter than typical cryoconite (Fig. 3) seems to be more stable on the ice. Also GMGAs penetrate few mm into ice due to radiation warming (Fig. 2b). These also prevent to wash this material to downward. Therefore, downward transportation of well-developed GMGAs seems unlikely happened.

During the biological growth season, supraglacial light conditions generally change based on depth of snow cover. In the early melt season seasonal snow is removed by melting to expose the glacial ice surface to sunlight at lower elevations only. The snow line then retreats until reaching equilibrium line altitudes at the end of melt season. This leaves the entire surface of the ablation zone snow-free, making it an available habitat for photosynthetic microorganisms. As a result, the biodiversity of glacier microbial communities changes with elevation [44,46–48].

In early February 2013, the entire glacier surface was covered by snow except for a steep slope at the glacier terminus, where ST1 is located. The snow cover near ST2 was 0.85 m deep, but near ST1, zero or a few centimeters of snow cover were observed. Because snow blocks radiation, these differences in snow depth cause variable light conditions and GMGA internal temperatures. Although the precise factors causing this difference in snow depth are unknown, slope angle and wind erosion likely to cause gradients in snow cover. We observed similar types of steep slopes on all edges of this glacier and other glaciers located beyond the ridge (Margarita Glacier: Fig. 2a); however, further observations and measurements must be necessary.

Direct ecological linkage between glacier and glacier foreland

In the glacier foreland immediately adjacent to the glacier terminus, we found an abundance of dried GMGAs on rock surfaces, which were likely left on the freshly bared subglacial rocks after glacier retreat. The temperature of these dried GMGAs on the rocks reaches approximately 20°C and conditions appear much drier than on the glacier, where water is supplied by melting (Fig. 5). Dried GMGAs create a soil-like structure on the abiotic rock surface with gametophyte of different dominant Bryophyta (*Bryum* sp.). The succession of dominant bryophyte species from *C. purpureus* gemmae to *Bryum* sp. shows that GMGAs had changed after leaving the glacier. Previous studies conceptually proposed the linkage between glacier and glacier foreland by nutrient connection [49] and outwash of cryoconite granules [2]. Otherwise, our findings indicate that GMGAs accumulate as a soil-like structure on the abiotic rock surface, directly linking the glacier and glacier foreland ecosystems.

Furthermore, recently another linkage between glacier and glacier foreland was found from subglacial environment [50,51]. In Canadian Arctic, varaeties of mosses regenerate from old populations, which had been entombed in subglacial environment from Little Ice Age (LIA) and recently released onto ice –free glacier foreland due to glacial retreat [50]. These evidences show releases of developed biological material from both supraglacier and subglacier supply the stable ecological substructure to glacier foreland ecology.

If the glacier disappears due to climate change and/or albedo reduction, this unique glacial ecosystem and its contribution to the glacier foreland will also disappear. Many other tropical glaciers that are expected to disappear in the near feature [52,53], which may also contain unique biota that are under threat. In this respect, the tropical glacial ecosystem is an urgent subject of study to understand the biodiversity.

Acknowledgments

The authors thank A. Wada of the Greenleaf Tourist Club for management of local transportation in Uganda. We also thank the guides and porters of the Rwenzori Mountaineering Service for guidance and transport of research equipment on the mountain, and K. Watanabe and M. Mori for assistance in laboratory experiments. I am also indebted to three reviewers (Dr. Peter Convey, Dr. Catherine La Farge and Dr. Nicoletta Cannone) for valuable suggestions, which greatly improved this article. This study was partially supported by Ministry of Education, Culture, Sports, Science and Technology Grant-in-Aid for Scientific Research (A) No. 22241005, and by Proposal for Seeds of Transdisciplinary Research from the Transdisciplinary Research Integration Center, and by an NIPR publication subsidy. Field research was supported by the Uganda Wildlife Authority, the Uganda National Council for Science and Technology, and Dr. S. Anguma of Mbarara University of Science and Technology.

Author Contributions

Conceived and designed the experiments: JU YT. Performed the experiments: JU ST KH YT DS. Analyzed the data: JU YT. Contributed reagents/materials/analysis tools: JU YT HM SI SK. Wrote the paper: JU YT DS SK.

References

1. Anesio AM, Hodson AJ, Fritz A, Psenner R, Sattler B (2009) High microbial activity on glaciers: importance to the global carbon cycle. Glob Chang Biol 15: 955–960.
2. Wharton RA, McKay CP, Simmons GM, Parker BC (1985) Cryoconite holes on glaciers. Bioscience 35: 499–503.
3. Takeuchi N, Nishiyama H, Li Z (2010) Structure and formation process of cryoconite granules on Ürümqi glacier No. 1, Tien Shan, China. Ann. Glaciol 51: 9–14.
4. Edwards A, Anesio AM, Rassner SM, Sattler B, Hubbard B, et al. (2011) Possible interactions between bacterial diversity, microbial activity and supraglacial hydrology of cryoconite holes in Svalbard. ISME J 5: 150–60.
5. Cameron KA, Hodson AJ, Osborn AM (2011) Structure and diversity of bacterial, eukaryotic and archaeal communities in glacial cryoconite holes from the Arctic and the Antarctic. FEMS Microbiol Ecol 82: 254–267.
6. Hamilton TL, Peters JW, Skidmore ML, Boyd ES (2013) Molecular evidence for an active endogenous microbiome beneath glacial ice. ISME J 7: 1402–1412.
7. Telling J, Anesio AM, Tranter M, Irvine-Fynn T, Hodson A, et al. (2011) Nitrogen fixation on Arctic glaciers, Svalbard. J Geophys Res 116: 2–9.
8. Eythórsson J (1951) Correspondence. Jökla-mýs. J Glaciol **1** (9): 503.
9. Benninghoff WS (1955) Correspondence. Jokla mys. J Glaciol 2: 514–515.
10. Heusser CJ (1972) Polsters of the moss Drepanocladus berggrenii on Gilkey Glacier, Alaska. Bull. Torrey Bot. Club 99: 34–36.
11. Porter PR, Evans AJ, Hodson AJ, Lowe AT, Crabtree MD (2008) Sediment-moss interactions on a temperate glacier: Falljokull, Iceland. Ann Glaciol 48 (1): 25–31.
12. Coulson SJ, Midgley NG (2012) The role of glacier mice in the invertebrate colonisation of glacial surfaces: the moss balls of the Falljökull, Iceland. Polar Biol 35: 1651–1658.
13. Hope GS, Peterson JA, Radok U, Allison I (1976) The Equatorial Glaciers of New Guinea: Results of the 1971–1973 Australian Universities' Expeditions to Irian Jaya: Survey, Glaciology, Meteorology, Biology and Palaeoenvironments, A.A. Balkema, Rotterdam, 81–92.
14. Taylor RG, Mileham L, Tindimugaya C, Majugu A, Muwanga A, et al. (2006). Recent glacial recession in the Rwenzori Mountains of East Africa due to rising air temperature. Geophys Res Lett 33: 2–5.
15. Mölg T, Rott H, Kaser G, Fischer A, Cullen NJ (2006) Comment on "Recent glacial recession in the Rwenzori Mountains of East Africa due to rising air temperature" by Richard G. Taylor, Lucinda Mileham, Callist Tindimugaya,

Abushen Majugu, Andrew Muwanga, and Bob Nakileza. Geophys Res Lett 33: 33–36.
16. Kaser G, Osmaston H (2002) Tropical glaciers. Cambridge University Press, Cambridge, UK. 63–116.
17. Quast C, Pruesse E, Yilmaz P, Gerken J, Schweer T, et al. (2013) The SILVA ribosomal RNA gene database project: improved data processing and web-based tools. Nucl Acids Res 41 (D1): D590–D596.
18. Schloss PD, Westcott SL, Ryabin T, Hall JR, Hartmann M, et al. (2009) Introducing mothur: open-source, platform-independent, community-supported software for describing and comparing microbial communities. Appl Environ Microbiol 75: 7537–7541.
19. Edgar RC, Haas BJ, Clemente JC, Quince C, Knight R (2011) UCHIME improves sensitivity and speed of chimera detection, Bioinformatics 27 (16): 2194–2200. Available at: doi:0.1093/bioinformatics/btr381.
20. Andersen R (2005) Algal Culturing Techniques. Elsevier, Amsterdam, Netherlands. 437.
21. Remias D, Schwaiger S, Aigner S, Leya T, Stuppner H, et al. (2012) Characterization of an UV- and VIS-absorbing, purpurogallin-derived secondary pigment new to algae and highly abundant in *Mesotaenium berggrenii* (Zygnematophyceae, Chlorophyta), an extremophyte living on glaciers. FEMS Microbiol Ecol 79: 638–648.
22. Taberlet P, Gielly L, Pautou G, Bouvet J (1991) Universal primers for amplification of three non-coding regions of chloroplast DNA. Plant Mol Bio 17: 1105–1109.
23. Nadot S, Bittar G, Carter L (1995) A Phylogenetic Analysis of Monocotyledons Based on the Chloroplast Gene rpsi4, Using Parsimony and a New Numerical Phenetics Method. Mol Phylogenet Evol 4: 257–282.
24. Souza-Chies T, Bittar G, Nadot S, Carter L, Besin E, et al. (1997) Phylogenetic analysis ofIridaceae with parsimony and distance methods using the plastid generps4. Plant Syst Evol 204: 109–123.
25. Chiang T, Schaal BA, Peng C (1998) Universal primers for amplification and sequencing a noncoding spacer between the *atpB* and *rbcL* genes of chloroplast DNA. Bot. Bull. Acad. Sin. 39: 245–250.
26. Shaw AJ, Cox CJ, Boles SB (2003) Polarity of peatmoss (*Sphagnum*) evolution: who says bryophytes have no roots? Am J Bot 90: 1777–1787.
27. Tanabe Y, Shitara T, Kashino Y, Hara Y, Kudoh S (2011) Utilizing the effective xanthophyll cycle for blooming of *Ochromonas smithii* and *O. itoi* (Chrysophy-

ceae) on the snow surface. PLoS One 6 (2), e14690. doi:10.1371/journal.pone.0014690.

28. McMinn A, Hegseth EN (2004) Quantum yield and photosynthetic parameters of marine microalgae from the southern Arctic Ocean, Svalbard. J Mar Biol Ass UK 84: 865–871.

29. Imura S, Kanda H (1986) The gemmae of the mosses collected from the Syowa Station area, Antarctica. Mem. Natl. Inst. Polar Res. 44: 241–246.

30. Takeuchi N, Matsuda Y, Sakai A, Fujita K (2005) A large amount of biogenic surface dust (cryoconite) on a glacier in the Qilian Mountains, China. Bull Glaciol Res 22: 1–8.

31. McDaniel SF, Shaw J (2005) Selective sweeps and intercontinental migration in the cosmopolitan moss *Ceratodon purpureus* (Hedw.) Brid. Mol. Ecol. 14: 1121–32.

32. Shaw J, Jules ES, Beer SC (1991) Effects of metals on growth, morphology, and reproduction of *Ceratodon purpureus*. Bryologist 94(3): 270–277.

33. Kanda H (1979) Regenerative development in culture of Antarctic plants of *Ceratodon purpureus* (HEDW.) BRID. Mem Natl Inst Polar Res Special issue 11: 58–69.

34. Imura S, Kanda H (1986) The gemmae of the mosses collected from the Syowa Station area, Antarctica. Mem Natl Inst Polar Res 24: 241–246.

35. Lewis Smith RI (1999) Biological and environmental characteristics of three cosmopolitan mosses dominant in continental Antarctica. J Veg Sci 10: 231–242.

36. Hauman L (1942) Les Bryophytes des hautes altitudes au Ruwenzori. Bulletin du Jardin botanique de l'État a Bruxelles 16: 311–353. (in French).

37. Oechel WC, Sveinbjörnsson B (1978) Primary Production Processes in Arctic Bryophytes at Barrow, Alaska. Vegetation and Production Ecology of an Alaskan Arctic Tundra, Ecological Studies Volume 29, Springer-Verlag, New Tork, USA. 269–298.

38. Hicklenton PR, Oechel WC (1976) Physiological aspects of the ecology of Dicranum fuscescens in the subarctic. I: Acclimation and acclimation potential of CO_2 exchange in relation to habitat, light, and temperature, Canadian Journal of Botany 54(10): 1104–1119.

39. Longton RE (1988) Biology of polar bryophytes and lichens. Cambridge University Press, Cambridge, UK. 141–210.

40. Sonoike K (1998) Various Aspects of Inhibition of Photosynthesis under Light/Chilling Stress: "Photoinhibition at Chilling Temperatures" versus "Chilling Damage in the Light". J Plant Res 111: 121–129.

41. Lentini G, Cristofanelli P, Duchi R, Marinoni A, Verza G, et al. (2011) Mount Rwenzori (4750 m a.s.l., Uganda): meteorological characterization and Air-Mass transport analysis. Geografia Fisica e Dinamica Quaternaria 34 (3): 183–193.

42. Higuchi S, Kawamura M, Miyajima I, Akiyama H, Kosuge K, et al. (2003) Morphology and phylogenetic position of a mat-forming green plant from acidic rivers in Japan. J Plant Res 116: 461–467.

43. Satake K, Nishikawa M, Shibata K (1990) A copper-rich protonemaI colony of the moss *Scopelophila cataractae*. J Bryol 1928: 109–116.

44. Takeuchi N, Uetake J, Fujita K, Aizen VB, Nikitin SD (2006) A snow algal community on Akkem glacier in the Russian Altai mountains. Ann Glaciol 43: 378–384.

45. Hodson A, Anesio AM, Tranter M, Fountain A, Osborn M, et al. (2008) Glacial Ecosystems Ecol Monogr 78: 41–67.

46. Takeuchi N (2001) The altitudinal distribution of snow algae on an Alaska glacier (Gulkana Glacier in the Alaska Range). Hydrol Process 15: 3447–3459.

47. Segawa T, Takeuchi N, Ushida K, Kanda H, Kohshima S (2010) Altitudinal Changes in a Bacterial Community on Gulkana Glacier in Alaska. Microbes Environ 25: 171–182.

48. Uetake J, Yoshimura Y, Nagatsuka N, Kanda H (2012) Isolation of oligotrophic yeasts from supraglacial environments of different altitude on the Gulkana Glacier (Alaska). FEMS Microbiol Ecol 82: 279–286.

49. Stibal M, Tranter M, Telling J, Benning LG (2008) Speciation, phase association and potential bioavailability of phosphorus on a Svalbard glacier. Biogeochemistry 90: 1–13.

50. La Farge C, Williams KH, England JH (2013) Regeneration of Little Ice Age bryophytes emerging from a polar glacier with implications of totipotency in extreme environments. Proc Natl Acad Sci USA 110: 9839–9844.

51. Thompson LG, Mosley-Thompson E, Davis ME, Zagorodnov VS, Howat IM, et al. (2013) Annually resolved ice core records of tropical climate variability over the past ~1800 years. Science 340: 945–50.

52. Rabatel A, Francou B, Soruco A, Gomez J, Cáceres B, et al. (2013) Current state of glaciers in the tropical Andes: a multi-century perspective on glacier evolution and climate change. Cryosph 7: 81–102.

53. Thompson LG, Brecher HH, Mosley-Thompson E, Hardy DR, Mark BG (2009) Glacier loss on Kilimanjaro continues unabated. Proc Natl Acad Sci USA 106: 19743–19744.

9

Spatial Diversity of Bacterioplankton Communities in Surface Water of Northern South China Sea

Jialin Li[1,9], **Nan Li**[1,9], **Fuchao Li**[2], **Tao Zou**[3], **Shuxian Yu**[1,4], **Yinchu Wang**[1,4], **Song Qin**[1]*, **Guangyi Wang**[5,6]*

1 Key Laboratory of Coastal Biology and Bioresource Utilization, Yantai Institute of Coastal Zone Research, Chinese Academy of Sciences, Yantai, China, 2 Key Laboratory of Experimental Marine Biology, Institute of Oceanology, Chinese Academy of Sciences, Qingdao, China, 3 Key Laboratory of Coastal Environmental Processes and Ecological Remediation, Yantai Institute of Coastal Zone Research, Chinese Academy of Sciences, Yantai, China, 4 Graduate University of Chinese Academy of Sciences, Beijing, China, 5 Tianjin University Center for Marine Environmental Ecology, School of Environmental Sciences and Engineering, Tianjin University, Tianjin, China, 6 Department of Microbiology, University of Hawaii at Manoa, Honolulu, Hawaii, United States of America

Abstract

The South China Sea is one of the largest marginal seas, with relatively frequent passage of eddies and featuring distinct spatial variation in the western tropical Pacific Ocean. Here, we report a phylogenetic study of bacterial community structures in surface seawater of the northern South China Sea (nSCS). Samples collected from 31 sites across large environmental gradients were used to construct clone libraries and yielded 2,443 sequences grouped into 170 OTUs. Phylogenetic analysis revealed 23 bacterial classes with major components *α*-, *β*- and *γ*-Proteobacteria, as well as *Cyanobacteria*. At class and genus taxon levels, community structure of coastal waters was distinctively different from that of deep-sea waters and displayed a higher diversity index. Redundancy analyses revealed that bacterial community structures displayed a significant correlation with the water depth of individual sampling sites. Members of *α*-Proteobacteria were the principal component contributing to the differences of the clone libraries. Furthermore, the bacterial communities exhibited heterogeneity within zones of upwelling and anticyclonic eddies. Our results suggested that surface bacterial communities in nSCS had two-level patterns of spatial distribution structured by ecological types (coastal VS. oceanic zones) and mesoscale physical processes, and also provided evidence for bacterial phylogenetic phyla shaped by ecological preferences.

Editor: Chih-hao Hsieh, National Taiwan University, Taiwan

Funding: This research work was financially supported by grants from the National Natural Science Foundation of China 41106100 and the CAS/SAFEA International Partnership Program for Creative Research Teams "Representative environmental processes and resources effects in coastal zone". The funders had no role in study design, data collection and analysis, decision to publish, or preparation of the manuscript.

Competing Interests: The authors have declared that no competing interests exist.

* Email: sqin@yic.ac.cn (SQ); gywang@tju.edu.cn (GW)

9 These authors contributed equally to this work.

Introduction

The oceans harbor more than 3×10^{28} bacteria, which are organized within an estimated 10^6 to 10^9 taxa [1,2]. These bacteria play vital roles in cycling nutrients and mediating climate on a global scale. However, bacterial communities in the oceans are structured by a variety of environmental factors, including currents, input of nutrients and pollutants, rising atmospheric carbon dioxide, and climate change [3,4]. Although many studies have focused on the spatial and temporal diversity of marine bacteria, their responses to environmental perturbation remain undiscovered in many oceanic regions. The International Census of Marine Microbes (ICOMM) and the Global Ocean Sampling program have made tremendous efforts to obtain both basic and global information on marine bacterial diversity. However, information gaps remain to be filled in many ocean regions, particularlyin continental shelf regions such as coastal and marginal seas [5]. Clearly, in order to understand how environmental variables alter the community structure of microbial flora, it is necessary to proceed with broader research on bacterial

communities and their distribution in relation to environmental conditions [6].

Marginal seas, major areas of biogeochemical cycling, are biologically more active regions compared with open oceans of the same latitude [7]. Meanwhile, these areas are characterized by their close relationships with adjacent terrestrial anthropogenic contaminations. The northern South China Sea (nSCS) is one part of the largest marginal sea located in the subtropical and tropical western North Pacific Ocean. It includes deep basins, with depths of over 5000 m, and the continental shelf; less than 100 mdeep. In summer, multi-scale physical processes mainly driven by the monsoon winds feature the nSCS with complex circulation as upwelling, costal currents, and cyclonic eddies [8]. The main body of nSCS water is oligotrophic, which is characterized with low nutrient concentrations, low phytoplankton biomass and low primary production [9]. Nutrient-rich fluvial input from the Pearl River discharges into the estuarine and adjacent waters, forming the sharp physical and chemical gradients over a small spatial scale [10]. Complex geographic and chemical marine systems make the nSCS sharing abundant biological diversity [11]. Thus, it provides an ideal area to investigate bacterial phylogenetic lineages shaped by environmental gradients from coast to open ocean, and from eutrophic and oligotrophic ecosystems [12,13]. To date, most microbiological studies in the nSCS have focused on microbial resources and their applications [11,14], and some studies on distribution of certain functional microbiota. Of microbial ecological studies in the nSCS [13,15,16], only a few reports investigated the relationship between bacterial abundance with water masses and nutrient status [17,18]. Particularly, reports on the molecular characterization of bacterial communities in the surface water of the nSCS are rare and no studies have been carried out to investigate the effect of regulation of hydrologic variables on bacterial populations in this region.

This work aimed to describe spatial distribution patterns of bacterial communities in the nSCS surface water through analysis of samples collected from 31 sites. The sampling sites covered major environmental features, including the Pearl River estuarine, coastal, offshore, deep-sea, upwelling, and prospective eddies areas. We aimed to answer the following questions related to bacterial communities in the nSCS. What is the spatial diversity in the region? What are the major environmental factors shaping the community structure? It represents the first report of surface water bacterial communities in the nSCS.

Materials and Methods

Description of Study Area

The nSCS is located south of the Tropic of Cancer and is heavily influenced by the East Asian monsoon system. It is connected with the East China Sea in the northeast through the Taiwan Strait, with the Pacific Ocean and the Sulu Sea in the east through Bashi Channel. The topography is characterized by a wide continental shelf and deep basins with maximum depth of 5,000 m at the center, and isobaths is parallel to the continental coastline. With large amounts of nutrient input from the Pearl River, and with fresh waters predominantly flowing along the coast via the coastal currents system, the nSCS features a gradient of P limitation in the estuary to N limitation in oceanic ocean [20]. Upwelling and eddies are common mesoscale phenomena mainly due to the southwest monsoon in summer. There are two strong upwelling regions in the inshore areas, Yuedong Upwelling from Shantou coast to the Nanri Islands (S30 site) and Qiongdong Upwelling in the east of Hainan Island (S52 site). The anomalous anticyclonic circulation is found along the 18°N latitude (S61 to S69 sites). Sampling sites represent most of the typical environments of the nSCS (Figure 1) and were classified into coastal and oceanic groups based on the 200 m water-depth contour.

Sample Collection and Environment Characteristics

Seawater samples were collected from a water depth of 4.0–4.8 m using a rosette of Niskin bottles attached to a CTD probe frame during an Open Cruise of R/V *Shiyan 3* in August of 2007 (Figure 1). No specific permissions were required for these locations and activities. No endangered or protected species were involved in the field work of this study. The specific location (i.e., GPS coordinates) of sampling sites is listed in Table S1. For bacterial analyses, 20 L of surface seawater were aseptically filtered through Millipore 0.22-µm Millipore filter. The resulting filtrate was sealed in airtight sterile plastic tubes and stored at −80°C until use.

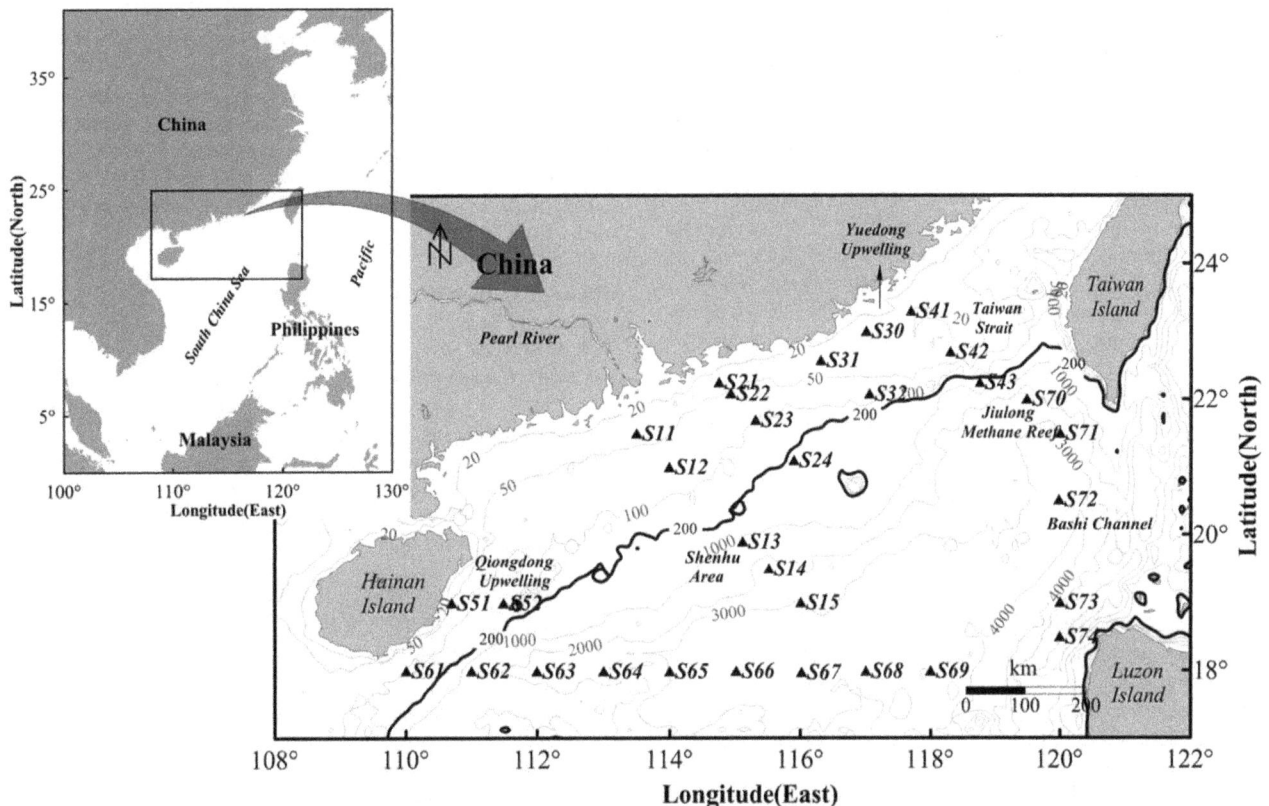

Figure 1. Location of the sampling sites in the nSCS (Isobath contour ploted by data from GEBCO website: www.gebco.net).

Temperature, salinity and depth were recorded by a Neil Brown MKIII CTD. Nutrient analyses were done in the South China Sea Institute of Oceanology, Chinese Academy of Sciences (Figure S1).

Sea level height anomaly data over the same sample-period was derived from the AVISO (Archiving, Validation, and Interpretation of Satellite Oceanographic data) website. A merged and gridded satellite product was generated based on TOPEX/Poseidon, Jason 1, ERS-1 and ERS-2 data [21]. The velocity field derived from SLA assuming geostrophic balance:

$$u = -\frac{g}{f}\frac{\partial h}{\partial x}, v = -\frac{g}{f}\frac{\partial h}{\partial y}$$

Where h is the SLA, g is gravitational acceleration, and f is the Coriolis parameter. Computational data were processed using MATLAB.

DNA Extraction and Clone Library Construction

Total genomic DNA was extracted from the membrane filters of individual sites using standard phenol-chloroform extraction procedure described previously for filtrate material [22]. DNA was treated with RNase and subjected to two rounds of ethanol precipitation.

Fragments of 16S rRNA gene were amplified in a Tprofessional standard thermal cycler (Biometra) using bacterial universal primers 27F/1390R, under previously described PCR conditions [23]. PCR products were gel-purified, cloned into pGEM-T easy vectors (Promega), and then transformed into *Escherichia coli* TOP10 competent cells. Approximately, 120 colonies were randomly selected for sequence analysis. Plasmids carrying insert of correct size were sequenced using the SP-6 and T7 primers on an ABI model 3730 sequencer at Chinese National Human Genome Center (Shanghai, China).

The resulting sequences were aligned using Muscle v3.8 [24], then imported into Mothur v1.29 to remove chimera prior to further diversity analysis [25]. Sequences were classified using the mother Bayesian classifier (80% confidence) with the mothur-formatted version of the Ribosomal Database Project (RDP) training set (v. 9). The stand-alone BLAST v2.2.28 was used for local alignment of sequence similarity search with 'env_nt' databases in NCBI GenBank. The 16S rRNA gene sequences from each library with a percentage sequence identity of ≥97% were placed in the same Operational Taxonomic Unit (OTU). One representative sequence for each OTU was chosen to build a more concise phylogenetic tree using Mothur v1.29. The maximum likelihood tree was implemented in program PhyML v3.0 [26], on the basis of the best-fit substitution model as determined by jModelTest v2.1 [27]. The 16S rRNA gene sequences were deposited in GenBank database under the accession numbers of KC872051–872789, KC872791–873358, KC873360–873759, and KC873761–874493.

Diversity Comparison and Statistical Analyses

Diversity within each bacterial community (α-diversity) was assessed by plotting a rarefaction curve and calculating diversity indices, including Chao (S_{Chao}) and the inverse Simpson index ($1/D$) using Mothur v1.29. In order to illustrate the scope of bacterial diversity, Good's coverage (C) was calculated as [1-(n/N)] where n is the number of OTUs that had been observed once and N is the total number of OTUs in the sample.

Community comparison of bacterial assemblages (β-diversity) was performed with Fast UniFrac environmental clustering and

principal coordinate analyses (PCoA) [28]. Diversity comparison matrix was generated into a heatmap based on the weighted UniFrac distance. Correlations between bacterial populations and environmental variables were determined by redundancy analysis (RDA) at class level by downweighting rare taxa in software Canoco v4.5 [29]. RDA was performed with the linear method because DCA (detrended correspondence analysis) on species variables revealed that the length of the first axis gradient was short (<2). Detrending was carried out in segments using the non-linear rescaling method. Prior to DCA and RDA, species values underwent square root transformation and environmental variables were normalized by z-score. The significance of the canonical axes was assessed using the permutation test with 499 unrestricted Monte Carlo permutations (P<0.05).

Results

Bacterial Diversity

To assess bacterial diversity, 2443 clones were selected from 31 bacterial libraries derived from surface water samples in the nSCS. Similarity of those sequences ranged from 65.7 to 100%. Non-redundant analyses identified 1,980 unique sequences, which were assigned into 310 OTUs. Of these sequences, 33% had less than 97% similarity with known sequences, which indicated that they were potential novel species. Moreover, three sequences (S51_38: KC823798, S51_47: KC873301, and S51_71: KC873325) had less than 95% identity with their best-matched reference sequences. About 67.3% of these sequences had their closest matches originally recovered from surface seawater collected along voyage from Eastern North American coast to the Eastern Pacific Ocean during the *Sorcerer II* Global Ocean Sampling Expedition [30].

The coverage of clone libraries ranged from 67.7 to 92.2% (Figure S2), suggesting that the selected sequences can reasonably represent bacterial communities of individual samples (Table 1). Of 31 sampling sites (Figure 1), sites S72 and S73 were observed with significantly higher bacterial diversity, S21 displaying the highest diversity found at site S21. Relatively low bacterial diversity was found at sites S32, S43 and S66 ($1/D$ <2). Meanwhile, spatial variation of bacterial diversity was observed in the study area.

Phylogeny of Bacterial Community

The vast majority (93.1%) of acquired sequences were affiliated with 4 bacterial phyla, *i.e.*, *Proteobacteria* (α-36.6%; β-16.6%; and γ-13.3%), *Cyanobacteria* (13.9%), *Bacilli* (6.9%), and *Actinobacteria* (5.8%). Overall, the identified sequences belonged to 23 bacterial classes, suggesting a great variety of bacterial communities in the study area (Figure 2). The phyla of *Proteobacteria*, *Acidobacteria*, *Deferribacteres*, *Planctomycetes*, and *Verrucomicobia* were shared with sediment samples from nSCS [31,32]. *Alphaproteobacteria* and *Actinobacteria*, were observed in all sampling sites, and interestingly, members of α-*Proteobacteria* constituted more than 70% of 4 clone libraries derived from 4 sites (S64, S66, S71 and S74). *Bacilli* were the most dominant class at sites S43 (81.0%) and S32 (73.3%), γ-*Proteobacteria* at sites S12 (53.4%) and S13 (62.0%), and β-*Proteobacteria* at site S22 (50.6%). Rare bacterial classes, whose sequences were observed only once from a library, were identified as members of *Acidobacteria_Gp4* (S15), *Aquificae* (S69), *Bacteroidia* (S31), *Nitrospira* (S12), *Anaerolineae* (S15), and *Opitutae* (S69).

Classification analysis (80% confidence threshold) revealed that 170 OTUs belonged to members of *Proteobacteria* (Figure 3 and Figure S3). Those OTUs were distinctively clustered with α-, β-,

Table 1. Biodiversity indices of bacterial communities obtained from the sampling sites of the nSCS.

Sites	Sequences	C (%)	OTUs	S_{Chao}	$1/D$
S11	67	88.06	17	22.6	5.91
S12	92	86.96	20	36.5	3.22
S13	92	88.04	19	37.3	3.76
S14	86	79.07	30	60.6	12.14
S15	71	76.06	26	53.2	7.22
S21	73	84.93	28	34.1	18.25
S22	89	87.64	28	33.0	11.32
S23	80	88.75	21	28.2	10.29
S24	88	92.05	13	23.5	4.64
S30	68	67.65	30	76.2	7.40
S31	73	80.82	24	46.8	9.25
S32	77	92.21	10	25.0	1.69
S41	83	86.75	23	32.2	8.49
S42	84	85.71	20	31.0	3.18
S43	83	91.57	10	31.0	1.41
S51	86	69.77	36	90.2	11.57
S52	66	78.79	22	67.5	9.84
S61	87	71.26	34	94.0	9.74
S62	74	74.32	29	57.5	9.19
S63	78	74.36	29	67.0	7.07
S64	80	88.75	14	23.0	1.83
S65	72	69.44	31	88.8	11.94
S66	82	91.46	11	21.5	1.54
S67	85	78.82	26	56.6	4.88
S68	70	74.29	25	76.0	9.66
S69	75	81.33	24	37.0	7.71
S70	62	79.03	19	58.0	7.95
S71	76	86.84	15	37.5	2.43
S72	79	69.62	30	168.0	9.22
S73	77	74.03	25	120.0	7.26
S74	88	90.91	15	24.3	2.27

γ- and δ- *Proteobacteria*. Of these OTUs, 83 fell into the class α-*Proteobacteria* and clustered with *Caulobacterales*, *Rhizobiales*, *Rickettsiales*, *Rhodospirillales*, *Rhodobacterales*, and *Sphingomonadales*. The OTUs, which were clustered with unclassified α-*Proteobacteria*, had close affiliations with sequences derived from Chesapeake Bay, coastal Delaware Bay and open sea Panama regions in the Pacific [33,34]. Furthermore, 21 OTUs, which were members of *Burkholderiales* in the phylume β-*Proteobacteria*, were mostly affiliated with accelerating utilization of organic nitrogen. Members of γ-*Proteobacteria* contributed to major components of 12 phylogenetic clades in the nSCS bacterial libraries. Their abundance and rich diversity supports their important ecological functions, including anaerobic sulfur and ammonia oxidation [35]. An abundance of *Enterobacteriaceae*, which were best-matched with sequences isolated from the human gut, suggests that anthropogenic influence brought non-marine origins into nSCS microbiota. Notably, the genus *Alteromonas* sp., whose presence substantially promots growth of toxic dinoflagellate *Alexandrium fundyense* [36], was present in extraordinarily high concentrations at sites S14 and S61. Those two sites were located at the regions

where algal blooms frequently occurred in summer [37,38]. The remaining minority OTUs were members of the orders *Bdellovibrionales* and *Desulfobacterales* in the class of δ-*Proteobacteria*, which were reported to play a fundamental role in sulfur and metal element biogeochemical cycling [39].

Thirty-one OTUs were affiliated with the phylum *Cyanobacteria* (Figure 3). They were detected in almost all libraries except the library of S13 and best matched with sequences from a wide range of aquatic samples in Chesapeake Bay, Coco's Island, Antarctica Lake Vida, Sargasso Sea and Panama Canal. GpIIa (*Synechococcu*) was the largest genus that clustered with culture-independent representative clones. GpXI (mostly members of *Microcystis* strains) and *Bacillariophyta* were minor components with located-specific distribution at Pearl River Estuary and continental area, respectively.

Forty OTUs belonged to *Bacteroidetes*. Those phylotypes fell into three classes (*Bacteroidia*, *Sphingobacteria* and *Flavobacteria*) and one unclassified group. The phylum *Bacteroidetes* has been implicated as a major utilizer of complex polymers and is abundant in marine ecosystem, especially in eutrophic waters

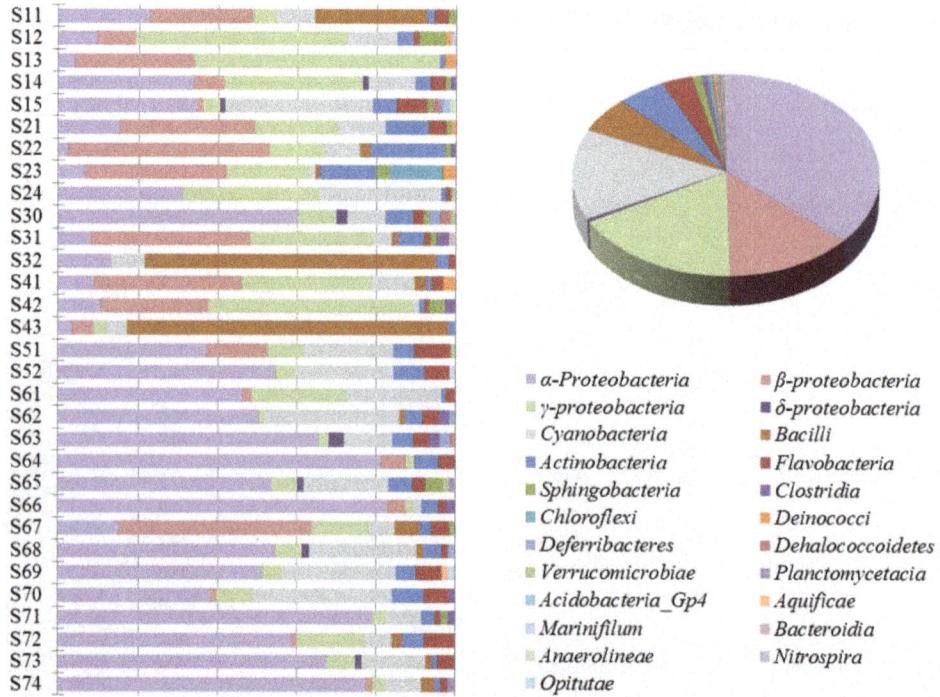

Figure 2. Phylotype distribution and comparison of the clone libraries from individual sampling sites (histogram) and entire nSCS (pie).

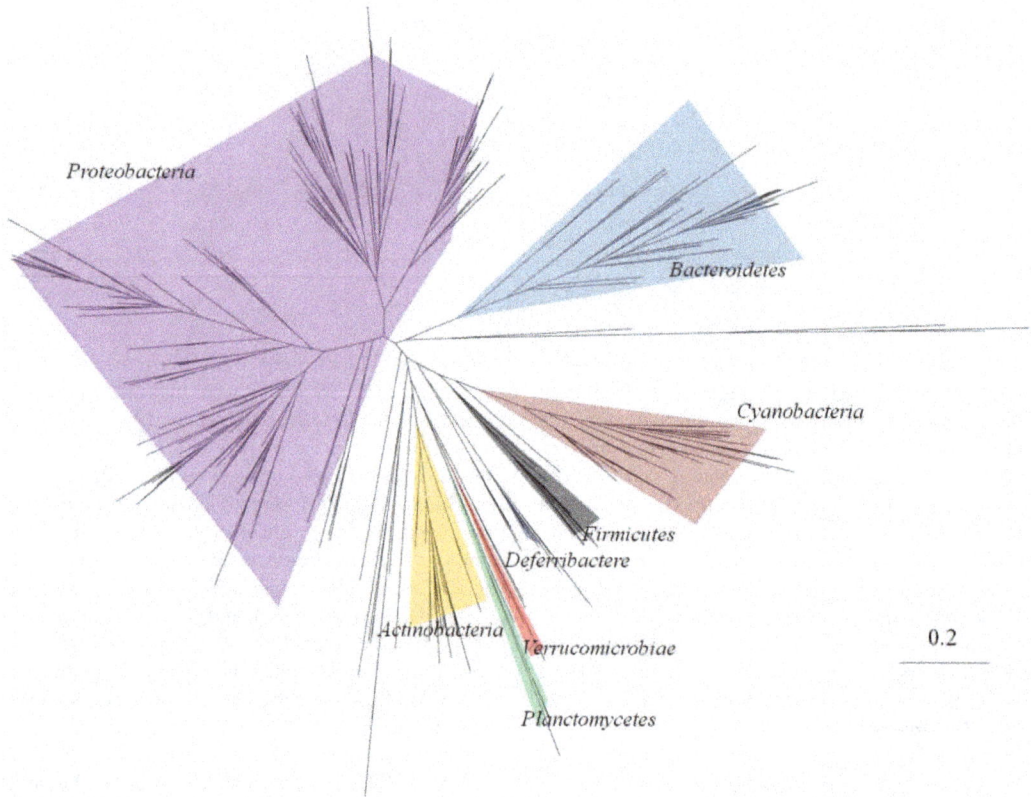

Figure 3. Maximum likelihood phylogenetic radual tree generated by 170 OTUs assembled of 2443 sequences derived from 31 bacterial clone libraries. The detailed polar tree was shown in Figure S3.

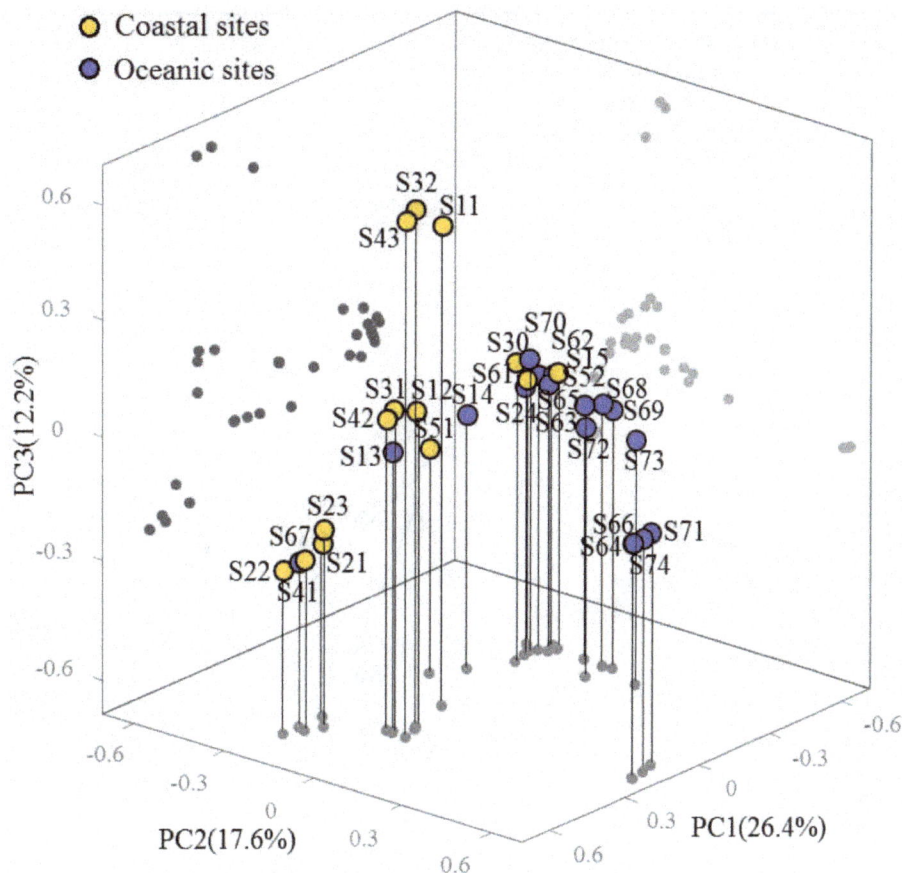

Figure 4. Ordination diagram for PCoA of the bacterial communities using weighted UniFrac with 16S rRNA sequences. Shown is the first three principal coordinate vectors (PC1, PC2 and PC3) and the distribution of bacterial libraries in response to these axes.

[40]. It was concentrated at shallow-water site S51, close to Hainan Island (accounting for 10.9% of sequences).

Members of the other 8 phyla were also identified in nSCS. Some clades were site-specific, for example *Chloroflexus* of *Chloroflexi* (at S23) and *Enterococcus* of *Firmicutes* (at sites S11 and S32). Four sequences (S51_38: KC873293, S51_47: KC873301, S51_71: KC873325 and S61_67: KC873469) were distantly related to others and represented the deep-rooted branch (Figure S3), indicating that they are divergent from other bacteria in the library. Eleven sequences (S14_12: KC872313, S24_13: KC872712, S30_16: KC872802, S30_35: KC872817, S30_40: KC872821, S62_02: KC873497; S63_02: KC873571, S63_20: KC873586, S63_27: KC873593, S68_41: KC874004 and S72_48: KC874293) formed a monophyletic clade (Figure S3) and were closely related to other uncultured sequences that derived from seawater of Panama and Northeast subarctic Pacific Ocean [34,41].

Spatial Distribution and Diversity Comparison

The maximum likelihood phylogeny was used to examine phylogenetic comparison between bacterial libraries using a UniFrac based method. The first three principal coordinates (PC) together accounted for 56.2% of the variation. Considering these primary vectors, the bacterial assemblages derived from sites close to continental shelf (water depth <200 m) were generally more similar amongst one another versus those from oceanic areas (water depth >200 m) with the exception of S30, S52 and S67 sites

(Figure 4). Comparison between two individual bacterial communities revealed overall high distances, suggesting an underrated and versatile role of bacteria within various marine environments with a highly niche-specific community structure (Figure S4). Spatial distribution of surface-water bacterial assemblages might be influenced by a variety of hydrological and physio-chemical factors, such as ocean currents, thermohaline background, and eutrophication condition. RDA of bacterial classes was used to reveal their relationship with environmental variables (Figure 5). The sum of all canonical eigenvalues indicated 30.0% of the total variation can be explained by environmental variations. Concerning the bacterial class data, the first two RDA axes explained 25.4% of the total variance in the bacterial composition and accounted for 84.8% of the cumulative variance of the bacteria-environment relationship. Correlations of bacterial classes and environment variables were 66.0% and 53.2% for axis RDA 1 and 2, respectively.

RDA1 represented a depth gradient and had a correlation coefficient of −0.5589. It distinguished the bacterial assemblages derived from sites of continental shelf from those of oceanic sites. RDA2 represented a silicate gradient caused by silicate and had a correlation coefficient of −0.2419. Based on the partial Monte Carlo permutation test (P<0.05), the variable of the depth alone contributed significantly ($P = 0.005$, $F = 5.28$) to the bacteria-environment relationship, providing 50.0% of the total CCA explanatory power. Although no other variables had statistically significant contribution to the relationship, thermohaline background of temperature and salinity provided more RDA

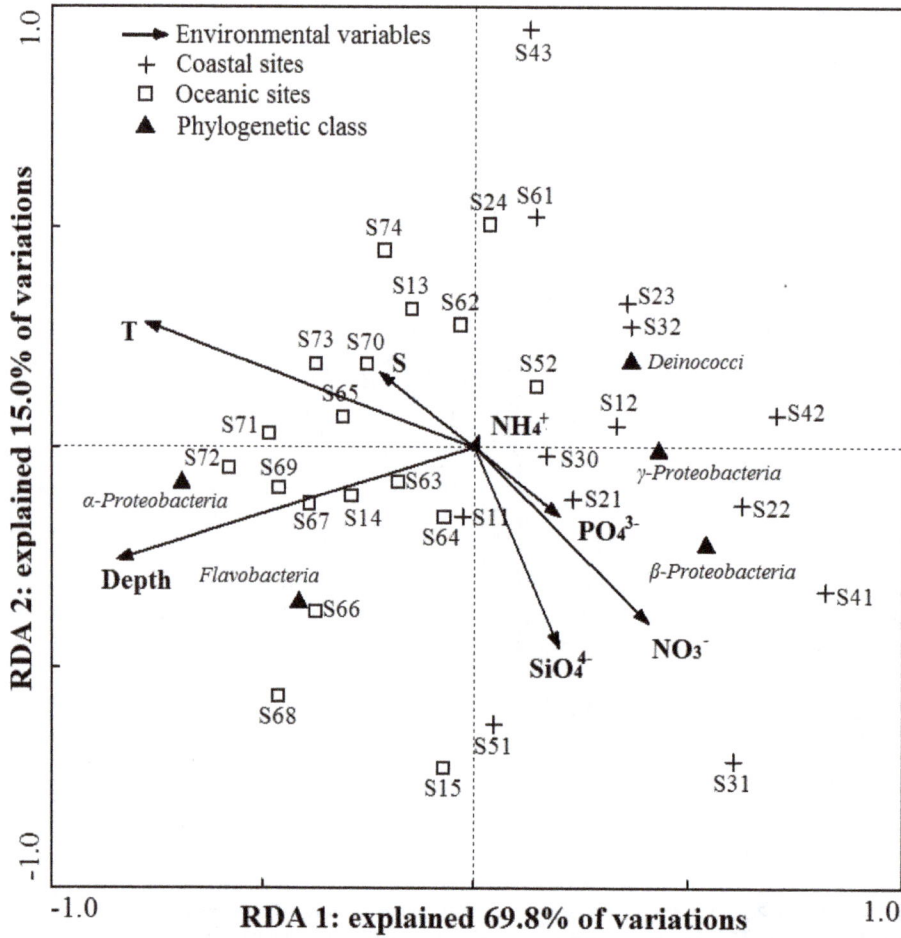

Figure 5. RDA ordination plots for the first two dimensions of the relationship among the sampling sites, environmental variables and bacterial phylogenetic classes. Environmental variables are represented by arrows with a cutoff of $r^2 = 0.2$. Correlations are indicated by the length and angle of arrows.

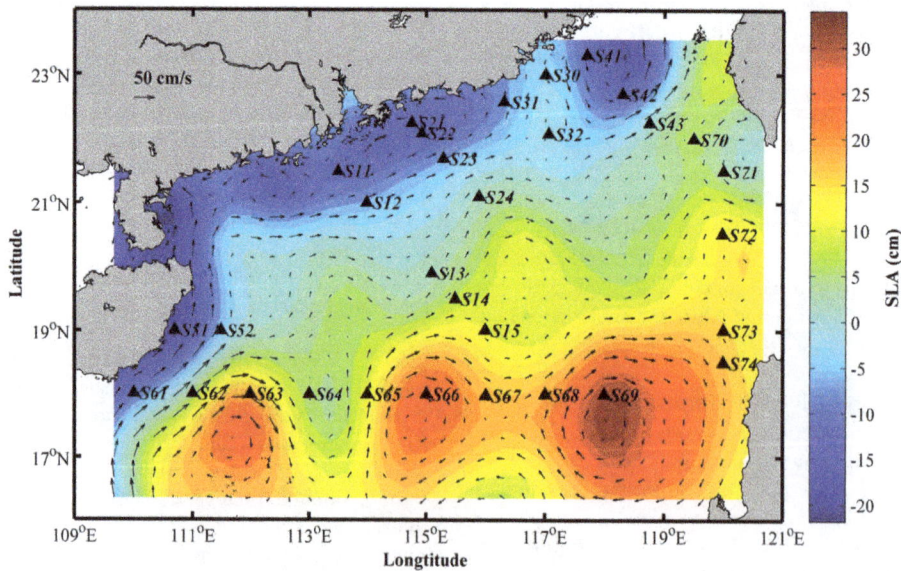

Figure 6. Sea level anomaly derived from the sea level anomaly maps during August 2007 provided by the AVISO.

explanatory power on bacterial composition than nutrition concentration. The correlation of the bacterial classes with environmental variables indicated that α- and β-Proteobacteria were major components that contributed to statistical difference of cluster analyses between sites of coastal and oceanic sites. The distribution of α-Proteobacteria was positively correlated with the depth of sampling sites, while the distribution of β-Proteobacteria was positively correlated with the nitrate concentration and negatively correlated with salinity.

Discussion

Studies on bacterial community composition in marine systems are nowadays routinely done by using culture-independent methods. For bacterial ecologists, it is tempting to correlate bacterial taxonomy and functions to particular environmental features. However, this relationship is far from conclusive because both samples and datasets are scanty relative to the vast bacterial categories and habitat types [42]. Particularly, the ocean is the largest contiguous environment and characterized by strong physical mixing of currents and storms, different nutrient factors, and occurrence of widely distributed microbes [2]. Characteristics of bacterial communities need to be approached and determined in more marine areas beyond the North Pacific [43], Arctic [44], and Mediterranean Sea [45]. Surface samples of nSCS water can be taken from diverse environmental habitats, such as coastal vs. oceanic, oligotrophication vs. eutrophication, and saline vs. freshwater [46]. This study is the first report of the surface seawater bacteria in a large environmental gradient, with variations in the oceanic province. Our results revealed that community structures appear to be spatial heterogeneity of distribution driven by habitat characteristics.

Diversity and Novelty of Bacterial Assemblages

Diverse and novel bacterial species were observed in this study (Figure 2). Bacterial communities showed higher diversity than previous reports in the same region, and shared several phyla with sediment samples from nSCS [30,31,47]. A great part of the 16S bacterial dataset had their closest matched sequences originally detected from the *Sorcerer II* Global Ocean Sampling expedition that has reported the most extensive dataset of microbiota in surface water consisting of 7.7 million sequences [30]. These results indicated that similar habitats may contain a similar genetic diversity of bacterial communities. Compared with the sequences collected from the *Sorcerer II* expedition, the major phylum were *Proteobacteria, Cyanobacteria, Firmicutes, Actinobacteria, Bacteroidetes*, and *Planctomycetes*. The results of this study also indicated that nSCS contained the dominant surface-seawater bacterial groups (α-*Proteobacteria*, γ-*Proteobacteria* and *Cyanobacteria*) commonly found in other regions [48]. Nevertheless, a relatively high proportion of β-*Proteobacteria* was detected in the clone library, which is generally found in small proportions (approximately <3%) in other oceanic surface seawaters [43,44,49].

Some sequences were found to be novel at species level and even at order level. Most of these species were collected from the coastal area of nSCS, especially at the S51site, which is the center of Qiongdong Upwelling. The anthropogenic activities of coastal urbanization, industrialization and economic growth have led to the current pollution through the increasing input of metal contaminants, nutrient substances and organic carbon in the last few decades [38]. Coupled with the upwelling system at southeast of Hainan Island, S51 was characterized by low temperature, high salinity, low dissolved oxygen, high chlorophyll *a* and primary production [50]. Furthermore, different from another Yuedong

Upwelling, seawater of Qiongdong Upwelling is also enriched with silicon [51]. The unique temperature, salinity and silicon concentration were also detected during sampling period (see Figure S1). These physiochemical variables may have contributed to the formation and evolution of new microbial species [52].

Environmental Influence on Bacterial Distribution Pattern

Based on the bacterial communities clustering analysis at species level, the bacterial communities of oceanic sites had more commonalities than those in coastal sites (Figure 4), which was further reflected at phylum level (Figure 5). Bacterial diversity index ($1/D$) revealed a generally inverse relationship to depth of sites along all transects (Table 1), suggesting that bacterial community was more diverse in coastal area. Spatial distances did not generate considerable differences in bacterial community composition, which likely resulted from contiguous environments due to physical mixing of currents and storms [46]. Furthermore, bacterial distribution patterns showed large-scale continuum and beta-diversity heterogeneity through intermediate habitat types across coastal and oceanic ecosystems. This is consistent with what has been reported from the synthesis of global and pole bacterial datasets [49,53].

Large proportion of sequences belonged to the members of α-*Proteobacteria*, supporting the dominance of α-*Proteobacteria* in saltwater [54]. Moreover, a significantly high percentage of α-*Proteobacteria* ($P = 0.000$, $F = 15.572$) in coastal water communities concurred with a previous report on global ocean sampling data [52]. *SAR11*, the most abundant free-living cluster, was also found in this study (Figure S5). The increase of SAR11 relative abundance in oceanic samples corresponded well with local oligotrophic conditions and also supported previous reports [55].

It was unexpected that the depth of sampling sites as the principal factor determines bacterial community structure in the surface water. Previous studies have revealed the existence of bacterial variation between coastal and oceanic seawaters [48]. As depth was unlikely to directly impact surface water, the most proper explanation was that the variation in bacterial populations was due to synergetic driving forces of environmental variables, which are involved in characteristics of coastal and oceanic waters. In other words, the different habitats (reflected in water depth) should account for variability in bacterial community composition. On the other hand, depth of sample sites seemed to have little impact on oceanic bacterial community distribution as it was an only factor, which was correlated with spatial distance [48]. Nutrients were originally expected to play a major role in the microbial composition based on the shift from P to N limitation in nSCS [10]. Nevertheless, 4 parameters of nutrients only explained 7% bacterial variability, which could be explained by the limited impacts of scale and scope on surface water transmitting by coastal currents from terrestrial input [56]. All the environmental variables can only explain 30% of the variability in the community composition. Thus, the composition of bacterial assemblages was additionally influenced by other environmental variables not investigated in the present study, such as residence time, availability of metal elements and bacterial competitors (e.g., protists, viruses and metazoans).

Previous studies have revealed that the physical oceanographic processes influence phytoplankton stocks and production by monsoon-driven circulation and upwelling in the SCS [18,57]. Moreover, bacterial assemblages were proved to be distinct in order to adapt for different oceanographic water masses in eastern Australian sea [38]. In our study, the bacterial community structures were apparently discrete in several sites of similar geochemical conditions. It is likely that the hydrological factors lead to this dissimilarity by stimulating the existence of clusters belonging to adjacent areas or layers through transportation (Figure 4). The abnormal community diversity was observed at the sites S30 and S52 with higher similarity to deep-sea samples, which might be influenced by the upwelling system of Qiongdong and Yuedong, respectively. The future study of bacterial diversity throughout water profile would demonstrate whether this difference was generated by insertion of components from lower layer. Moreover, community structure appeared discrete at the sites S61 to S69. The existence of mesoscale local anticyclonic circulations in summer has been reported [8]. The data from AVISO also proved their existence during sampling period (Figure 6). It is likely that mass transport or extraordinarily hydrology of special current systems could result in the opportunistic taxa and the ecological shift of bacteria in surface seawater.

In conclusion, the composition of bacterial communities exhibited remarkable biogeographic differences between coastal and oceanic ecological systems in surface seawater of nSCS. Similar to other coastal environments, bacterial communities were dominated by members of *Proteobacteria*, *Cyanobacteria*, and *Bacteroidetes*. Moreover, bacterial communities derived from upwelling and mesoscale anticyclonic eddy sites displayed abnormal compositions compared with those of adjacent sites (Figure 4 and Figure S4). Our finding of spatial heterogeneity in marine contiguous environment implied that environmental factors other than dispersal (?) were the drivers of the distribution of bacterial compositions. This study demonstrated that bacterial composition at class level was influenced by the depth of sampling sites. Further investigation to define biomes for underlying patterns of marine bacteria should focus on what common rules of natural selection impact the bacterial communities and how bacteria change the functional biogeochemical cycle.

Supporting Information

Figure S1　Contour maps of environmental variables in nSCS.
(TIF)

Figure S2　Rarefaction curve of 16S rRNA clone libraries derived from nSCS. Phylogenetic diversity is represented by branch length.
(TIF)

Figure S3　Maximum likelihood phylogenetic polar tree generated using 170 OTUs.
(TIF)

Figure S4　Heatmap showing the bacterioplankton diversity comparison among different sites. The scale at the bottom of the heatmap indicates the similarity level between each comparison. The darker the color is, the more different the two comparing bacterioplankton communities are.
(TIF)

Figure S5　Phylogenetic tree of the 16S rRNA clusters affiliated with the α-Proteobacteria lineage, constructed from an alignment of OTUs from nSCS in bold. Reference sequences were selected from GenBank with accession numbers are in parentheses. The OTU names were labeled with the numbers of contained sequences, while were designated as sequence name when containing only one sequence.
(TIF)

Table S1　Coordinates and characteristics of the sampling sites.
(DOCX)

Acknowledgments

The seawater samples used in this study were collected during the 2007 South China Sea Open Cruise by R/V *Shiyan* 3. We thank the South China Sea Institute of Oceanology (SCSIO), CAS, Guangzhou for the opportunity to join their open cruise and share the environmental variables from South China Sea Ocean Database. Our special thanks go to Prof. Dongxiao Wang for providing CTD data and Jia Yue for excellent technical assistance in this project.

Author Contributions

Conceived and designed the experiments: NL SQ FL. Performed the experiments: NL. Analyzed the data: JL GW TZ SY YW. Contributed reagents/materials/analysis tools: NL. Contributed to the writing of the manuscript: NL GW. Revision and software use: GW JL.

References

1. Copley J (2002) All at sea. Nature 415: 572–574.
2. Pedrós-Alió C (2006) Marine microbial diversity: can it be determined? Trends Microbiol 14: 257–263.
3. Doney SC (2010) The growing human footprint on coastal and open-ocean biogeochemistry. Science 328: 1512–1516.
4. Sarmento H, Montoya JM, Vázquez-Domínguez E, Vaqué D, Gasol JM (2010) Warming effects on marine microbial food web processes: how far can we go when it comes to predictions? Philos T Roy Soc B 365: 2137–2149.
5. Zehr JP, Robidart J, Scholin C (2011) Global environmental change demands a deeper understanding of how marine microbes drive global ecosystems. Microbe 6: 169–175.
6. Whitman WB, Coleman DC, Wiebe WJ (1998) Prokaryotes: the unseen majority. P Natl Acad Sci USA 95: 6578–6583.
7. Lee K, Sabine CL, Tanhua T, Kim TW, Feely RA, et al. (2011) Roles of marginal seas in absorbing and storing fossil fuel CO$_2$. Energ Environ Sci 4: 1133–1146.
8. Hu J, Kawamura H, Hong H, Qi Y (2000) A review on the currents in the South China Sea: seasonal circulation, South China Sea warm current and Kuroshio intrusion. J Oceanogr 56: 607–624.
9. Chen YL, Chen HY, Karl DM, Takahashi M (2004) Nitrogen modulates phytoplankton growth in spring in the South China Sea. Cont Shelf Res 24: 527–541.
10. Yin K, Qian PY, Wu MC, Chen JC, Huang L, et al. (2001) Shift from P to N limitation of phytoplankton growth across the Pearl River estuarine plume during summer. Mar Ecol Prog Ser 221: 17–28.
11. Zhou MY, Chen XL, Zhao HL, Dang HY, Luan XW, et al. (2009) Diversity of both the cultivable protease-producing bacteria and their extracellular proteases in the sediments of the South China Sea. Microbial Ecol 58: 582–590.
12. Ishida Y, Eguchi M, Kadota H (1986) Existence of obligately oligotrophic bacteria as a dominant population in the South China Sea and the West Pacific Ocean. Mar Ecol-Prog Ser 30: 197–203.
13. Moisander PH, Beinart RA, Voss M, Zehr JP (2008) Diversity and abundance of diazotrophic microorganisms in the South China Sea during intermonsoon. ISME J 2: 954–967.
14. Wang J, Li Y, Bian J, Tang SK, Ren B, et al. (2010) *Prauserella marina* sp. nov., isolated from ocean sediment of the South China Sea. Int J Syst Evol Micr 60: 985–989.
15. Cai H, Jiao N (2008) Diversity and abundance of nitrate assimilation genes in the northern South China Sea. Microbial Ecol 56: 751–764.
16. Dang H, Yang J, Li J, Luan X, Zhang Y, et al. (2013) Environment-dependent distribution of the sediment *nif*H-harboring microbiota in the northern South China Sea. Appl Environ Microbiol 79: 121–132.
17. Ning X, Li WK, Cai Y, Shi J (2005) Comparative analysis of bacterioplankton and phytoplankton in three ecological provinces of the northern South China Sea. Mar Ecol-Prog Ser 293: 17–28.
18. Yuan X, He L, Yin K, Pan G, Harrison PJ (2011) Bacterial distribution and nutrient limitation in relation to different water masses in the coastal and northwestern South China Sea in late summer. Cont Shelf Res 31: 1214–1223.
19. Seymour JR, Doblin MA, Jeffries TC, Brown MV, Newton K, et al. (2012) Contrasting microbial assemblages in adjacent water masses associated with the East Australian Current. Env Microbiol Rep 4: 548–555.

20. Xu J, Yin K, He L, Yuan X, Ho AY, et al. (2008) Phosphorus limitation in the northern South China Sea during late summer: influence of the Pearl River. Deep-Sea Res PT I 55: 1330–1342.

21. Ducet N, Le Traon PY, Reverdin G (2000) Global high-resolution mapping of ocean circulation from TOPEX/Poseidon and ERS-1 and −2. J Geophys Res, 105: 19477–19498.

22. Urakawa H, Martens-Habbena W, Stahl DA (2010) High abundance of ammonia- oxidizing archaea in coastal waters, determined using a modified DNA extraction method. Appl Environ Microbiol 76: 2129–2135.

23. Eckburg PB, Bik EM, Bernstein CN, Purdom E, Dethlefsen L, et al. (2005) Diversity of the human intestinal microbial flora. Science 308: 1635–1638.

24. Edgar RC (2004) MUSCLE: multiple sequence alignment with high accuracy and high throughput. Nucleic Acids Res 32: 1792–1797.

25. Schloss PD, Westcott SL, Ryabin T, Hall JR, Hartmann M, et al. (2009) Introducing mothur: open-source, platform-independent, community-supported software for describing and comparing microbial communities. Appl Environ Microbiol 75: 7537–7541.

26. Guindon S, Gascuel O (2003) A simple, fast and accurate algorithm to estimate large phylogenies by maximum likelihood. Syst Biol 52: 696–704.

27. Darriba D, Taboada GL, Doallo R, Posada D (2012) Jmodeltest 2: more models, new heuristics and parallel computing. Nat Methods 9: 772–772.

28. Lozupone C, Knight R (2005) UniFrac: A new phylogenetic method for comparing microbial communities. Appl Environ Microbiol 71: 8228–8235.

29. Song H, Li Z, Du B, Wang G, Ding Y (2012) Bacterial communities in sediments of the shallow lake Dongping in China. J Appl Microbiol 112: 79–89.

30. Rusch DB, Halpern AL, Sutton G, Heidelberg KB, Williamson S, et al. (2007) The Sorcerer II global ocean sampling expedition: northwest Atlantic through eastern tropical Pacific. PLoS Biol 5: e77.

31. Jiang H, Dong H, Ji S, Ye Y, Wu N (2007) Microbial diversity in the deep marine sediments from the Qiongdongnan Basin in South China Sea. Geomicrobiol J 24: 505–517.

32. Wang G, Dong J, Li X, Sun H (2010) The bacterial diversity in surface sediment from the South China Sea. Acta Oceanol Sin 29: 98–105.

33. Kan J, Evans SE, Chen F, Suzuki MT (2008) Novel estuarine bacterioplankton in rRNA operon libraries from the Chesapeake Bay. Aquat Microb Ecol 51: 55–66.

34. Shaw AK, Halpern AL, Beeson K, Tran B, Venter JC, et al. (2008) It's all relative: ranking the diversity of aquatic bacterial communities. Environ Microbiol 10: 2200–2210.

35. Bowman JP, McCammon SA, Dann AL (2005) Biogeographic and quantitative analyses of abundant uncultivated γ-proteobacterial clades from marine sediment. Microbial Ecol 49: 451–460.

36. Ferrier M, Martin JL, Rooney-Varga JN (2002) Stimulation of Alexandrium fundyense growth by bacterial assemblages from the Bay of Fundy. J Appl Microbiol 92: 706–716.

37. Liu H, Song X, Huang L, Tan Y, Zhang J (2011) Phytoplankton biomass and production in northern South China Sea during summer: influenced by Pearl River discharge and coastal upwelling. Acta Ecol Sin 31: 133–136.

38. Wang SF, Tang DL, He FL, Fukuyo Y, Azanza RV (2008) Occurrences of harmful algal blooms (HABs) associated with ocean environments in the South China Sea. Hydrobiologia 596: 79–93.

39. North NN, Dollhopf SL, Petrie L, Istok JD, Balkwill DL, et al. (2004) Change in bacterial community structure during in situ biostimulation of subsurface sediment cocontaminated with uranium and nitrate. Appl Environ Microbiol 70: 4911–4920.

40. Lydell C, Dowell L, Sikaroodi M, Gillevet P, Emerson D (2004) A population survey of members of the phylum Bacteroidetes isolated from salt marsh sediments along the East Coast of the United States. Microbial Ecol 48: 263–273.

41. Allers E, Wright JJ, Konwar KM, Howes CG, Beneze E, et al. (2012) Diversity and population structure of Marine Group A bacteria in the Northeast subarctic Pacific Ocean. ISME J 7: 256–268.

42. Azam F, Malfatti F (2007) Microbial structuring of marine ecosystems. Nat Rev Microbiol 5: 782–791.

43. Brown MV, Philip GK, Bunge JA, Smith MC, Bissett A, et al. (2009) Microbial community structure in the North Pacific Ocean. ISME J 3: 1374–1386.

44. Bowman JS, Rasmussen S, Blom N, Deming JW, Rysgaard S, et al. (2011) Microbial community structure of Arctic multiyear sea ice and surface seawater by 454 sequencing of the 16S RNA gene. ISME J 6: 11–20.

45. Pinhassi J, Gómez-Consarnau L, Alonso-Sáez L, Sala MM, Vidal M, et al. (2006) Seasonal changes in bacterioplankton nutrient limitation and their effects on bacterial community composition in the NW Mediterranean Sea. Aquat Microb Ecol 44: 241–252.

46. Morton B, Blackmore G (2001) South China Sea. Mar Pollut Bull 42: 1236–1263.

47. Lai X, Cao L, Tan H, Fang S, Huang Y, et al. (2007) Fungal communities from methane hydrate-bearing deep-sea marine sediments in South China Sea. ISME J 1: 756–762.

48. Zinger L, Amaral-Zettler LA, Fuhrman JA, Horner-Devine MC, Huse SM, et al. (2011) Global patterns of bacterial beta-diversity in seafloor and seawater ecosystems. PLoS ONE 6: e24570.

49. Yin Q, Fu B, Li B, Shi X, Inagaki F, et al. (2013) Spatial variations in microbial community composition in surface seawater from the ultra-oligotrophic center to rim of the South Pacific Gyre. PLoS ONE 8: e55148.

50. Ning X, Chai F, Xue H, Cai Y, Liu C, et al. (2004) Physical-biological oceanographic coupling influencing phytoplankton and primary production in the South China Sea. J Geophys Res 109: C10005, doi:10010.11029/12004JC002365

51. Li L, He J, Wang H (2014) Factors affecting the abundance and community structure of the phytoplankton in northern South China Sea in the summer of 2008: a biomarker study. Chin Sci Bull 59: 981–991.

52. Schäfer H, Bernard L, Courties C, Lebaron P, Servais P, et al. (2001) Microbial community dynamics in mediterranean nutrient-enriched seawater mesocosms: changes in the genetic diversity of bacterial populations. FEMS Microbiol Ecol 34: 243–253.

53. Ghiglione JF, Galand PE, Pommier T, Pedrós-Alió C, Maas EW, et al. (2012) Pole-to-pole biogeography of surface and deep marine bacterial communities. P Natl Acad Sci USA 109: 17633–17638.

54. Cottrell MT, Kirchman DL (2000) Community composition of marine bacterioplankton determined by 16S rRNA gene clone libraries and fluorescence in situ hybridization. Appl Environ Microbiol 66: 5116–5122.

55. Chen CTA, Wang SL, Wang BJ, Pai SC (2001) Nutrient budgets for the South China Sea basin. Mar Chem 75: 281–300.

56. Cottrell MT, Kirchman DL (2000) Community composition of marine bacterioplankton determined by 16 s rRNA gene clone libraries and fluorescence in situ hybridization. Appl Environ Microbiol 66: 5116–5122.

57. Liu F, Tang S, Chen C (2013) Impact of nonlinear mesoscale eddy on phytoplankton distribution in the northern South China Sea. J Marine Syst 123–124: 33–40.

10

High Phylogenetic Diversity of Glycosyl Hydrolase Family 10 and 11 Xylanases in the Sediment of Lake Dabusu in China

Guozeng Wang[1,2☺], **Xiaoyun Huang**[1☺], **Tzi Bun Ng**[3], **Juan Lin**[1,2]*, **Xiu Yun Ye**[1,2]*

1 College of Biological Science and Technology, Fuzhou University, Fuzhou 350108, P.R. China, **2** National Engineering Laboratory for High-efficiency Enzyme Expression, Fuzhou 350002, P. R. China, **3** School of Biomedical Sciences, Faculty of Medicine, The Chinese University of Hong Kong, Shatin, New Territories, Hong Kong, China

Abstract

Soda lakes are one of the most stable naturally occurring alkaline and saline environments, which harbor abundant microorganisms with diverse functions. In this study, culture-independent molecular methods were used to explore the genetic diversity of glycoside hydrolase (GH) family 10 and GH11 xylanases in Lake Dabusu, a soda lake with a pH value of 10.2 and salinity of 10.1%. A total of 671 xylanase gene fragments were obtained, representing 78 distinct GH10 and 28 GH11 gene fragments respectively, with most of them having low homology with known sequences. Phylogenetic analysis revealed that the GH10 xylanase sequences mainly belonged to Bacteroidetes, Proteobacteria, Actinobacteria, Firmicutes and Verrucomicrobia, while the GH11 sequences mainly consisted of Actinobacteria, Firmicutes and Fungi. A full-length GH10 xylanase gene (*xynAS10-66*) was directly cloned and expressed in *Escherichia coli*, and the recombinant enzymes showed high activity at alkaline pH. These results suggest that xylanase gene diversity within Lake Dabusu is high and that most of the identified genes might be novel, indicating great potential for applications in industry and agriculture.

Editor: Chunxian Chen, USDA/ARS, United States of America

Funding: This research was supported by the National Natural Science Foundation of China (31301406), the Oceanic Public Welfare Industry Special Research Project of China (201305015) and the Priming Scientific Research Foundation of Fuzhou University (XRC-1326). The funders had no role in study design, data collection and analysis, decision to publish, or preparation of the manuscript.

Competing Interests: The authors have declared that no competing interests exist.

* Email: ljuan@fzu.edu.cn (JL); xiuyunye@fzu.edu.cn (XYY)

☺ These authors contributed equally to this work.

Introduction

Xylan is the major hemicellulose component of the plant cell wall, which is the second most abundant polysaccharide on earth after cellulose [1]. Xylan is composed of a homopolymeric backbone chain of β-1, 4-linked xylopyranose units with substituted side chains at different positions, and its complete hydrolysis requires a group of enzymes including endo-1,4-β-D-xylanase, β-D-xylosidase, α-D-glucuronidase, α-L-arabinofuranosidase, acetyl xylanesterase, and arylesterase [2,3]. Endo-1,4-β-D-xylanase (EC 3.2.1.8) is a crucial component because it catalyzes the hydrolysis of xylan to short xylooligosaccharides of varying lengths [2,3]. Xylanases have been classified into glycosyl hydrolase (GH) families (http://www.cazy.org/fam/acc_GH.html; [4]) 5, 7, 8, 10, 11 and 43 [3] based on sequence similarities of the catalytic domain. Among these, GH10 and 11 xylanases are the most abundant, which have distinct three-dimensional structures [5], mechanisms of action [6] and substrate specificity to xylan.

Although xylanases are produced by diverse organisms, including bacteria, algae, fungi, protozoans, gastropods and anthropods [3], microbial xylanases are the focus of intense research owing to their significant application in various industrial processes. Specifically, they are used to improve the digestibility of animal feedstock, enhance filterability in brewing, increase dough volume and improve the textural and staling properties of bread, and for fruit juice and wine clarification, the bioconversion of lignocellulosic materials into fermentative products, and facilitation of the release of lignin from the pulp [1,2,7,8]. Many microbial xylanases have been purified and characterized, and the genes encoding xylanases have been cloned and expressed in heterologous systems, and the structures of a number of enzymes have been determined [3,8].

Most microbial xylanases reported to date have acidic or neutral pH optima. Alkaline xylanases are of great interest because xylan is more readily soluble under alkaline pH than neutral pH. Additionally, the application of alkaline xylanases in the paper and pulp industry can reduce the use of chlorine, which is very attractive from an economical and technical point of view [9,10]. Consequently, studies are continually conducted in attempts to identify novel xylanases with potential applications in the pulp and paper industries. Although alkaline xylanases have been reported from microorganisms isolated from nonalkaline environments [11,12], xylanases produced by microorganisms isolated from extreme alkaline environments have attracted increasing attention. These organisms are exposed to hostile environments, resulting in their evolution and accumulation of a variety of adaptive features for activity and stability under these conditions [13–15].

Natural alkaline environments are not common, and soda lakes and deserts are the most stable naturally occurring alkaline ecosystems on earth [16,17]. Soda lakes have high carbonate alkalinity, a pH of 9 to 11, and moderate to extremely high salinity [17]. Despite their extreme environmental conditions, soda lakes harbor extremely productive microbial communities. A number of alkaliphilic microorganisms that play key metabolic roles in soda lakes have previously been studied, and many novel alkaliphilic microorganisms have been isolated from these unique ecosystems [18]. Molecular biological techniques that do not depend on culture have revealed a high diversity and novelty of microbial communities in soda lake environments [19–21].

Although the microbial diversity of soda lakes has been studied extensively, the functional diversity of genes in such systems has not. Lin et al. analyzed the methane monooxygenase genes in Mono Lake and suggested that increased methane oxidation activity was correlated with changes in methanotroph community structure [22]. Kovaleva et al. explored the diversity of RuBisCO and ATP citrate lyase genes in the sediments from six soda lakes of the Kulunda Steppe [23]. However, few studies have focused on the functional gene diversity of plant material dehygrolysis in soda lake environments. We explored the genetic diversity of GH10 xylanase in diverse soil environments using culture-independent molecular methods and found that pH was one of the most important factors influencing the xylan degrading microbial community [24].

In this study, we focused on the genetic diversity of GH10 and GH11 xylanases in the natural Lake Dabusu, an alkaline (pH 10.2) and saline (101 g/liter) lake situated in northeastern China. Partial xylanase genes were amplified directly from the metagenomic DNA of the soda lake sediment. Sequence analysis showed that most xylanase gene fragments had low homologies to known xylanases, suggesting a large number of uncharacterized xylanase genes in the soda lake. Additionally, phylogenetic diversity analysis suggested a surprising diversity of xylanase genes in this harsh environment. A novel GH10 full-length xylanase gene was directly cloned from the metagenomic DNA and expressed in *Escherichia coli*. The recombinant xylanases showed high activity at alkaline pH. Our study provides new insight into the genetic diversity and distribution of microbial xylanases in the soda lake ecosystem, which will help us understand their roles in this microenvironment.

Materials and Methods

Ethics statement

We declare that no living animals were used in this research. No specific permissions were required for the described field studies. The sediment sample was collected from the location that is not privately-owned or protected in any way. The field studies did not involve endangered or protected species because this study only concentrated on the sediment sample.

Sample site

Lake Dabusu is located in the southwestern part of the Qian'an County (Jilin Province, China), in the center of the depressed belt of Songliao Basin. Lake Dabusu is a closed inland alkaline lake located 122 meters above sea level, with an average water depth of about 0.90 meters and an area of approximately 38 km^2 in the rainy season. Because of strong evaporation and a lack of outflow within the closed basin, alkaline materials accumulate in the lake water. Lake Dabusu is a typical soda lake with salinity of 62.34 g L^{-1} to 347.34 g L^{-1} and a pH of 10 to 11 depending on season [25].

Table 1. Physicochemical characters of the alkaline saline lake sediment and xylanase fragment sequences obtained.

Location	T (°C)	pH	Total organic carbon (mg/g)	Total nitrogen (mg/g)	C/N ratio	GH family	Clones sequenced	Sequences recovered	OTUs[a]
44 48 20 N	20	10.2	1.10	0.35	3.14	GH10	550	467	78
123 40 32 E						GH11	250	204	28

[a]5% dissimilarity as cutoff.

A

GH 10

B

GH 11

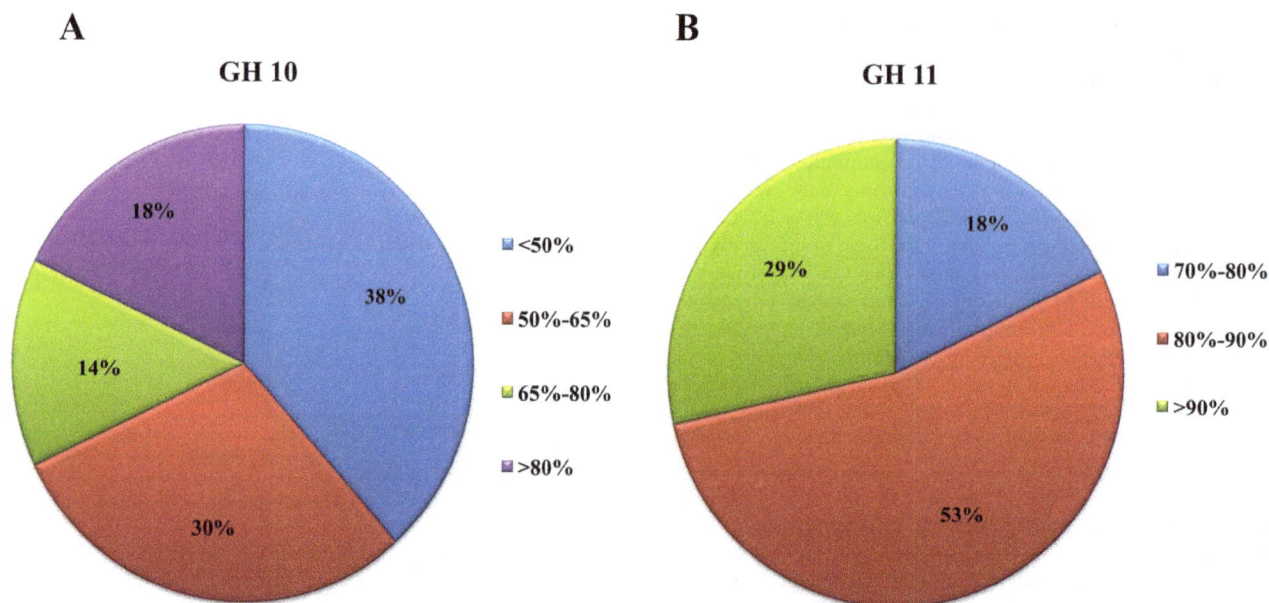

Figure 1. Amino acid sequence homologies of GH10 and GH11 xylanase gene fragments from sediment metagenomic DNA to known xylanases. Each sequence was analyzed by NCBI BLASTp (version 2.2.29) against the GenBank nr database. An E-score (expect value) cutoff of 10^{-10} (default) was applied and the top BLASTp hit to the known xylanases was collected.

Sample collection and DNA extraction

On June 12, 2013, superficial (0 to 10 cm depth) sediment samples were collected and transported into a sterile sampling bag and immediately shipped to the laboratory on ice packs. Upon arrival, portions of the samples were stored at 4°C for physicochemical characterization and −80°C for metagenomic DNA extraction. Details regarding the location and physicochemical properties of the lake sediment are provided in Table 1. By using a modified protocol specific for high molecular weight DNA from environmental samples [26], the metagenomic DNA of the sediment was extracted as described in detail previously [27]. The sediment metagenomic DNA was purified using an Omega Gel Extraction Kit (Norcross, GA) and stored at −20°C until use.

Xylanase gene fragments amplification, library construction and sequencing

GH10 and GH11 xylanase gene fragments were amplified with the purified metagenomic DNA as a template by touchdown PCR using the CODEHOP primers X10-F and X10-R specific for GH10 xylanase and X11-F and X11-R specific for GH11 xylanase [27], respectively. PCR products were then visualized on an agarose gel and purified using a Qiaquick gel extraction kit (Qiagen, Valencia, CA). The purified PCR products were then ligated into the PMD 19-T vector (TaKaRa, Tokyo, Japan) and electroporated into *Escherichia coli* DH5a (TaKaRa, Tokyo, Japan) following the procedure recommended by the manufacturer to construct the clone library for each xylanase family. Positive transformants (white colonies) from each library were randomly picked for further confirmation by PCR with primers M13F (GTAAAACGACGGCCAGT) and M13R (GGATAA-CAATTTCACACAGGA). They were then sequenced by Life Technologies using the Sanger method with an ABI-3730 automatic sequencer (Life Technologies, Carlsbad, CA).

Phylogenetic analysis

By using the Figaro software [28] (http://sourceforge.net/apps/mediawiki/amos/index.php?title=Figaro), vector sequences introduced by automated Sanger sequencing machines were removed. The sequences were analyzed by NCBI BLASTx (version 2.2.29) searches against the GenBank nr database, with an E-score (expect value) cutoff of 10^{-10}. Nucleotide sequences identified as xylanase gene fragments were translated into amino acids by EMBOSS Transeq (http://www.ebi.ac.uk/Tools/st/emboss_transeq/). Then, they were aligned with known sequences in the GenBank database at the protein level using ClustalW. Redundant amino acid sequences were removed using Cd-hit [29] with a 95% sequence identity cutoff.

The protein sequence similarities were assessed using the BLASTp program (http://www.ncbi.nlm.nih.gov/BLAST/; until January 15, 2014). Phylogenetic trees were constructed with MEGA 4.1 [30] using the neighbor-joining method [31]. Confidence for tree topologies was estimated by bootstrap values based on 1,000 replicates. A total of 39 and 17 representative sequences were selected and used as references for GH10 and GH11 phylogenetic tree constructions, respectively.

Abundance analysis

The abundance of each GH family was estimated using the distance-based operational taxonomic unit and richness determination (DOTUR) software [32]. By using PHYLIP software (http://evolution.genetics.washington.edu/phylip.html), distance matrices of the fragment sequences were calculated at the protein level with the default parameters of protdist. Based on UPGMA (average linkage clustering) implemented in DOTUR with default parameters of precision (0.01) and 1000 bootstrap replicates, sequences were then assigned to OTUs.

Cloning of full-length xylanase gene

Fragment AS10-66, which showed phylogenetic novelty (Figure 1, Table S1) and was most abundant in the GH10 library (see

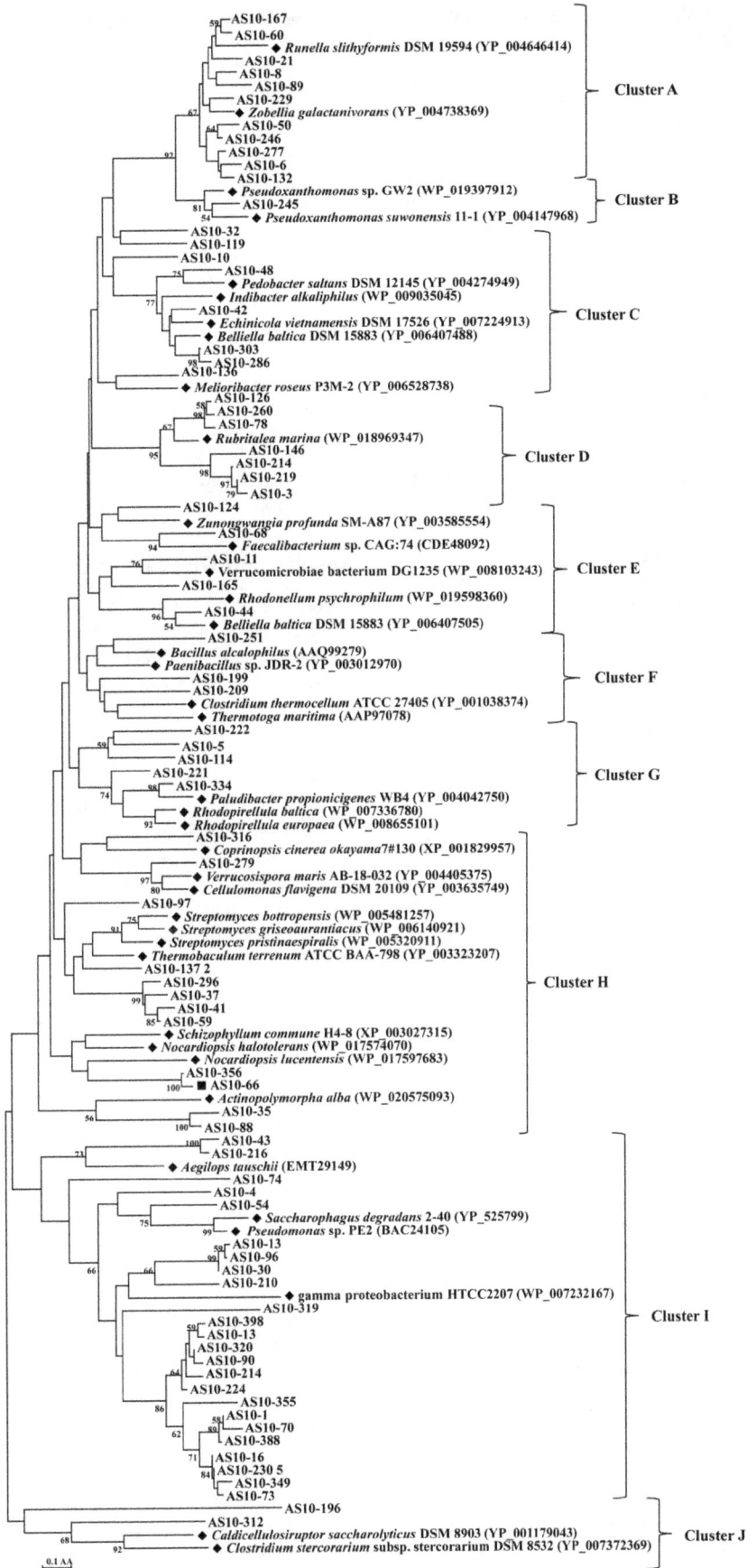

Figure 2. Phylogenetic analysis based on the partial amino acid sequences of GH10 xylanase genes detected in the Lake Dabusu sediment metagenomic DNA. The tree was constructed using the neighbor joining method (MEGA 4.1). The lengths of the branches indicate the relative divergence among amino acid sequences. The reference sequences are marked with a closed diamond (♦) with source strains and GenBank accession numbers in parentheses. The gene fragments (AS10-66) used for full length cloning were marked with a solid square (■). The numbers at the nodes indicate bootstrap values based on 1,000 bootstrap replications and bootstrap values (>50) are displayed. The scale bar represents 0.1 amino acid substitutions per position.

Table S1), was subjected to full-length gene cloning. The flanking regions of the xylanase gene fragments were cloned by using a modified TAIL-PCR with eight specific primers (Table S3) following the protocol [33]. PCR products of the expected size appeared between the third and fourth rounds of amplification were purified, cloned into the PMD 19-T vectors, sequenced, and then assembled with the known fragment sequence. The full-length xylanase gene was designated as *xynAS10-66*. The signal peptide sequence of XynAS10-66 was predicted with SignalP [34] (http://www.cbs.dtu.dk/services/SignalP/). The DNA and protein sequence identities/similarities were assessed with the BLASTn and BLASTp programs (http://www.ncbi.nlm.nih.gov/BLAST/), respectively.

Xylanase expression and activity assay

Using two primer sets (Table S3), the coding sequence of the xylanase gene *xynAS10-66* without the signal peptide was amplified and cloned into vector pET-22b(+), and then transformed into *E. coli* BL21 (DE3) competent cells for recombinant expression. The positive transformant harboring pET-*xynAS10-66* was grown in LB medium containing 100 μg mL^{-1} ampicillin at 37°C to an A$_{600}$ of 0.6. Protein expression was induced by addition of isopropyl-β-D-1-thiogalactopyranoside (IPTG) at a final concentration of 1 mM, and the culture was incubated at 30°C for additional 12 h.

Xylanase activity was determined by measuring the release of reducing sugar from substrate by the 3, 5-dinitrosalicylic acid method [35]. To accomplish this, reactions containing 0.1 ml of appropriately diluted enzyme and 0.9 ml of 1% (w/v) beechwood xylan as substrate. After incubation at 55°C for 10 min, the reaction was stopped with 1.5 ml DNS reagent and boiled for 5 min, and the absorbance at 540 nm (A$_{540}$) was measured. Finally, using a standard curve generated with D-xylose, the absorbance was converted into moles of reducing sugars produced. One unit (U) of xylanase activity was defined as the amount of enzyme that released 1 μmol of reducing sugar equivalent to xylose per minute.

Purification and partial characterization of recombinant XynAS10-66

To purify the His-tagged recombinant proteins (rXynAS10-66), culture supernatant was collected after centrifugation (12,000×g, 4°C for 15 min). Then the culture supernatant was concentrated with an ultrafiltration membrane (PES5000; Sartorius Stedim Biotech, Germany) and then loaded onto a Ni^{2+}-NTA agarose gel column (Qiagen, Germany) with an imidazole gradient of 20–200 mM in Tris-HCl buffer (20 mM Tris-HCl, 500 mM NaCl, 10% glycerol, pH 7.6). Sodium dodecyl sulfate-polyacrylamide gel electrophoresis (SDS-PAGE) was used to determine the purity and apparent molecular mass of rXynAS10-66.

The optimal pH for xylanase activity of the purified rXynAS10-66 was determined at 37°C with pH ranging from 4.0 to 11.0. The buffers used were McIlvaine buffer (0.2 M Na$_2$HPO$_4$/0.1 M citric acid) for pH 4.0–7.0, 0.1 M Tris-HCl for pH 7.0–9.0, and 0.1 M glycine-NaOH for pH 9.0–11.0. The optimal temperature for purified rXynAS10-66 activity was determined over the range of

40–90°C in Tris-HCl buffer (pH 9.0). The K_m and V_{max} values for rXynAS10-66 were determined in Tris-HCl buffer (pH 7.0) containing 1–10 mg mL^{-1} beechwood xylan at 55°C, respectively. K_m and V_{max} were determined from a Lineweaver-Burk plot using the non-linear regression computer program GraFit (Erithacus, Horley, Surrey, UK).

Nucleotide sequence accession numbers

The GH10 and GH11 xylanase gene fragments were deposited into the GenBank database under accession numbers KJ463250–KJ463327 and KJ463328–KJ463355, respectively. Accession number KJ463356 was assigned to the full-length xylanase gene *xynAS10-66*.

Results and Discussion

Soda lakes are one of the most stable naturally occurring alkaline and saline environments that harbor numerous novel microorganisms [16,17,21]. Although the microbial diversity of this unique ecosystem has been thoroughly investigated, only a few studies have considered the functional diversity [22,23]. Functional gene diversity based on the metagenomic sequences can provide insight into functional diversity and metabolic potential at the community level. Xylanases play key roles in the initial steps of plant cell wall breakdown and have great potential for industrial and agricultural applications. Thus xylanase genes were targeted for diversity analysis in this study. Because many microorganisms cannot be cultured in the laboratory owing to the extremely harsh conditions of soda lakes, culture-independent molecular approaches were used to explore the xylanase gene diversity of Lake Dabusu.

Abundance of GH10 xylanase in Lake Dabusu sediment

Using the CODEHOP primers X10-F and X10-R specific for GH10 xylanases [27], PCR product of about 260 bp was amplified from the metagenomic DNA of the sediment. The product was purified and used to construct a clone library. The positive transformants were picked and then confirmed by PCR with primers M13F and M13R. Overall, 550 clones were sequenced, 467 sequences of which were identified as GH10 xylanase gene fragments based on BLASTx analysis and presence of the Asn residue in the protein sequence, which is conserved among GH10 xylanases [36].

After removing redundant sequences using the CD-hit program [29], 78 sequences showed divergence (<95% homology, Table S1). Based on BLASTp analysis, about 68% (53/78) of the sequences had low similarities (<65%) with known xylanases in GenBank (Figure 1), implying that they may be novel xylanases. Abundance analysis using the distance based operational taxonomic unit and richness determination (DOTUR) software [32] showed that AS10-66 was the predominant operational taxonomic unit (OTU), representing 117 sequences. Forty OTUs contained only one sequence (Table S1).

Culture dependent and culture-independent methods revealed the presence of numerous novel microorganisms in soda lake environments [16,21]. In the present study, GH10 xylanase gene

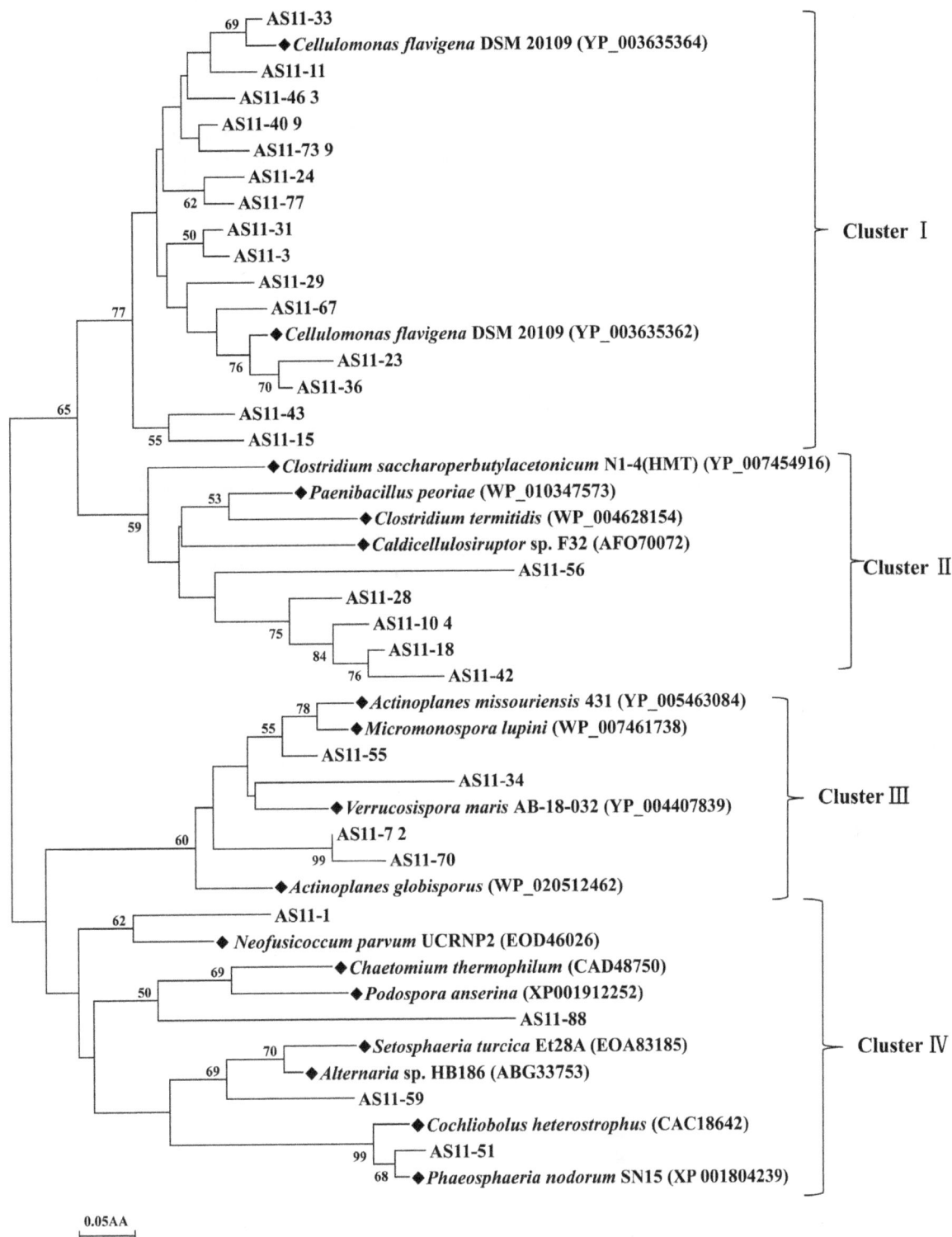

Figure 3. Phylogenetic analysis based on the partial amino acid sequences of GH11 xylanase genes detected in the Lake Dabusu sediment metagenomic DNA. The tree was constructed using the neighbor joining method (MEGA 4.1). The lengths of the branches indicate the relative divergence among amino acid sequences. The reference sequences are marked with a closed diamond (◆) with source strains and GenBank accession numbers in parentheses. The numbers at the nodes indicate bootstrap values based on 1,000 bootstrap replications and bootstrap values (>50) are displayed. The scale bar represents 0.05 amino acid substitutions per position.

fragment sequences obtained directly from the sediment metage-nomic DNA had low homology with known xylanases. Specifical-ly, none of the xylanase fragments amplified in this study had greater than 90% homology with known xylanases in the

A

B

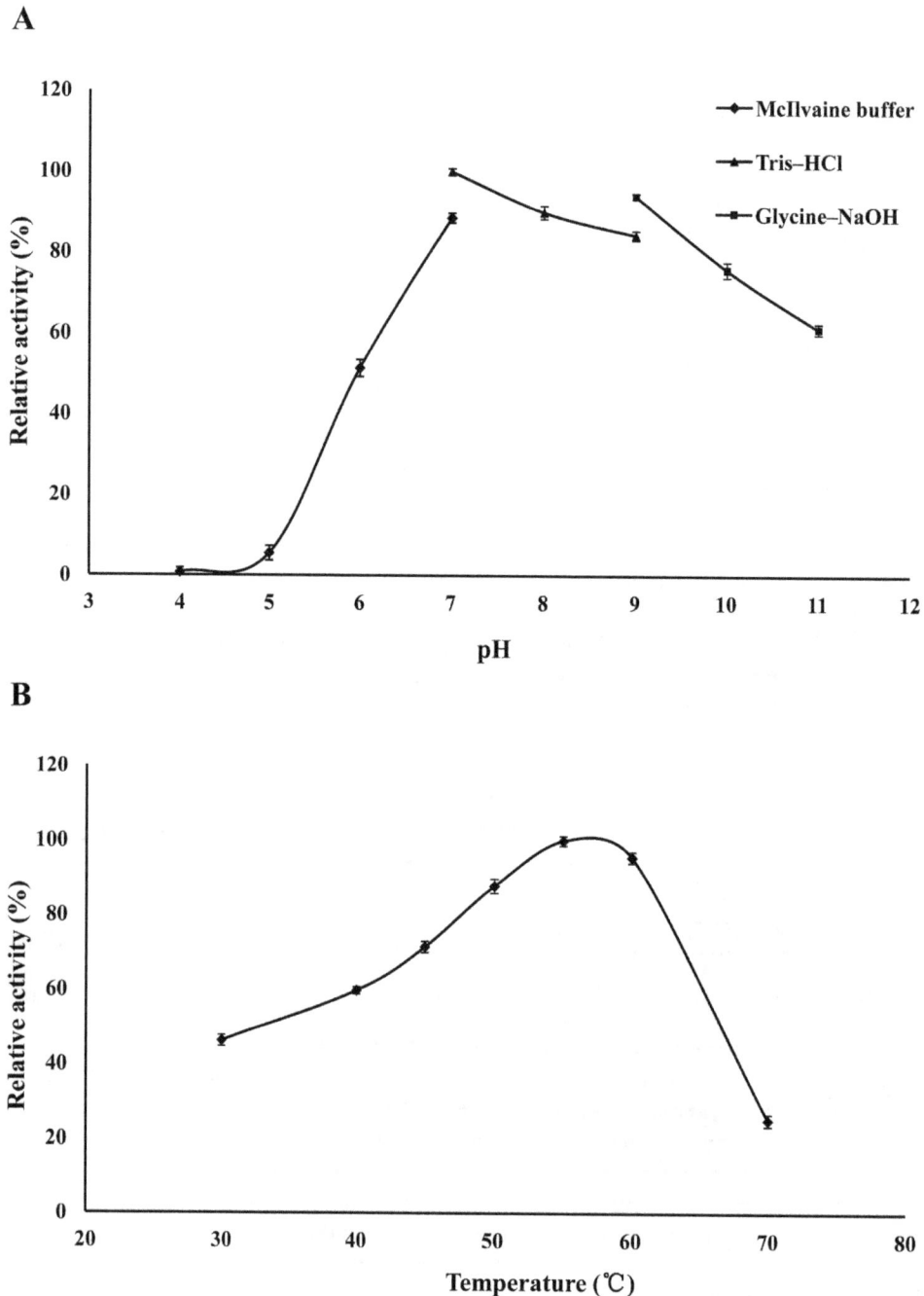

Figure 4. pH and temperature activity profiles of purified recombinant XynAS10-66. A Effect of pH on XynAS10-66 activity. Activities at various pHs were assayed at 30°C in different buffer. **B** Effect of temperature on XynAS10-66 activity in Tris-HCl buffer (pH 9.0).

GenBank database (Table S1). The lowest similarities were observed for AS10-35 and AS10-88, which both had 35% homology with xylanase from *Nocardiopsis* sp. CNS639 (WP_019610727), while AS10-8 showed the greatest homology of 87% with xylanase from *Cellulophaga algicola* DSM 14237 (YP_004163134). Moreover, almost 70% of the obtained sequences had low similarity (<65%) with known xylanases. This is much higher than that from the other soil environments we previously investigated [24]. Overall, these findings suggested that there may be abundant novel xylan-degrading microorganisms in this environment.

High genetic diversity of GH10 xylanase in the alkaline soda lake sediment

Using 78 divergent sequences from the GH10 clone library and 39 reference sequences from the GenBank Database, an unrooted protein-level phylogenetic tree of GH10 xylanases was constructed. All sequences were confined to ten clusters, denoted as Cluster A to Cluster J, indicating substantial diversity among GH10 xylanases in the sediment (Figure 2). The presence of many clades without close relatives suggests their novelty, which might be because of the large portion of unidentified microorganisms in this environment.

Cluster A, C, E and G contained a total of 28 sequences from the soda lake sediment and 14 reference sequences of genera belonging to Bacteroidetes including *Runella slithyformis* DSM 19594, *Pedobacter saltans*, *Indibacter alkaliphilus* and *Rhodopirellula baltica*. Cluster B and cluster I contained the sequences of 23 clones and reference sequences from the phylum Proteobacteria, including gamma proteobacterium HTCC2207, *Pseudomonas* sp. PE2, *Saccharophagus degradans* 2–40 and *Pseudoxanthomonas suwonensis* 11-1. Twelve sequences and 11 reference sequences from different genera in the phylum Actinobacteria, including *Actinopolymorpha*, *Streptomyces* and *Nocardiopsis*, formed cluster H. Three sequences in cluster F and two in cluster J were closely related to xylanase produced by members of the phylum Firmicutes, including *Bacillus alcalophilus*, *Paenibacillus* sp. JDR-2, *Clostridium thermocellum* ATCC 27405, *Caldicellulosiruptor saccharolyticus* DSM 8903 and *Clostridium stercorarium* subsp. stercorarium DSM 8532. Seven sequences in cluster D shared the highest identity with xylanase from *Rubritalea marina*.

As shown in Fig. 2, phylogenetic analysis revealed that Bacteroidetes, Proteobacteria, Actinobacteria, Firmicutes and Verrucomicrobia are the main xylanolytic bacteria to produce xylanases in the sediment of Lake Dabusu. These findings are consistent with the conclusions of previous studies [16,17,20,37]. However, the composition of the xylanolytic community differs from that of the microbial community. Microbial community analysis suggested that Firmicutes are the most abundant microorganisms in soda lake environments, followed by Proteobacteria and Actinobacteria. In the present study, xylanases were found to primarily belong to Bacteroidetes, which harbored more than 35% (28/78) of all GH10 fragment sequences. One reason for this may be that the xylantic microbial community differs from that of the total microbial community. Soda lakes harbor abundant and diverse microorganisms which are critical in decomposition of organic matter and cycling of carbon, nitrogen, phosphorus, and sulphur [38]. Second, the physicochemical properties of Lake Dabusu differ from those of other soda lakes [16], which play important role in the composition of the microbial community [20,39].

Xylanase fragment sequences related to those from Proteobacteria were the second most abundant, accounting for about 30% (23/78) of all GH10 xylanase sequences. All reference sequences of gamma proteobacterium HTCC2207, *Pseudomonas* sp. PE2, *Saccharophagus degradans* 2–40 and *Pseudoxanthomonas suwonensis* 11-1 belonged to the Gammaproteobacteria. This is concurrent with the finding that Gammaproteobacteria is one of the most abundant and diverse groups in soda lakes [37]. Moreover, we found that although these sequences were grouped in a cluster, they were not closely related to each other (Figure 2), suggesting that these sequences might be novel.

Amplification and sequence analysis of GH11 xylanase gene fragments

PCR product with a size of about 210 bp was obtained from the metagenomic DNA of sediment using CODEHOP primers X11-F and X11-R specific for GH11 xylanases [24]. Overall, 250 clones were randomly selected from the clone library constructed using the PCR products and sequenced. As a result, 204 sequences showed 70–98% amino acid identity with known GH11 xylanases (Table S2). Additionally, Glu (catalytic residue), which is highly conserved in GH11 xylanases, was found in all protein sequences (Table S2). Thus, these sequences were considered to be partial GH11 xylanases. After removing the redundant sequences using the CD-hit program, 28 sequences showed divergence (sharing < 95% identity) (Table S2). Abundance analysis using DOTUR

software showed that AS11-42 was the predominant OTU, occurring in 39 sequences. Nine OTUs contained only one sequence (Table S2).

Phylogenetic diversity of GH11 xylanase in lake sediment

The 28 distinct partial sequences of GH11 xylanases were used to construct an unrooted phylogenetic tree with 17 reference sequences (Figure 3). Four clusters (I, II, III, and IV) were formed based on high bootstrap values. Some clades formed without closely related references, suggesting that these sequences differed from known xylanases. Nineteen sequences shared the highest identity with xylanases from *Cellulomonas flavigena* DSM 20109, *Actinoplanes missouriensis* 431, *A. globisporus Micromonospora lupine* and *Verrucosispora maris* AB-18-032, and fell into clusters I and III. Five sequences and those of *Clostridium saccharoperbutylacetonicum* N1-4(HMT), *C. Paenibacillus peoriae*, and *Caldicellulosiruptor* sp. F32 were grouped into cluster II. Four sequences closely related to xylanases from *Neofusicoccum parvum* UCRNP2, *Podospora anserine*, *Setosphaeria turcica* Et28A, *Alternaria* sp. HB186 and *Phaeosphaeria nodorum* SN15 fell into cluster IV.

Unlike GH10 family xylanase genes, GH11 xylanase genes from Lake Dabusu were less diverse and showed a different microbial distribution. Sequences of the GH11 xylanase fragments were found to be related to Actinomycetes, Firmicutes and Fungi. Actinomycetes are important degraders of organic matter in most habitats. An alkaline xylanase was purified and characterized from alkaliphilic *Micrococcus* sp. AR-135, which was isolated from an alkaline soda lake in Ethiopia [14]. Upon phylogenetic analysis, more than 67% (19/28) of the sequences were grouped into a cluster containing reference sequences from different genera in Actinobacteria. Along with the 12 sequences of GH10 xylanase, a total 31 fragment sequences were found to be associated with Actinobacteria, indicating the high diversity of xylanolytic actinobacteria in Lake Dabusu.

Using culture and culture-independent methods, the bacterial community in the soda lake was found to be dominated by clones affiliated with Firmicutes. Moreover, several xylanases were characterized from the isolated Firmicutes with high activity at alkaline pH [13,15]. In this study, xylanase fragments related to xylanases of different genera Firmicutes were also found. Specifically, five GH10 xylanase fragment sequences were found to be closely related to *Bacillus alcalophilus*, *Clostridium thermocellum* ATCC 27405 and *Caldicellulosiruptor saccharolyticus* DSM 8903, while five GH11 fragments were closely related to *Caldicellulosiruptor* sp. F32 (AFO70072), *Clostridium termitidis* (WP_004628154) and *Clostridium saccharoperbutylacetonicum* N1-4(HMT) (YP_007454916), suggested the diversity of xylanolytic Firmicutes in this environment.

When compared with Bacteria and Archaea, little is known about the diversity, abundance and activity of micro-eukaryotes in soda lakes. Recently, a high diversity of micro-eukaryotes in soda lakes located in the Ethiopian Rift Valley was revealed by high-throughput sequencing [21]. However, no functional genetic diversity has been reported regarding fungi in soda lake environments. In the present study, four sequences were found to be related to xylanases from Ascomycota, the dominant fungi in soda lakes, suggesting that xylantic fungi are present in these systems.

Direct cloning and novelty of xylanase gene XynAS10-66

Until recently, the majority of known xylanases were obtained from pure cultures. Only a few were cloned using culture-independent approaches, such as construction and screening of metagenomic libraries and PCR-based methods. In our lab, we developed a fast and efficient modified TAIL-PCR method to obtain full-length genes directly from metagenomic DNA based on fragment sequences [33]. In this study, the fragment sequence AS10-66 was selected to obtain the full-length gene because it was the most abundant of all the sequences, representing about one quarter of all sequences (117/467). Moreover, AS10-66 has been reported to have low homology with known xylanases (Table S1).

Based on the fragment sequences of AS10-66 and using modified TAIL-PCR [33], a full-length xylanase gene (*xynAS10-66*) was direct cloned from metagenomic DNA of the soda lake sediment. The complete sequence of *xynAS10-66* contained an open reading frame of 1092 bp encoding a 363-residue polypeptide with a typical signal peptide (residues 1–24). The calculated mass of XynAS10-66 was 38.7 kDa, and the theoretical isoelectric point was 4.74. Sequence similarity searches showed that deduced XynAS10-66 shared the highest homology (44%) with the thermostable GH10 xylanase from *Thermomonospora alba* ULJB1, followed by *Thermobispora bispora* DSM 43833 (44%), an isolate recovered from decaying manure and *Nocardiopsis halotolerans* (43%), an isolate recovered from salt marsh soil in Kuwait [40].

Partial biochemical characterization of purified recombinant XynAS10-66

The gene encoding the mature proteins was expressed in *E. coli* BL21 (*DE3*). Following induction with IPTG at 30°C for 12 h, substantial xylanase activity was detected in the culture supernatant of recombinant cells. Using beechwood xylan as the substrate, rXynAS10-66 showed the highest activity at pH 7.0, while it retained more than 95% activity at pH 9.0, and >60% of the maximum activity at pH 7.0–11.0 (Figure 4A). Using beechwood xylan as the substrate, the K_m and V_{max} values were 1.5 ± 0.04 mg $\cdot$ mL^{-1}, 1102 ± 9.54 µmol $\cdot$ mg^{-1} $\cdot$ min^{-1}, respectively. Although the optimum temperature of the recombinant XynAS10-66 is neutral, it has substantial activity at alkaline pH, with more than 80% and 60% activity being retained at pH 10 and pH 11, respectively. This characterization differs from those of alkaline xylanases obtained from *Bacillus* and *Micrococcus* sp. isolated from other soda lakes, which showed optimum activity at pH 8 to 9 [14,15]. The optimal temperature for the enzyme activity of XynAS10-66 was 55°C (Figure 4B), while more than 60% activity was retained at 45°C to 65°C. The recombinant XynAS10-66 had the same optimum temperature as the alkaline xylanase from alkaliphilic *Micrococcus* sp. AR-135 [14]. However, it had a lower

activity than that of xylanases from *Bacillus halodurans* S7 [13] and two thermostable alkaline xylanases from an alkaliphilic *Bacillus* sp. [15]. BLAST analysis revealed that XynAS10-66 shared high similarity with a thermostable endo-beta-1,4-xylanase of *Thermomonospora alba* ULJB1 [41], which showed good activity at up to 95°C. Sequence alignment revealed that XynAS10-66 lacks a carbohydrate binding module (CBM) at the C terminal, which may have resulted in the lower temperature optimum of XynAS10-66. Future studies will be conducted to investigate this phenomenon. Moreover, other fragment sequences will be used to generate more full-length genes directly from the metagenomic DNA, and any genes generated will be characterized in subsequent studies.

In conclusion, culture-independent molecular methods led to recovery of abundant GH10 and GH11 xylanase genes from the sediment of the Lake Dabusu. Similarities among these sequences with known xylanases were low, and they were distantly related based on phylogenetic analysis. These results suggest that xylanase gene diversity within Lake Dabusu is high and most of them might be novel. Our study provides new insight into the genetic diversity and distribution of xylanases in soda lake environments and a rapid culture-independent molecular method for the retrieval of the source of xylanase genes with potential industrial applications.

Supporting Information

Table S1 GH10 xylanase gene fragments detected in the sediment of Lake Dabusu and their closest relative based on amino acid sequence identity and similarity.
(PDF)

Table S2 GH11 xylanase gene fragments detected in the sediment of Lake Dabusu and their closest relative based on amino acid sequence identity and similarity.
(PDF)

Table S3 Primers used for gene cloning and expression.
(PDF)

Author Contributions

Conceived and designed the experiments: JL XY. Performed the experiments: GW XH. Analyzed the data: GW TBN JL XY. Contributed reagents/materials/analysis tools: JL XY. Contributed to the writing of the manuscript: GW XH TBN JL XY.

References

1. Kulkarni N, Shendye A, Rao M (1999) Molecular and biotechnological aspects of xylanases. FEMS Microbiol Rev 23: 411–456.
2. Sunna A, Antranikian G (1997) Xylanolytic enzymes from fungi and bacteria. Crit Rev Biotechnol 17: 39–67.
3. Collins T, Gerday C, Feller G (2005) Xylanases, xylanase families and extremophilic xylanases. FEMS Microbiol Rev 29: 3–23.
4. Cantarel BL, Coutinho PM, Rancurel C, Bernard T, Lombard V, et al. (2009) The Carbohydrate-Active EnZymes database (CAZy): an expert resource for Glycogenomics. Nucl Acids Res 37: D233–238.
5. Biely P, Vrsanska M, Tenkanen M, Kluepfel D (1997) Endo-beta-1,4-xylanase families: differences in catalytic properties. J Biotechnol 57: 151–166.
6. Jeffries TW (1996) Biochemistry and genetics of microbial xylanases. Curr Opin Biotechnol 7: 337–342.
7. Beg QK, Kapoor M, Mahajan L, Hoondal GS (2001) Microbial xylanases and their industrial applications: a review. Appl Microbiol Biotechnol 56: 326–338.
8. Polizeli ML, Rizzatti AC, Monti R, Terenzi HF, Jorge JA, et al. (2005) Xylanases from fungi: properties and industrial applications. Appl Microbiol Biotechnol 67: 577–591.
9. Martin-Sampedro R, Rodriguez A, Ferrer A, Garcia-Fuentevilla LL, Eugenio ME (2012) Biobleaching of pulp from oil palm empty fruit bunches with laccase and xylanase. Bioresour Technol 110: 371–378.

10. Khandeparkar R, Bhosle NB (2007) Application of thermoalkalophilic xylanase from *Arthrobacter* sp. MTCC 5214 in biobleaching of kraft pulp. Bioresour Technol 98: 897–903.
11. Verma D, Kawarabayasi Y, Miyazaki K, Satyanarayana T (2013) Cloning, expression and characteristics of a novel alkalistable and thermostable xylanase encoding gene (Mxyl) retrieved from compost-soil metagenome. PLoS ONE 8: e52459.
12. Simkhada JR, Yoo HY, Choi YH, Kim SW, Yoo JC (2012) An extremely alkaline novel xylanase from a newly isolated Streptomyces strain cultivated in corncob medium. Appl Biochem Biotechnol 168: 2017–2027.
13. Mamo G, Delgado O, Martinez A, Mattiasson B, Hatti-Kaul R (2006) Cloning, sequence analysis, and expression of a gene encoding an endoxylanase from *Bacillus halodurans* S7. Mol Biotechnol 33: 149–159.
14. Gessesse A, Mamo G (1998) Purification and characterization of an alkaline xylanase from alkaliphilic *Micrococcus* sp AR-135. Journal of Industrial Microbiology and Biotechnology 20: 210–214.
15. Gessesse A (1998) Purification and properties of two thermostable alkaline xylanases from an alkaliphilic bacillus sp. Appl Environ Microbiol 64: 3533–3535.
16. Antony CP, Kumaresan D, Hunger S, Drake HL, Murrell JC, et al. (2013) Microbiology of Lonar Lake and other soda lakes. ISME J 7: 468–476.

17. Jones BE, Grant WD, Duckworth AW, Owenson GG (1998) Microbial diversity of soda lakes. Extremophiles 2: 191–200.

18. Zhao B, Chen S (2012) Alkalitalea saponilacus gen. nov., sp. nov., an obligately anaerobic, alkaliphilic, xylanolytic bacterium from a meromictic soda lake. Int J Syst Evol Microbiol 62: 2618–2623.

19. Rees HC, Grant WD, Jones BE, Heaphy S (2004) Diversity of Kenyan soda lake alkaliphiles assessed by molecular methods. Extremophiles 8: 63–71.

20. Foti MJ, Sorokin DY, Zacharova EE, Pimenov NV, Kuenen JG, et al. (2008) Bacterial diversity and activity along a salinity gradient in soda lakes of the Kulunda Steppe (Altai, Russia). Extremophiles 12: 133–145.

21. Lanzen A, Simachew A, Gessesse A, Chmolowska D, Jonassen I, et al. (2013) Surprising prokaryotic and eukaryotic diversity, community structure and biogeography of Ethiopian soda lakes. PLoS ONE 8: e72577.

22. Lin JL, Joye SB, Scholten JC, Schafer H, McDonald IR, et al. (2005) Analysis of methane monooxygenase genes in mono lake suggests that increased methane oxidation activity may correlate with a change in methanotroph community structure. Appl Environ Microbiol 71: 6458–6462.

23. Kovaleva OL, Tourova TP, Muyzer G, Kolganova TV, Sorokin DY (2011) Diversity of RuBisCO and ATP citrate lyase genes in soda lake sediments. FEMS Microbiol Ecol 75: 37–47.

24. Wang G, Meng K, Luo H, Wang Y, Huang H, et al. (2012) Phylogenetic diversity and environment-specific distributions of glycosyl hydrolase family 10 xylanases in geographically distant soils. PLoS ONE 7: e43480.

25. Shen J, Cao JT, Wu YH (2001) Paieoclimatic changes in Dabusu Lake. Chin J Oceanol Limnol 19: 91–96.

26. Brady SF (2007) Construction of soil environmental DNA cosmid libraries and screening for clones that produce biologically active small molecules. Nat Protoc 2: 1297–1305.

27. Wang G, Wang Y, Yang P, Luo H, Huang H, et al. (2010) Molecular detection and diversity of xylanase genes in alpine tundra soil. Appl Microbiol Biotechnol 87: 1383–1393.

28. White JR, Roberts M, Yorke JA, Pop M (2008) Figaro: a novel statistical method for vector sequence removal. Bioinformatics 24: 462–467.

29. Li W, Godzik A (2006) Cd-hit: a fast program for clustering and comparing large sets of protein or nucleotide sequences. Bioinformatics 22: 1658–1659.

30. Tamura K, Dudley J, Nei M, Kumar S (2007) MEGA4: Molecular Evolutionary Genetics Analysis (MEGA) Software Version 4.0. Mol Biol Evol 24: 1596–1599.

31. Saitou N, Nei M (1987) The neighbor-joining method: a new method for reconstructing phylogenetic trees. Mol Biol Evol 4: 406–425.

32. Schloss PD, Handelsman J (2005) Introducing DOTUR, a computer program for defining operational taxonomic units and estimating species richness. Appl Environ Microbiol 71: 1501–1506.

33. Huang H, Wang G, Zhao Y, Shi P, Luo H, et al. Direct and efficient cloning of full-length genes from environmental DNA by RT-qPCR and modified TAIL-PCR. Appl Microbiol Biotechnol 87: 1141–1149.

34. Petersen TN, Brunak S, von Heijne G, Nielsen H (2011) SignalP 4.0: discriminating signal peptides from transmembrane regions. Nat Methods 8: 785–786.

35. Miller GL, Blum R, Glennon WE, Burton AL (1960) Measurement of carboxymethylcellulase activity. Analytical Biochemistry 1: 127–132.

36. Solomon V, Teplitsky A, Shulami S, Zolotnitsky G, Shoham Y, et al. (2007) Structure-specificity relationships of an intracellular xylanase from *Geobacillus stearothermophilus*. Acta Crystallographica Section D 63: 845–859.

37. Ma Y, Zhang W, Xue Y, Zhou P, Ventosa A, et al. (2004) Bacterial diversity of the Inner Mongolian Baer Soda Lake as revealed by 16S rRNA gene sequence analyses. Extremophiles 8: 45–51.

38. Joshi AA, Kanekar PP, Kelkar AS, Shouche YS, Vani AA, et al. (2008) Cultivable bacterial diversity of alkaline Lonar lake, India. Microb Ecol 55: 163–172.

39. Pagaling E, Wang H, Venables M, Wallace A, Grant WD, et al. (2009) Microbial biogeography of six salt lakes in Inner Mongolia, China, and a salt lake in Argentina. Appl Environ Microbiol 75: 5750–5760.

40. Al-Zarban SS, Abbas I, Al-Musallam AA, Steiner U, Stackebrandt E, et al. (2002) *Nocardiopsis halotolerans* sp. nov., isolated from salt marsh soil in Kuwait. Int J Syst Evol Microbiol 52: 525–529.

41. Blanco J, Coque JJ, Velasco J, Martin JF (1997) Cloning, expression in Streptomyces lividans and biochemical characterization of a thermostable endo-beta-1,4-xylanase of *Thermomonospora alba* ULJB1 with cellulose-binding ability. Appl Microbiol Biotechnol 48: 208–217.

11

Toxin-Antitoxin Systems in the Mobile Genome of *Acidithiobacillus ferrooxidans*

Paula Bustamante[1], Mario Tello[2], Omar Orellana[1]*

1 Programa de Biología Celular y Molecular, ICBM, Facultad de Medicina, Universidad de Chile, Santiago, Chile, 2 Centro de Biotecnología Acuícola, Departamento de Biología, Facultad de Química y Biología, Universidad de Santiago de Chile, Santiago, Chile

Abstract

Toxin-antitoxin (TA) systems are genetic modules composed of a pair of genes encoding a stable toxin and an unstable antitoxin that inhibits toxin activity. They are widespread among plasmids and chromosomes of bacteria and archaea. TA systems are known to be involved in the stabilization of plasmids but there is no consensus about the function of chromosomal TA systems. To shed light on the role of chromosomally encoded TA systems we analyzed the distribution and functionality of type II TA systems in the chromosome of two strains from *Acidithiobacillus ferrooxidans* (ATCC 23270 and 53993), a Gram-negative, acidophilic, environmental bacterium that participates in the bioleaching of minerals. As in other environmental microorganisms, *A. ferrooxidans* has a high content of TA systems (28-29) and in twenty of them the toxin is a putative ribonuclease. According to the genetic context, some of these systems are encoded near or within mobile genetic elements. Although most TA systems are shared by both strains, four of them, which are encoded in the active mobile element ICE*Afe*1, are exclusive to the type strain ATCC 23270. We demostrated that two TA systems from ICE*Afe*1 are functional in *E. coli* cells, since the toxins inhibit growth and the antitoxins counteract the effect of their cognate toxins. All the toxins from ICE*Afe*1, including a novel toxin, are RNases with different ion requirements. The data indicate that some of the chromosomally encoded TA systems are actually part of the *A. ferrooxidans* mobile genome and we propose that could be involved in the maintenance of these integrated mobile genetic elements.

Editor: Finbarr Hayes, University of Manchester, United Kingdom

Funding: This work was supported by grants from Fondecyt Chile 1110203 to OO (http://www.conicyt.cl/fondecyt/) and Proyecto Bicentenario PDA20 (www.conicyt.cl) and Proyecto FIA PYT20120056 to MT (www.fia.cl). PB was the recipient of a graduate studies fellowship and supporting fellowship AT-24100112 from Conicyt (www.conicyt.cl), Chile. The funders had no role in study design, data collection and analysis, decision to publish, or preparation of the manuscript.

Competing Interests: The authors have declared that no competing interests exist.

* Email: oorellan@med.uchile.cl

Introduction

Toxin-antitoxin (TA) systems are small genetic modules widely distributed in bacteria and archaea [1] that are comprised of a pair of genes encoding a stable toxin and an unstable antitoxin capable of inhibiting toxin activity [1,2]. In contrast to bacteriocins [3] and toxins from contact-dependent inhibition systems [4], TA toxins are not secreted and inhibit cell growth by targeting key molecules in essential cellular processes such as DNA replication, mRNA stability or protein, cell-wall or ATP biosynthesis [1].

TA systems were first discovered as systems that contribute to plasmid maintenance by a phenomenon denoted as "post-segregational killing" or "addiction" [5,6]. When a plasmid encoding a TA system is lost from a cell, the toxin is released from the existing TA complex as the unstable antitoxin decays, resulting in cell growth inhibition and eventually death [7]. In addition to plasmids, TA systems are also found in bacterial chromosomes, particularly in free-living prokaryotic cells [8,9], but their function is not well understood [10]. Although chromosomal TA systems are not essential for normal cell growth [11], it is believed that they play key roles in stress response [12], persister phenotype [13] and stabilization of horizontally acquired genetic elements [14].

Five types of TA systems have been proposed to date. All of them comprise a toxic protein (toxin) and an antitoxin that can be either a small non-coding RNA (type I and type III [15,16]) or a low molecular weight protein (types II, IV and V [17–19]). Recent studies have identified an ever-increasing number of experimentally defined, or putative, type I, type II and type III TA systems [8,9,15,16]. On the other hand, type IV and type V TA systems were recently discovered and to date have only a few representatives [17,18,20,21].

Type II TA systems, the most well known and the interest of this work, are encoded in operons consisting of genes that overlap (or are a few bases apart); the toxin and its cognate antitoxin form a stable protein TA complex that prevents the toxic effect [1]. Type II TA systems are diverse and are classified in 12 toxin and 20 antitoxin super-families based on sequence similarity [19]. Targets of type II toxins are also diverse, most frequently acting to cleave mRNA at specific sequences to inhibit translation in a ribosome-dependent or independent manner [22,23].

Type II systems are thought to move from one genome to another by horizontal gene transfer (HGT) [9]. In fact, some TA systems (besides plasmidial TA) are localized within mobile genetic elements (MGEs) such as transposons and superintegrons [24,25].

Chromosomally encoded TA systems have also been shown to have a role in the stabilization of large genomic fragments and integrative-conjugative elements (ICEs) [14,26]. Thus, it is possible that TA systems considered to be chromosomally encoded could actually be associated with active or inactive integrated genetic elements.

The number of type II TA systems in an organism varies greatly, not only from one bacterial species to another, but also between isolates from the same species [9,19]. Most of the organisms that have many TA systems grow in nutrient-limited environments and/or are chemolithoautotrophs (although a high TA content is observed in some obligate intracellular bacterial genomes [19]), leading to the proposal that these systems might be beneficial for this type of slow-growing microorganisms [9].

Acidithiobacillus ferrooxidans is an environmental acidophilic, chemolithoautotrophic Gram-negative γ-proteobacterium (although some discrepancies exist concerning its classification in this bacterial class [27]) that obtains its energy from the oxidation of ferrous ions or reduced sulfur compounds [28]. It belongs to the consortium of microorganisms that participate in the bioleaching of minerals, being a model organism for the study of bioleaching, metabolic and genomic studies of acidophilic bacteria [28,29]. Although no genetic system has been developed for this microorganism, the genome sequences of two strains are available in public databases (ATCC 23270 and ATCC 53993 strains). A number of MGE-related DNA sequences have been described in its genome as insertion sequence elements, transposons and plasmids [28,30,31], including a large genomic island [32] and an actively excising integrative-conjugative element (ICE*Afe*1) [33]. As these MGEs are stably integrated into the chromosome of *A. ferrooxidans* and a number of TA-related proteins have been annotated in the genome of the two sequenced strains [28], it is possible that this environmental bacterium relies on TAs to avoid the loss of these mobile elements.

To shed light into the role of chromosomally encoded TA systems from *A. ferrooxidans* and their relation with MGEs, we studied the distribution of type II TA systems in the two available sequenced genomes in public databases. We also studied the functionality of the systems encoded in the actively excising ICE*Afe*1. Based on our data we propose that type II TA systems from *A. ferrooxidans* could be part of its mobile genome and might be involved in the maintenance of its MGEs.

Materials and Methods

Bioinformatic analysis

In silico screening for type II TA systems in *A. ferrooxidans* ATCC 23270 (NCBI RefSeq NC_011761) and ATCC 53993 (NCBI RefSeq NC_011206) was conducted using the web-based search tool TADB (http://bioinfo-mml.sjtu.edu.cn/TADB/) [34], an online resource of type II TA loci-relevant data from 'wet' experimental data as well as information garnered by bioinformatics analyses. We also used the data from RASTA-Bacteria (http://genoweb1.irisa.fr/duals/RASTA-Bacteria/) [35], an automated method allowing identification of TA loci in sequenced prokaryotic genomes, whether they are annotated open reading frames or not.

The classification of putative toxin and antitoxins in super-families was according to Leplae et al. [19]. Using BLASTP, each putative toxin and antitoxin from *A. ferrooxidans* was compared against the sequences of toxins and antitoxins from the different super-families described by Leplae et al. [19], either 'original', 'similar' or validated sequences. An E-value score threshold of 0.001 and 50% query residues aligned were used to select

candidates. Each toxin or antitoxin was assigned to the super-family with the best hit and a name was given according to the best protein hit. Protein structure predictions were assayed by Phyre 2.0 server [36].

The Integrated Microbial Genomes platform (IMG, http://img.jgi.doe.gov/cgi-bin/w/main.cgi) [37] was used for the visualization of genome contexts and characteristics of each gene and protein.

Phylogenetic analysis

Multi-alignment between nucleotide sequences encoding TA toxins was performed using ClustalW [38]. The parameters were set up to align codons using Gonnet as substitution matrix [39]. The evolutionary history was inferred using the Neighbor-Joining method [40]. The optimal tree with the sum of branch length = 380.7 is shown. The confidence probability (multiplied by 100) that the interior branch length is greater than 0, as estimated using the bootstrap test (1000 replicates), is shown next to the branches [41]. The evolutionary distances were computed using the Maximum Composite Likelihood method [42]. The rate of variation among sites was modeled with a gamma distribution (shape parameter = 1). The analysis involved 72 nucleotide sequences. All ambiguous positions were removed for each sequence pair. There were a total of 618 positions in the final dataset. Multialignment and evolutionary analyses were conducted in MEGA5 [43].

Bacterial strains and growth conditions

Escherichia coli JM109 strain was used for cloning and plasmid maintenance. *E. coli* BL21(DE3)pLysS strain was used for recombinant protein expression and BL21(DE3) strain for plasmid maintenance tests. The strains were grown in Luria-Bertani (LB) or on LB agar at 37°C with 1% glucose. When appropriate, media were supplemented with ampicillin (100 μg/ml) or chloramphenicol (34 μg/ml). When both antibiotics were used together, they were added to half of the concentration.

Cloning of TA systems

Toxin and antitoxin genes were amplified by PCR using PfuUltra II Fusion HS DNA Polymerase (Agilent Technologies), *A. ferrooxidans* ATCC 23270 chromosomal DNA as a template and the oligonucleotides indicated in Table 1. The pETDuet-1 expression vector (Novagen) was used for cloning. The amplified genes and vector DNA were double digested with *Bam*HI/*Hind*III or *Nde*I/*Xho*I according to the protocols indicated by the manufacturer (ThermoScientific), ligated with T4 DNA Ligase (New England Biolabs), and used to transform *E. coli* JM109 by a chemical method [44]. Three types of recombinant vectors were constructed: pETDuet-T, with toxin genes cloned into multiple cloning site-1 (MCS1) so that the toxins are expressed as N-terminal (His)₆-tagged proteins; pETDuet-A, with the antitoxin genes cloned into MCS1; and pETDuet-TA, corresponding to pETDuet-T vectors with the cognate antitoxin genes cloned into MCS2. Transformants were selected with ampicillin and checked by colony PCR with the oligonucleotides indicated in Table 1. Cloned genes were analyzed by DNA sequencing (Macrogen, USA). Recombinant plasmids were used for transformation of *E. coli* BL21(DE3)pLysS cells by a chemical method [44].

For the plasmid maintenance test (see below) we constructed pACYCDuet-A plasmids, corresponding to the pACYCDuet-1 vector (Novagen) with antitoxin genes cloned into its MCS2. DNA fragments containing the antitoxin genes were obtained from the corresponding pETDuet-TA plasmids double digested with *Nde*I/*Xho*I. The fragments were ligated to pACYCDuet-1 double

Table 1. Oligonucleotides used.

Name	Sequence 5'-3'	Use
HtoxinDuet-F	AGA TCT TCT GAT GGG CGC TGC	Forward oligonucleotide for cloning the toxin gene from MazEF-1 system on pGEM-T Easy and further sub-cloning into the MCS1 from pETDuet-1.
HtoxinDuet-R	AAG CTT CTC CCA ATA GCT ATG CC	Reverse oligonucleotide for cloning the toxin gene from MazEF-1 system on pGEM-T Easy and further sub-cloning into the MCS1 from pETDuet-1.
AntitoxinDuet-F	ACC ATA TGC GGG TGA TTG TG	Forward oligonucleotide for cloning the toxin gene from MazEF-1 system on pGEM-T Easy and further sub-cloning into the MCS2 from pETDuet-1.
AntitoxinDuet-R	ATC TCG AGC GCC CAT CAG AG	Reverse oligonucleotide for cloning the toxin gene from MazEF-1 system on pGEM-T Easy and further sub-cloning into the MCS1 from pETDuet-1.
AFE1361_NdeI	GCC AGA GGC ATA TGA TTA CAA TG	Forward oligonucleotide for cloning of the antitoxin gene from StbC/VapC-3 system into the MCS2 from pETDuet-1
AFE1361_XhoI	GGT CTC GAG CAA AAT CAT GC	Reverse oligonucleotide for cloning of the antitoxin gene from StbC/VapC-3 system into the MCS2 from pETDuet-1
AFE1362_BamHI	ATA GGA TCC CAT GAT TTT GCT GG	Forward oligonucleotide for cloning of the toxin gene from StbC/VapC-3 system into the MCS1 from pETDuet-1
AFE1362_HindIII	CAT TAA GCT TGT CTC ATG TCT C	Reverse oligonucleotide for cloning of the toxin gene from StbC/VapC-3 system into the MCS1 from pETDuet-1
AFE1367_NdeI	TGT GCA TAT GCT TGA TAA GC	Oligonucleotide forward for cloning of the antitoxin gene from TA system number 9 into the MCS2 from pETDuet-1
AFE1367_XhoI	TCT CTC GAG TTG CGC ATC AAC	Reverse oligonucleotide for cloning of the antitoxin gene from TA system number 9 into the MCS2 from pETDuet-1
AFE1368_BamHI	GGG GAT CCG AAA TTT TTA GTT G	Forward oligonucleotide for cloning of the toxin gene from TA system number 9 into the MCS1 from pETDuet-1
AFE_1368_HindIII	CGA TAA GCT TCT TCA CTG ATG G	Reverse oligonucleotide for cloning of the toxin gene from TA system number 9 into the MCS1 from pETDuet-1
AFE1383_NdeI	CAT CCA TAT GAG CGG TGG CAA TG	Forward oligonucleotide for cloning of the antitoxin gene from EcoA1/EcoT1-1 system into the MCS2 from pETDuet-1
AFE1383_XhoI	CGA TCT CGA GTC ATA GCG CAC	Reverse oligonucleotide for cloning of the antitoxin gene from EcoA1/EcoT1-1 system into the MCS2 from pETDuet-1
AFE1384_BamHI	GCA GGA TCC TTT GCT CTG GGT G	Forward oligonucleotide for cloning of the toxin gene from EcoA1/EcoT1-1 system into the MCS1 from pETDuet-1
AFE1384_HindIII	GAC ATA AGC TTC GCT CAT CTC G	Reverse oligonucleotide for cloning of the toxin gene from EcoA1/EcoT1-1 system into the MCS1 from pETDuet-1
pET Upstream Primer	ATG CGT CCG GCG TAG A	Oligonucleotide for sequencing genes inserted into MCS1 from pETDuet-1
DuetDOWN-1 Primer	GAT TAT GCG GCC GTG TAC AA	Oligonucleotide for sequencing genes inserted into MCS1 from pETDuet-1
DuetUP2 Primer	TTG TAC ACG GCC GCA TAA TC	Oligonucleotide for sequencing genes inserted into MCS2 from pETDuet-1 and pACYCDuet-1
T7 Terminator Primer	GCT AGT TAT TGC TCA GCG G	Oligonucleotide for sequencing genes inserted into MCS2 from pETDuet-1 and pACYCDuet-1

digested with the same enzymes and the constructs were used to transform *E. coli* JM109 by a chemical method [44]. Transformants were selected with chloramphenicol and checked by colony PCR with the oligonucleotides indicated in Table 1. Recombinant plasmids were used for transformation of *E. coli* BL21(DE3) cells by a chemical method [44].

Evaluation of toxicity in *E. coli*

The toxicity of toxin proteins in *E. coli* BL21(DE3)pLysS was determined by the growth pattern of cultures on liquid and solid media in the presence or absence of the inducer IPTG. Overnight cultures of *E. coli* BL21(DE3)pLysS cells with plasmids containing toxin, antitoxin or both genes of each TA system were diluted 100-fold and grown in LB broth until an OD_{600} of 0.2-0.3. At this point, 1 mM IPTG was added and growth was monitored by measuring OD_{600} of the cultures in a microplate spectrophotometer (Epoch). Three hours after the induction, aliquots of each culture were 10-fold serial diluted, and 5 µl of each dilution were

spotted on LB agar without IPTG and growth at 37°C for 16 hours. In addition, following the induction with IPTG, a viability assay was performed. At different time intervals culture samples were serially diluted (10-fold) and aliquots were seeded on LB plates to determine the number of colony-forming units (CFU/ml).

Plasmid maintenance test

E. coli BL21(DE3) was double-transformed with the corresponding pETDuet-T and pACYCDuet-A or pETDuet-1 and pACYCDuet-A vectors. With these cultures a plasmid maintenance test was performance as in [45].

Protein expression and purification of (His)$_6$-toxins

Toxins were expressed in *E. coli* BL21(DE3)pLysS carrying the corresponding plasmids after induction with 1 mM IPTG for three hours and purified by Ni^{+2}-affinity chromatography.

(His)$_6$-MazF-1 was purified under native conditions from *E. coli* carrying pETMazF-1. The cells were harvested by centrifugation

Table 2. Putative type II TA systems encoded on A. ferrooxidans[§].

TA	Antitoxin locus	Accession gi	Hits in CDD	Antitoxin super-family[a] (name)	Toxin locus	Accession gi	Hits in CDD	Toxin super-family[a] (name)	Chromosomal or in a MGE[b]
1	AFE_0085	218866122	Phd super family [cl18766]; COG4118	Phd (Phd-1)	AFE_0086	218666107	PIN_MT3492 [cd09874]	NI (tox1)	Chr
	Lferr_0087	198282235	Phd super family [cl18766]; COG4118	Phd (Phd-1)	Lferr_0088	198282236	PIN_MT3492 [cd09874]	NI (tox1)	Chr
2	AFE_0089	218666717	HTH_XRE [cd00093]	HigA (HigA-1)	AFE_0088	218665352	Plasmid_killer super family [cl01422]	RelE/ParE (HigB-1)	Chr
	Lferr_0091	198282239	HTH_XRE [cd00093]	HigA (HigA-1)	Lferr_0090	198282238	Plasmid_killer super family [cl01422]	RelE/ParE (HigB-1)	Chr
3	AFE_0414	218665317	Phd super family [cl18766]; COG4118	Phd (Phd-2)	AFE_0413	218666088	PIN_SlI0205 [cd09872]	VapC (VapC-1)	Chr
	Lferr_0577	198282717	Phd super family [cl18766]; COG4118	Phd (Phd-2)	Lferr_0576	198282716	PIN_SlI0205 [cd09872]	VapC (VapC-1)	Chr
4	AFE_0477	218665855	VagC super family [cl18787]	VapB (VapB-1)	AFE_0478	218667984	PIN_VapC-FitB [cd09881]	VapC (VapC-2)	Chr
	Lferr_0637	198282777	VagC super family [cl18787]	VapB (VapB-1)	Lferr_0638	198282778	PIN_VapC-FitB [cd09881]	VapC (VapC-2)	Chr
5	AFE_0869	218667933	PhdYeFM_antitox [pfam02604]	NI (antitox5)	AFE_0870	218667345	Plasmid_Txe super family [cl17389]	RelE/ParE (YoeB-1)	Chr
	Lferr_0994	198283126	PhdYeFM_antitox [pfam02604]	NI (antitox5)	Lferr_0995	198283127	Plasmid_Txe super family [cl17389]	RelE/ParE (YoeB-1)	Chr
6	AFE_0889	218667173	PhdYeFM_antitox super family [cl09153]	Phd (StbD-1)	AFE_0890	218665915	RelE [COG2026]	RelE/ParE (StbE-1)	Chr
	Lferr_1011	198283139	PhdYeFM_antitox super family [cl09153]	Phd (StbD-1)	Lferr_1012	198283140	RelE [COG2026]	RelE/ParE (StbE-1)	Chr
7	AFE_1413	218667280	PRK09974; AbrB [COG2002]	NI (antitox7)	AFE_1412	218665529	Toxin_YhaV [pfam11663]	NI (tox7)	Chr
	Lferr_1133	198283260	PRK09974; AbrB [COG2002]	NI (antitox7)	Lferr_1132	198283259	Toxin_YhaV [pfam11663]	NI (tox7)	Chr
8	AFE_1418	218665082	COG4453	NI (antitox8)	AFE_1417	218665278	Acetyltransf_1 [pfam00583]	NI (tox8)	Chr
	Lferr_1137	198283264	COG4453	NI (antitox8)	Lferr_1136	198283263	Acetyltransf_1 [pfam00583]	NI (tox8)	Chr
9	AFE_1560	218667662	DUF4415 [pfam14384]	NI (antitox9)	AFE_1559	218665905	NI	NI (tox9)	ICEAfe2-23270
	Lferr_0133	198282280	DUF4415 [pfam14384]	NI (antitox9)	Lferr_0132	198282279	NI	NI (tox9)	GI
10	AFE_1579	218667623	DUF2191 [pfam09957]	RelB (RelB-1)	AFE_1578	218666342	VapC [COG1487]	NI (tox10)	ICEAfe2-23270
	Lferr_0230	198282374	DUF2191 [pfam09957]	RelB (RelB-1)	Lferr_0229	198282373	VapC [COG1487]	NI (tox10)	ICEAfe2-53993
	Lferr_1290	198283414	DUF2191 [pfam09957]	RelB (RelB-1)	Lferr_1289	198283413	VapC [COG1487]	NI (tox10)	GI
11	AFE_1614	218666352	Excise [TIGR01764]	NI (antitox11)	AFE_1613	218668111	PIN_3 [pfam13470]	NI (tox11)	ICEAfe2-23270
	Lferr_0234	198282378	Excise [TIGR01764]	NI (antitox11)	Lferr_0233	198282377	PIN_3 [pfam13470]	NI (tox11)	GI
12	AFE_1631	218665941	HTH_XRE [cd00093]	NI (antitox12)	AFE_1630'		upstrm_HI1419 [TIGR02683]	NI (tox12)	ICEAfe2-23270
	Lferr_1332	198283452	HTH_XRE [cd00093]	NI (antitox12)	Lferr_1331	198283451	upstrm_HI1419 [TIGR02683]	NI (tox12)	ICEAfe2-53993
13	AFE_1700	218667519	NI	RelB (Paa1-1)	Pseudo		ParE [COG3668]	NI (tox13)	ICEAfe2-23270
	Lferr_0263	198282407	NI	RelB (Paa1-1)	Lferr_0264	198282408	ParE [COG3668]	NI (tox13)	ICEAfe2-53993
	Lferr_1399	198283518	NI	RelB (Paa1-1)	Lferr_1400	198283519	ParE [COG3668]	NI (tox13)	GI

Table 2. Cont.

TA	Antitoxin locus	Accession gi	Hits in CDD	Antitoxin super-family[a] (name)	Toxin locus	Accession gi	Hits in CDD	Toxin super-family[a] (name)	Chromosomal or in a MGE[b]
14	AFE_1732	218667149	YcfA super family [cl00752]	NI (antitox14)	AFE_1733	218666062	UPF0150 [pfam03681]	NI (tox14)	ICEAfe2-23270
	Lferr_1422	198283540	YcfA super family [cl00752]	NI (antitox14)	Lferr_1423	198283541	UPF0150 [pfam03681]	NI (tox14)	ICEAfe2-53993
15	AFE_1779	218665129	StbC super family [cl01921]	NI (antitox15)	AFE_1780	218665399	PIN_VapC-FitB [cd09881]	VapC (VapC-4)	Chr
	Lferr_1455	198283572	StbC super family [cl01921]	NI (antitox15)	Lferr_1456	198283573	PIN_VapC-FitB [cd09881]	VapC (VapC-4)	Chr
16	AFE_2130	218665450	VagC [COG4456]	VapB (MvpA-1)	AFE_2129	218668039	PIN_VapC-FitB [cd09881]	VapC (VapC-5)	Chr
	Lferr_1789	198283896	VagC [COG4456]	VapB (MvpA-1)	Lferr_1788	198283895	PIN_VapC-FitB [cd09881]	VapC (VapC-5)	Chr
17	AFE_2415	218666500	DUF4415 [pfam14384]	NI (antitox17)	AFE_2414	218666978	DUF497 super family [cl01108]	NI (tox17)	Chr
	Lferr_1314	198283435	DUF4415 [pfam14384]	NI (antitox17)	Lferr_1315	198283436	DUF497 super family [cl01108]	NI (tox17)	ICEAfe2-53993
	Lferr_2046	198284147	DUF4415 [pfam14384]	NI (antitox17)	Lferr_2045	198284146	DUF497 super family [cl01108]	NI (tox17)	Chr
18	AFE_2658	218666707	DUF4415 [pfam14384]	NI (antitox18)	AFE_2657'		NI	NI (tox18)	Chr
	Lferr_2284	198284371	DUF4415 [pfam14384]	NI (antitox18)	Lferr_2283	198284370	NI	NI (tox18)	Chr
19	AFE_2771	218665366	VagC super family [cl18787]	VapB (VapB-2)	Pseudo			NI (tox19)	Chr
	Lferr_2392	198284478	VagC super family [cl18787]	VapB (VapB-2)	Lferr_2391	198284477	PIN_VapC-FitB [cd09881]	NI (tox19)	Chr
20	AFE_2886	218668170	Antitoxin-MazE super family [cl00877]	VapB (MazE-2)	AFE_2885	218665972	PemK [pfam02452]	CcdB/MazF (PemK-1)	Chr
	Lferr_2506	198284586	Antitoxin-MazE super family [cl00877]	VapB (MazE-2)	Lferr_2505	198284585	PemK [pfam02452]	CcdB/MazF (PemK-1)	Chr
21	AFE_2889	218666996	Antitoxin-MazE super family [cl00877]	VapB (VapB-3)	AFE_2888	218666880	PIN_VapC-Smg6-like [cd09855]	VapC (NspT2-1)	Chr
	Lferr_2509	198284589	Antitoxin-MazE super family [cl00877]	VapB (VapB-3)	Lferr_2508	198284588	PIN_VapC-Smg6-like [cd09855]	VapC (NspT2-1)	Chr
22	AFE_2981	218665144	NI	NI (antitox22)	AFE_2982	218667184	NI	RelE/ParE (CcrT1-1)	Chr
	Lferr_2595'		NI	NI (antitox22)	Lferr_2595"		NI	RelE/ParE (CcrT1-1)	Chr
23	AFE_2983	218666488	Phd [COG4118]	NI (antitox23)	AFE_2984	218665493	PIN_MT3492 [cd09874]	NI (tox23)	Chr
	Lferr_2596	198284676	Phd [COG4118]	NI (antitox23)	Lferr_2597	198284677	PIN_MT3492 [cd09874]	NI (tox23)	Chr
24	AFE_3174	218665847	Phd super family [cl18766]	NI (antitox24)	AFE_3173	218665111	PIN_2 [pfam10130]	NI (tox24)	Chr
	Lferr_2770	198284847	Phd super family [cl18766]	NI (antitox24)	Lferr_2769	198284846	PIN_2 [pfam10130]	NI (tox24)	Chr
25	AFE_3268	218667137	PhdYefM_antitox [pfam02604]	Phd (YefM-1)	AFE_3269	218665388	PIN_3 super family [cl17397]	NI (tox25)	Chr
	Lferr_2866	198284937	PhdYefM_antitox [pfam02604]	Phd (YefM-1)	Lferr_2867	198284938	PIN_3 super family [cl17397]	NI (tox25)	Chr
26	AFE_1098	218667753	MazE [COG2336]	VapB (MazE-1)	AFE_1099	218666923	PemK super family [cl00995]	CcdB/MazF (MazF-1)	ICEAfe1
27	AFE_1361	218667301	StbC super family [cl01921]	NI (antitox27)	AFE_1362	218667390	PIN_VapC-FitB [cd09881]	VapC (VapC-3)	ICEAfe1
28	AFE_1367	218667849	DUF433 [pfam04255]	NI (antitox28)	AFE_1368	218665288	COG4634 super family [cl18792]	NI (tox28)	ICEAfe1
29	AFE_1383	218666557	HTH_XRE [cd00093]; HipB [COG1396]	HigA (EcoA1-1)	AFE_1384	218667318	Gp49 super family [cl01470]	RelE/ParE (EcoT1-1)	ICEAfe1

[s]Locus, accession gi and hits in CDD are according to the NCBI.

[a] according to the classification by Leplae et al, [19]; NI: not identified.

[b]Chr: chromosomal TA II; when a TA II is encoded in a MGE, the name of the element is indicated; in the case of ICEAfe2, the name is followed by the number of A. ferrooxidans ATCC strain where the TA is present.

at 3800 g at 4°C for 10 minutes, resuspended in native lysis buffer (50 mM NaH_2PO_4, 300 mM NaCl, 20 mM imidazole, pH 8.0) with 1 mM PMSF and subjected to lysis by sonication. The protein extract was cleared by centrifugation at 15350 g at 4°C for 30 minutes and the supernatant was applied to a column containing 500 µl of Ni^{+2}-Sepharose resin (GE Healthcare). The resin was washed with 30 column volumes of washing buffer (50 mM NaH_2PO_4, 300 mM NaCl, 150 mM imidazole, pH 8.0) and the retained proteins were eluted with the same buffer containing 250 mM imidazole. The purified proteins were dialyzed against storage buffer (25 mM Tris-HCl pH 8.0, 100 mM NaCl, 20% glycerol, 0.5 mM dithiothreitol) at 4°C for 16 hours followed by a second dialysis for 4 hours against fresh storage buffer and stored at −20°C.

$(His)_6$-VapC-3, $(His)_6$-tox28 and $(His)_6$-EcoT1-1 were purified from E. coli carrying the corresponding pETDuet-TA plasmids. Cells were harvested by centrifugation at 3800 g at 4°C for 10 minutes, resuspended in denaturing lysis buffer (100 mM NaH_2PO_4, 10 mM Tris-HCl, 6 M GuHCl, pH 8.0) and incubated at ambient temperature for 1 h with agitation to achieve the TA complexes dissociation. Protein extracts were cleared by centrifugation at 15350 g at ambient temperature for 30 minutes and the supernatant applied to a column containing 500 µl of Ni^{+2}-Sepharose resin (GE Healthcare). The resin was washed with 30 column volumes of denaturing wash buffer (100 mM NaH_2PO_4, 10 mM Tris-HCl, 8 M urea, 20 mM imidazole, pH 8.0). The elution of bound proteins was achieved by increasing the imidazole concentration in the buffer to 50 mM (for $(His)_6$-VapC-3 and $(His)_6$-tox28) or 100 mM (for $(His)_6$-EcoT1-1). The purified proteins were refolded by dialysis against storage buffer as before and stored at -20°C.

All proteins were quantified by the method of Bradford (Bio-Rad Protein Assay) in a microplate spectrophotometer (Epoch), analyzed by Tricine-SDS-PAGE [46] and visualized by staining with Coomassie brilliant blue.

RNase activity

The digestion reaction mixture (20 µl) consisted of 1.6 µg of MS2 RNA substrate (Roche) in 10 mM Tris-HCl (pH 7.8) with or without 10 mM $MgCl_2$ or $MnCl_2$, 40 U RNase inhibitor Ribolock (ThermoScientific) and 100 pmol of each purified toxin. Parallel reactions with 12 mM EDTA were used as controls. The reactions were incubated for 15 or 30 minutes at 37°C and stopped by adding 4 µl of 6X electrophoresis loading buffer. The reaction products were run on a 1% agarose gel (in 1X TAE) and visualized by staining with GelRed (Biotium).

Results

Content of type II TA systems in A. ferrooxidans

To further understand the role of chromosomally encoded TA systems in environmental microorganisms, we searched for type II TA systems (hereafter named as TA) in the publicly available genome sequences from two strains (ATCC 23270 and ATCC 53993) of the bioleaching bacterium A. ferrooxidans.

To identify shared TA between both strains, BLASTP searches were conducted, using toxin and antitoxin protein sequences from one strain as query (based on the information available in TADB) to search the proteins encoded by the other strain. From this analysis, 29 TA are encoded in A. ferrooxidans ATCC 23270 (including TA 13 and 19 in which the toxin gene corresponds to a pseudogene; Table 2 and Figure 1) and 28 in ATCC 53993 (including TA 10, 13 and 17 which have two identical copies; Table 2 and Figure 1). A total of 13 new putative TA were

identified that were either not assigned or erroneously assigned by TADB (Supporting information S1). In support of this, we note that A. ferrooxidans has a high TA content and it is expected that this characteristic is shared with other bioleaching bacteria. When we analyzed the TA content of other sequenced acidophilic bioleaching bacteria with the RASTA-Bacteria platform (because they are not available in TADB), we found that A. caldus SM-1, Leptospirillum ferriphilum ML-04, L. ferrooxidans C2-3 and A. ferrivorans SS3 encode at least, 30, 16, 29 and more than 50 putative TA, respectively (data not shown).

As is described in some TA (mainly in higBA family) [47–49], an organization opposite to the classical gene arrangement (toxin gene encoded after the antitoxin gene) was found in six systems from A. ferrooxidans (TA 2, 9, 12, 17, 18 and 29).

All TA from A. ferrooxidans ATCC 53993 have counterparts in the other strain (sharing 94-100% amino acid identity; Figure 1, TA 1-25). Strikingly, type strain ATCC 23270 contains four exclusive TA (TA 26-29), encoded in a MGE as discussed below (Figure 1, highlighted in red). As TA 1 to 25 are the same in both strains, we will refer only to TA from ATCC 23270 strain hereafter (if not otherwise indicated).

Nowadays, TA systems are classified as independent toxin and antitoxin super-families instead of TA families as before [19]. Based on this classification, we assigned 14 antitoxins and 13 toxins to a given super-family according to amino acid sequence similarity (Table 2). The prevalent antitoxin super-families in A. ferrooxidans are Phd and VapB (4 and 5 representative of each respectively), whereas the toxins that we could assign to a super-family belong to VapC, RelE/ParE and CcdB/MazF super-families, with 6, 5 and 2 representatives each, respectively. Specifically, toxins containing PIN domains are the most abundant in A. ferrooxidans (thirteen TA, 48% of the toxins). It is known that TA toxins show limited sequence similarity, despite having common folds [50] and this might explain why sixteen toxins from A. ferrooxidans could not be assigned in the current classification. Indeed, there are seven PIN domain toxins in A. ferrooxidans (TA 1, 10-11, 19 and 23-25, Table 2) that do not show a suitable sequence similarity with VapC super-family proteins (those containing PIN domains). Nonetheless, these toxins show high structural homology with characterized VapC toxins (Supporting information S2) and are clustered within the VapC super-family in a phylogenetic analysis (Figure 2, green squared). Using the same phylogenetic approach, the rest of the unclassified toxins grouped within different super-families, with some of them forming a different clade (e. g. TA 1, 9, 10, 23 and 28; Figure 2, shown in open symbols).

Based on the conserved domain database (CDD [51]) hits and the super-families of each toxin identified in A. ferrooxidans, we predicted that twenty toxins might be ribonucleases that possibly function as translation inhibitors. Functional analysis of some of these ribonucleases associated with MGEs is described below.

TA encoded in MGEs

TA may be associated with MGEs allowing their movement between microorganisms by HGT [25,52]. To elucidate whether chromosomal TA systems from A. ferrooxidans form part of MGEs and to predict whether they have been acquired or they have the potential to be mobilized by HGT, we analyzed their genetic context.

Recently, we identified and characterized ICEAfe1, an active 291-kbp ICE from A. ferrooxidans type strain ATCC 23270; this element is excised from the chromosome of the bacterium and has the potential to be transferred by conjugation [33]. A. ferrooxidans ATCC 53993 also encodes a 164-kbp genomic island (GI) that

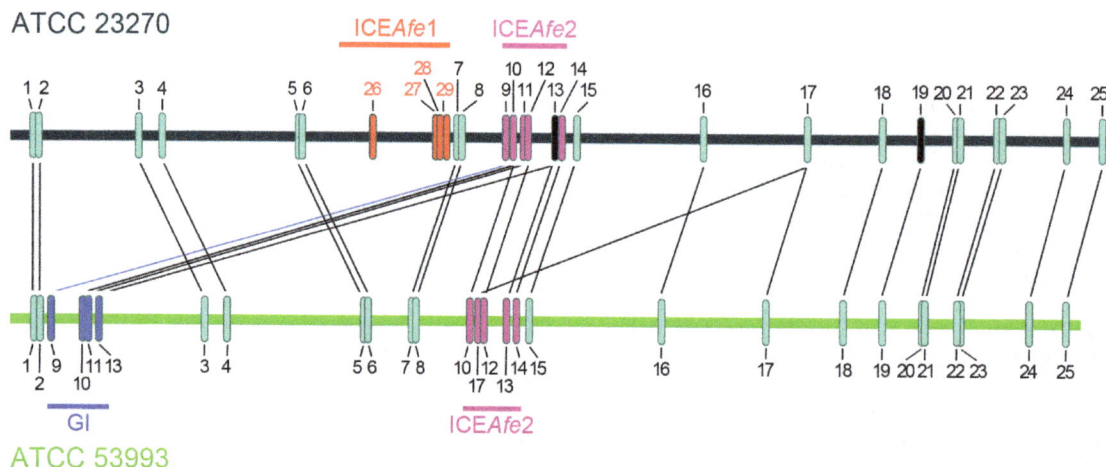

Figure 1. Comparison of the relative genomic locations of *A. ferrooxidans* TA systems. Using BLASTP, TA from each *A. ferrooxidans* genome were paired according to protein similarity. TA encoded in MGEs are shown in red (ICE*Afe*1), pink (ICE*Afe*2) and blue (Genomic island, GI). In black are shown TA in which the gene that must encode the toxin are pseudo genes. Black lines link TA that have 94-100% amino acid identity between the two strains. The blue line links a TA that has 49% (antitoxin) and 52% (toxin) amino acid identity with its counterpart in the other strain. Numbers of the TA are according to Table 2.

provides additional copper resistance to the bacterium [32]. Further other putative ICE, ICE*Afe*2, is shared by both strains, although these elements are not identical (176-kbp and 159-kbp in ATCC 23270 and ATCC 53993, respectively). A detailed analysis of TA encoded in these MGEs revealed that ICE*Afe*1, ICE*Afe*2 and the GI contain four, five (six in ATCC 23270 including a pseudo gene) and four of them, respectively (Figure 1 highlighted in red, pink and blue). Remarkably, TA from ICE*Afe*1 (TA 26-29, Figure 1 highlighted in red) are exclusive to ATCC 23270, consistent with the unique presence of this ICE in this strain. On the other hand, ICE*Afe*2 is present in both strains and thus TA encoded within this MGE are shared (TA 9-14 and 17, Figure 1 highlighted in pink). Some of these TA are also encoded in the GI from ATCC 53993 (TA 9-11 and 13, Figure 1, highlighted in blue) suggesting a duplication of these TA systems in this strain. Interestingly, TA encoded in the GI are close to transposon-related sequences and integrases genes and have a different genomic context to their counterparts in the ICE*Afe*2 (not shown). These findings reinforce the notion that certain chromosomal TA in *A. ferrooxidans* have the potential to be mobilized by HGT and to form part of its mobile genome. Other TA in *A. ferrooxidans* are also encoded near to transposases or transposon related-genes (e.g. TA systems from ICE*Afe*1, Figure S1).

TA from ICE*Afe*1 are functional and their toxins are ribonucleases

Because TA systems have been proposed to participate in the maintenance of MGEs, we hypothesize that TAs encoded in ICE*Afe*1, ICE*Afe*2 and the GI might contribute to prevent the loss of these elements from the *A. ferrooxidans* chromosome. Indeed, although we do not know yet the function of ICE*Afe*1 or the advantage for strain ATCC 23270 to carry it, it is stably maintained in laboratory conditions despite being unique to this strain among the other 12 strains that we have analyzed [33]. We therefore carried out a functional analysis of TA from ICE*Afe*1.

Three out of four TA from ICE*Afe*1 (TA 26, 27 and 29) share sequence similarity to well-known super-families (Table 2) and are grouped within their corresponding super-families on phylogenetic trees (Figure 2).

TA 26 (MazEF-1) is similar to the MazEF system from *E. coli* [53–56]. The putative toxin (MazF-1) is 51.8% identical (65.8% similar) with its counterpart from *E. coli* (Figure S2A). On the other hand, the putative antitoxin (MazE-1) is 42.7% identical (65.9% similar) to the orthologue from *E. coli* (Figure S2B). A number of conjugation genes and genes from a transposon are encoded both upstream and downstream to this TA system, respectively (Figure S1A).

TA 27 has conserved domains similar to StbC antitoxins and VapC toxins (Table 2). The toxin, VapC-3, has low sequence identity with VapC proteins but it conserves the three acidic residues from PIN-domains that are important for toxin activity (Figure S3). It is encoded near a cluster of genes that are involved in the biosynthesis and export of exopolysaccharides (Figure S1B).

TA 29 is encoded by AFE_1383/AFE_1384 genes. This system is encoded near to two other TA and close to transposition-related sequences (Figure S1B). The antitoxin is encoded by AFE_1383 and has a HTH_XRE conserved domain present on HigA and VapB antitoxins [2,35]. Similar to the classical TA loci *higBA* [43], this TA is unusual because the toxin-encoding gene is located upstream of the antitoxin-encoding gene. The toxin encoded by AFE_1384 has a Gp49 super family conserved domain and amino acid similarity with RelE/ParE super-family. The highest amino acid identity found is with a new toxin EcoT1$_{EDL933}$ identified and validated by Leplae et al [19], thus we named it EcoT1-1.

Phylogenetic data revealed that TA toxins 26, 27 and 29 clustered within their corresponding toxin super-families, but in a different clade from their chromosomal counterparts (Figure 2). These data reinforces the fact that these systems are part of the mobile genome from *A. ferrooxidans*.

TA 28, encoded by AFE_1367/AFE_1368 genes, has no orthologue described to date and it was ascribed as a TA system based on the characteristic of the operon by RASTA-Bacteria (Supporting information S1). On the phylogenetic analysis this toxin is closer to a putative RelE/ParE toxin from TA 22 but within a heterogeneous clade involving CcdB and not classified toxins (Figure 2). According to the information from the Integrated Microbial Genomes platform, tox28 contains a Mut7-C domain (pfam01927), which corresponds to a C-terminal RNase domain with a PIN fold [57]. Indeed, structural homology

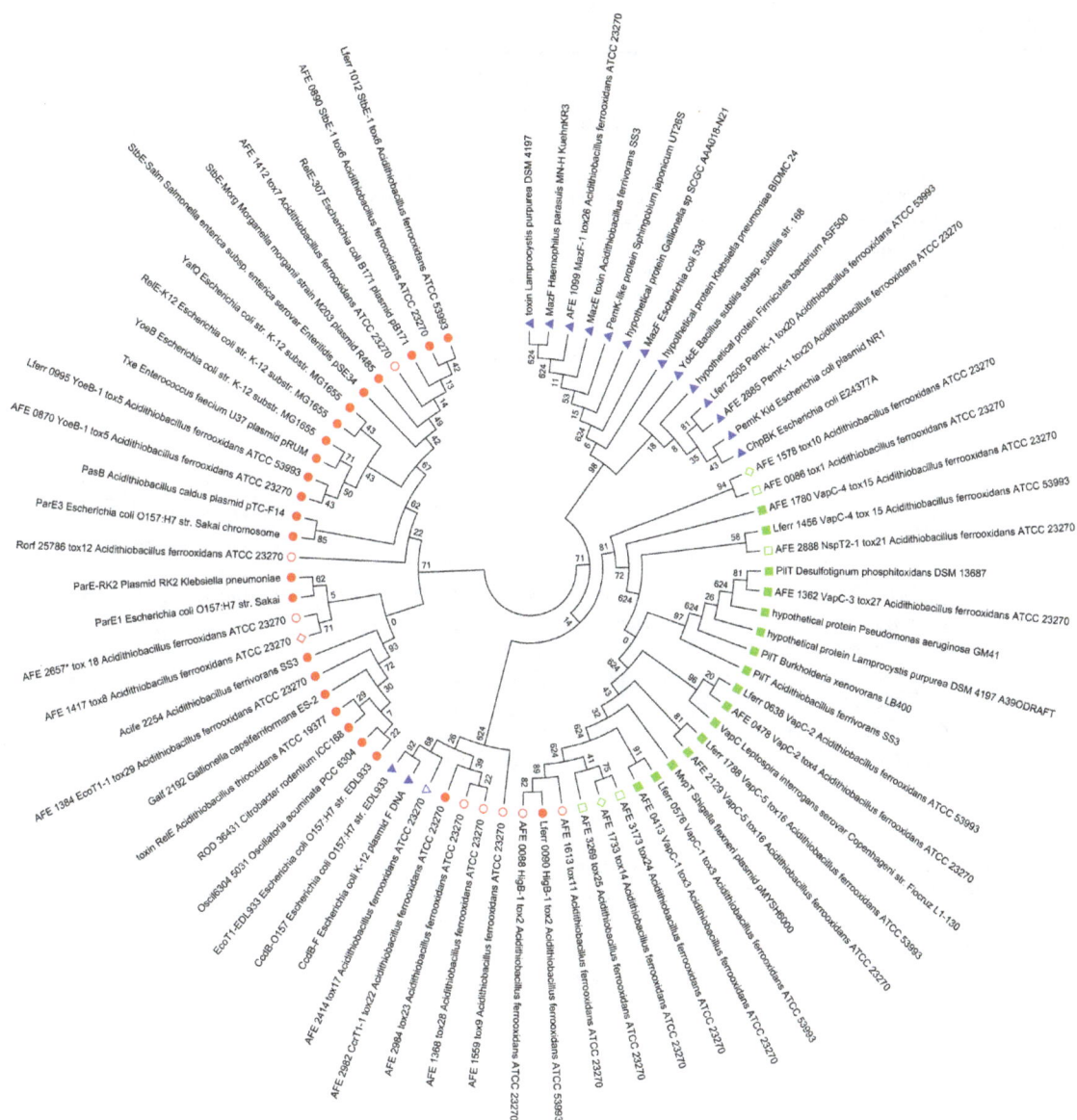

Figure 2. Phylogenetic relationship between TA toxins of *A. ferrooxidans* **ATCC 23270.** Circular unrooted dendogram built using Neighbor-Joining method. Scale shows the evolutionary distance in number of base substitutions per site. Toxins described by Leplae et al [19] belonging to RelE/ParE (red full-filled circle), CcdB/MazF (blue full-filled triangles) and VapC (green full-filled squared) super-families were introduced in the analysis as reference. Toxin classifications performed according the homologs with lower evolutionary distance (Table S1) are show in open symbols. The sequences whose homologs with lower evolutionary distance correspond to a non previously classified toxin are show in open rhomboid. The accession numbers of the sequences used in the analysis are in Supporting information S3.

searches predicted in tox28 the presence of a putative fold like a 3-phosphoglycerate dehydrogenase (PDB ID: 2EKL; 77% coverage) and low structural homology with PIN domain proteins (PDB ID: 1O4W, 39% coverage; PDB ID: 3H87, 26% coverage). According to BLASTP results, this toxin is conserved in different species, especially in cyanobacteria and Gram-positive bacteria. Therefore, this TA system might be a novel system with an RNase toxin related to PIN domain proteins. In *A. ferrooxidans* TA 28 is encoded very close to TA 27 and transposon-related sequences (Figure S1B).

The functionality of TA from ICE*Afe*1 was tested by transformation of *E. coli* BL21(DE3)pLysS with a multicopy plasmid (pETDuet-1 derivatives) carrying genes encoding either the toxin, the antitoxin or both (see Table 3 and Methods for a

description of each plasmid). Induction of VapC-3 and tox28 expression caused cell growth arrest of *E. coli* (Figure 3, blue lines). As expected, the co-induction of cognate antitoxins restored cell growth (Figure 3, green lines). Growth of cells expressing only the antitoxins was not affected (Figure 3, red lines). Overexpression of MazF-1 did not caused cell growth arrest of *E. coli* (Figure 3A). Conversely, overexpression of VapC-3 and tox28 seems to be bactericidal. In these cases the growth is not restored when the cells are transferred to a non-inducer medium (Figure 3B and C, lower panels). These results are consistent with a decrease in the CFU count (Figure S4). On the other hand, EcoT1-1 could not be cloned in the absence of its cognate antitoxin gene, which sheds light on its high toxicity in *E. coli* cells. These results indicated that, as all these toxins are RNases (bellow), they could target

Table 3. Plasmids used.

Plasmid	Characteristics	Reference
pGEM-T Easy	*E. coli* cloning vector. Ampicillin resistance.	Promega
pETDuet-1	*E. coli* expression vector. Ampicillin resistance. This vector is designed for the co-expression of two target genes. The vector contains two multiple cloning site (MCS1 and MCS2), each of which is preceded by a T7 promoter/lac operator and a ribosome binding site. ColE1 replicon.	Novagen
pACYCDuet-1	*E. coli* expression vector. Chloramphenicol resistance. This vector is designed for the co-expression of two target genes. The vector contains two multiple cloning site (MCS1 and MCS2), each of which is preceded by a T7 promoter/lac operator and a ribosome binding site. P15A replicon.	Novagen
pETMazE-1	pETDuet-1 derivative. Expressing the antitoxin gene from TA 26 system.	This work
pETMazF-1	pETDuet-1 derivative. Expressing the toxin gene from TA 26 system with a His$_6$ tag at the N-terminal.	This work
pETMazEF-1	pETDuet-1 derivative. Expressing TA 26 system. In this construction the toxin gene has a His$_6$ tag at the N-terminal.	This work
pETantitox27	pETDuet-1 derivative. Expressing the antitoxin gene from the TA 27 system.	This work
pETVapC-3	pETDuet-1 derivative. Expressing the toxin gene from TA 27 system with a His$_6$ tag at the N-terminal.	This work
pETStbC-VapC-3	pETDuet-1 derivative. Expressing the TA 27 system. In this construction the toxin gene has a His$_6$ tag at the N-terminal.	This work
pETantitox28	pETDuet-1 derivative. Expressing the antitoxin gene from TA 28 system.	This work
pETtox28	pETDuet-1 derivative. Expressing the toxin gene from TA 28 system with a His$_6$ tag at the N-terminal.	This work
pETTA28	pETDuet-1 derivative. Expressing the TA 28 system. In this construction the toxin gene has a His$_6$ tag at the N-terminal.	This work
pEcoA1/EcoT1-1	pETDuet-1 derivative. Expressing the TA 29 system. In this construction the toxin gene has a His$_6$ tag at the N-terminal.	This work
pACYCantitox27	pACYCDuet-1 derivative with the antitoxin27 gene cloned at its MCS2.	This work
pACYCantitox28	pACYCDuet-1 derivative with the antitoxin28 gene cloned at its MCS2.	This work

different cellular RNAs and/or have different sequence specificities. VapC-3, tox28 and EcoT1-1 probably target important RNAs in *E. coli* cells, while no target (or underrepresented targets) is present for MazF-1 in this host.

According to bioinformatic data all these toxins were predicted to be ribonucleases. To test the ability of these proteins to

hydrolyze RNA, we performed *in vitro* cleavage of viral MS2 RNA assays with purified toxins (Figure S5). MazF-1 digested the RNA only in the absence of Mg^{+2} ions (Figure 4, lane 2). Surprisingly, RNA cleavage by MazF-1 apparently is blocked by Mg^{+2} and it appears to be insensitive to Mn^{+2} ions (Figure 4B, lane 2-4), a phenomenon not yet described for a MazF toxin to our

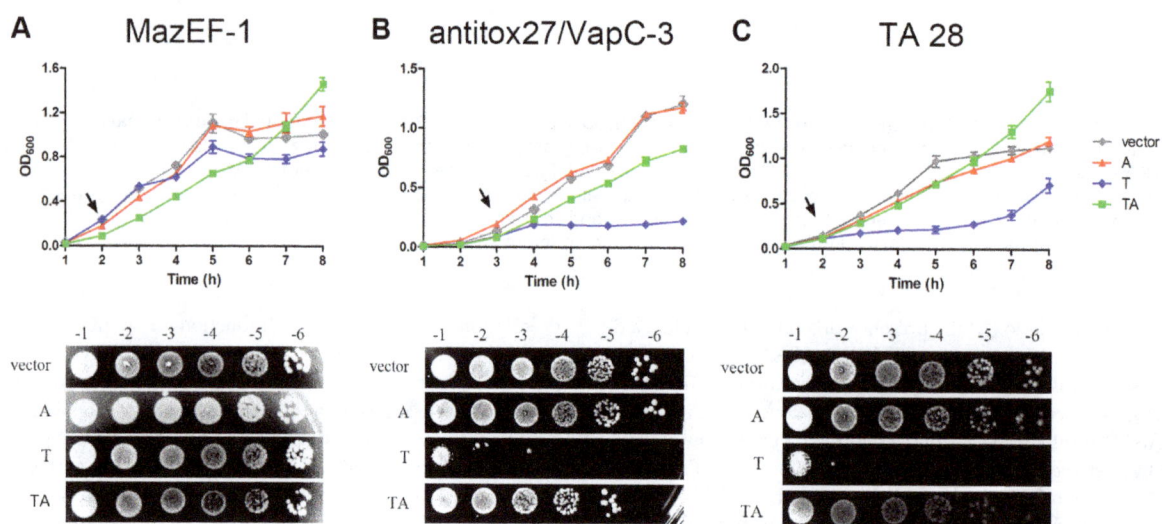

Figure 3. Effect of ICE*Afe*1 TA systems expression in *E. coli* growth. Cellular growth of *E. coli* BL21(DE3)pLysS cells harboring plasmids containing toxin (T, blue curves), antitoxin (A, red curves) or both (TA, green curves) genes of TA 26 (A), TA 27 (B) and TA 28 (C) was monitored by measuring the OD_{600}. Cells containing the empty vector (gray curves) were used as a control. The arrows indicate the moment when 1 mM IPTG was added to each culture. 3 hours after the induction 10-fold serial dilutions of each culture were spotted on LB plates without IPTG (panels below each graph). The means and standard deviation of three different experiments are plotted.

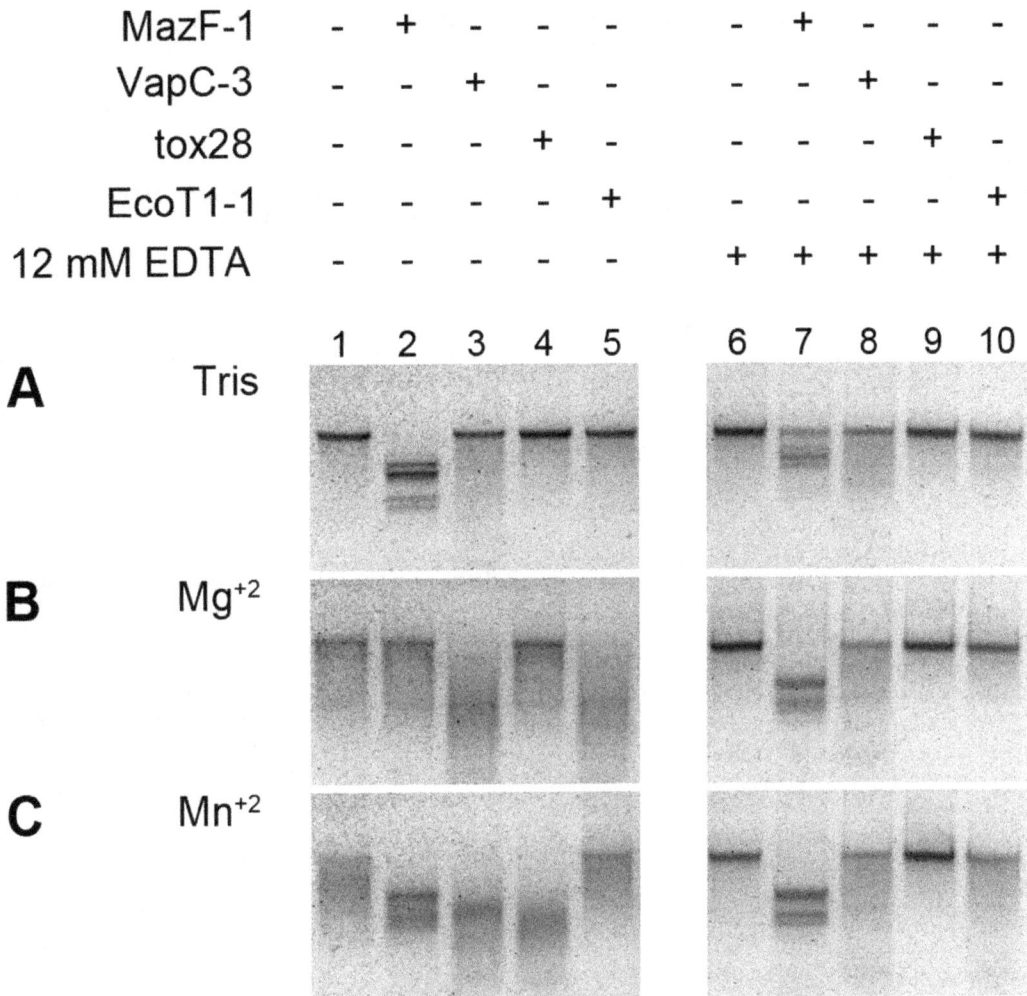

Figure 4. *In vitro* **RNase assay of ICE*Afe*1 toxins.** 1.6 µg of MS2 RNA was incubated with (+) or without (−) the purified toxins in 10 mM Tris-HCl (pH 7.8) in the absence of divalent ions (A) or with 10 mM MgCl$_2$ (B) or MnCl$_2$ (C). The reactions were incubated at 37°C for 15 (A and C) or 30 minutes (B). 12 mM EDTA was added to some reactions as a control (lanes 6-10).

knowledge. Consistent with this, EDTA addition restored the RNase activity (Figure 4B, lane 7). Conversely, VapC-3 exhibited RNase activity only when Mg^{+2} or Mn^{+2} ions were present (Figure 4B and C, lane 3). Strikingly, under our tested conditions tox28 was active only in the presence of Mn^{+2} ions (Figure 4C, lane 4). According to our knowledge there are no TA toxins that required Mn^{+2} ions for their RNase activity. Some VapC toxins

bind Mg^{+2} and/or Mn^{+2} [58] or have been crystallized with bound Mn^{+2}, as in the case of VapC toxin PAE0151 from *Pyrobaculum aerophilum* [59], but its activity has been assayed only with Mg^{+2}. In contrast to tox28, in the analyzed conditions EcoT1-1 was an active RNase only in the presence of Mg^{+2} ions (Figure 4B, lane 5). All Mg^{+2}/Mn^{+2}-dependent RNase activities

Table 4. Plasmid maintenance test. *E. coli* BL21(DE3) was double transformed with the plasmids indicated.

Culture	Plasmid maintenance (%)[a]			
	15 days	20 days	26 days	30 days
pETVapC-3+pACYCantitox27	70.87±37.93	88.10±16.83	89.03±13.51	76.67±33.00
pETDuet-1+pACYCantitox27	52.77±33.56	44.67±27.15	44.97±22.13	35.03±31.94
pETtox28+pACYCantitox28	62.23±34.59	55.42±37.93	55.83±41.05	51.43±40.88
pETDuet-1+pACYCantitox28	64.70±44.87	60.83±43.03	40.73±29.02	28.20±21.25

[a]percentage of chloramphenicol-resistant bacteria (resistance gene encoded on pACYCDuet-1) when cultured on ampicillin-containing media (resistance gene encoded on pETDuet-1) for the days indicated. The data are expressed as the means of three independent cultures ± standard deviation.

were inhibited in the presence of EDTA (Figure 4, lane 6-10), confirming that the divalent ions are necessary for the activity.

Thus, three out of four TA encoded in ICE*Afe*1 (with the exception of MazF-1) are active in a heterologous system, i.e. the expression of toxins in *E. coli* arrest growth and the cognate antitoxins suppress this toxicity. Furthermore, these results show that all ICE*Afe*1 TA toxins are RNases with different ions requirements that could inhibit translation via RNA cleavage. Cellular targets and cleavage sites of these toxins are our future interest to study.

TA systems from ICE*Afe*1 might allow plasmid maintenance

Given that TA from ICE*Afe*1 are functional and in general TA systems have been proposed to participate avoiding the loss of MGEs from their hosts, we hypothesize that these TA might be responsible of the stable maintenance of ICE*Afe*1 in *A. ferrooxidans* chromosome. To determine whether TA systems from ICE*Afe*1 mediate plasmid stability, we assay the capacity of TA 27 and TA 28 to avoid the loss of pACYCDuet-1, a vector carrying chloramphenicol resistance. We could not assay TA 26 and TA 29 given that MazF-1 (TA 26) did not affect the growth of *E. coli* (Figure 3A) and, on the other hand, the gene encoding EcoT1-1 (TA 29) could not be cloned in *E. coli*.

The assay consisted on monitoring the stability of pACYCDuet-A vectors during cultivation in the presence of selective antibiotic pressure towards pETDuet-T vectors [45]. After 30 days of culture in the absence of chloramphenicol, although they were not fully maintained (and that we obtained high standard deviations in the experiment), pACYCantitox27 and pACYCantitox28 were lost to a lower level in the absence of their cognate pETDuet-T vectors in comparison to cultures with the pETDuet-1 empty vector (Table 4). As *E. coli* BL21(DE3) contains its native chromosomal TA systems, the failure to pACYCantitox27 and pACYCantitox28 to be fully maintained, may be due to cross-interaction between ICE*Afe*1 toxins with cognate antitoxins from *E. coli*. This may have made the presence of ICE*Afe*1 antitoxins dispensable and thus the pACYCDuet-A vectors lost. Functional interaction between chromosomal and plasmidial TA systems has been demonstrated before [60]. Nevertheless, these data demonstrate that TA 27 and TA 28 encoded on a MGE from *A. ferrooxidans* could have a plasmid-stabilizing role.

Discussion

In this study, the content of type II TA systems and their relationship with MGEs in the environmental bacterium *A. ferrooxidans* were investigated. According to the data presented here, *A. ferrooxidans* encodes at least 29 and 28 TA in ATCC 23270 and ATCC 53993 strains, respectively (representing 1.8 and 2% of total number of CDSs, respectively). Given this content of TA and considering the number of putative TA proteins that we could not classify within a super-family, it is expected that this microorganism could be a source of novel systems, expanding the repertoire currently known. It seems to be the case for TA 28 characterized in this work; a TA present in a MGE encoding a novel toxin that causes a bactericidal state in *E. coli* and has a Mn^{+2}-dependent RNase activity. The described number of TA in *A. ferrooxidans* could be underestimated because of the several hypothetical proteins encoded in a TA-like gene organization as well as putative orphan toxins. Additionally, we must consider that type I, III, IV and V TA systems, that were not the subject of this study, might contribute to the total number of TA systems in *A. ferrooxidans*.

Putative roles of ICE*Afe*1 TA systems

In the ICE SXT from *V. cholera*, the MosAT system promotes the maintenance of the element. The mRNA levels of MosAT system are enhanced when the element is excised, preventing its loss [14]. Since the levels of TA mRNAs from ICE*Afe*1 do not increase upon their excision (based on qRT-PCR, data not shown), it seems that these systems do not behave in the same way as the MosAT system. We cannot rule out that there might be changes at a protein level in different growth conditions.

Plasmid-encoded TA systems prevent the proliferation of plasmid-free progeny and thereby contribute to the maintenance of their replicons. By a similar mechanism, chromosomal genes closely linked to a TA locus could have a selective advantage; as a consequence the maintenance of specific genes (like TA genes) might have an effect on the stability and spread of MGEs. Here we have shown that two TA from *A. ferrooxidans* ICE*Afe*1 (TA 27 and TA 28) seem to have a role in the maintenance of MGEs (Figure 4). Thus, the presence of TA on MGEs in *A. ferrooxidans* could explain why they are stably maintained in this bacterium.

Based on the genomic contexts, we hypothesize that TA from ICE*Afe*1 might be involved in the conjugal transfer and/or biofilm formation, putative roles ascribed to this MGE [33]. The MazEF-1 system (TA 26) is encoded very close to the conjugation cluster probably responsible for the horizontal transfer of this element (Figure S1A). On the other hand, TA 27, TA 28 and TA 29 are encoded near to a cluster of genes predicted to be involved in the biosynthesis and export of exopolysaccharides which could be linked to biofilm formation in this bacterium (Figure S1B). To determine whether TA systems from ICE*Afe*1 contribute to biofilm formation is to be further analyzed. However, we can not rule out that each system might have a different function under different physiological conditions.

Elucidation of the sequence specificity and thus the cellular targets of each toxin from ICE*Afe*1 might be crucial to determine their role in the physiology of *A. ferrooxidans*.

Type II TA systems, chromosomal or mobile TA systems?

Hitherto TA systems are classified as chromosomal (stable) or plasmid encoded (mobile). All TA in ICE*Afe*1 are encoded near to transposon-related sequences (Figure S1). Similar distribution occurs with TA from the genomic island of *A. ferrooxidans* ATCC 53993. As it is known that ICEs are modular elements [61], it is possible that TA are part of modules that have been acquired by HGT and contributed to the creation of ICE*Afe*1. Thus we propose that most *A. ferrooxidans* TA systems may belong to either active or inactive MGEs that are inserted in the bacterial chromosome. A similar case has been reported in *Acidithiobacillus caldus*, another acidophilic bacterium [62].

Supporting Information

Figure S1 Genetic overview of ICE*Afe*1 TA and the flanking DNA regions. The genetic contexts of TA 26 (A), TA 27, TA 28 and TA 29 (B) are indicated. Each gene is colored by COG according to the information on the Integrated Microbial Genomes platform (IMG, http://img.jgi.doe.gov/cgi-bin/w/main.cgi [37]). Color codes of function category for COGs are indicated in the insert below the images.
(TIF)

Figure S2 Alignment of MazEF-1 system from ICE*Afe*1 with its ortholog from *E. coli*. Protein sequences from toxin (A) and antitoxin (B) were aligned using ClustalW. GenBank accession numbers: MazF_Ec, BAA03918.1; MazF ICE*Afe*1, YP_002425571.1; MazE_Ec, BAA41177.1; MazE ICE*Afe*1,

YP_002425570.1. Identical and similar amino acids are shown in black and grey, respectively. Functionally important conserved regions [51] are indicated below the MazF and MazE sequences by black lines.
(TIF)

Figure S3 Alignment of VapC toxins from *A. ferrooxidans* ATCC 23270. Protein sequences were aligned using ClustalW. Identical and similar amino acids are shown in black and grey, respectively. The three conserved acidic residues of the PIN-domain are highlighter in green. GenBank accession numbers: VapC-1, YP_002424909; VapC-2, YP_002424974; VapC-3, YP_002425797; VapC-4, YP_002426198; and VapC-5, YP_002426529.
(TIF)

Figure S4 Effect of ICE*Afe*1 toxins expression on *E. coli* CFU. Cellular growth of *E. coli* BL21(DE3)pLysS cells harboring plasmids containing toxin (T, blue curves) of TA 26 (A), TA 27 (B) and TA 28 (C) post IPTG addition was monitored by measuring the CFU/ml. Cells containing the empty vector (gray curves) were used as a control. The means and standard deviation of two different experiments are plotted.
(TIF)

Figure S5 ICE*Afe*1 toxins purification. Tricine-SDS-PAGE of (His)$_6$-tagged toxin proteins purified as it is described at Materials and Methods. The proteins were visualized by staining with Coomassie brilliant blue. The molecular weights of some reference bands (M, PageRuler Unstained Broad Range Protein Ladder, Thermo Scientific) are indicated at the left of the figure.
(TIF)

Figure S6 *In vitro* RNase assay of MazEF-1 system. 1.6 µg of MS2 RNA was incubated with (+) or without (−) 50

picomoles of the purified MazF-1 toxin and/or MazE-1 antitoxin in 10 mM Tris-HCl (pH 7.8). The reactions were incubated at 37°C for 15 minutes. Lanes 5-7: the reactions contain 100, 150 and 200 picomoles of MazF-1, respectively. Lane 8-10: the reactions contain 100, 150 and 200 picomoles of MazF-1, respectively.
(TIF)

Table S1 Evolutionary distances among toxins from *A. ferrooxidans*.
(XLS)

Supporting Information S1 Identification of new TA II not describe in TADB.
(DOCX)

Supporting Information S2 Structural homologous of PIN domain toxins from *A. ferrooxidans*.
(DOCX)

Supporting Information S3 Gene ID or locus tag of nucleotide sequences used in the phylogenetic analysis.
(DOCX)

Acknowledgments

We thank to Dr. Jonathan Iredell from the University of Sydney to review and to correct the manuscript and to María Catalina Aranda for her generous assistance in some experiments.

Author Contributions

Conceived and designed the experiments: PB MT OO. Performed the experiments: PB MT. Analyzed the data: PB MT OO. Wrote the paper: PB OO.

References

1. Yamaguchi Y, Park J-H, Inouye M (2011) Toxin-antitoxin systems in bacteria and archaea. Annu Rev Genet 45: 61–79.
2. Gerdes K, Christensen SK, Løbner-Olesen A (2005) Prokaryotic toxin-antitoxin stress response loci. Nat Rev Microbiol 3: 371–382.
3. Riley MA, Wertz JE (2002) Bacteriocins: evolution, ecology, and application. Annu Rev Microbiol 56: 117–137.
4. Hayes CS, Aoki SK, Low DA (2010) Bacterial contact-dependent delivery systems. Annu Rev Genet 44: 71–90.
5. Gerdes K, Bech FW, Jørgensen ST, Løbner-Olesen A, Rasmussen PB, et al. (1986) Mechanism of postsegregational killing by the hok gene product of the parB system of plasmid R1 and its homology with the relF gene product of the E. coli relB operon. EMBO J 5: 2023–2029.
6. Gerdes K, Rasmussen PB, Molin S (1986) Unique type of plasmid maintenance function: postsegregational killing of plasmid-free cells. Proc Natl Acad Sci U S A 83: 3116–3120.
7. Yamaguchi Y, Inouye M (2011) Regulation of growth and death in Escherichia coli by toxin-antitoxin systems. Nat Rev Microbiol 9: 779–790.
8. Makarova KS, Wolf YI, Koonin EV (2009) Comprehensive comparative-genomic analysis of type 2 toxin-antitoxin systems and related mobile stress response systems in prokaryotes. Biol Direct 4: 19.
9. Pandey DP, Gerdes K (2005) Toxin-antitoxin loci are highly abundant in free-living but lost from host-associated prokaryotes. Nucleic Acids Res 33: 966–976.
10. Magnuson RD (2007) Hypothetical functions of toxin-antitoxin systems. J Bacteriol 189: 6089–6092.
11. Tsilibaris V, Maenhaut-Michel G, Mine N, Van Melderen L (2007) What is the benefit to Escherichia coli of having multiple toxin-antitoxin systems in its genome? J Bacteriol 189: 6101–6108.
12. Norton JP, Mulvey MA (2012) Toxin-antitoxin systems are important for niche-specific colonization and stress resistance of uropathogenic Escherichia coli. PLoS Pathog 8: e1002954.
13. Fasani RA, Savageau MA (2013) Molecular mechanisms of multiple toxin-antitoxin systems are coordinated to govern the persister phenotype. Proc Natl Acad Sci U S A 110: E2528–2537.
14. Wozniak RAF, Waldor MK (2009) A toxin-antitoxin system promotes the maintenance of an integrative conjugative element. PLoS Genet 5: e1000439.

15. Fozo EM, Makarova KS, Shabalina SA, Yutin N, Koonin EV, et al. (2010) Abundance of type I toxin-antitoxin systems in bacteria: searches for new candidates and discovery of novel families. Nucleic Acids Res 38: 3743–3759.
16. Blower TR, Short FL, Rao F, Mizuguchi K, Pei XY, et al. (2012) Identification and classification of bacterial Type III toxin-antitoxin systems encoded in chromosomal and plasmid genomes. Nucleic Acids Res 40: 6158–6173.
17. Masuda H, Tan Q, Awano N, Wu K-P, Inouye M (2012) YeeU enhances the bundling of cytoskeletal polymers of MreB and FtsZ, antagonizing the CbtA (YeeV) toxicity in Escherichia coli. Mol Microbiol 84: 979–989.
18. Dy RL, Przybilski R, Semeijn K, Salmond GPC, Fineran PC (2014) A widespread bacteriophage abortive infection system functions through a Type IV toxin-antitoxin mechanism. Nucleic Acids Res 42: 4590–4605.
19. Leplae R, Geeraerts D, Hallez R, Guglielmini J, Drèze P, et al. (2011) Diversity of bacterial type II toxin-antitoxin systems: a comprehensive search and functional analysis of novel families. Nucleic Acids Res 39: 5513–5525.
20. Sala A, Bordes P, Genevaux P (2014) Multiple toxin-antitoxin systems in Mycobacterium tuberculosis. Toxins 6: 1002–1020.
21. Wang X, Lord DM, Cheng H-Y, Osbourne DO, Hong SH, et al. (2012) A new type V toxin-antitoxin system where mRNA for toxin GhoT is cleaved by antitoxin GhoS. Nat Chem Biol 8: 855–861.
22. Cook GM, Robson JR, Frampton RA, McKenzie J, Przybilski R, et al. (2013) Ribonucleases in bacterial toxin-antitoxin systems. Biochim Biophys Acta 1829: 523–531.
23. Yamaguchi Y, Inouye M (2009) mRNA interferases, sequence-specific endoribonucleases from the toxin-antitoxin systems. Prog Mol Biol Transl Sci 85: 467–500.
24. Cambray G, Guerout A-M, Mazel D (2010) Integrons. Annu Rev Genet 44: 141–166.
25. Guérout A-M, Iqbal N, Mine N, Ducos-Galand M, Van Melderen L, et al. (2013) Characterization of the phd-doc and ccd toxin-antitoxin cassettes from Vibrio superintegrons. J Bacteriol 195: 2270–2283.
26. Szekeres S, Dauti M, Wilde C, Mazel D, Rowe-Magnus DA (2007) Chromosomal toxin-antitoxin loci can diminish large-scale genome reductions in the absence of selection. Mol Microbiol 63: 1588–1605.
27. Williams KP, Gillespie JJ, Sobral BWS, Nordberg EK, Snyder EE, et al. (2010) Phylogeny of gammaproteobacteria. J Bacteriol 192: 2305–2314.

28. Valdés J, Pedroso I, Quatrini R, Dodson RJ, Tettelin H, et al. (2008) Acidithiobacillus ferrooxidans metabolism: from genome sequence to industrial applications. BMC Genomics 9: 597.

29. Bonnefoy V, Holmes DS (2012) Genomic insights into microbial iron oxidation and iron uptake strategies in extremely acidic environments. Environ Microbiol 14: 1597–1611.

30. Holmes DS, Zhao HL, Levican G, Ratouchniak J, Bonnefoy V, et al. (2001) ISAfe1, an ISL3 family insertion sequence from Acidithiobacillus ferrooxidans ATCC 19859. J Bacteriol 183: 4323–4329.

31. Oppon JC, Sarnovsky RJ, Craig NL, Rawlings DE (1998) A Tn7-like transposon is present in the glmUS region of the obligately chemoautolithotrophic bacterium Thiobacillus ferrooxidans. J Bacteriol 180: 3007–3012.

32. Orellana LH, Jerez CA (2011) A genomic island provides Acidithiobacillus ferrooxidans ATCC 53993 additional copper resistance: a possible competitive advantage. Appl Microbiol Biotechnol 92: 761–767.

33. Bustamante P, Covarrubias PC, Levicán G, Katz A, Tapia P, et al. (2012) ICE Afe 1, an actively excising genetic element from the biomining bacterium Acidithiobacillus ferrooxidans. J Mol Microbiol Biotechnol 22: 399–407.

34. Shao Y, Harrison EM, Bi D, Tai C, He X, et al. (2011) TADB: a web-based resource for Type 2 toxin-antitoxin loci in bacteria and archaea. Nucleic Acids Res 39: D606–611.

35. Sevin EW, Barloy-Hubler F (2007) RASTA-Bacteria: a web-based tool for identifying toxin-antitoxin loci in prokaryotes. Genome Biol 8: R155.

36. Kelley LA, Sternberg MJE (2009) Protein structure prediction on the Web: a case study using the Phyre server. Nat Protoc 4: 363–371.

37. Markowitz VM, Chen I-MA, Palaniappan K, Chu K, Szeto E, et al. (2012) IMG: the Integrated Microbial Genomes database and comparative analysis system. Nucleic Acids Res 40: D115–122.

38. Thompson JD, Higgins DG, Gibson TJ (1994) CLUSTAL W: improving the sensitivity of progressive multiple sequence alignment through sequence weighting, position-specific gap penalties and weight matrix choice. Nucleic Acids Res 22: 4673–4680.

39. Gonnet GH, Cohen MA, Benner SA (1992) Exhaustive matching of the entire protein sequence database. Science 256: 1443–1445.

40. Saitou N, Nei M (1987) The neighbor-joining method: a new method for reconstructing phylogenetic trees. Mol Biol Evol 4: 406–425.

41. Dopazo J (1994) Estimating errors and confidence intervals for branch lengths in phylogenetic trees by a bootstrap approach. J Mol Evol 38: 300–304.

42. Tamura K, Nei M, Kumar S (2004) Prospects for inferring very large phylogenies by using the neighbor-joining method. Proc Natl Acad Sci U S A 101: 11030–11035.

43. Tamura K, Peterson D, Peterson N, Stecher G, Nei M, et al. (2011) MEGA5: molecular evolutionary genetics analysis using maximum likelihood, evolutionary distance, and maximum parsimony methods. Mol Biol Evol 28: 2731–2739.

44. Chung CT, Niemela SL, Miller RH (1989) One-step preparation of competent Escherichia coli: transformation and storage of bacterial cells in the same solution. Proc Natl Acad Sci U S A 86: 2172–2175.

45. Bukowski M, Lyzen R, Helbin WM, Bonar E, Szalewska-Palasz A, et al. (2013) A regulatory role for Staphylococcus aureus toxin-antitoxin system PemIKSa. Nat Commun 4: 2012.

46. Schägger H (2006) Tricine-SDS-PAGE. Nat Protoc 1: 16–22.

47. Budde PP, Davis BM, Yuan J, Waldor MK (2007) Characterization of a higBA toxin-antitoxin locus in Vibrio cholerae. J Bacteriol 189: 491–500.

48. Christensen-Dalsgaard M, Gerdes K (2006) Two higBA loci in the Vibrio cholerae superintegron encode mRNA cleaving enzymes and can stabilize plasmids. Mol Microbiol 62: 397–411.

49. Tian QB, Ohnishi M, Tabuchi A, Terawaki Y (1996) A new plasmid-encoded proteic killer gene system: cloning, sequencing, and analyzing hig locus of plasmid Rts1. Biochem Biophys Res Commun 220: 280–284.

50. Blower TR, Salmond GPC, Luisi BF (2011) Balancing at survival's edge: the structure and adaptive benefits of prokaryotic toxin-antitoxin partners. Curr Opin Struct Biol 21: 109–118.

51. Marchler-Bauer A, Zheng C, Chitsaz F, Derbyshire MK, Geer LY, et al. (2013) CDD: conserved domains and protein three-dimensional structure. Nucleic Acids Res 41: D348–352.

52. Wozniak RAF, Fouts DE, Spagnoletti M, Colombo MM, Ceccarelli D, et al. (2009) Comparative ICE genomics: insights into the evolution of the SXT/R391 family of ICEs. PLoS Genet 5: e1000786.

53. Amitai S, Kolodkin-Gal I, Hananya-Meltabashi M, Sacher A, Engelberg-Kulka H (2009) Escherichia coli MazF leads to the simultaneous selective synthesis of both "death proteins" and "survival proteins." PLoS Genet 5: e1000390.

54. Kamada K, Hanaoka F, Burley SK (2003) Crystal structure of the MazE/MazF complex: molecular bases of antidote-toxin recognition. Mol Cell 11: 875–884.

55. Mittenhuber G (1999) Occurrence of mazEF-like antitoxin/toxin systems in bacteria. J Mol Microbiol Biotechnol 1: 295–302.

56. Zhang Y, Zhang J, Hoeflich KP, Ikura M, Qing G, et al. (2003) MazF cleaves cellular mRNAs specifically at ACA to block protein synthesis in Escherichia coli. Mol Cell 12: 913–923.

57. Anantharaman V, Koonin EV, Aravind L (2002) Comparative genomics and evolution of proteins involved in RNA metabolism. Nucleic Acids Res 30: 1427–1464.

58. Arcus VL, McKenzie JL, Robson J, Cook GM (2011) The PIN-domain ribonucleases and the prokaryotic VapBC toxin-antitoxin array. Protein Eng Des Sel PEDS 24: 33–40.

59. Bunker RD, McKenzie JL, Baker EN, Arcus VL (2008) Crystal structure of PAE0151 from Pyrobaculum aerophilum, a PIN-domain (VapC) protein from a toxin-antitoxin operon. Proteins 72: 510–518.

60. Wilbaux M, Mine N, Guérout A-M, Mazel D, Melderen LV (2007) Functional Interactions between Coexisting Toxin-Antitoxin Systems of the ccd Family in Escherichia coli O157:H7. J Bacteriol 189: 2712–2719.

61. Wozniak RAF, Waldor MK (2010) Integrative and conjugative elements: mosaic mobile genetic elements enabling dynamic lateral gene flow. Nat Rev Microbiol 8: 552–563.

62. Acuña LG, Cárdenas JP, Covarrubias PC, Haristoy JJ, Flores R, et al. (2013) Architecture and gene repertoire of the flexible genome of the extreme acidophile Acidithiobacillus caldus. PloS One 8: e78237.

12

Oligoribonucleotide (ORN) Interference-PCR (ORNi-PCR): A Simple Method for Suppressing PCR Amplification of Specific DNA Sequences Using ORNs

Naoki Tanigawa, Toshitsugu Fujita, Hodaka Fujii*

Chromatin Biochemistry Research Group, Combined Program on Microbiology and Immunology, Research Institute for Microbial Diseases, Osaka University, Suita, Osaka, Japan

Abstract

Polymerase chain reaction (PCR) amplification of multiple templates using common primers is used in a wide variety of molecular biological techniques. However, abundant templates sometimes obscure the amplification of minor species containing the same primer sequences. To overcome this challenge, we used oligoribonucleotides (ORNs) to inhibit amplification of undesired template sequences without affecting amplification of control sequences lacking complementarity to the ORNs. ORNs were effective at very low concentrations, with IC_{50} values for ORN-mediated suppression on the order of 10 nM. DNA polymerases that retain $3'-5'$ exonuclease activity, such as KOD and *Pfu* polymerases, but not those that retain $5'-3'$ exonuclease activity, such as *Taq* polymerase, could be used for ORN-mediated suppression. ORN interference-PCR (ORNi-PCR) technology should be a useful tool for both molecular biology research and clinical diagnosis.

Editor: Richard C. Willson, University of Houston, United States of America

Funding: This study was supported by Research Institute for Microbial Diseases, Osaka University (TF, HF). The funder had no role in study design, data collection and analysis, decision to publish, or preparation of the manuscript.

Competing Interests: The authors filed a patent application on ORNi-PCR. The patent details are as follows: Name: Method for suppressing amplification of specific nucleic acid sequences; Number: Japanese Patent Application No. 2014-176018.

* Email: hodaka@biken.osaka-u.ac.jp

Introduction

Polymerase chain reaction (PCR) is an essential method for molecular biological research. Amplification of multiple templates using common primers is widely used for cloning of members of gene families [1,2] and detection of microorganisms in clinical diagnosis. In such applications, amplicons of abundant templates that contain the primer sequences can often predominate over those of less abundant species, making it difficult to detect amplification of minor species. Therefore, it is necessary to develop methods that efficiently suppress amplification of abundant templates but allow amplification of minor or desired templates containing the same primer sequences.

Several methods have been developed to achieve this goal. Blocking oligodeoxyribonucleotides (ODNs) or locked nucleic acids (LNAs) complementary to the specific sequences of known genes have been used to inhibit their amplification [3–7]. In these procedures, the 3′ termini of ODNs and LNAs must be modified so that they are not able to serve as primers for PCR reactions. The 3′ modifications used for this purpose include dideoxidization [4], nonbase-pairing tails [5,6], and addition of a phosphate group [7]; however, these approaches are not feasible when used with DNA polymerases that have 3′–5′ exonuclease activity, because such enzymes can remove the 3′ modifications and thus allow the ODNs or LNAs to serve as primers. Therefore, these methods are applicable only when 3′–5′ exonuclease-deficient DNA polymerases such as Stoffel fragment or *Taq* polymerases are used [4–7].

Because 3′–5′ exonuclease activity is important for proofreading activity of DNA polymerases, the inability to use DNA polymerases that retain this activity may result in higher error rates in the amplification process.

Previous work showed that long RNAs (ca. 750 bases) could specifically inhibit amplification of targets by PCR [8]; however, this approach requires synthesis of RNA by *in vitro* transcription, which is expensive and time-consuming. Moreover, it remains unclear how tightly specificity of inhibition can be controlled using this strategy.

To overcome the aforementioned problems with conventional methods, we developed oligoribonucleotide (ORN) interference-PCR (ORNi-PCR) technology, in which ORNs are used to inhibit amplification of specific templates. This strategy has several advantages, namely, ORNs can be chemically synthesized easily and inexpensively, and because of their short length, it is straightforward to design ORNs for specific inhibition. Furthermore, ORNs can be easily removed by RNases for downstream applications and thus ORNi-PCR technology should be a useful tool for various applications in molecular biology research and clinical diagnosis.

Results and Discussion

Specific inhibition of PCR amplification using ORNs

We reasoned that ORNs could suppress PCR amplification of specific sequences by annealing to the template, and sought to

A

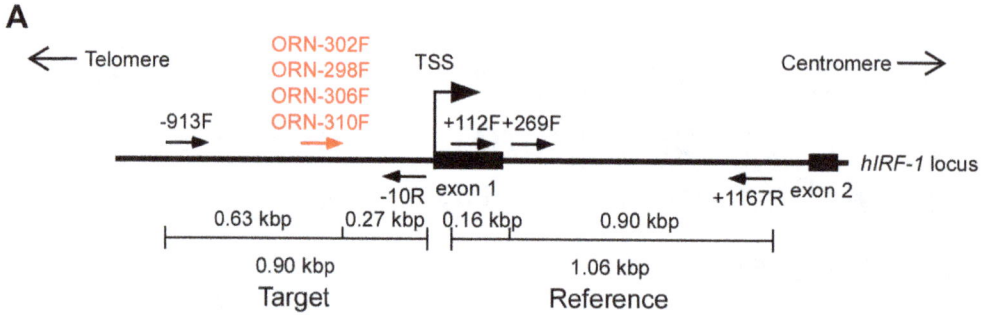

B

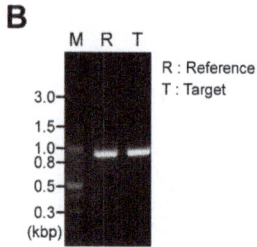

R : Reference
T : Target

C

ORN-302F	ccgggggcgcugggcuguccc (21 bases)
ORN-298F	gggcgcugggcuguccc (17 bases)
ORN-306F	ggggccggggcgcugggcuguccc (25 bases)
ORN-310F	ggcugggggccggggcgcugggcuguccc (29 bases)

D

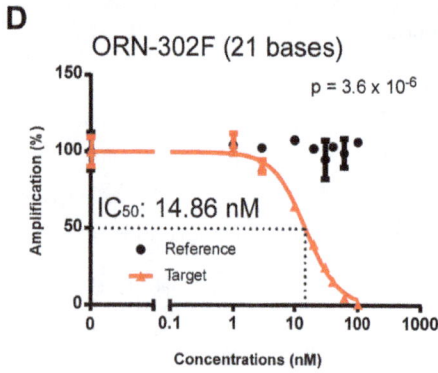

E

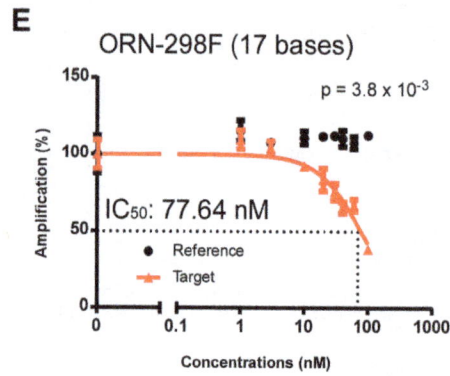

F

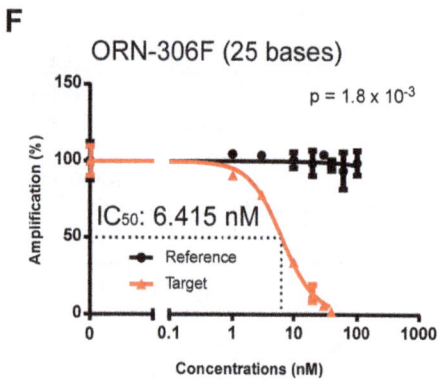

G

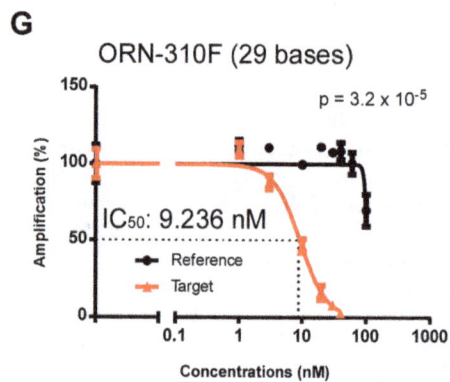

H

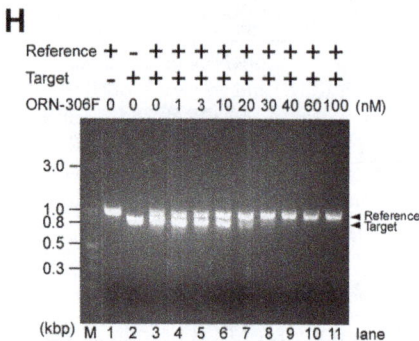

I

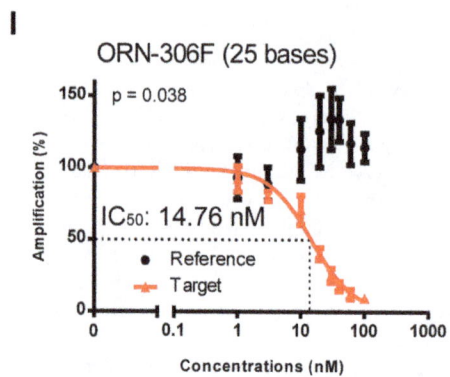

Figure 1. Inhibition of PCR amplification by ORNs of various lengths. (**A**) Positions of oligoribonucleotides (ORNs) in the target region. Primers were designed against the human *IRF-1* locus to amplify two neighboring regions. (**B**) Comparable amplification of the target and reference regions in the *IRF-1* locus. (**C**) Alignment of ORNs (5′–3′) of various lengths. (**D–G**) Dose responses of ORNs of various lengths. Real-time PCR amplification in the presence of various concentrations of ORNs was normalized against the value in the absence of ORNs. Dose-response curves were generated using the Prism software; Prism was not able to generate curves of the references for ORN-302F (**D**) and −298F (**E**). Data are represented as means ± S.D. (n = 3). (**H**) Suppression of amplification of the target template by ORN-306F in a reaction in which the target and reference primers were mixed. (**I**) Dose responses calculated for the condition used in (**H**). P-values for the differences in amplification between the reference and the target at an ORN concentration of 10 nM were calculated using Student's t-test (**D–G, I**).

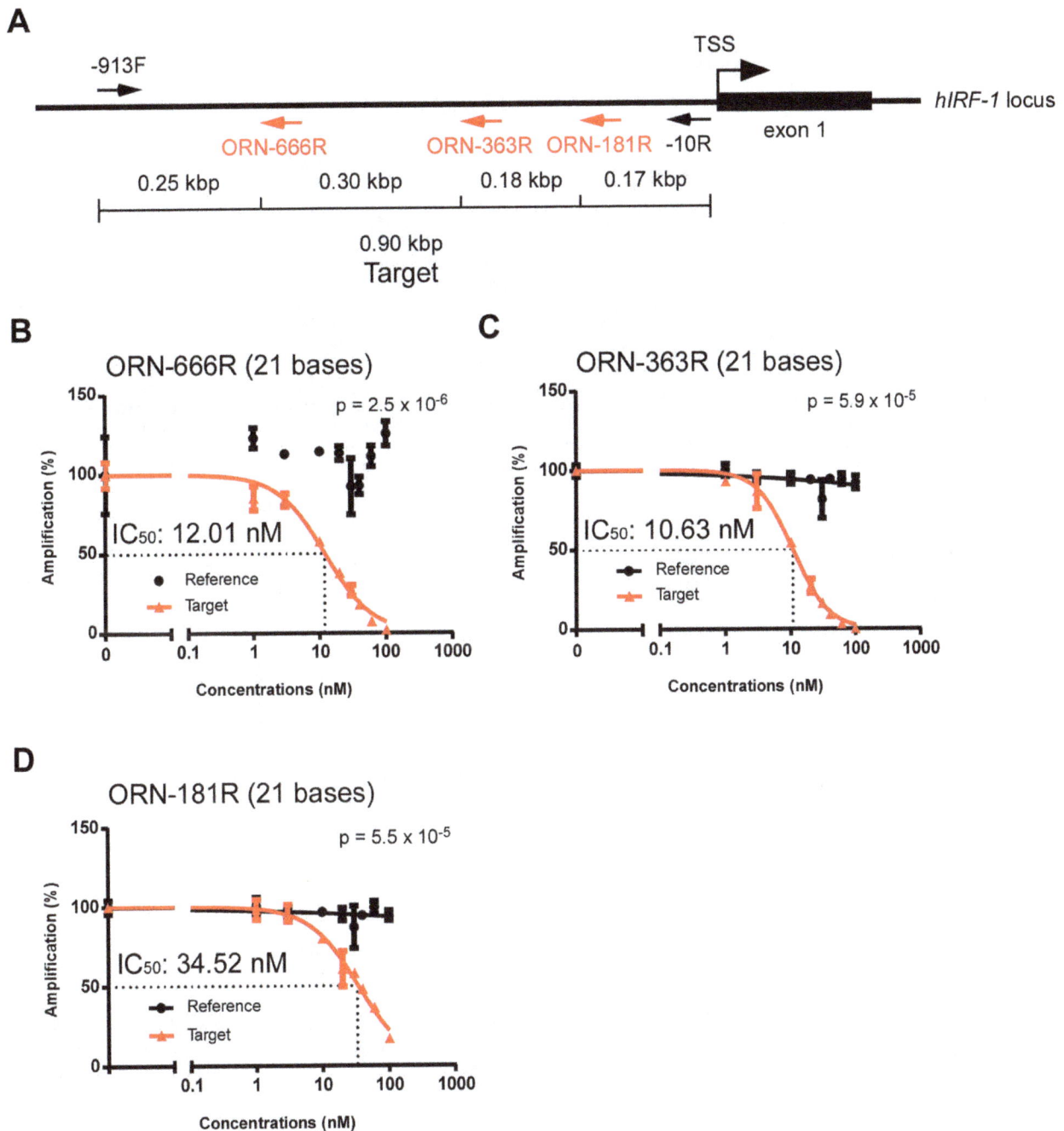

Figure 2. Effects of ORN position on inhibition of PCR amplification. (**A**) Positions of ORNs in the target region. (**B–D**) Dose responses of ORNs. Real-time PCR amplification in the presence of various concentrations of ORNs was normalized against the value in the absence of ORNs. Dose-response curves were generated using the Prism software; Prism was not able to generate a curve of the reference for ORN-666R (**B**). Data are represented as means ± S.D. (n = 3). P-values for the differences in amplification between the reference and the target at an ORN concentration of 10 nM were calculated using Student's t-test (**B–D**).

A

B

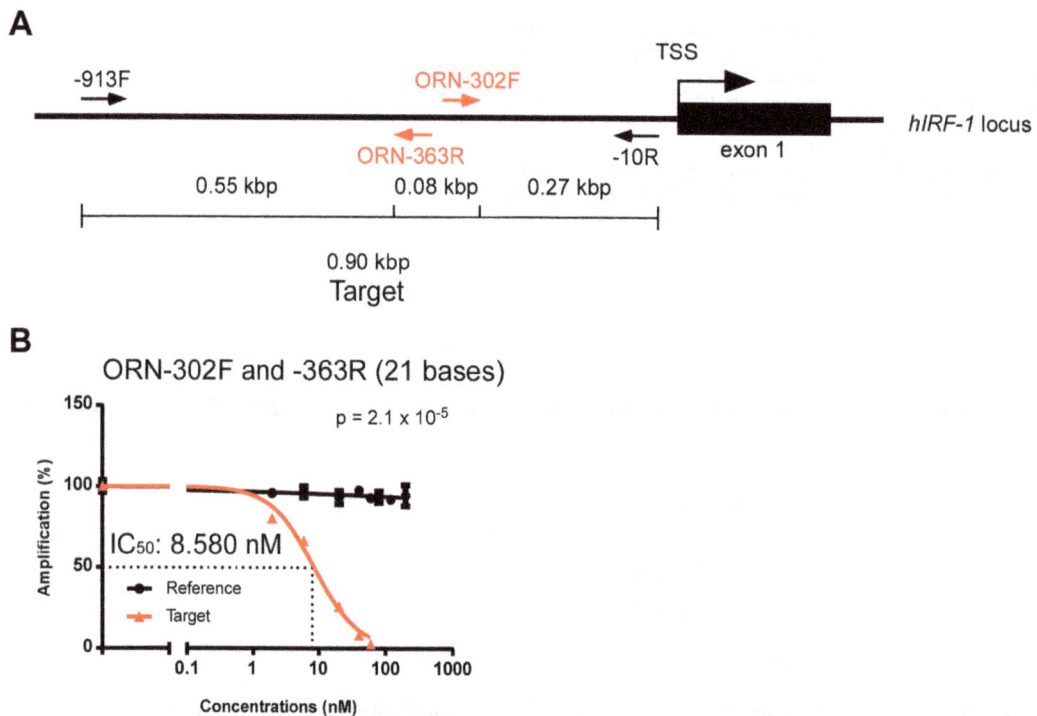

Figure 3. Inhibitory effects of multiple ORNs on PCR amplification. (**A**) Positions of ORNs in the target region. (**B**) Dose responses of multiple ORNs (ORN-302F and -363R). Real-time PCR amplification in different concentrations of ORNs was normalized against the value in the absence of ORNs. Dose-response curves were generated using the Prism software. Data are represented as means ± S.D. (n = 3). P-values for the differences in amplification between the reference and the target at a total ORN concentration of 20 nM (i.e., 10 nM for each ORN) were calculated using Student's t-test.

determine whether such ORN-mediated suppression occurs. As a model for testing, we designed primers complementary to the human *IRF-1* (*hIRF-1*) locus that amplify two neighboring regions 900 bp in size, namely, the reference region (hIRF1+269F and hIRF1+1167R) and the target region (hIRF1-913F and hIRF1-10R) (Figure 1A). The efficiency of amplification was comparable between the two regions (Figures 1B and S1). We then designed a 21-base ORN, ORN-302F, complementary to the target region (Figure 1A, C). As shown in Figure 1D, amplification of the target region was inhibited by increasing concentrations of ORN-302F, whereas amplification of the reference region was not inhibited. The IC_{50} of ORN-302F-mediated inhibition was 14.86 nM (Figure 1D). This result demonstrated the feasibility of ORN-mediated specific inhibition of PCR. In mechanistic terms, ORNs may hybridize to the target template and prevent DNA polymerases from extending the DNA strand from the primer, thereby suppressing amplification of the target template.

Optimal lengths of ORNs for suppression of PCR

Next, we tested ORNs of various lengths (Figure 1C) and evaluated the efficiency with which they suppressed PCR. As shown in Figure 1E, the slightly shorter ORN-298F (17 bases) specifically suppressed amplification of the target; however, its IC_{50} value (77.64 nM) was markedly higher than that of the 21-base ORN-302F. By contrast, the longer ORN-306F (25 bases) also specifically suppressed amplification of the target, but with a slightly lower IC_{50} than that of ORN-302F (6.415 nM) (Figure 1F). An even longer ORN, the 29-base ORN-310F, had an IC_{50} (9.236 nM) comparable to that of ORN-306F, and also suppressed amplification of the reference template at a higher concentration (100 nM) (Figure 1G). These results indicated that

ORNs with lengths of 21–25 bases can specifically inhibit amplification of target sequences with optimal IC_{50} values.

We validated the suppression of amplification of the target template by ORN-306F in a reaction in which the target and reference primers were mixed (Figure 1H and I). To distinguish the bands of the reference (1.06 kbp) and target (0.90 kbp) amplicons by agarose gel electrophoresis, we used the hIRF1+112F primer instead of hIRF1+296F (Figure 1A). Suppression efficiency was calculated semi-quantitatively by image-processing analysis. The calculated IC_{50} from this experiment (14.76 nM) (Figure 1I) was comparable to that determined by real-time PCR (6.415 nM) (Figure 1F).

Selection of ORNs for effective suppression

Next, we examined how easily optimal ORNs can be found in the target regions by designing 21-base ORNs to hybridize to various positions within the target sequence (ORN-666R, -363R and -181R) (Figure 2A). All of them efficiently suppressed amplification of the target sequence, although the IC_{50} of ORN-181R (34.52 nM) was higher than those of the other two (12.01 and 10.63 nM, respectively) (Figure 2B–D). These results suggested that ORNs can be designed flexibly in target regions, and that optimal ORNs can be easily found empirically.

Potential synergistic or additive effects of multiple ORNs

Next, we investigated whether the use of multiple ORNs had a synergistic or additive effect on suppression of amplification. When a combination of ORN-302F and -363R was used, which are oriented in opposite directions (Figure 3A), the IC_{50} was 8.580 nM (Figure 3B). This value was slightly lower than that of

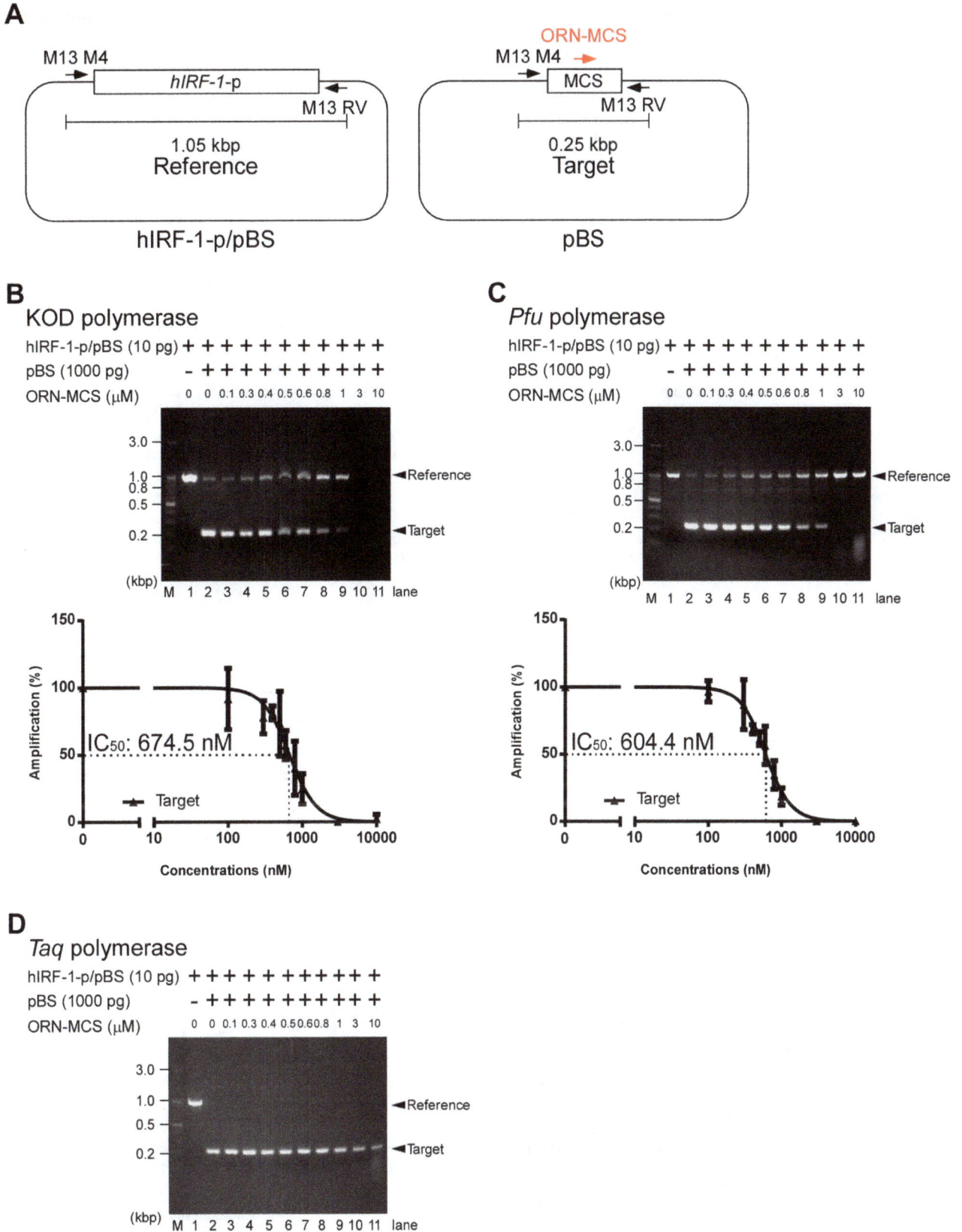

Figure 4. ORN-mediated inhibition of PCR amplification in a competitive setting. (A) Scheme for ORNi-PCR in a competitive setting. ORNi-PCR can be used for specific inhibition of target sequences while allowing amplification of reference sequences containing the same primer sequences. We designed an ORN targeting the multiple cloning site (MCS) of pBluescript-SK+ (pBS) (ORN-MCS). PCR amplification of the target and reference sequence in the presence or absence of different concentrations of ORNs, using KOD polymerase (B), *Pfu* polymerase (C), or *Taq*

polymerase (**D**). Gel images (upper panels) are representative of three independent experiments. The lower panels show dose-response curves of ORN-mediated inhibition.

ORN-302F (Figure 1D) or -363R (Figure 2C) used separately, but did not represent a significant improvement. This result suggested that, in practice, a single ORN would be sufficient for near-optimal suppression of PCR amplification.

Specific inhibition of target sequences by ORNi-PCR in competitive settings

Next, we investigated whether ORNi-PCR can be used for specific inhibition of target sequences while allowing amplification of non-target templates containing the same primer sequences. In these experiments, we used the template plasmids pBluescript-SK+ (pBS) and a pBS-derived plasmid containing *hIRF-1* promoter sequence (hIRF-1-p/pBS) (Figure 4A). Because the sequence of *hIRF-1* promoter was inserted into *Kpn* I and *Sac* I sites of pBS, hIRF-1-p/pBS lacks the multiple cloning site (MCS) of pBS. hIRF-1-p/pBS served as an efficient template for PCR amplification using the M13 primers with the KOD DNA polymerase (Figure 4B, upper panel, lane 1). When pBS and hIRF-1-p/pBS were mixed in a 100:1 ratio, PCR reactions resulted in selective amplification of the sequence derived from pBS (Target) but poor amplification of the sequence from hIRF-1-p/pBS (Reference) (Figure 4B, upper panel, lane 2). This selective amplification might have been caused by the lower concentration of hIRF-1-p/pBS, as

well as the longer insert size of the *hIRF-1*-p sequence. Therefore, we designed an ORN targeting the multiple cloning site of pBS (ORN-MCS) (Figure 4A) and tested whether it could inhibit amplification of the pBS sequence. As shown in Figure 4B (upper panel, lane 9), ORN-MCS effectively inhibited amplification of the target sequence at a concentration of 1 μM, and amplification of the reference sequence in the same reaction increased. At higher concentrations (3 and 10 μM), ORN-MCS non-specifically suppressed amplification of the reference sequence, as well as the target sequence, when KOD polymerase was used (Figure 4B, upper panel, lanes 10 and 11). We speculate that KOD polymerase might be non-specifically inhibited by the ORN through direct binding to the active site. These results suggested that ORNi-PCR can be used for selective amplification of minor templates that are otherwise more difficult to amplify.

Next, we examined whether other DNA-dependent DNA polymerases can be used for ORNi-PCR. PCR reactions using the *Pfu* DNA polymerase in the same setting described above suppressed amplification of the target sequence, but augmented amplification of the *hIRF-1* promoter region in hIRF-1-p/pBS (Figure 4C, upper panel). The calculated IC_{50} (604.4 nM) was comparable to that obtained using KOD polymerase (674.5 nM) (Figure 4C, lower panel). The IC_{50} values in these experimental

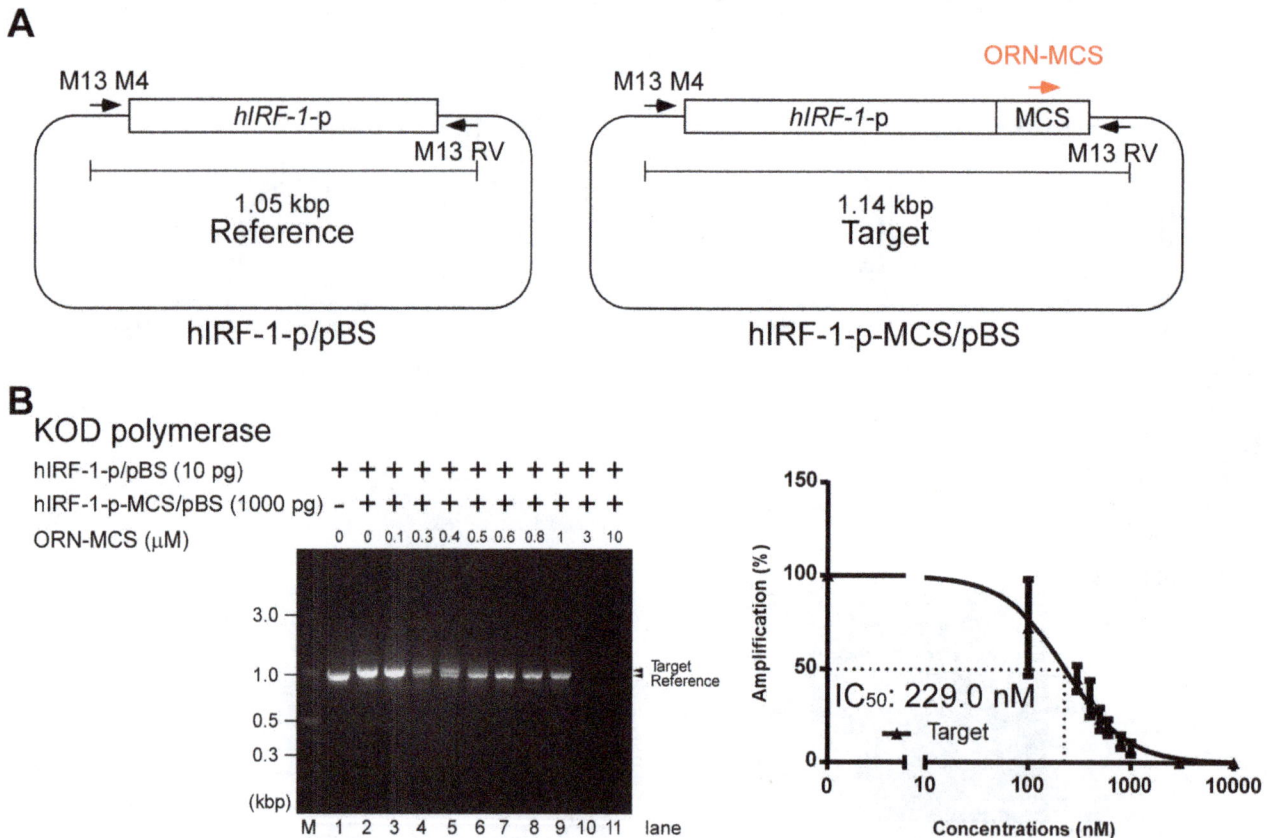

Figure 5. ORN-mediated inhibition of PCR amplification of a target sequence similar to the reference sequence, in a competitive setting. (**A**) Scheme for ORNi-PCR using a target sequence similar to the reference sequence in a competitive setting. (**B**) PCR amplification of the target sequence and reference sequence in the absence or presence of different concentrations of ORNs using the KOD polymerase. The gel image (left panel) is representative of three independent experiments. The right panel shows a dose-response curve of ORN-mediated inhibition.

A

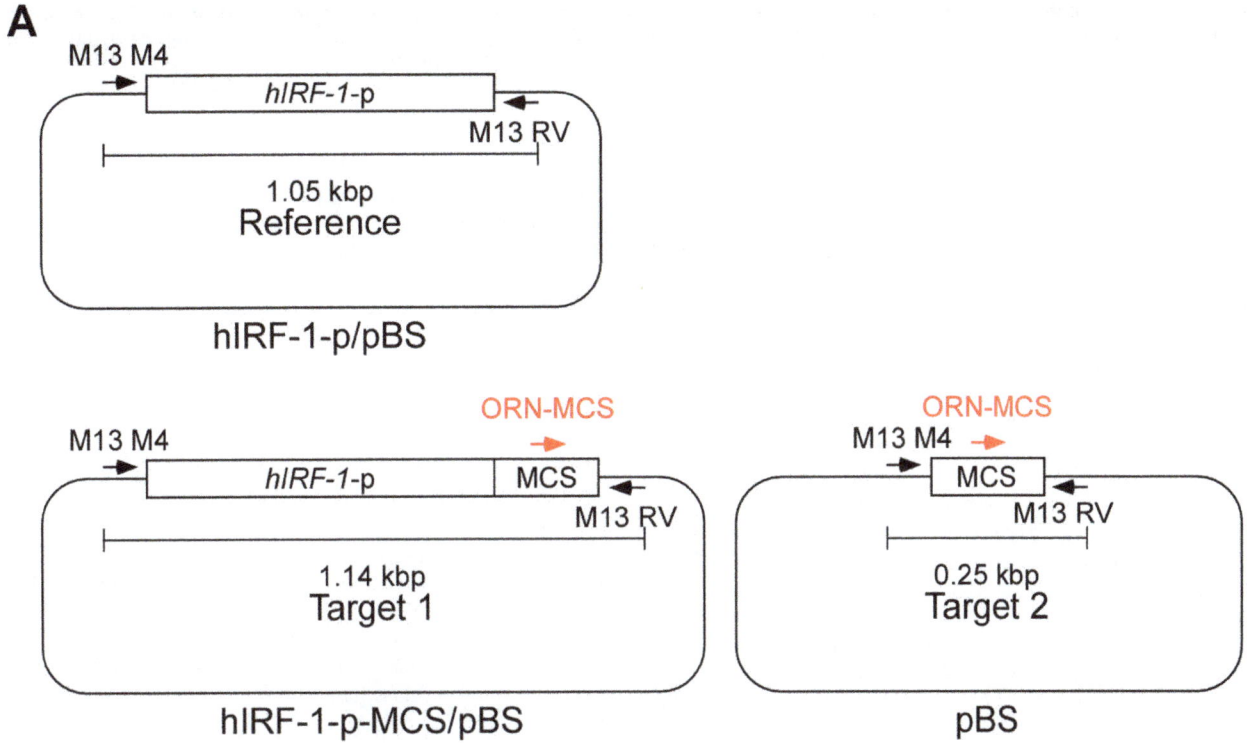

B

KOD polymerase

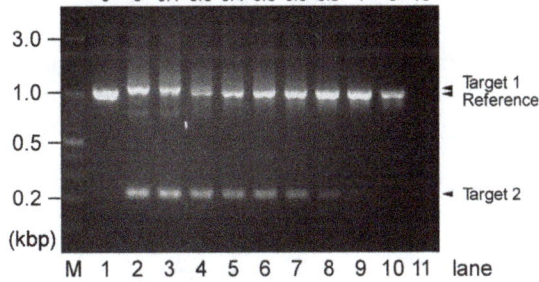

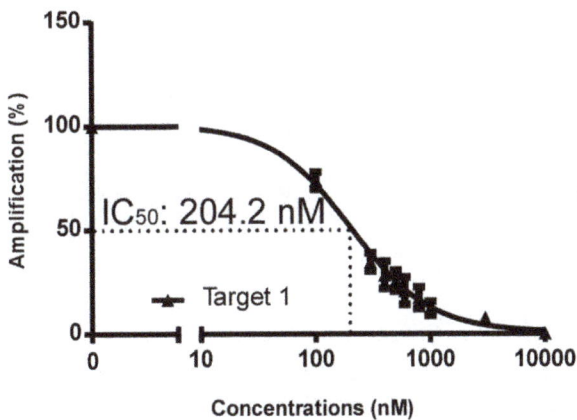

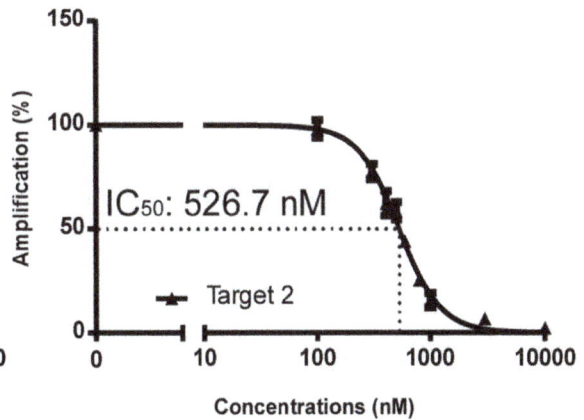

Figure 6. ORN-mediated inhibition of PCR amplification of multiple target sequences in a competitive setting. (A) Scheme for ORNi-PCR using multiple target sequences in a competitive setting. **(B)** PCR amplification of target sequences and the reference sequence in the presence or absence of various concentrations of ORNs, using KOD polymerase. The gel image (upper panel) is representative of three independent experiments. The lower right and left panels show dose-response curves of ORN-mediated inhibition for Targets 1 and 2, respectively.

settings, using plasmid DNAs as templates (Figure 4B and C), were much higher than those for ORNi-PCR using genomic DNA as the template (6–78 nM) (Figures 1–3). Because plasmid DNA is of much lower complexity than genomic DNA, PCR amplification of plasmid DNA is much more efficient; consequently, higher concentrations of ORNs may be required for effective suppression.

On the other hand, in PCR reactions using *Taq* DNA polymerase, increasing concentrations of ORN-MCS (up to 10 µM) gradually suppressed amplification of the target sequence, but the suppression was never complete (Figure 4D). In contrast to PCR reactions using the KOD and *Pfu* polymerases, we failed to detect amplification of the reference sequence when *Taq* was used; thus these results showed that KOD and *Pfu* are suitable for ORNi-PCR, whereas *Taq* is not. There are several salient differences between these enzymes, namely, the KOD and *Pfu* DNA polymerases are α-type DNA polymerases that retain $3'–5'$ exonuclease activity but not $5'–3'$ exonuclease activity [9–11], whereas *Taq* DNA polymerase belongs to the Pol I–type DNA polymerase family and retains $5'–3'$ exonuclease activity but not $3'–5'$ exonuclease activity [12]. Our results suggested that α-type DNA polymerases, but not Pol I-type DNA polymerases, might be compatible with ORNi-PCR. It is possible that $5'–3'$ exonuclease activity may remove ORNs bound to the target sequence, as DNA polymerase I does in prokaryotic DNA replication [13]. The fact that α-type DNA polymerases can be used for ORNi-PCR is advantageous, because $3'–5'$ exonuclease activity is important for proofreading [14]. In this regard, other methods such as blocking ODNs or LNAs are not compatible with α-type DNA polymerases, because 3' modifications of blocking ODNs or LNAs may be removed by $3'–5'$ exonuclease activity [3]. Therefore, ORNi-PCR might enable more accurate amplification while effectively suppressing target amplification.

Next, we investigated whether ORN-MCS could specifically inhibit a target sequence similar to the reference sequence. To this end, we constructed hIRF-1-p-MCS/pBS (Target), which is

distinguished from hIRF-1-p/pBS (Reference) by the presence of the pBS MCS (Figure 5A). When hIRF-1-p-MCS/pBS and hIRF-1-p/pBS were mixed in a 100:1 ratio, PCR reactions selectively amplified the reference sequence (Figure 5B, left panel, lane 2). ORN-MCS inhibited amplification of the target sequence, whereas amplification of the reference sequence was elevated in the same reaction mixture (Figure 5B, left panel, lanes 4–9). The calculated IC_{50} was 229.0 nM (Figure 5B, right panel). As shown in Figure 4B, ORN-MCS non-specifically suppressed amplification of the reference sequence at higher concentrations (Figure 5B, left panel, lanes 10 and 11). This result suggested that ORNi-PCR can be used for selective amplification of minor templates that are similar to abundant templates.

Finally, we investigated whether ORN-MCS can specifically inhibit multiple target sequences. In these experiments, we used the template plasmids pBS and hIRF-1-p-MCS/pBS and the reference plasmid hIRF-1-p/pBS (Figure 6A). When hIRF-1-p-MCS/pBS (Target 1), pBS (Target 2), and hIRF-1-p/pBS (Reference) were mixed in a 10:10:1 ratio, PCR reactions selectively amplified Targets 1 and 2 but only marginally amplified the Reference (Figure 6B, upper panel, lane 2). ORN-MCS could inhibit amplification of both target sequences (Figure 6B, upper panel, lanes 4–10). By contrast, amplification of the Reference was elevated in the same reaction mixture (Figure 6B, upper panel, lanes 4–10). The calculated IC_{50} values were 204.2 nM and 526.7 nM for Targets 1 and 2, respectively (Figure 6B, lower panels). This result suggested that ORNi-PCR can be used for selective amplification of minor templates in the presence of multiple abundant templates.

Conclusions

In this study, we demonstrated that ORNs specifically inhibit PCR amplification of target templates containing sequences complementary to the ORNs. For optimal suppression, ORNs should be around 21–25 bases long. α-type DNA-dependent

Table 1. ORNs used in this study and their efficiency of suppression in ORNi-PCR.

Number	Name	ORN sequence (5'–3')	Length(base)	IC_{50} (nM)	Experiments
R4	ORN-302F	ccgggggcgcugggcuguccc	21	14.86	Figures 1D and 3B
R6	ORN-298F	gggcgcugggcuguccc	17	77.64	Figure 1E
R5	ORN-306F	ggggccgggggcgcugggcuguccc	25	6.415	Figure 1F
				14.76	Figure 1I
R7	ORN-310F	ggcuggggccgggggcgcugggcuguccc	29	9.236	Figure 1G
R10	ORN-666R	ggccgcugcuggcacagcccc	21	12.01	Figure 2B
R9	ORN-363R	cacccuccuggcggggcgggg	21	10.63	Figures 2C and 3B
R8	ORN-181R	cacccucuccggccgggcgcc	21	34.52	Figure 2D
R11	ORN-MCS	agagcggccgccaccgcggug	21	674.5	Figure 4B
				604.4	Figure 4C
				229.0	Figure 5B
				204.2	Figure 6B
				526.7	Figure 6B

polymerases retaining $3'-5'$ exonuclease activity, such as the KOD and *Pfu* polymerases, can be used for ORNi-PCR. Regarding the mechanism, ORNs may hybridize to the target template and prevent DNA polymerases from extending the DNA strand from the primer, thereby suppressing amplification of the target template. ORNi-PCR technology will be a useful tool in both molecular biology research and clinical diagnosis. In addition, the ORN-mediated suppression strategy might also be applied to other types of template-dependent synthesis of nucleic acids, e.g., DNA-dependent DNA synthesis, transcription, and reverse transcription.

Materials and Methods

Oligonucleotides

Primers and ORNs used in this study were chemically synthesized (Greiner), and their sequences are provided in Table 1, Figure 1C, and Table S1.

Genomic DNA was purified from the 293T cell line [15,16]. For analysis of the amplification efficiencies of target and reference templates, PCR reactions were performed in mixtures containing 40 ng of 293T genomic DNA, 1× Buffer provided with KOD FX, 0.32 mM dNTPs, 0.25 μM of each primer, and 0.2 μL of KOD FX (Toyobo) in a 10 μL volume. The reactions were carried out with an initial denaturation at 94°C for 2 min, followed by 28 cycles of denaturation at 98°C for 10 sec, primer annealing and extension at 68°C for 1 min, and final extension at 68°C for 2 min. Real-time PCR reaction mixtures contained 40 ng of 293T genomic DNA, 0.2 μM of each primer, different concentrations of ORN, 1× ROX reference dye, and 5 μL of KOD SYBR qPCR Mix (Toyobo) in 10 μL reactions. The reactions were carried out with an initial denaturation at 98°C for 2 min, followed by 40 cycles of denaturation at 98°C for 10 sec, primer annealing at 60°C for 10 sec, and extension at 68°C for 1 min. The specificity of PCR products was confirmed by performing a dissociation curve analysis at 95°C for 15 sec, 60°C for 15 sec, and at 99°C for 15 sec. PCR amplification was quantitated on a 7900HT Fast Real-Time PCR System (Applied Biosystems). PCR products were subsequently electrophoresed on 2% agarose gels to confirm the presence of a unique amplicon of the expected size. All samples were amplified in triplicate, and IC_{50} values were calculated from the means using the Prism software (GraphPad).

For endpoint analysis, PCR reactions were performed with 40 ng of 293T genomic DNA, 1× Buffer provided with KOD -Plus- Ver.2, 0.2 mM dNTPs, 1.5 mM MgSO₄, 0.3 μM of each primer, and 0.2 μL of KOD -Plus- Ver.2 (Toyobo) in a 10 μL volume. The reactions were carried out with an initial denaturation at 94°C for 2 min, followed by 29 cycles of denaturation at 98°C for 10 sec, primer annealing at 60°C for 30 sec, and extension at 68°C for 1 min. The PCR products were subsequently electrophoresed on 2% agarose gels.

Statistical analysis

P-values were calculated using the Excel software (Microsoft) using Student's t-test.

Densitometric analysis

Densitometric analyses were performed using the ImageJ software (National Institutes of Health), and IC_{50} values were calculated from the means using the Prism software.

Plasmid construction

To construct hIRF-1-p/pBS, the sequence of the *hIRF-1* promoter was amplified using the hIRF1-913F_KpnI and hIRF1-10R_SacI primers with 293T genomic DNA as the template. The resulting amplicon was cleaved with *Kpn*I and *Sac*I, and ligated into pBluescript-SK+ (Stratagene) digested with the same enzymes.

To generate hIRF-1-p-MCS/pBS, the sequence of the *hIRF-1* promoter was amplified using the hIRF1-913F_KpnI and hIRF1-10R_XhoI primers with 293T genomic DNA as template. The resulting amplicon was cleaved with *Kpn*I and *Xho*I, and ligated into pBluescript-SK+ (Stratagene) digested with the same enzymes.

ORNi-PCR using plasmid DNA as templates

PCR was performed using KOD -Plus- Ver.2 (Toyobo) or *TaKaRa Taq* (Takara) in the presence of 1× Buffer provided with enzymes, 0.2 mM dNTPs, 0.3 μM each of M13 Primer M4 and M13 Primer RV, and various concentrations of ORN-MCS. When KOD was used, 1.5 mM MgSO₄ was added. Templates were mixtures of 100 or 1,000 pg of the target plasmids (pBS and/or hIRF-1-p-MCS/pBS) and 10 pg of the reference plasmid (hIRF-1-p/pBS). The reactions were carried out with an initial denaturation at 94°C for 2 min, followed by 30 cycles of denaturation at 98°C for 10 sec, primer annealing at 55°C for 30 sec, and extension at 68°C for 1 min. When *TaKaRa Taq* was used, an additional step of 68°C for 2 min was added at the end of the reaction.

PCR was also performed with Pfu-X (Greiner) in the presence of 1× Buffer provided with Pfu-X, 0.2 mM dNTPs, 5% DMSO, 0.4 μM each of M13 Primer M4 and M13 Primer RV, and various concentrations of ORN-MCS. The reactions were carried out with an initial denaturation at 95°C for 2 min, followed by 30 cycles of denaturation at 95°C for 20 sec, primer annealing at 55°C for 20 sec, extension at 68°C for 30 sec, and a final extension at 68°C for 30 sec.

Supporting Information

Figure S1 Comparable amplification of target and reference regions in the *IRF-1* locus by real-time PCR.
(PDF)

Table S1 PCR primers used in this study.
(PDF)

Acknowledgments

We thank M. Yuno for technical assistance.

Author Contributions

Conceived and designed the experiments: NT TF HF. Performed the experiments: NT TF HF. Analyzed the data: NT TF HF. Contributed reagents/materials/analysis tools: NT TF HF. Contributed to the writing of the manuscript: NT HF. Initiated, directed and supervised the study: HF.

References

1. Mack DH, Sninsky JJ (1988) A sensitive method for the identification of uncharacterized viruses related to known virus groups: hepadnavirus model system. Proc Natl Acad Sci U S A 85: 6977–6981.

2. Pytela R, Suzuki S, Breuss J, Erle DJ, Sheppard D (1994) Polymerase chain reaction cloning with degenerate primers: homology-based identification of adhesion molecules. Methods Enzymol 245: 420–451.

3. Vestheim H, Deagle BE, Jarman SN (2011) Application of blocking oligonucleotides to improve signal-to-noise ratio in a PCR. Methods Mol Biol 687: 265–274.

4. Seyama T, Ito T, Hayashi T, Mizuno T, Nakamura N, et al. (1992) A novel blocker-PCR method for detection of rare mutant alleles in the presence of an excess amount of normal DNA. Nucleic Acids Res 20: 2493–2496.

5. Yu D, Mukai M, Liu QL, Steinman CR (1997) Specific inhibition of PCR by non-extendable oligonucleotides using a 5′ to 3′ exonuclease-deficient DNA polymerase. Biotechniques 23: 714–720.

6. Dominguez PL, Kolodney MS (2005) Wild-type blocking polymerase chain reaction for detection of single nucleotide minority mutations from clinical specimens. Oncogene 24: 6830–6834.

7. Carlson CM, Dupuy AJ, Fritz S, Roberg-Perez KJ, Fletcher CF, et al. (2003) Transposon mutagenesis of the mouse germline. Genetics 165: 243–256.

8. Yuen PS, Brooks KM, Li Y (2001) RNA: a method to specifically inhibit PCR amplification of known members of a multigene family by degenerate primers. Nucleic Acids Res 29: E31.

9. Takagi M, Nishioka M, Kakihara H, Kitabayashi M, Inoue H, et al. (1997) Characterization of DNA polymerase from Pyrococcus sp. strain KOD1 and its application to PCR. Appl Environ Microbiol 63: 4504–4510.

10. Lundberg KS, Shoemaker DD, Adams MW, Short JM, Sorge JA, et al. (1991) High-fidelity amplification using a thermostable DNA polymerase isolated from Pyrococcus furiosus. Gene 108: 1–6.

11. Uemori T, Ishino Y, Toh H, Asada K, Kato I (1993) Organization and nucleotide sequence of the DNA polymerase gene from the archaeon Pyrococcus furiosus. Nucleic Acids Res 21: 259–265.

12. Innis MA, Myambo KB, Gelfand DH, Brow MA (1988) DNA sequencing with Thermus aquaticus DNA polymerase and direct sequencing of polymerase chain reaction-amplified DNA. Proc Natl Acad Sci U S A 85: 9436–9440.

13. Kornberg A, Baker T (1992) DNA replication. Sausalito, California: University Science Books.

14. Blanco L, Bernad A, Blasco MA, Salas M (1991) A general structure for DNA-dependent DNA polymerases. Gene 100: 27–38.

15. DuBridge RB, Tang P, Hsia HC, Leong PM, Miller JH, et al. (1987) Analysis of mutation in human cells by using an Epstein-Barr virus shuttle system. Mol Cell Biol 7: 379–387.

16. Pear WS, Nolan GP, Scott ML, Baltimore D (1993) Production of high-titer helper-free retroviruses by transient transfection. Proc Natl Acad Sci U S A 90: 8392–8396.

13

Comparison of Immunity in Mice Cured of Primary/Metastatic Growth of EMT6 or 4THM Breast Cancer by Chemotherapy or Immunotherapy

Reginald M. Gorczynski[1,2]*, Zhiqi Chen[1], Nuray Erin[3], Ismat Khatri[1], Anna Podnos[1]

1 University Health Network, Toronto General Hospital, Toronto, Canada, **2** Department of Immunology, Faculty of Medicine, University of Toronto, and Institute of Medical Science, University of Toronto, Toronto, Ontario, Canada, **3** Department of Medical Pharmacology, Akdeniz University, School of Medicine, Antalya, Turkey

Abstract

Purpose: We have compared cure from local/metastatic tumor growth in BALB/c mice receiving EMT6 or the poorly immunogenic, highly metastatic 4THM, breast cancer cells following manipulation of immunosuppressive CD200:CD200R interactions or conventional chemotherapy.

Methods: We reported previously that EMT6 tumors are cured in CD200R1KO mice following surgical resection and immunization with irradiated EMT6 cells and CpG oligodeoxynucleotide (CpG), while wild-type (WT) animals developed pulmonary and liver metastases within 30 days of surgery. We report growth and metastasis of both EMT6 and a highly metastatic 4THM tumor in WT mice receiving iv infusions of Fab anti-CD200R1 along with CpG/tumor cell immunization. Metastasis was followed both macroscopically (lung/liver nodules) and microscopically by cloning tumor cells at limiting dilution in vitro from draining lymph nodes (DLN) harvested at surgery. We compared these results with local/metastatic tumor growth in mice receiving 4 courses of combination treatment with anti-VEGF and paclitaxel.

Results: In WT mice receiving Fab anti-CD200R, no tumor cells are detectable following immunotherapy, and CD4+ cells produced increased TNFα/IL-2/IFNγ on stimulation with EMT6 in vitro. No long-term cure was seen following surgery/immunotherapy of 4THM, with both microscopic (tumors in DLN at limiting dilution) and macroscopic metastases present within 14 d of surgery. Chemotherapy attenuated growth/metastases in 4THM tumor-bearers and produced a decline in lung/liver metastases, with no detectable DLN metastases in EMT6 tumor-bearing mice-these latter mice nevertheless showed no significantly increased cytokine production after restimulation with EMT6 in vitro. EMT6 mice receiving immunotherapy were resistant to subsequent re-challenge with EMT6 tumor cells, but not those receiving curative chemotherapy. Anti-CD4 treatment caused tumor recurrence after immunotherapy, but produced no apparent effect in either EMT6 or 4THM tumor bearers after chemotherapy treatment.

Conclusion: Immunotherapy, but not chemotherapy, enhances CD4+ immunity and affords long-term control of breast cancer growth and resistance to new tumor foci.

Editor: Fabrizio Mattei, Istituto Superiore di Sanità, Italy

Funding: Supported by a grant (RG-11) to RMG from the Canadian Cancer Society (www.cancer.ca). The funders had no role in study design, data collection and analysis, decision to publish, or preparation of the manuscript.

Competing Interests: The authors have declared that no competing interests exist.

* Email: rgorczynski@uhnres.utoronto.ca

Introduction

The immunoregulatory molecule CD200 has been reported to regulate growth of human solid tumors [1,2] and hematological tumors [3–5]. Using a transplantable EMT6 mouse breast cancer line CD200 expression, by tumor cells or host, increased local tumor growth and metastasis to DLN [6,7], which was abolished by neutralizing antibody to CD200, or following growth in mice lacking the primary inhibitory receptor for CD200 (CD200R1KO mice). In contrast to these observations, growth of the highly metastatic 4THM breast tumor (derived from a 4T1 parent line) was increased in CD200R1KO mice, with somewhat diminished growth in CD200[tg] animals [8]. Surgical resection in CD200R1KO EMT6 tumor-bearing mice, followed by immunization with CpG as adjuvant, cured CD200R1KO mice of breast cancer recurrence in the absence of lung/liver metastases, and of micro metastases (defined by limiting dilution cloning in vitro) in DLN [9].

Multiple factors both intrinsic to tumor cells themselves and host associated elements are implicated in tumor metastasis [10–14]. Many such factors are associated with altering trafficking of either host inflammatory-type cells to the local tumor environment where they can facilitate metastasis through a variety of mechanisms [15–17], including regulation of host resistance mechanisms [18–21]. Metastatic tumor cells are known to undergo changes in gene expression profile leading to increased cancer stem cell- like properties and the ability to survive, establish and grow in a foreign environment [22–24]. Like CD200, an inhibitory member of the B7 family of T cell co stimulation, expression of another such molecule, B7× (B7-H4) has been reported to influence metastasis using 4T1 tumor cells and B7KO mice [25]. B7KO mice with 4T1 tumors, like CD200R1KO with EMT6, showed enhanced survival and a memory response to tumor re-challenge, which was correlated with decreased infiltration of immunosuppressive cells, including tumor-associated neutrophils, macrophages, and regulatory T cells, into tumor-bearing metastatic lung tissue [25]. CD200R1KO mice showed increased growth of 4THM tumors [24].

The studies below compared protection seen in surgically treated/immunized EMT6 or 4THM tumor injected WT mice with/without manipulation of CD200:CD200R interactions using Fab anti-CD200R, with attenuation of disease after surgical resection followed by chemotherapy.

Materials and Methods [9]

Ethics approval and animal use guidelines

This study was carried out in strict accordance with the recommendations of the Canadian council for Animal Care (CCAC). The protocol was approved by the Committee on the Ethical use of Animals for experimentation at the University Health Network (Permit Number:AUP.1.5). All surgery was performed under sodium pentobarbital anesthesia, and all efforts were made to minimize suffering.

Mice

CD200KO and CD200R1 knockout mice are described elsewhere [9]. WT BALB/c mice were from Jax Labs. All mice were housed 5/cage in an accredited facility at UHN. Female mice were used at 8 wk of age.

Monoclonal antibodies, and CpG deoxyoligonucleotide for adjuvant use, are described elsewhere [6,9,26]

Rabbit Fab anti-CD200R1 antibody was prepared using a commercial kit (Pierce Protein Products, Rockford, IL, USA) and rabbit IgG isolated by Cedarlane Labs (Hornby, Ontario, Canada), following immunization of rabbits with 500 µg mouse CD200R1 emulsified in Freund's Adjuvant. In independent studies (not shown) this antibody (1:1000 dilution) inhibited binding (FACS analysis) of FITC-labeled mouse CD200 to Hek cells transduced to over-express murine CD200R1.

EMT6 breast tumor cells, induction of tumor growth in BALB/c mice, and limiting dilution cultures to establish frequency of metastasis to draining lymph nodes (DLN) were as described earlier [9,26]

4THM tumors, a highly metastatic variant of 4T1, were derived by Erin et al as reported elsewhere [24].

Surgical resection and immunotherapy/chemotherapy of tumor-bearing mice [9]

Mice receiving 5×10^5 EMT6 or 1×10^5 4THM tumor cells injected into the mammary fat pad in 100 µl PBS underwent surgical resection 14–16 d later. For immunotherapy, mice received intraperitoneal immunization with 3×10^6 EMT6 (or 4THM) tumor cells (irradiated with 2500Rads) mixed with 100 ug CpG ODN (see above) in 100 µl PBS, emulsified with an equal volume of Incomplete Freund's adjuvant, 2 days after surgery. Mice treated with chemotherapy post surgical resection, received 4 injections of paclitaxil intraperitoneally in 0.15 ml PBS (Taxol: 10 mg/Kg), beginning on the day of surgery, and at 21 day intervals thereafter. In addition, beginning on the day following surgery, and at 14 day intervals for a total of 6 injections, the same mice also received anti-VEGF (30 mg/Kg) iv in 0.3 ml PBS.

All animals were monitored ×3/week for weight loss and general health and sacrificed at the times indicated in individual experiments (>10% weight loss), with visible tumor colonies in the lung/liver enumerated. DLN cell suspensions were prepared from individual mice and cloned under limiting dilution in 96-well flat-bottomed microtitre plates to assess tumor colony formation [7]. Important variables measured were time post treatment to sacrifice, and tumor growth-note that aggressive uncontrolled tumor growth in some groups in individual experiments led to certain groups being sacrificed before others (see text).

Preparation of cells and cytotoxicity, proliferation and cytokine assays: see [9,26]

In brief, 5×10^6 splenocytes from mice treated as described in the text were stimulated in vitro in triplicate with 2×10^5 irradiated (2500Rads) tumor cells in 2 ml αMEM with 10% fetal calf serum. 100 µl aliquots of supernatants were assayed at 48 hr for various cytokines using commercial kits (BioLegend, San Diego, USA). Cells were harvested from cultures at 6 d, washed ×2, and incubated for 18 hr with 1×10^3 ^{3}HTdR-labelled tumor target cells at varying effector:target ratios to determine direct anti-tumor cytotoxicity.

Statistics

Cloneable tumor cell frequency was determined as before [6]. Within experiments, comparison between groups used ANOVA, with subsequent paired Student's t-tests as indicated.

Results

Surgical resection followed by immunization along with Fab anti-CD200R, or chemotherapy alone, prevents metastasis of EMT6, but not 4THM, in BALB/c mice

Surgical resection of a primary tumor in CD200R1KO mice followed by immunization prevented macroscopic lung/liver metastases enumerated at 90 d post tumor inoculation, compared with surgery alone [9]. As shown in Figure 1 (data pooled from 2 independent studies) no protection was seen in wild type (WT) mice Figure 1, panel a), but WT mice were cured if given Fab anti-CD200R following surgery/immunization (panel b). Note that aggressive tumor growth led to WT control mice having to be sacrificed within 18 d or 21 d of surgery (panels a/b), unlike immunotherapy-treated mice receiving anti-CD200R (panel b) where mice were able to be followed for ≥90 d post surgery. When mice in this latter group were sacrificed earlier (18–21 d post surgery) again no lung/liver colonies were observed (not shown, but note no colonies at 90 d). Both CD200R1KO and WT mice showed no evidence of macroscopic metastases following chemotherapy instead of immunotherapy post surgery (Figure 1, panels c/d respectively). Again note that addition of chemotherapy treatment allowed mice to be monitored for tumor metastases (90 d post surgery) much longer than non-chemotherapy controls (21 and 18 d in panels c, d respectively-however, in studies where chemotherapy mice were deliberately sacrificed early, no metastases were observed on days 18/21 (not shown-but note data for 90 d). In mice receiving 4THM tumors, attenuation of lung/liver metastasis was achieved using surgery+chemotherapy, but not by surgery followed by immunotherapy (see Figure 1, panels e and f respectively). Failure of immunotherapy to protect from 4THM tumors again led to these mice (panel e) being sacrificed much earlier (10 d post surgery) than with EMT6 mice (panels a–d) or 4THM mice receiving chemotherapy (panel f). Once again, in studies where chemotherapy-treated 4THM injected mice were sacrificed at 10 d post surgery, no metastases were seen (not

shown-but seen marked attenuation of metastases even at 90 d in panel f).

DLN cell suspensions of mice sacrificed at the times shown in Figure 1 were cultured under limiting dilution conditions with cultures monitored over a 21-day period for colony growth, to enumerate the frequency of tumor cells in the initial DLN samples (Figure 2: panel a shows data for EMT6 tumors, panel b for 4THM) [7]. Data to the far left in each panel show the frequency of tumor cells cloned from DLN of mice sacrificed on the day of

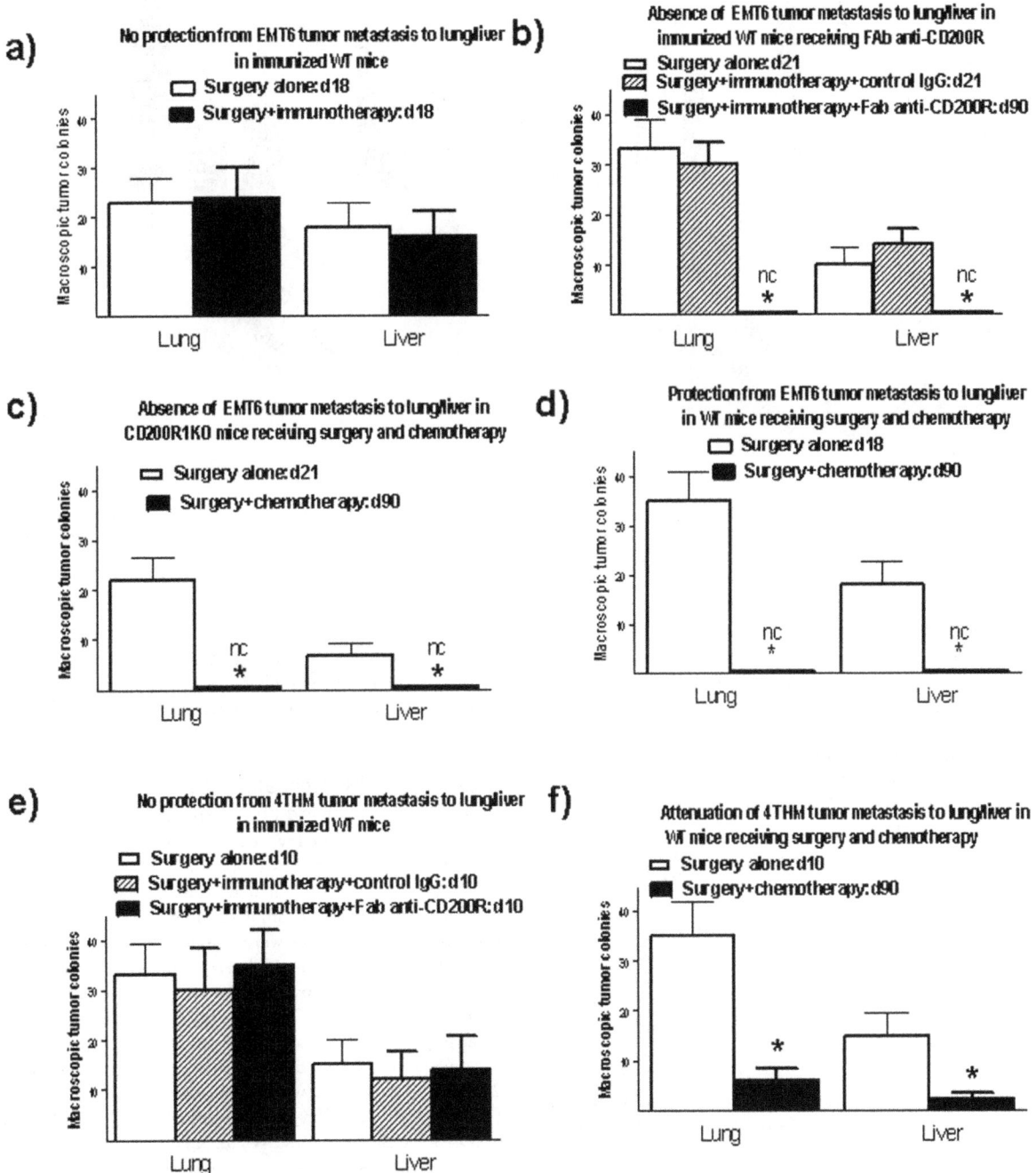

Figure 1. Comparison of lung and liver metastases of tumor cells in WT BALB/c mice receiving EMT6 or 4THM tumor cells and subsequently treated with surgical resection and chemotherapy/immunotherapy (see)Methods. 4 mice were used per group, with mice sacrificed at the times show post surgery (number above histogram bars) to measure macroscopic tumor metastases in the lung/liver. All data represent arithmetic means (±SD) for each group. nc indicates no metastatic colonies detected; *, p<0.05 relative to similar group receiving either immunotherapy or chemotherapy.

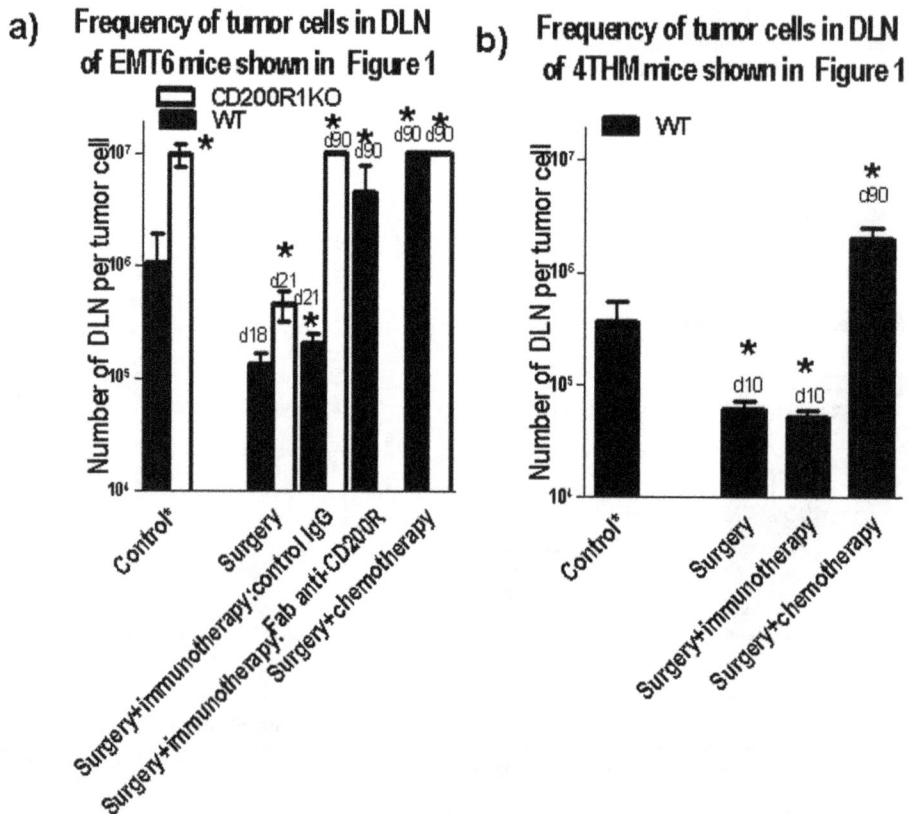

Figure 2. Attenuation of outgrowth of tumor from DLN of mice shown in Figure 1 as assessed by limiting dilution frequency (see Methods). DLN cells from separate mice were also cloned alone at the time of surgery (data to far left of each panel-control*). All frequencies were calculated based on the input numbers of cells from DLN of control mice only. *, p<0.05 compared with control* mice

tumor resection. Cells in all clones were stained (~100% positive) with anti-BTAK (anti-tumor) antibody (data not shown-see [7]).

The frequency of tumor cells cloned from DLN of both WT and CD200R1KO EMT6-injected mice treated only by surgical resection increased over 18–21 d post resection, relative to the frequency seen in DLN at the time of surgical resection (panel a). Surgical resection followed by immunotherapy and control IgG led to little decrease in the DLN tumor frequency in WT mice sacrificed at 21 d post surgery. Fab anti-CD200R along with surgery/immunization resulted in a marked decrease (>7x) in tumor cells cloned from DLN of WT mice (d90). In similarly treated CD200R1KO mice no tumor cells were detected (detection limits in assay ~1 in 1×10^7) at 90 d post surgery. No detectable tumor cells could be cloned from DLN of either WT or CD200R1KO mice 90 d post surgery if animals received chemotherapy following surgical resection (data to far right in Figure 2a). In 4THM tumor-bearers (panel b), sacrifice of mice 10 d after surgery with either no additional treatment, or immunotherapy (CpG+ irradiated 4THM), indicated an increase (~8x) in frequency of cloned tumor cells in DLN compared with the numbers present at the time of surgery. Surgery followed by chemotherapy decreased the number of cloned tumor cells at d90 (far right in Figure 2b).

In separate studies (not shown), no WT or CD200R1KO mice survived following treatment with surgery and anti-VEGF alone, and survival with paclitaxil as the sole chemotherapeutic agent was ≤50% of that seen using the combination shown, in both CD200R1KO and WT mice with each tumor used. Combined surgery and chemotherapy "cured" WT mice of EMT6 tumor

growth, as defined by an absence of macroscopic metastases at 300 d post surgery, and undetectable tumor cells cloned from DLN of mice at this time (limits of detection ~1 in 2×10^7 DLN cells)-see also [9]. All 4THM mice treated in this fashion died before110days post surgery (data not shown).

Absence of cells attenuating ability to clone tumor from DLN of mice receiving chemotherapy

Figure S1 investigated whether DLN of either immunotherapy- or chemotherapy-treated WT mice contained populations of cells which non-specifically attenuated growth of tumor cells, leading to inaccurate estimation of tumor cell frequency in limiting dilution [9]. Groups of 5WT mice were treated as in Figure 1 with EMT6 or 4THM tumor cells, followed by surgical resection and combined chemotherapy with anti-VEGF and paclitaxil. Mice were sacrificed 90 days post surgery. DLN cells from WT mice receiving either EMT6 or 4THM tumor cells 14d earlier (WT* in Figure S1) were cultured under limiting dilution conditions (from 2×10^3 to 1×10^5 cells/well) alone, or with a five-fold excess of DLN cells from the 90d chemotherapy-treated mice (from 1×10^4 to 5×10^5). Cells from these WT or CD200R1KO mice were also cloned alone. All tumor cells frequencies were subsequently calculated based on the input numbers of control cells only. Data shown in this Figure are pooled from 3 separate studies.

The frequency of detected tumor cells in the mice at 90 d post combined surgery/chemotherapy was below the limits of detection in this assay (see data to far right in each of the EMT6/4THM groups of Figure S1). Addition of a 5-fold excess of cells from the

DLN of these populations **did not** alter the measured frequency of cloneable tumor cells from DLN of WT* mice sacrificed at 14 d post tumor injection.

CD4$^+$ cells in immunotherapy-treated, but not in chemotherapy-treated mice, are responsible for decreased metastasis

Protection (in CD200KO or CD200R1KO mice) was not related to a direct immune response from recipient mice to CD200 expressed on tumor cells themselves [9,25]. CD200/CD200R is not expressed on 4THM tumors, and thus an immune response to such tumor-bearing epitopes could not explain the differences observed above. Immunotherapy of EMT6 tumor growth was abolished by infusion of anti-CD4 mAb [9]. To investigate whether an active CD4-dependent immune process was implicated in protection afforded by (surgery + chemotherapy) we performed the following study.

Groups of 30 WT mice received EMT6 or 4THM cells into the mammary fat pad, followed by surgical resection. 5 mice/group received no further treatment. Two subgroups of 15 mice each then received either combination chemotherapy, or immunotherapy with irradiated tumor cells, CpG and Fab anti-CD00R. 10 d after immunotherapy/chemotherapy was initiated 5mice/group began a course of anti-CD4mAb or control IgG injections (3 injections of 75 µg in 300 µlPBS at 72 hr intervals iv). Mice were monitored for overall health, with sacrifice of all mice when there was evidence of respiratory distress and/or weight loss (10%) in any individual. Note that in the case of 4THM mice not receiving chemotherapy, this necessitated sacrifice at 10 d post surgery, while for EMT6 control mice, or EMT6 mice receiving immunotherapy and anti-CD4 treatment, this necessitated sacrifice at 18, 26 d post surgery respectively (see also text to Figure 1 above). All surviving mice were terminated at 90 d post surgery, and macroscopic liver/lung metastases determined, along with frequency of tumor cells in DLN (see Figure 2). In addition (see Figure S2), splenocytes from individual mice were stimulated in vitro with irradiated tumor cells for 6 d, with cytokine production measured (48 hr) and CTL assayed at 6 days, as described in the Methods. Data for 1 of 3 such studies are shown in Figure 3.

Macroscopically visible metastases in lung/liver (Figure 3a), along with increased frequency of tumor cells cloned from DLN (Figure 3b), was seen in EMT6 tumor injected mice receiving immunotherapy and anti-CD4 relative to mice receiving control Ig (see also [9])-as noted in Figure 1, where other immunotherapy-treated (but no anti-CD4) EMT6 groups were sacrificed at d18/26 (not 90 d as shown) there were, as expected, no metastases seen. Also as noted in Figure 1, immunotherapy afforded no protection from 4THM growth, regardless of subsequent anti-CD4 treatment, and these mice had to be sacrificed early in the study (10 d post surgery, by comparison to chemotherapy-treated mice, sacrificed at 90 d post surgery). In contrast to these data, following both EMT6 and 4THM tumor injection, the protection from macroscopic (lung/liver) and microscopic (DLN) metastases afforded by chemotherapy was apparently resistant to anti-CD4mAb therapy (Figure 3a/b). In separate studies (not shown) no affect was seen after infusion of anti-CD8 mAb into chemotherapy treated mice either. These in vivo studies need to be seen in the context of data from Figure S2, showing elevated cytotoxicity (CD4$^+$-dependent) only using splenocytes from immunotherapy-treated EMT6 tumor-injected mice (panel b), while in turn CD4$^+$ cells from these same mice produced increased cytokines (TNFα, IL-2 and IFNγ) relative to mice receiving surgery alone. Note that in the cytotoxicity assay used in Figure

S2b, killing itself was a function of CD8$^+$ cells in all groups (data not shown).

Resistance to implantation of fresh EMT6, but not 4THM, tumor in immunotherapy-treated EMT6-injected mice, but not in chemotherapy-treated EMT6/4THM-injected mice

The data in Figure 3 show that cure of both EMT6- and 4THM-injected mice of macroscopic and microscopic (DLN) tumor metastases following surgical resection and chemotherapy is resistant to anti-CD4 treatment, unlike mice cured of EMT6 tumor following surgery and immunotherapy. We next investigated resistance to fresh tumor implants of the same or different tumor in mice cured following immunotherapy/chemotherapy.

Groups of mice receiving EMT6/4THM tumors underwent surgical resection, followed by either chemotherapy (for all of 15 4THM- and 15 EMT6-injected mice) or immunotherapy (15 EMT6- injected mice). 90 d post surgical resection, with all animals free of obvious tumor growth and gaining weight, 5 mice/group, and 5 fresh mice, received either 5×10^5 EMT6 or 1×10^5 4THM tumors in the contralateral mammary fat pad to that used previously. Primary tumor growth was followed daily for all mice, and animals sacrificed 20 d later, with DLN harvested to assess tumor cells by limiting dilution. Data in Figure 4 show results (1 of 2 studies) for this experiment. None of the mice not receiving further tumor inoculation developed overt tumor recurrence in this time-data not shown to retain clarity.

Figure 4a shows that mice which undergo surgical eradication of EMT6, followed by immunotherapy, are refractory to re-challenge with EMT6 as monitored over 20 d by either visible tumor (panel a) or microscopic DLN metastases (panel b). There was no such protection seen if re-challenge was with 4THM tumor cells. Growth of either EMT6 or 4THM in mice receiving EMT6 followed by surgery/chemotherapy was equivalent to that seen in naive mice. Mice receiving primary injections with 4THM, and subsequently treated with chemotherapy, showed no resistance to re-challenge with either EMT6 or 4THM (Figure 4b). These data were mirrored by analysis of tumor cells frequencies in DLN of treated/re-challenged mice (Figure 4c). Only EMT6 tumor bearers cured by immunotherapy showed decreased DLN micro-metastasis after re-challenge with EMT6, but not 4THM, tumors. Note however, that in these mice (and mice cured of 4THM and re-challenged with EMT6) we cannot discern whether tumor cells measured were of EMT6 or 4THM origin.

Further evidence suggesting that immunotherapy, but not chemotherapy, treatment of EMT6-injected mice resulted in protective immunity to re-challenge with the same tumor came from studies using splenocytes pooled from 4mice/group 90 d post either surgical resection of primary tumors followed by either chemotherapy or immunotherapy. 50×10^6 of these cells were infused iv into fresh mice initially receiving 5×10^5 EMT6, or 1×10^5 4THM, tumor cells (Figure 5) 15 d earlier, and surgically removed 1 d before spleen cell transfer. Lung tumor colonies were enumerated in all groups at 15 days after surgery (14 d after spleen cell transfer), and DLN used to estimate tumor cell frequency by limiting dilution. Data for 1 of 2 studies are shown in Figure 5.

In this independent assay, protection from metastatic tumor colony growth, either macroscopic (to lung) or microscopic (DLN metastases assayed by limiting dilution), was afforded only by transfer of splenocytes from mice cured of EMT6 by surgical resection and immunotherapy, and not from mice cured by chemotherapy. Furthermore, no protection from growth of 4THM tumors was observed.

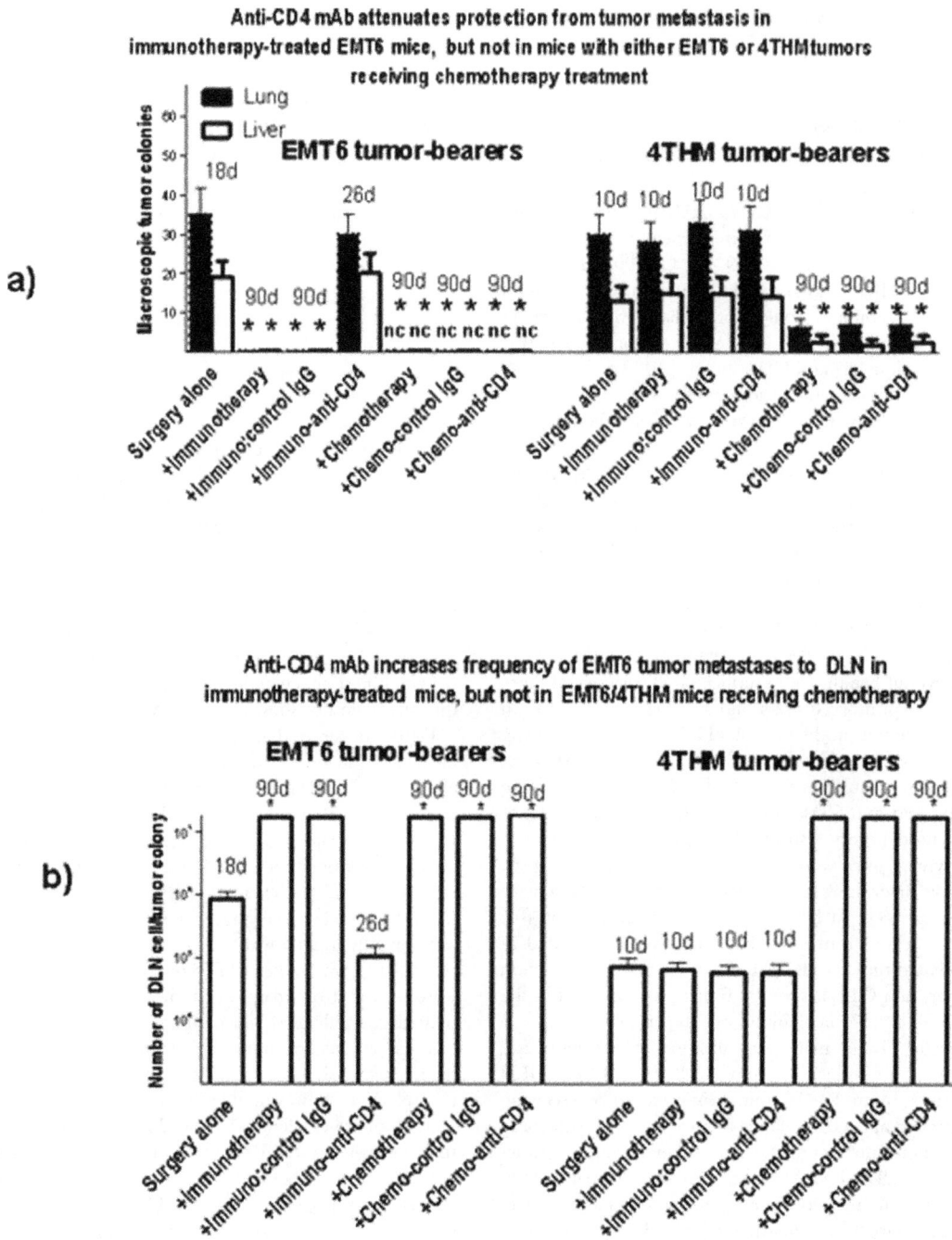

Figure 3. Effect of anti-CD4 mAb on lung/liver (panel a) or DLN (panel b) metastases in mice receiving EMT6 or 4THM tumor cells and treatment as in Figure 1. 5 mice were used per group for sacrifice at the time post surgery points shown (numbers above histogram bars). Data show means for macroscopic tumor colonies/group; nc= no visible tumor colonies. * indicates p<0.05, compared with control treated with surgery alone;

Discussion

Breast cancer cells are thought to be continuously monitored by host resistance mechanisms (immunosurveillance [27]), as evidenced by linkage of MHC expression (Class I) with breast cancer growth [28–30], as well as analysis of the role of other immune parameters on disease incidence/progression [31–34]. Included amongst such studies are several reporting on the possible importance of regulation of inflammation by T lymphocytes

[35–37]. Consistent with these concepts, lymphocyte infiltration into breast tumors is correlated with improved overall survival [38], and peripheral blood of breast cancer patients show evidence at both the cellular and humoral level of immunity to antigens (MUC-1 and Her-2/neu) associated with human breast cancer [39,40]. This in turn is reflected in the moderate success seen using Her-2/neu peptides, and other antigenic moieties, as a cancer vaccine [41,42]. While there remains controversy concerning whether development of CD4 or CD8 immunity will best predict

Specific host resistance to fresh EMT6 reinjection in mice cured of EMT6, but not 4THM, tumors by immuno- but not chemo-therapy

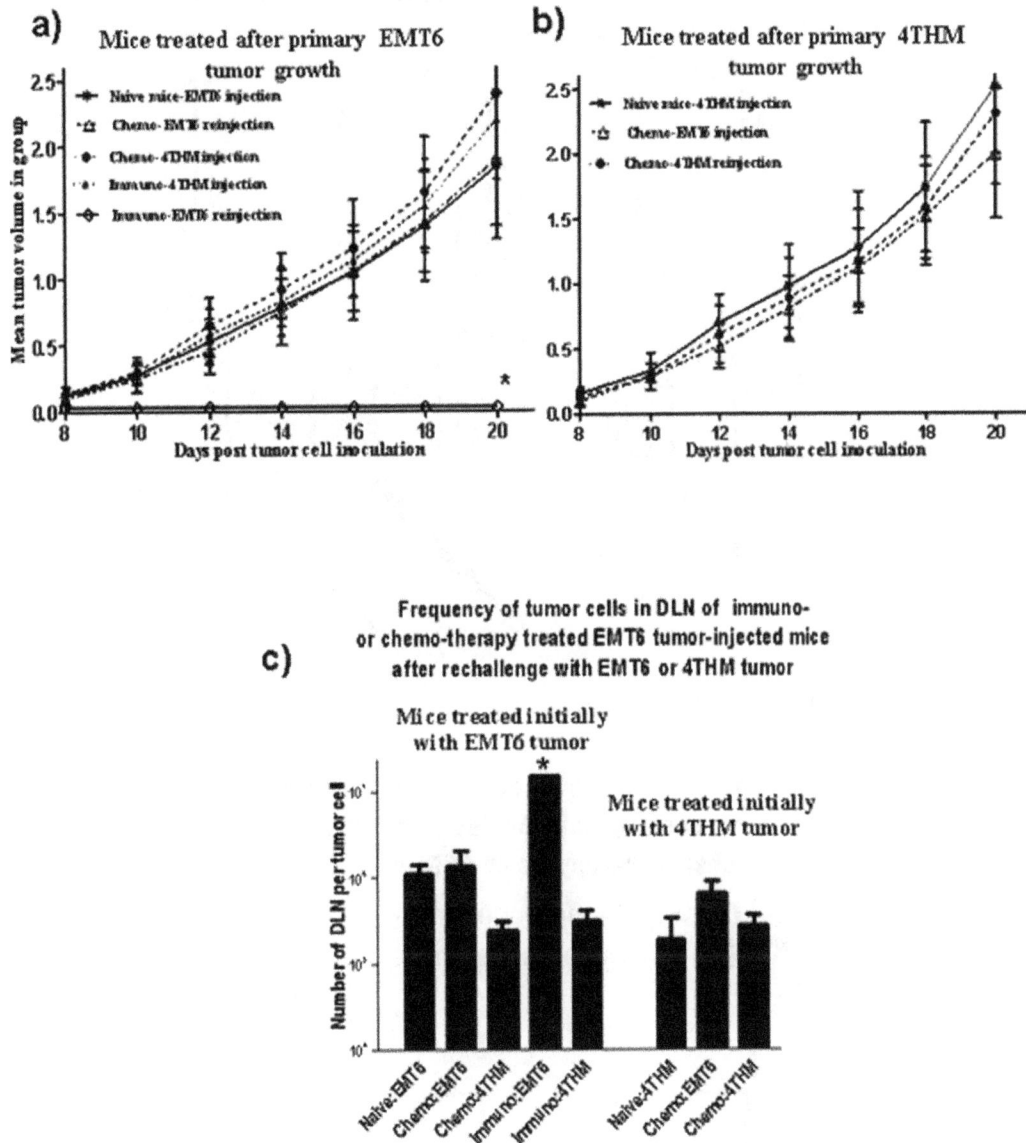

Figure 4. Specific protection from re-challenge with EMT6, but not 4THM, assaying either local tumor growth (panel a) or DLN metastases (panel c) in mice treated 90 d earlier by surgical tumor resection and immunotherapy. Naïve mice had had no previous EMT6 or 4THM tumor implants. All mice were sacrificed at 20 d post re-challenge. Data represent means for group. No protection was seen in mice initially treated with 4THM tumors before treatment/re-challenge (panel b). *, $p < 0.05$ compared with equivalent fresh control mice.

host-resistance [43,44], there is also concern that vaccination may augment induction of Tregs to block effective tumor immunity [45,46]. Compounding the complexity of understanding the role of immunotherapy in breast cancer treatment is the potential effect of concomitant chemotherapy on the immune system of the tumor host. Conventional cyclophosphamide-methotrexate-5-fluorouracil (CMF) chemotherapy decreases both NK cell activity [47]. In contrast, in studies of taxane-based chemotherapy in 30 women with advanced breast cancer, increased NK and LAK cell activity and increased IL-6, GM-CSF, and IFNγ levels with decreased IL-1 and TNFα levels were reported in cancer patients following chemotherapy, and correlated with clinical responses [48].

Similarly, cyclophosphamide which is known to suppress T reg cells, has been incorporated into some vaccine *HER2/neu* vaccine trials [39].

Anti-CD200 mAb protects mice from micro-metastasis of EMT6 to DLN, while EMT6 over-expressing a CD200 transgene, or growing in CD200[tg] hosts, grew more aggressively and metastasized at higher frequency [7]. CD200RKO mice were more resistant both to primary and metastatic growth of tumor [25]. In CD200R1KO mice cured (tumor-free for >300 d) by surgical tumor resection and immunotherapy, CD4[+] cells, rather than effector CD8[+] cells, were critical for protection [9]. Growth and metastasis of a highly aggressive metastatic variant (4THM) of

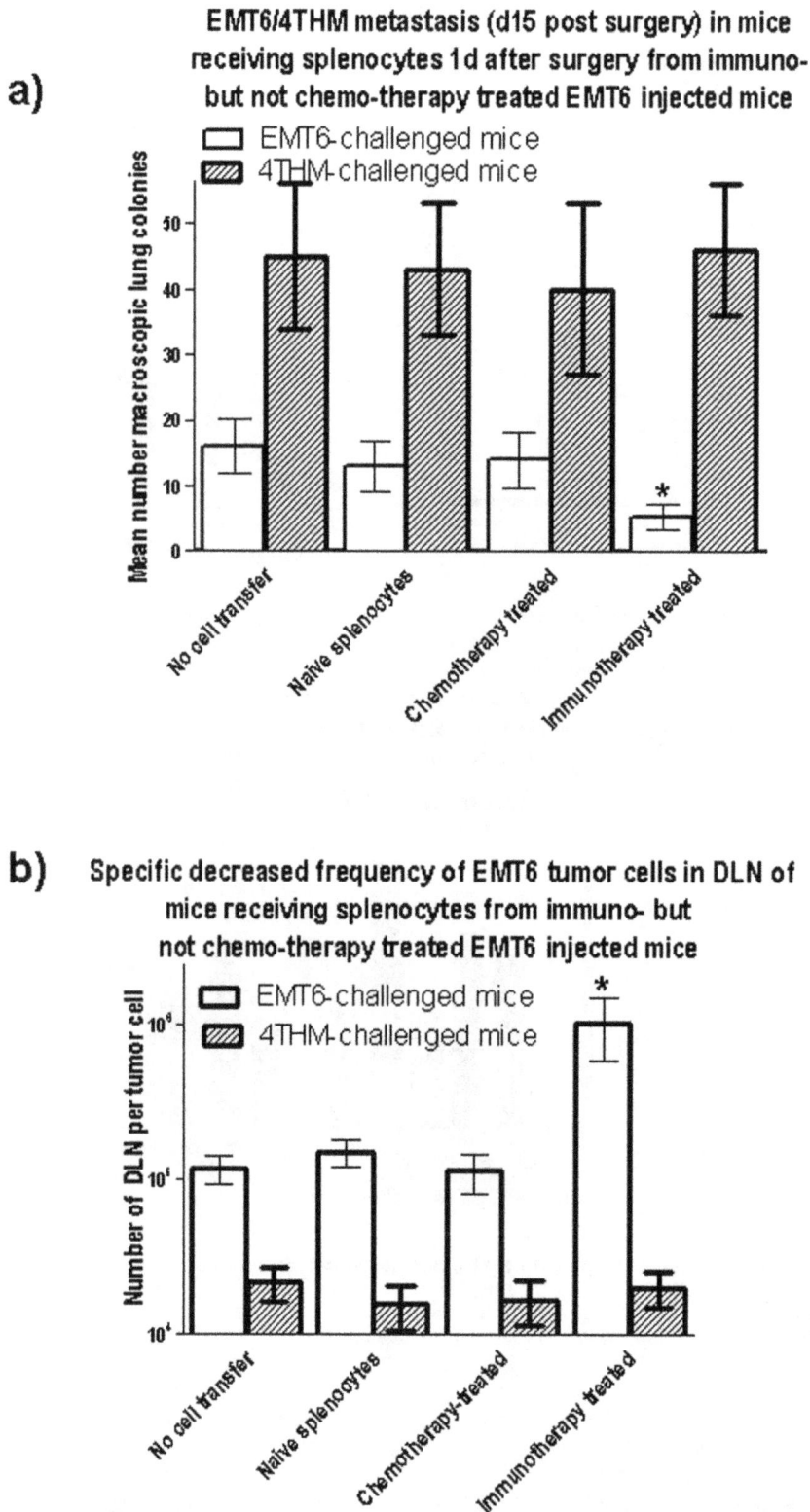

Figure 5. Adoptive transfer of splenocytes from immune- but not chemo-therapy treated mice receiving EMT6 tumors can decrease lung (panel a) and DLN (panel b) metastases in mice which had previously received EMT6 but not 4THM tumors. The tumors in the latter mice were surgically removed 1 d before spleen transfer, and all mice sacrificed 14 d after spleen cell transfer. Data show means ($\pm$SD). *, p< 0.01 relative to control (no cell transfer).

the breast tumor 4T1 was reported to be refractory to attenuation of CD200:CD200R interactions in CD200R1KO mice [8].

The current studies have extended our understanding of host resistance to EMT6 tumors using WT mice as tumor recipients, and, following surgical resection of tumor, by augmenting immunization with tumor cells (with CpG as adjuvant) with infusion of Fab anti-CD200R to block CD200:CD200R interactions. We compared this treatment with a more conventional approach using surgery followed by chemotherapy with anti-VEGF and paclitaxel, and compared results with EMT6 and the less immunogenic tumor, 4THM. 4THM mice were not effectively treated with immunotherapy, as was evident from the different times at which mice were sacrificed to measure tumor metastases endpoints in Figures 1–3. In contrast, chemotherapy was effective for both EMT6 and 4THM tumors, allowing us to study mice up to 90 d post surgery (Figures 1–3). Data in Figures 3–5, show that: (i) cure following chemotherapy in both tumor models is not abolished by anti-CD4 treatment, unlike cure of EMT6 tumors by immunotherapy (Figure 3-see also [9]). Immunotherapy in the EMT6 tumor model led to increased induction of direct killing (by CD8$^+$ effector cells) using splenocytes from treated mice, along with increased cytokine production in vitro-both effects were attenuated in mice receiving anti-CD4 treatment in vivo (Figure S2). (ii) following chemotherapy, mice initially cured of either 4THM or EMT6 tumors were not resistant to re-challenge with the same tumor, though immunotherapy of EMT6 tumors afforded resistance to re-challenge with the same tumor, but not with 4THM (Figure 4); and finally, (iii) only splenocytes from immuno- but not chemo-therapy treated EMT6 mice, could adoptively transfer protection from macroscopic/microscopic metastases to surgically treated WT mice (Figure 5) previously injected with the same tumor. Again no protection was afforded against 4THM tumors. Thus we were able to induce a tumor-protective immune response in WT mice with EMT6 tumors, but not mice with the more aggressive 4THM tumors. Additional features differentiating host inflammatory responses to EMT6 and 4THM have been described elsewhere by Erin et al (8). Given that the sensitivity of detection of metastases from DLN in our limiting dilution assay is ~1:10^7 cells, and that anti-CD4 treatment of immunotherapy-treated EMT6 tumor injected mice reveals increased metastases in mice otherwise "cured" of disease, we speculate that such mice may harbor quiescent tumor cells, whose growth is held in check by mechanisms which are CD4-dependent.

The nature of the resistant mechanism(s) in mice undergoing chemotherapy in the regimen prescribed is not yet clear. Preliminary data show a difference in intra-tumoral cytokine profiles in such animals, and a difference in phenotype of cells infiltrating the re-challenged EMT6 tumor in WT mice compared with those infiltrating a primary tumor challenge, with increased CD4$^+$ cells. This in itself is of interest given the data of Figure S2a, showing a CD4$^+$-dependent augmented cytokine production

(TNFα, IL-2 and IFNγ) in mice receiving immunotherapy, but not chemotherapy. Infusion of exogenous soluble CD200 into mice undergoing chemotherapy treatment did not attenuate cure or increase metastasis (RMG-unpublished), confirming the independence of this protection from an effect mediated by CD200:CD200R interactions, which is clearly implicated in the immunotherapy described. Our data suggest that optimal treatment of breast cancer should take into consideration the importance in "trade-off" between cancer cell sterilization by immunosuppressive drug treatment and the potential benefit of enhancing immune resistance by manipulation of co-inhibitory (CD200) pathways.

Supporting Information

Figure S1 DLN cell from (surgery+chemotherapy) treated WT mice do not antagonize outgrowth of tumor clones from DLN of WT mice sacrificed 14d post EMT6/4THM tumor cell injection. DLN cells from 5/group WT mice were harvested at 90 d post tumor resection and chemotherapy treatment (see Figures 1 and 2), and from separate groups of WT mice 14 d post EMT6/4THM injection-WT* in Figure). Cells from the latter were cultured under limiting dilution conditions (from 2×10^3 to 1×10^5 cells/well) alone, or with a 5-fold excess of cells from the 90 d treated mice. DLN cells from the latter were also cloned alone (data to far left in each subgroup in the Figure). All tumor cell frequencies cloned were calculated based on the input numbers of cells from DLN of WT* only. (TIF)

Figure S2 Cytokine production (panel a) and CD8$^+$-dependent antigen specific lyses of ^{3}HTdR tumor target cells (panel b), using splenocytes from mice described in Figure 3. Control mice in each panel received no tumor cells-in this case only data are pooled for groups stimulated with either EMT6 or 4THM cells. Other mice shown were injected with EMT6 (left side of each panel) or 4THM tumor (right side of each panel), and received surgery alone, or followed by chemotherapy/immunotherapy. For all these studies splenocytes were harvested at 90 d post surgery, or earlier as necessary for groups where tumor growth was not controlled (see Figure 3), and re-stimulated in vitro with the same tumor cells (EMT6 or 4THM). Data show mean ($\pm$SD) for triplicate cultures, with a minimum of 4 individual spleen cells assayed/group. * p<0.05 compared with a surgery-only control group. (TIF)

Author Contributions

Conceived and designed the experiments: RMG. Performed the experiments: RMG ZC IK AP. Analyzed the data: RMG NE IK. Contributed reagents/materials/analysis tools: RMG IK. Wrote the paper: RMG.

References

1. Petermann KB, Rozenberg GI, Zedek D (2007) CD200 is induced by ERK and is a potential therapeutic target in melanoma. J Clin Invest 117: 3922–3929.
2. Siva A, Xin H, Qin F, Oltean D, Bowdish KS, et al. (2008) Immune modulation by melanoma and ovarian tumor cells through expression of the immunosuppressive molecule CD200 Cancer Immunol Immunother 57: 987–996.
3. Moreaux J, Veyrune JL, Reme T, DeVos J, Klein B (2008) CD200: A putative therapeutic target in cancer. Biochem Biophys Res Commun 366: 117–122.
4. McWhirter JR, KretzRommel A, Saven A (2006) Antibodies selected from combinatorial libraries block a tumor antigen that plays a key role in immunomodulation. Proc Nat Acad Sci Usa 103: 1041–1046.
5. Tonks A (2007) CD200 as a prognostic factor in acute myeloid leukemia. Leukemia 21: 566–571

6. Gorczynski RM, Chen Z, Diao J (2010) Breast cancer cell CD200 expression regulates immune response to EMT6 tumor cells in mice. Breast Cancer Res Treat 123: 405–415.
7. Gorczynski RM, Clark DA, Erin N, Khatri I (2011) Role of CD200 in regulation of metastasis of EMT6 tumor cells in mice. Breast Cancer Res Treatment 130: 49–60.
8. Erin N, Podnos A, Tanriover G, Duymus O, Cote E, Khatri I, et al. (2014) Bidirectional effect of CD200 on breast cancer development and metastasis, with ultimate outcome determined by tumor aggressiveness and a cancer-induced inflammatory response Oncogene: in press

9. Gorczynski RM, Chen Z, Khatri I, Podnos A, Yu K (2013) Cure of metastatic growth of EMT6 tumor cells in mice following manipulation of CD200:CD200R signaling. Breast Cancer Res Treatment 142: 271–282.

10. Pandit TS, Kennette W, MacKenzie L (2009) Lymphatic metastasis of breast cancer cells is associated with differential gene expression profiles that predict cancer stem cell- like properties and the ability to survive, establish and grow in a foreign environment. Int J Oncol 35: 297–308.

11. Pfeffer U, Romeo F, Noonan DM, Albini A (2009) Prediction of breast cancer metastasis by genomic profiling: where do we stand? Clin Exp Metastas 26: 547–558.

12. Pollard JW (2008) Macrophages define the invasive microenvironment in breast cancer. J Leukocyte Biol 84: 623–630.

13. Olkhanud PB, Baatar D, Bodogai M (2009) Breast Cancer Lung Metastasis Requires Expression of Chemokine Receptor CCR4 and Regulatory T Cells. Cancer Res 69: 5996–6004.

14. Lu X, Kang YB (2009) Chemokine (C-C Motif) Ligand 2 Engages CCR2(+) Stromal Cells of Monocytic Origin to Promote Breast Cancer Metastasis to Lung and Bone. J Biol Chem 284: 29087–29096.

15. Liang ZX, Yoon YH, Votaw J, Goodman MM, Williams L, et al. (2005) Silencing of CXCR4 blocks breast cancer metastasis. Cancer Res 65: 967–971.

16. Takahashi M, Miyazaki H, Furihata M (2009) Chemokine CCL2/MCP-1 negatively regulates metastasis in a highly bone marrow-metastatic mouse breast cancer model. Clin Exp Metastas 26: 817–828.

17. Ma XR, Norsworthy K, Kundu N (2009) CXCR3 expression is associated with poor survival in breast cancer and promotes metastasis in a murine model. Mol Cancer Ther 8: 490–498.

18. Huang B, Pan PY, Li QS (2006) Gr-1(+)CD115(+) immature myeloid suppressor cells mediate the development of tumor-induced T regulatory cells and T-cell anergy in tumor-bearing host. Cancer Res 66: 1123–1131.

19. Yang L, Debusk LM, Fukuda K (2004) Expansion of myeloid immune suppressor GR1+CD11b+ cells in tumor-bearing host directly promotes tumor angiogenesis. Cancer Cell 6: 409–421.

20. Qin FXF (2009) Dynamic Behavior and Function of Foxp3(+) Regulatory T Cells in Tumor Bearing Host. Cell Mol Immunol 6: 3–13.

21. Yang L, Huang JH, Ren XB (2008) Abrogation of TGF beta signaling in mammary carcinomas recruits Gr- 1+CD11b+ myeloid cells that promote metastasis. Cancer Cell 13: 23–35.

22. Pandit TS, Kennette W, MacKenzie L, Zhang GH, AlKatib W, et al. (2009) Lymphatic metastasis of breast cancer cells is associated with differential gene expression profiles that predict cancer stem cell- like properties and the ability to survive, establish and grow in a foreign environment. Int J Oncol. 35: 297–308

23. Pakala SB, Rayala SK, Wang R, Ohshiro K, Mudvari P, et al. (2013) MTA1 Promotes STAT3 Transcription and Pulmonary Metastasis in Breast Cancer. Cancer Res. 73: 3761–3770

24. Erin N, Zhao W, Bylander J, Chase G, Clawson G (2006) Capsaicin-induced inactivation of sensory neurons promotes a more aggressive gene expression phenotype in breast cancer cells. Breast Cancer Res Treat 99: 351–364.

25. Abadi YM, Jeon H, Ohaegbulam KC, Scandiuzzi L, Ghosh K, et al. (2013) Host B7× Promotes Pulmonary Metastasis of Breast Cancer. J Immunol 190: 3806–3814

26. Podnos A, Clark DA, Erin N, Yu K, Gorczynski RM (2012) Further evidence for a role of tumor CD200 expression in breast cancer metastasis: decreased metastasis in CD200R1KO mice or using CD200-silenced EMT6. Breast Cancer Res Treatment 136: 117–127.

27. Standish LJ, Sweet ESND, Novack J, Wenner CA, Bridge C, et al. (2008) Breast Cancer and the Immune System. J Soc Integr Oncol. 6: 158–168.

28. Chaudhuri S, Cariappa A, Tang M (2000) Genetic susceptibility to breast cancer: HLA DQB*03032 and HLA DRB1*11 may represent protective alleles. Proc Natl Acad Sci USA 97: 11451–11454.

29. Marincola FM, Jaffee EM, Hicklin DJ (2000) Escape of human solid tumors from T-cell recognition: molecular mechanisms and functional significance. Adv Immunol 74: 181–273.

30. Camploi M, Changg CC, OLdford SA (2004) HLA antigen changes in malignant tumors of mammary epithelial origin: molecular mechanisms and clinical implications. Breast Dis 2004: 105–125.

31. Hamilton G, Reiner A, Teleky B (1988) Natural killer cell activities of patients with breast cancer against different target cells. J Cancer Res Clin Oncol. 114: 191–196.

32. Jarnicki AG, Lysaght J, Todryk S, Mills KH (2006) Suppression of antitumor immunity by IL-10 and TGF-beta-producing T cells infiltrating the growing tumor: influence of tumor environment on the induction of CD4+ and CD8+ regulatory T cells. J Immunol. 177: 896–904.

33. Ramsey-Goldman R, Mattai SA, Schilling E (1998) Increased risk of malignancy in patients with systemic lupus erythematosus. J Investig Med. 46: 217–222.

34. Calogero RA, Cordero F, Forni G, Cavallo F (2007) Inflammation and breast cancer. Inflammatory component of mammary carcino-genesis in ErbB2 transgenic mice. Breast Cancer Res. 9: 211–212.

35. Denardo DG, Coussens LM (2007) Inflammation and breast cancer. Balancing immune response: crosstalk between adaptive and innate immune cells during breast cancer progression. Breast Cancer Res. 9: 212–213.

36. Tan TT, Coussens LM (2007) Humoral immunity, inflammation and cancer. Curr Opin Immunol. 19: 209–216.

37. Einav U, Tabach Y, Getz G (2005) Gene expression analysis reveals a strong signature of an interferon-induced pathway in childhood lymphoblastic leukemia as well as in breast and ovarian cancer. Oncogene. 24: 6367–6375.

38. Menard S, Tomasic G, Casalini P (1997) Lymphoid infiltration as a prognostic variable for early onset breast carcinomas. Clin Cancer Res 3: 817–819.

39. Disis ML, Calenoff E, McLaughlin G (1994) Existent T cell and antibody immunity to Her-2/neu protein in patients with breast cancer. Cancer Res 54: 16–20.

40. Jerome KR, Domenech N, Finn OJ (1993) Tumor-specific cytotoxic T cell clones from patients with breast and pancreatic adenocarcinoma recognize EBV-immortalized B cells transfected with polymorphic epithelial mucin complementary DNA. J Immunol 151: 1654–1662.

41. Baxevanis CN, Sotiriadou NN, Gritzapis AD (2006) Immunogenic HER-2/neu peptides as tumor vaccines. Cancer Immunol Immunother 55: 85–95.

42. Anderson KS (2009) Tumor vaccines for Breast Cancer. Cancer Invest 27: 361–368.

43. Assudani DP, Horton RBV, Mathieu MG, McArdle SEB, Rees RC (2007) The role of CD4(+) T cell help in cancer immunity and the formulation of novel cancer vaccines. Cancer Immunol Immunother 56: 70–80.

44. Beyer M, Karbach J, Mallmann MR (2009) Cancer Vaccine Enhanced, Non-Tumor-Reactive CD8(+) T Cells Exhibit a Distinct Molecular Program Associated with "Division Arrest Anergy". Cancer Res 69: 4346–4354.

45. Zhou G, Drake CG, Levitsky HI (2006) Amplification of tumor-specific regulatory T cells following therapeutic cancer vaccines. Blood 107: 628–636.

46. Duraiswamy J, Kaluza KM, Freeman GJ, Coukos G (2013) Dual Blockade of PD-1 and CTLA-4 Combined with Tumor Vaccine Effectively Restores T-Cell Rejection Function in Tumors. Cancer Research 73: 3591–3603.

47. Tichatschek E, Zielinski CC, Muller C (1988) Long-term influence of adjuvant therapy on natural killer cell activity in breast cancer. Cancer Immunol Immunother. 27: 278–282.

48. Tsavaris N, Kosmas C, Vadiaka M (2002) Immune changes in patients with advanced breast cancer undergoing chemotherapy with taxanes. Br J Cancer. 87: 21–27.

14

Amyotrophic Lateral Sclerosis-Linked Mutant VAPB Inclusions Do Not Interfere with Protein Degradation Pathways or Intracellular Transport in a Cultured Cell Model

Paola Genevini[1], Giulia Papiani[1¤a], Annamaria Ruggiano[1¤b], Lavinia Cantoni[2], Francesca Navone[1*], Nica Borgese[1,3*]

1 Institute of Neuroscience, Consiglio Nazionale delle Ricerche, and Department of Medical Biotechnology and Translational Medicine (BIOMETRA), Università degli Studi di Milano, Milano, Italy, 2 Department of Molecular Biochemistry and Pharmacology, Istituto di Ricerche Farmacologiche "Mario Negri", Milan, Italy, 3 Department of Health Science, Magna Graecia University of Catanzaro, Catanzaro, Italy

Abstract

VAPB is a ubiquitously expressed, ER-resident adaptor protein involved in interorganellar lipid exchange, membrane contact site formation, and membrane trafficking. Its mutant form, P56S-VAPB, which has been linked to a dominantly inherited form of Amyotrophic Lateral Sclerosis (ALS8), generates intracellular inclusions consisting in restructured ER domains whose role in ALS pathogenesis has not been elucidated. P56S-VAPB is less stable than the wild-type protein and, at variance with most pathological aggregates, its inclusions are cleared by the proteasome. Based on studies with cultured cells overexpressing the mutant protein, it has been suggested that VAPB inclusions may exert a pathogenic effect either by sequestering the wild-type protein and other interactors (loss-of-function by a dominant negative effect) or by a more general proteotoxic action (gain-of-function). To investigate P56S-VAPB degradation and the effect of the inclusions on proteostasis and on ER-to-plasma membrane protein transport in a more physiological setting, we used stable HeLa and NSC34 Tet-Off cell lines inducibly expressing moderate levels of P56S-VAPB. Under basal conditions, P56S-VAPB degradation was mediated exclusively by the proteasome in both cell lines, however, it could be targeted also by starvation-stimulated autophagy. To assess possible proteasome impairment, the HeLa cell line was transiently transfected with the ERAD (ER Associated Degradation) substrate CD3δ, while autophagic flow was investigated in cells either starved or treated with an autophagy-stimulating drug. Secretory pathway functionality was evaluated by analyzing the transport of transfected Vesicular Stomatitis Virus Glycoprotein (VSVG). P56S-VAPB expression had no effect either on the degradation of CD3δ or on the levels of autophagic markers, or on the rate of transport of VSVG to the cell surface. We conclude that P56S-VAPB inclusions expressed at moderate levels do not interfere with protein degradation pathways or protein transport, suggesting that the dominant inheritance of the mutant gene may be due mainly to haploinsufficiency.

Editor: Yanmin Yang, Stanford University School of Medicine, United States of America

Funding: This work was supported by the CARIPLO Foundation (http://www.fondazionecariplo.it/it/index.html) project 2007-5098 (NB), PNR-CNR Aging Program 2012–2014, and Università Statale di Milano. The funders had no role in study design, data collection and analysis, decision to publish, or preparation of the manuscript.

Competing Interests: The authors have declared that no competing interests exist.

* Email: f.navone@in.cnr.it (FN); n.borgese@in.cnr.it (NB)

¤a Current address: Oligomerix, Inc., New York, New York, United States of America
¤b Current address: Cell and Developmental Biology Programme, Centre for Genomic Regulation (CRG), Barcelona, Spain

Introduction

VAPB, and its homologue VAPA, are members of the highly conserved and ubiquitously expressed VAP (_Vesicle-Associated Membrane Protein (VAMP)-Associated Protein_) family of ER tail-anchored transmembrane proteins. The cytosolic N-terminal region, consists of a domain that is homologous to the nematode major sperm protein (MSP), followed by a central coiled-coil domain; the transmembrane segment is close to the C-terminus, and the last four C-terminal residues are probably exposed to the ER lumen [1].

By interacting with FFAT (two phenylalanines in an acidic tract) motif-containing polypeptides, VAPs are able to recruit a wide spectrum of proteins, and are thus implicated in a variety of physiological functions (reviewed in ref 1), including membrane trafficking [2,3], lipid transport and metabolism [4,5], membrane contact site formation [6,7,8,9,10], Ca^{2+} homeostasis [9], ER-cytoskeleton interactions [11], participation in the unfolded protein response [12], neurotransmitter release and neurite extension [13,14]. Specific roles that functionally distinguish the two mammalian VAP isoforms have not been identified so far.

The identification of a dominant missense mutation in the VAPB gene in patients affected by a slowly progressing form of familial motor neuron disease (ALS8) [15] greatly increased the interest in VAP proteins. The mutation, which causes substitution of proline 56 with serine in the MSP domain (P56S mutation), disrupts VAPB's three-dimensional structure and favors its aggregation [16,17,18]. Initially identified in eight Brazilian families with a shared Portuguese ancestor [19], the same mutation was subsequently detected in an unrelated German patient, carrying a haplotype distinct from the one linked to the mutation in the Brazilian families [20]. Three additional mutations of VAPB have since been identified in familial Amyotrophic Lateral Sclerosis (ALS) patients [21,22,23], however, in these cases, the segregation of the mutation with the disease was not demonstrated.

Like many proteins linked to neurodegenerative diseases, mutant VAPB forms intracellular inclusions. Work from our laboratory, however, revealed important differences between P56S-VAPB inclusions and other inclusion bodies. More specifically, we showed that, after insertion into the ER membrane, P56S-VAPB rapidly clusters to generate paired ER cisternae that give rise to a profoundly restructured ER domain and not to a cytosolic protein aggregate, as is generally the case [24]. Moreover, we demonstrated that, at variance with other inclusion bodies linked to neurodegenerative diseases, ER-derived ubiquitinated P56S-VAPB inclusions can be easily cleared by the proteasome, with no apparent involvement of basal macroautophagy (here referred to as autophagy) [25].

Although protein misfolding and aggregation are a common feature of several neurodegenerative diseases, including ALS, their precise pathogenic role is poorly understood, and both a toxic gain of function as well as loss of function by dominant negative effects are thought to be involved. In the case of ALS8, studies in transfected mammalian cells and in fly models have revealed that wild-type VAPB, as well as VAPA and other functionally important interactors, are sequestered within the VAPB inclusions, leading to the hypothesis that the dominant inheritance of ALS8 is due to a dominant negative effect of the mutant protein [12,16,26,27,28,29].

In addition to the loss of function mechanism, driven by sequestration of potentially functional proteins into inclusion bodies, evidence for a toxic gain-of-function of mutant VAPB has also been reported. P56S-VAPB inclusions are ubiquitin-positive both in transfected cells [25] and in motor neurons of transgenic animals [30], and both wild-type and P56S-VAPB, when overexpressed, have been observed to impair the activity of the proteasome [31]. These observations suggest that VAPB inclusions may disturb proteostasis, and are in line with the many studies pointing to alteration in protein degradation pathways as an important pathogenic mechanism underlying aggregated misfolded protein toxicity both in sporadic and familial ALS (reviewed in refs [32–35]).

One limitation of most of the studies on the mechanism of P56S-VAPB pathogenicity in mammalian systems has been the use of strongly overexpressing transfected cells, which may be inadequate to unravel the effects of the mutant protein expressed from a single allele, as in patients' cells. In our previous work, we developed a cell line inducibly expressing P56S-VAPB at levels comparable to those of the endogenous protein, and used this cell line to investigate the genesis, nature and clearance of the P56S-VAPB-containing aggregates [24,25]. In the present study, we have investigated whether the presence of P56S-VAPB-containing inclusions, generated by mutant VAPB expressed at levels comparable to those of the endogenous protein, interferes with

physiological protein degradation pathways or impairs normal protein transport from the ER to the plasma membrane. We find that the inclusions neither interfere with general proteostasis nor with the intracellular transport of a model secretory membrane protein. We also confirm that P56S-VAPB inclusions are exclusively cleared by the proteasome under basal conditions both in neuronal and non-neuronal cells, but find that they can be degraded by stimulated autophagy. Our results are consistent with the idea that haploinsufficiency alone may underlie the dominant inheritance of P56S-VAPB.

Materials and Methods

Plasmids

The pTre Tight vectors (Clontech), coding for *myc*-wt VAPB or *myc*-P56S-VAPB have been described [24,25].

pGEX vectors coding for fragments 132–225 or 1–225 of VAPB fused to GST were provided by C.C. Hoogenraad (Utrecht University, NL). VAPA-pGEX2T coding for full-length VAPA fused to GST was generated from the rat VAPA sequence amplified from a pGEM4 recombinant plasmid. The VAPA clone was provided by Stephen Kaiser [36]. Specific restriction sites for subcloning in the pGEX2T vector were introduced into the PCR primers: upper 5′ AT CCCGGGA ATGGCGAAACACGAGC 3′ (SmaI restriction site underlined) and lower 5′ TA GAATTCG-CAGGTCGACTCTAGAC 3′ (EcoRI restriction site underlined).

pTK-Hyg and pEGFP-N1 were from Clontech; pCINeoHA-CD3δ and pCDM8.1-ts045VSVG-EGFP were generously provided by A.M. Weissman (National Institutes of Health) and J. Lippincott-Schwartz (National Institutes of Health, Bethesda, MD) respectively.

All constructs generated in the laboratory were checked by sequencing.

Antibodies

The following primary antibodies were obtained from the indicated sources: anti-*myc* monoclonals (clone 9E10), Santa Cruz or Sigma; monoclonal anti-tubulin (clone B-5-1-2), monoclonal anti-actin, and polyclonal anti-LC3 (L8918), Sigma; polyclonal anti-p62 (ab91526), Abcam; monoclonal anti-VSVG (clone IE9F9), keraFAST; polyclonal anti-HA, Invitrogen (71-5500) or Santa Cruz (SC-805); polyclonal anti-GFP (ab290), Abcam. Polyclonal anti-giantin serum and anti-GM130 were kindly provided by Dr. M. Renz (Institute of Immunology and Molecular Genetics, Karlsruhe, Germany) [37] and A. de Matteis (Telethon Institute of Genetics and Medicine, Naples, Italy) [38], respectively.

Anti-VAPB polyclonal antibodies were produced in the laboratory as follows. The VAPB 132–225 fragment fused to GST was expressed in E. coli BL21 by induction with 0.5 mM Isopropyl β-D-1-thiogalactopyranoside (IPTG), following standard procedures. The expressed protein was purified with glutathione-Sepharose 4B resin (GE Healthcare) according to the manufacturer's protocol. A rabbit was immunized with the VAPB fragment excised from GST by thrombin digestion. The sera were first tested against lysates of E.coli BL21 induced to express either full-length VAPA-GST or VAPB 1-225-GST. Cross-reactive anti-VAPA antibodies were then eliminated by adsorption of 3 ml of sera with 1.60 mg of VAPA-GST immobilized on glutathione-sepharose beads. Finally, anti-VAPB antibodies were purified from the adsorbed sera using 1 mg of 132–225 VAPB fragment coupled to CNBr-activated Sepharose 4B as affinity ligand (see Fig. S1).

Peroxidase-conjugated anti-rabbit and anti-mouse IgG were from Sigma, anti-mouse IRDye 680 and anti-rabbit IRDye 800 from LI-COR Bioscience, Alexa Fluor 488 anti-rabbit and Alexa Fluor 568 anti-mouse IgG from Invitrogen, DyLight 549 or 633 anti-mouse and anti-rabbit IgG from Pierce.

Cell culture, transfection, and P56S-VAPB expression analysis

HeLa Tet-Off cell lines expressing *myc*-P56S-VAPB [24,25] were maintained in DMEM supplemented with 10% FBS Tet-free (Hyclone), 1% Pen/Strep, 1% L-Glut, G418 (100 μg/ml), Hygromycin (100 μg/ml), and doxycycline (Dox) (500 ng/ml). Expression of P56S-VAPB was induced by transferring the cells to Dox-free medium. Degradation of VAPB was followed after re-addition of Dox to the medium, as previously described [25]. Briefly, four days after removal of Dox, equal numbers of cells were seeded onto 35 mm Petri dishes containing a coverslip, and incubation in the absence of Dox was continued for another two days. At this time, the coverslips were fixed and stained with DAPI; nuclei from random fields were counted to assess that each dish contained an equal number of cells. Dox was then added back to the samples, and cells were collected after treatment with the indicated drugs and at the time intervals indicated in the figures. The collected cells were lysed with SDS-lysis buffer [2% SDS, 50 mM Tris-HCl, pH 8, plus Complete (Roche) protease inhibitors] and all samples were brought to the same volume. Equal aliquots were then analyzed by SDS-PAGE-Immunoblotting. The levels of VAPB were corrected for minor variations in the number of plated cells.

NSC34 Tet-Off cell lines were generated in the laboratory of L. Cantoni [39] and were maintained in DMEM supplemented with 10% FBS, 1% P/S, 1% L-Glut, 1% Na^+Pyruvate and G418 (250 μg/ml). For most experiments, NSC34 Tet-Off cells were plated on Matrigel (BD Biosciences)-coated wells.

All transfections were carried out with JetPei (Polyplus transfection) according to the manufacturer's protocol. For the transient transfection of induced P56S-VAPB-HeLaTet-Off cells, incubation with JetPei DNA complexes was carried out in the presence of FBS from Gibco. After 24 h, the medium was replaced with complete medium supplemented with Tet-free serum, and the cells were treated as indicated in the figure legends.

To generate NSC34 Tet-Off VAPB clones, cells were co-transfected with pTK-hyg and pTre vector coding either for *myc*-wt VAPB or *myc*-P56S-VAPB. After transfection cells were selected with 150 μg/ml hygromycin. After approximately four weeks of growth in selection medium, individual clones were collected, amplified and induced to express the transgene by growth in the absence of Dox for 4–5 days. Increased expression was obtained by addition of 10 mM Na^+butyrate for 12 h. For *myc*-P56S-VAPB, five positive clones were identified out of 41 tested, while for *myc*-wt-VAPB, two out of 23.

To investigate clearance of P56S-VAPB inclusions from the NSC34 lines, cells were induced to express P56S-VAPB by growth for 4 days in Dox-free medium followed by treatment with 10 mM Na^+butyrate for 12 h. Cells were then transferred to Na^+butyrate-free, Dox (0.5 μg/ml) -containing medium, and P56S-VAPB degradation was followed as described for the HeLa Tet-Off clones.

Drug treatments and starvation

Lactacystine and Torin 1 were from Cayman Chemical; MG132 was from Calbiochem. Other drugs were from Sigma. 3-Methyladenine (3-MA) and Cycloheximide (CHX) were dissolved in water and used at final concentration of 10 mM and 50 μg/ml, respectively. Na^+butyrate was dissolved in complete medium and used at 10 mM final concentration. The following

drugs were dissolved in DMSO and used at the final concentrations indicated between brackets: Bafilomycin (200 nM), MG132 (10 μM), Lactacystin (10 μM) and Torin1 (250 nM). Control cells received equal volumes of the vector.

Cells were starved by replacing culture media with EBSS (Earle's Balanced Salt Solution).

SDS-PAGE and Immunoblotting

SDS-PAGE and blotting were performed by standard procedures. Protein content was assayed with the BCA Protein Assay Kit (Thermo Scientific). Before immunostaining, blots were stained for total protein with Ponceau S (Sigma); they were then incubated with antibodies diluted in TBS+5% milk+0.1% Tween. Peroxidase-conjugated secondary antibodies were revealed by ECL (Perkin Elmer). The films were digitized, and band intensities were determined with ImageJ software (National Institutes of Health) after calibration with the optical density calibration step table (Stouffer Graphics Arts). Alternatively, Infrared dye-conjugated secondary antibodies were used. In this case, blots were scanned with the Odyssey CLx Infrared Imaging System (LI-COR Biosciences), and band intensities were determined with Image Studio software (LI-COR Biosciences).

Fluorescence Microscopy

Cells grown on coverslips were fixed with 4% paraformaldehyde (PFA)+4% sucrose and processed for immunofluorescence as described previously [25]. Images were acquired with the Zeiss LSM 510 Meta confocal system equipped with a 405/488/543/633 dichroic (Carl Zeiss, Oberkochen, Germany) and using a 63xPlanApo lens. Alexa Fluor 488 and GFP were acquired using the 488 line of the Argon/2 laser, and a 505–550 band pass emission filter. For Alexa Fluor 568 and DyLight 549, the 544 line of the He/Ne laser was used in combination with a 560–615 band pass emission filter. For DyLight 633, the 633 line of the He/Ne laser was used in combination with a 650 long pass emission filter. DAPI was imaged using the 405 diode laser and a 420–480 band pass emission filter. Wide-field imaging was performed with an Axioplan microscope (Carl Zeiss, Oberkochen, Germany), using the 40× PlanNeofluar lens equipped with a phase contrast ring.

Image analysis was performed with ImageJ software.

VSVG transport

Cells induced or not induced to express P56S-VAPB were transfected with ts045VSVG-EGFP and immediately placed at 39.3°C. After 24 h, cells were brought to 32° in the presence of CHX and incubated for the times indicated in the figures.

To evaluate the amount of VSVG in the Golgi area at each time point, coverslips were fixed and processed for immunofluorescence with anti-giantin and anti-*myc* antibodies. 1.2 μm thick z-stacks (~20 cells for each condition and time point) were acquired centered around the plane with maximum giantin staining (x–y sections). For each section, a ROI corresponding to giantin staining was outlined; the integrated EGFP fluorescence intensity of this region was determined, and summed over the entire stack. This value was normalized to that of the entire cell, determined in each section in ROIs drawn around the periphery of the cell.

For determination of surface VSVG, cells were placed on ice, medium was replaced with pre-chilled PBS+0.5 mM $CaCl_2$+ 1 mM $MgCl_2$ and then samples were transferred to the cold room. After two washes, cells were blocked with 0.1% BSA in the same buffer, and then incubated with anti-VSVG primary antibody diluted in blocking buffer for 1 h. Cells were washed 3 times, fixed with chilled PFA (see above: *Fluorescence microscopy*) first at 4° for 10 min, then at RT for an additional 10 min. After blocking with

17% goat serum, the non-permeabilized cells were exposed to secondary anti-mouse antibody for 50 min at room temperature. After 5 washes, the cells were fixed again for 5 min with PFA, permeabilized with Triton-X100 and processed for immunofluorescence with polyclonal anti-VAPB and secondary anti-rabbit antibodies under standard conditions [25]. Z-stacks (15–30 cells for each condition and time point) comprising the total height of the cells were acquired (X–Y sections at 0.5 μm intervals) to measure EGFP and anti-VSVG fluorescence as described for the Golgi analysis.

For both the Golgi and the surface quantification of VSVG, images were acquired with identical parameters, taking care to remain below saturation in the EGFP and anti-VSVG channels.

Statistical Analyses

Significance of the difference in VAPB levels between treated and untreated cells and possible differences in the intracellular distribution of VSVG in cells induced or not induced to express P56S-VAPB were evaluated by Student's unpaired two-tailed t test. Two-way matched Anova, followed by Bonferroni's post-test, was used to simultaneously evaluate the effects of cycloheximide (CHX) treatment and P56S-VAPB induction on CD3δ levels, or of autophagocytosis stimulation and P56S-VAPB induction on the LC3-II/LC3-I ratio. To compensate for different absolute values of band intensities in different experiments, values were either normalized to the band intensities before drug treatment, or converted to logarithms. p values are given in the figure legends.

Results

P56S-VAPB is cleared exclusively by the proteasome under basal conditions, but can be degraded by stimulated autophagy

To investigate the mechanism of P56S-VAPB clearance, we used the previously characterized HeLa Tet-Off cell line [24,25], in which expression of mutant, myc-tagged, VAPB is repressed by tetracycline or Dox, and induced by removal of the antibiotic from the medium (compare lanes 1 and 6 of Fig. 1A with lanes 2 and 7). We previously showed that mutant VAPB in these cells is expressed at levels close to those of the endogenous protein, and that the expressed protein is detected exclusively within inclusions ([24,25], Fig. 2C). When induced cells were shifted to Dox containing medium, ~2/3 of P56S-VAPB was degraded within 9–10 h (Fig. 1A, B). Degradation was prevented by two different proteasomal inhibitors, MG132, used in our previous study [25], and lactacystin (Fig. 1A, B). In contrast, autophagy inhibitors (3-MA and the proton pump blocker Bafilomycin) were without effect on myc-P56S-VAPB (which we will refer to here as P56S-VAPB) clearance. We verified that Bafilomycin was active by evaluating its capability to inhibit the lysosomal degradation of the autophagosomal ubiquitin receptor p62/SQSTM1 (to which we refer here as p62). As shown in Fig. 1C, we found that indeed p62 levels were higher in Bafilomycin-treated cells compared to controls.

Our previous work demonstrated that P56S-VAPB is less stable than the wild-type protein [25]; furthermore, we found that the levels of endogenous VAPB are not affected by expression of the mutant protein, indicating that, although native VAPB may be sequestered within P56S-VAPB-generated inclusions [12,16], this sequestration does not result in an alteration of its rate of degradation. To extend these findings, we probed the levels of endogenous VAPB with anti-VAPB antibodies in experiments like the one illustrated in Fig. 1A. Using anti-VAPB antibodies, we could simultaneously visualize the transfected P56S-VAPB and the endogenous wt protein. As shown in Fig. S2, endogenous VAPB levels were not affected either by P56S-VAPB expression or by proteasomal inhibitors, confirming its higher stability compared to the mutant protein as well as its insensitivity to the presence of the inclusions.

The results illustrated in Fig. 1A–C confirm and extend our previous results that indicated that under basal conditions P56S-VAPB is degraded exclusively by the proteasomal pathway, and that autophagy is not involved. We then asked whether the inclusions could become substrate for induced autophagy. To this end, we compared the rate of degradation of P56S-VAPB under normal or starvation conditions (Fig. 1D, E). Nine h after Dox addition, P56S-VAPB levels in starved cells were reduced to less than one half those of non-starved cells. Under starvation conditions, MG132 was less effective than under basal conditions in protecting mutant VAPB from degradation suggesting that the enhanced degradation was due to autophagy. This was confirmed by the observation that Bafilomycin rescued the excess degradation observed in starved cells, so that Bafilomycin-treated starved cells had P56S-VAPB levels similar to non-starved cells. Thus, whereas degradation of the P56S-VAPB is exclusively by the proteasomal pathway under basal conditions, the mutant protein may become an autophagosomal substrate under conditions that activate autophagy. The results of this biochemical analysis are in agreement with our previous morphological observations, showing close proximity of P56S-VAPB inclusions to p62 and LC3-positive autophagosomes in starved cells [25].

Neither proteasome-mediated degradation nor autophagic flux are altered by P56S-VAPB inclusions

Disturbance of proteostasis due to alterations in proteasomal function or autophagosomal flux represents an important mechanism of proteotoxicity of pathogenic aggregates [40]. Furthermore, interference with both these mechanisms by P56S-VAPB overexpressing cells has been reported [31,41]. We therefore investigated whether induction of the expression of P56S-VAPB inclusions in the HeLa Tet-Off cell line interferes with one or both of these pathways.

To investigate a possible interference with the proteasome, we analyzed the clearance of a substrate of ER associated degradation (ERAD), a pathway involving extraction of substrates from the ER, coupled to their ubiquitination and delivery to the proteasome [42]. Cells, induced or not to express P56S-VAPB, were transiently transfected with the CD3 complex δ chain (CD3δ), which, when expressed in the absence of the other subunits of the complex, is recognized by the quality control system of the ER and degraded by ERAD [43]. To follow CD3δ degradation, cells, grown in the presence or absence of Dox and co-transfected with EGFP and HA-tagged CD3δ, were treated with cycloheximide (CHX) for three h. This treatment did not affect EGFP nor tubulin (Fig. 2A), but strongly reduced CD3δ levels (Fig. 2A, B). Importantly, after CHX treatment, CD3δ levels were comparable in cells induced or not induced to express P56S-VAPB.

Since not all induced cells have detectable P56S-VAPB inclusions, we were concerned that the non-expressing cells might be preferentially transfected with CD3δ/EGFP, so that the results of Fig. 2A,B would be reporting on the situation in inclusion-negative cells. We therefore quantified the distribution of P56S-VAPB inclusions in transfected and non-transfected cells (Fig. 2C). In two separate experiments inclusions were detected in 52 and 44% of total cells and in 64 and 47% of EGFP-positive cells (~300 cells from random fields analyzed in each experiment). Thus, there is no bias towards P56S-VAPB low-expressing or negative cells in the efficiency of the transient transfection.

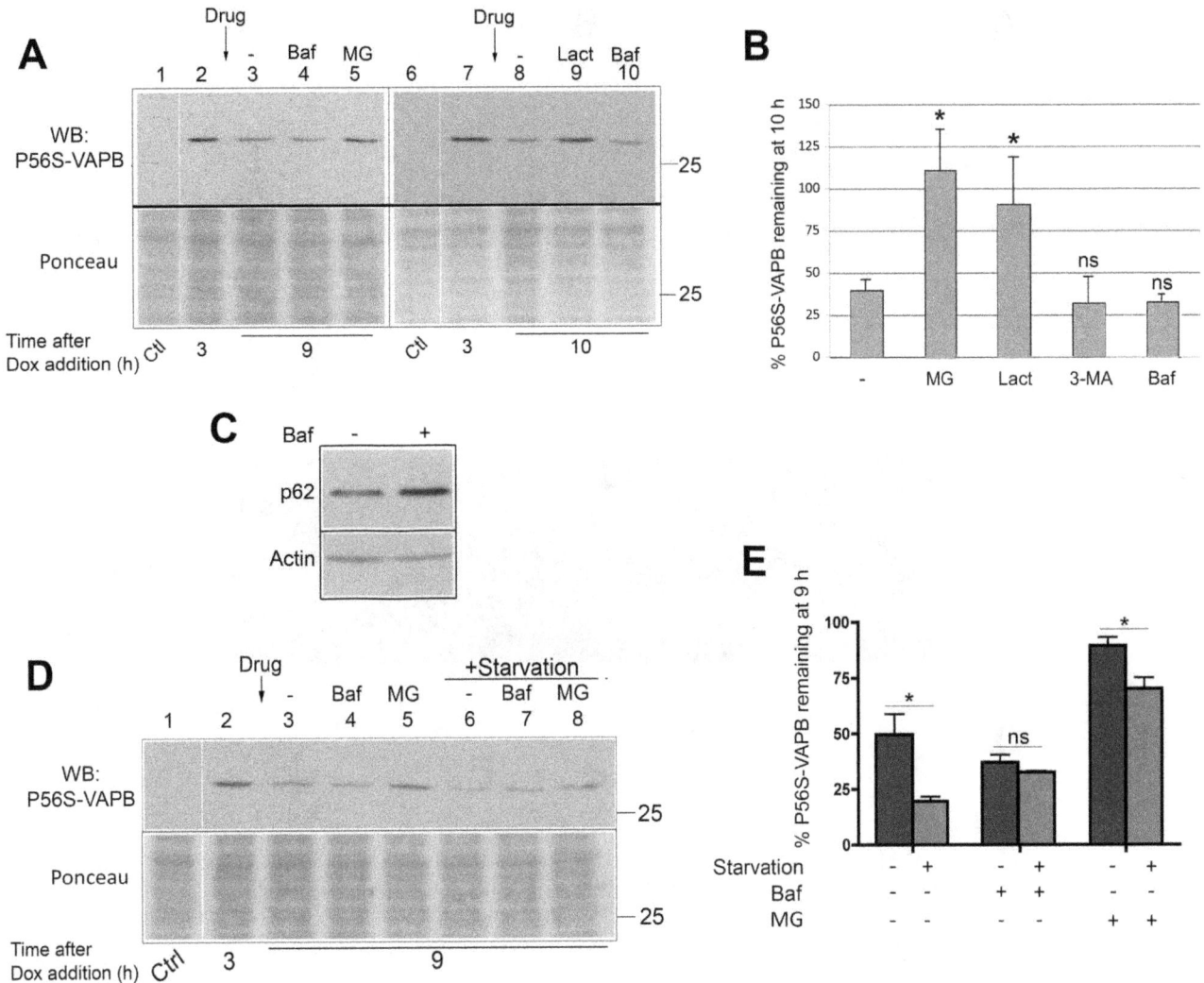

Figure 1. P56S-VAPB is degraded by the proteasome and by activated, but not basal, autophagy. A: Immunoblotting analysis of degradation of P56S-VAPB in the presence or absence of proteasome or autophagy inhibitors. 3 h after the inhibition of transcription of the P56S-VAPB transgene by addition of Dox to the media (lanes 2 and 7), cells were either left untreated (lanes 3 and 8), treated with the autophagy inhibitor Bafilomycin (Baf) or with the proteasome inhibitors MG132 (MG) or Lactacystin (Lact) for 6–7 h, as indicated. Control (Ctl) cells were grown in the presence of Dox. Equal aliquots of each sample were loaded (see Methods). The lower panel shows Ponceau staining of the blotted gel region, as loading control. The vertical white line (here and in panel D) juxtaposes lanes deriving from the same blot exposure. The position of the 25 kDa size marker is indicated. **B:** Quantification (means from 2–5 experiments +SEM) of P56S-VAPB remaining at 10 h after Dox addition in the presence or absence of drugs, as indicated, compared to levels measured at 3 h *: p = 0.013 and 0.025 for MG132 and lactacystin treated samples *vs* untreated by Student's t test. respectively. The difference between 3-MA or bafilomycin-treated samples and untreated was non-significant (ns). **C:** Equal amounts of protein of the samples of lanes 3 and 4 of panel A were analyzed for p62 by immunoblotting, to control for inhibition of autophagy by bafilomycin. Actin was probed as loading control. **D:** Effect of starvation on clearance of P56S-VAPB. 3 h after addition of Dox to the media (lane 2), cells were either left untreated (lane 3), or treated with bafilomycin (Baf) or MG132 (MG), as indicated, for 6 h; the samples of lanes 6–8 were also starved during the incubation with or without the drugs. Control (Ctl) cells were cultured in presence of Dox. Ponceau staining of the blotted region is shown in the lower panel. **E:** Quantification of three experiments (means +S.E.M.) of P56S-VAPB remaining 9 h after Dox addition under the indicated conditions compared to levels measured before drug treatment and/or starvation at 3 h after Dox addition. *: p = 0.036 by Student's t test; ns, non significant.

To investigate autophagosomal flux, we analyzed the behavior of two autophagosome markers after either pharmacological (torin 1) or starvation-induced autophagy [44]. The ubiquitin receptor p62 is degraded in autolysosomes; thus, its levels decrease under conditions of increased autophagy [44]. Analysis of autophagy-driven decrease of endogenous p62 levels in cells grown in the absence or presence of Dox showed that induction of P56S-VAPB expression did not interfere with p62 degradation (Fig. 2D, top). In similarly treated cells, we examined the generation of the lipidated form of LC3 (LC3-II), a reaction that occurs when LC3 is recruited to nascent autophagosomes [45]. In our HeLa cell line, the lipidated (LC3-II) form predominated already under basal conditions (Fig. 2D, lanes 1 and 4); the non-lipidated form (LC3-I) decreased both after torin 1 treatment and after starvation and differences between the ratio of the two forms were not detected between cells grown in the presence or absence of Dox (Fig. 2E).

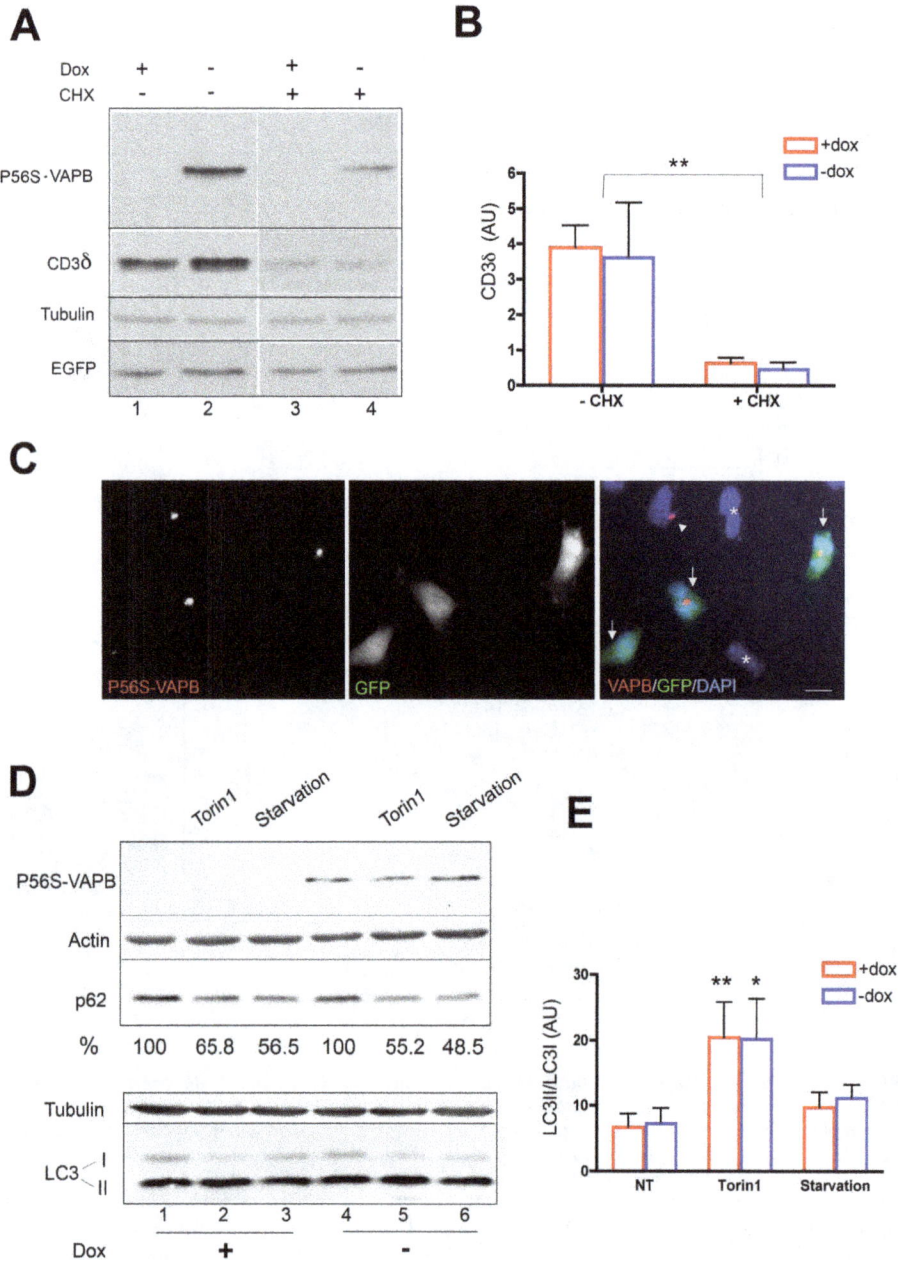

Figure 2. Lack of interference of P56S-VAPB inclusions with general proteostasis. A: Immunoblotting analysis of the degradation of the ERAD substrate CD3δ. Induced or not induced cells, co-transfected with plasmids specifying HA-CD3δ and EGFP, were treated with CHX for 3 h as indicated. Equal amounts of protein (30 μg) were loaded. **B:** Quantification of three experiments (means+SEM) of CD3δ remaining 3 h after CHX addition compared to untreated samples. Values were normalized to EGFP. By two-way Anova, the presence of Dox had no significant effect on CD3δ, while the effect of CHX was very significant (p = 0.0014). **C:** Immunofluorescence analysis of induced P56S-VAPB-Tet-Off cells co-transfected with HA-CD3δ and EGFP. The arrows in the merge panel indicate EGFP positive cells containing P56S-VAPB inclusions, revealed with anti-*myc* antibodies (left panel). Approximately equal proportions of cells with or without detectable inclusions were transfected (see text). The arrowhead indicates a non-transfected cell positive for P56S-VAPB. Asterisks indicate non-transfected cells negative for VAPB. Nuclei were stained with DAPI (blue). Scale bar, 10 μm. **D:** Immunoblotting analysis of the effect of P56S-VAPB inclusions on autophagic flux. Cells expressing or not expressing P56S-VAPB where either left untreated or treated for 3 h with Torin1 or starvation medium (EBSS), as indicated. The levels of p62, as percentage of the values in untreated cells are indicated below the lanes. Values were normalized to actin content. **E:** Quantification of three experiments (means+SEM) of LC3II/LC3I ratio of cells treated either with Torin 1 or with starvation medium, in comparison to untreated cells. Two-way Anova analysis reported that the source of variation between samples was due to autophagocytosis induction (non-treated *vs* Torin 1: p<0.01 and <0.05 for non-induced and induced cells, respectively) and not to P56S-VAPB expression.

P56S-VAPB inclusions in a model motoneuronal cell line are degraded by the proteasome and not by basal autophagocytosis

To extend our findings to a cell line with characteristics closer to motor neurons, we created NSC34 cell lines stably expressing wild-type or P56S-VAPB under the tetracycline-repressible promoter. NSC34 is a mouse cell line created by fusion of a neuroblastoma line with spinal cord primary motor neurons, and currently represents the best characterized available cell line with motoneuronal characteristics [46]. As shown in Fig. 3A, wt *myc*-VAPB in these cells was distributed throughout the cytoplasm, in a dense reticular network, as expected for an ER protein, whereas the P56S mutant formed inclusions similar to those of HeLa cells and of transiently transfected NSC34 cells [24,28]. We then investigated the mechanism of degradation of mutant VAB by adding Dox to the medium in the presence of MG132 or Bafilomycin, as done for the HeLa cell line. As shown in Fig. 3B, the decrease of P56S-VAPB levels observed between three and ten h after exposure to Dox was nearly completely reversed by MG132, while Bafilomycin was without effect. The efficacy of Bafilomycin treatment was confirmed by the increase of p62 content. The degradation of P56S-VAPB determined by western blot correlated with the decrease in number and size of VAPB-positive inclusions visualized by immunofluorescence (Fig. 3C). Thus, under basal conditions, P56S-VAPB inclusions in NSC34 cells are cleared by the same proteasome-mediated mechanism as observed in HeLa cells.

Close relationship between P56S-VAPB inclusions and the Golgi Complex

Inspection of the localization of P56S-VAPB inclusions revealed that in most cases they were close to the nucleus, in a position similar to that of the Golgi apparatus. Since disruption of the Golgi in neurons is a hallmark of many neurodegenerative diseases, including ALS [47,48], we investigated the relationship of the inclusions to the Golgi, comparing their distribution with the one of two different Golgi markers, GM130, which is preferentially localized to the *cis* face of the Golgi ribbon, and giantin, which is present on Golgi vesicles. Remarkably, the inclusions appeared to be embedded within the Golgi complex (Fig. 4). The intricate relationship between the inclusions and the Golgi is better appreciated in the 3D reconstructions obtained from confocal stacks (Video S1).

P56S-VAPB inclusions do not interfere with the intracellular transport of Vesicular Stomatitis Virus Glycoprotein (VSVG)

The above observations suggested that the tight relationship between P56S-VAPB inclusions and the Golgi complex might underlie interference of the inclusions with transport through the secretory pathway, as reported in cells transiently transfected with mutant VAPB [49]. To investigate the functionality of the secretory pathway in cells expressing moderate levels of P56S-VAPB, we transfected the Tet-Off HeLa cell line with cDNA coding for the ts045 version of the secretory membrane protein VSVG. This protein presents the advantage of accumulating in the ER at 39°C, so that a synchronized wave of transport through the secretory pathway can be followed after release of the high temperature transport block [50]. We first compared the time course of accumulation in the Golgi of transfected VSVG in cells induced and not induced to express mutant VAPB. Random cells were imaged and Golgi localization was evaluated by superposition on the giantin-positive area of the cells. In the case of the

induced sample, cells lacking visible inclusions were not considered. As shown in Fig. 5, VSVG accumulated rapidly in the Golgi, with maximum accumulation at 30 min after release of the temperature block, with similar time course in induced and non-induced cells. At later times, Golgi fluorescence decreased, with concomitant appearance of surface staining.

The experiment of Fig. 5 indicates that transport of VSVG from the ER to the Golgi is not impaired by the presence of mutant VAP inclusions. To quantify transport to the cell surface, we incubated non-permeabilized cells with an antibody that recognizes the lumenal/extracellular domain of VSVG and determined cell surface fluorescence at various times after release of the temperature block. As shown in Fig. 6, arrival of VSVG at the cell surface was not delayed in the induced, compared to the non-induced cells, indicating that the intracellular transport of this model glycoprotein is not affected by the presence of P56S-VAPB inclusions in tight association with the Golgi complex.

Discussion

ALS is a rapidly progressive and devastating neurodegenerative disease characterized by loss of motor neurons from the brain and spinal cord and consequent fatal respiratory failure. Only 10% of ALS cases are inherited (Familial ALS, or FALS), but understanding the pathogenic mechanism of each of the over ten identified FALS-linked mutations [51,52] represents an important step towards unraveling the molecular basis of the much more common sporadic form of this fatal disease. Among the identified ALS-linked genes, the one coding for VAPB is rare and perhaps the least understood. Nevertheless, the observation that VAP levels are decreased in sporadic ALS patients [16,53] is consistent with a more general role of the VAPs in motor neuron pathophysiology, and suggests that clarification of the cellular effects of the mutant gene will bring important insights into the molecular pathogenesis of ALS.

Because of its interaction with many different protein partners, VAPB is involved in a variety of functions [1]; accordingly, a number of possible, not mutually exclusive, pathogenic mechanisms of mutant VAPB have been proposed. Many of these are based on the observation that P56S-VAPB forms intracellular inclusions that sequester both the wild-type protein and, to a lesser extent, VAPA [12,16,26,27,28], suggesting that loss of function by a dominant negative mechanism underlies mutant VAPB's mode of inheritance. In addition, it has been hypothesized that cellular dysfunction is caused by the sequestration within the inclusions of functionally important VAPB interactors, such as the ER-Golgi recycling protein Yif1A, involved in transport within the early secretory pathway [3], and the phosphoinositide phosphatase Sac1 [29]. The VAPs have also been implicated in modulation of the ER Unfolded Protein Response (UPR), and overexpression of P56S-VAPB is reported both to attenuate UPR signaling [12,54], and to increase ER stress in animal disease models [41,55,56].

In addition to these cellular dysfunctions attributable to specific interactions of the VAPs, mutant VAPB inclusions have been reported to inhibit the proteasome [31], possibly leading to a general dysregulation of proteostasis, as is the case for other ALS-linked mutant genes [33]. Thus, a combination of dominant negative effects and general proteotoxicity could act together to cause the reduction in cell viability that has been observed in a number of transfected cell models [16,17,21,28,57,58].

As pointed out in the Introduction, the different mechanisms proposed for P56S-VAPB pathogenicity have been based mainly on studies on cultured cells acutely overexpressing mutant VAPB, and are thus not clearly related to the situation in cells chronically

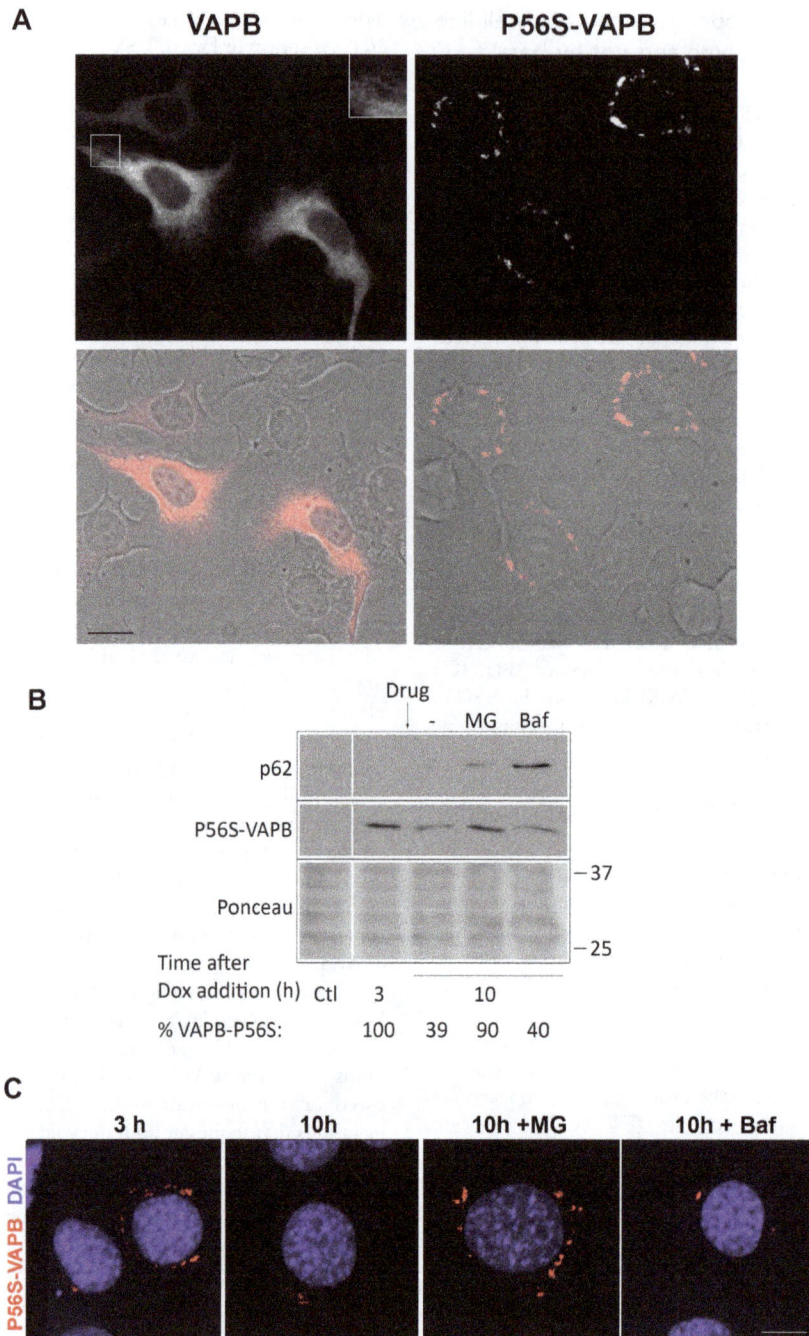

Figure 3. P56S-VAPB inclusions in a model motoneuronal cell line are degraded by the proteasome. A: Immunofluorescence analysis of NSC34 Tet-Off cells induced to express *myc*-wt-VAPB (left) or *myc*-P56S-VAPB (right). The upper panel shows anti-*myc* immunofluorescence, the lower one the superposition of *myc* staining with phase contrast. The inset of the upper left panel shows a 2 fold enlargement of the boxed area, and illustrates the web-like distribution of wt VAPB typical of an ER protein. Scale bar: 15 μm. **B:** Degradation of P56S-VAPB stably expressed in NSC34 cells. Induced cells were supplemented with Dox; 3 h thereafter the cells were either left untreated or treated with MG132 (MG) or Bafilomycin (Baf) for 7 h. Control (Ctl) cells were grown in the presence of Dox. Equal aliquots of each sample were loaded. The lower panel shows Ponceau staining of the blotted gel region; the positions of the 25 and 37 kDa size marker are indicated. The vertical white line indicates removal of irrelevant lanes form the image. The levels of P56S-VAPB, as percentage of values in untreated cells at 3 h after Dox addition, are indicated below the lanes. p62 immunoblotting was performed to check the efficacy of bafilomycin to inhibit autophagy (upper). **C:** Confocal analysis (single sections are shown) of P56S-VAPB inclusions stained with anti-*myc* antibody (red) at 3 h after Dox addition (left) and 7 h later in the presence or absence of the indicated drugs. Nuclei were stained with DAPI. The number and size of the inclusions decreased in the absence of drugs or in the presence of Bafilomycin, but remained similar to the 3 h cells when MG132 was present. Scale bar, 10 μm.

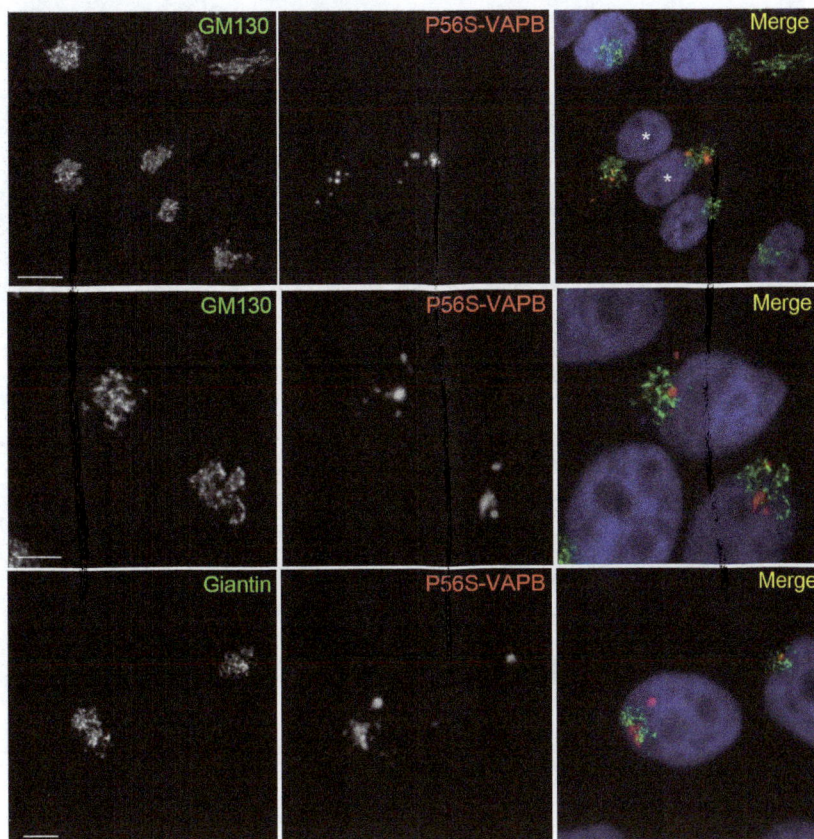

Figure 4. Close relationship between P56S-VAPB inclusions and the Golgi Complex. Induced HeLa Tet-Off cells were doubly immunostained with anti-*myc* antibodies, to reveal P56S-VAPB, and antibodies against the Golgi proteins GM130 or giantin, as indicated. Nuclei, stained with DAPI, are shown in the merge panel. Shown are maximum intensity projections of z-stacks. Scale bars: upper row, 10 µm; middle and lower row 5 µm.

expressing the mutant protein from a single allele. To investigate the effects of P56S-VAPB when expressed chronically at moderate levels, we turned to cell lines expressing mutant VAPB under the control of a Tet-repressible promoter. In Dox-free medium, these cells express P56S-VAPB at levels 2–3 fold higher than the endogenous protein ([24,25] and Figs. S1 and S2 of this study), and reach this steady state condition gradually over a period of several days after removal of Dox from the medium (unpublished results). Using these cells, we previously demonstrated that P56S-VAPB is unstable in comparison to the wt protein, and that its degradation is mediated by the proteasome and involves the participation of a key ERAD player, the AAA ATPase p97 [25]. Here, we have continued our investigation on the mechanism of degradation of P56S-VAPB inclusions as well as on their possible toxic effects on the cells.

First, we confirmed that under basal conditions mutant VAPB inclusions are cleared by the proteasome, both in HeLa and in a model motoneuronal cell line, but we also showed that autophagy, when stimulated, can further enhance degradation of the mutant protein. Thus, P56S-VAPB inclusions are available to degradation by both the major degradative pathways of the cell, and our results predict that, under conditions in which the cell potentiates autophagy, mutant VAPB inclusions will not become overrepresented in comparison to other compartments targeted by autophagy.

We then investigated whether P56S-VAPB inclusions interfere with two fundamental processes: (i) protein degradation mediated by the proteasome and by autophagy; and (ii) protein transport through the secretory pathway.

Moumen et al. [31] reported that transient overexpression of wild-type and mutant VAPB results in an increase of polyubiquitinated proteins and stabilization of three different proteasomal substrates, among which the classical ERAD substrate CD3δ. However, in our cells, clearance of CD3δ, whose degradative pathway shares with the one of P56S-VAPB the involvement both of the proteasome and of p97, was unaffected by the expression of the mutant protein.

Autophagic dysfunction has been described in ALS, and both a significant autophagy upregulation and/or impairment with abnormal accumulation of autophagosomes have been observed (reviewed in ref 32). However, to our knowledge, the effect of P56S-VAPB inclusions on autophagic flow had yet not been investigated. We found that autophagy, stimulated either pharmacologically or by starvation, was unaffected by P56S-VAPB expression. Thus, it appears that cells can adjust the capacity of their degradative machinery to cope with moderate levels of mutant VAPB without consequent disturbances in proteostasis.

A second fundamental process in which the VAPs are implicated is intracellular transport through the secretory pathway, but contrasting results have been reported on the effect of P56S-VAPB expression on intracellular transport. In CHO cells, Prosser and collaborators [49] found a strong interference of overexpressed P56S-VAPB (and also of overexpressed wt VAPA) with VSVG transport, while no delay of the transport of the same

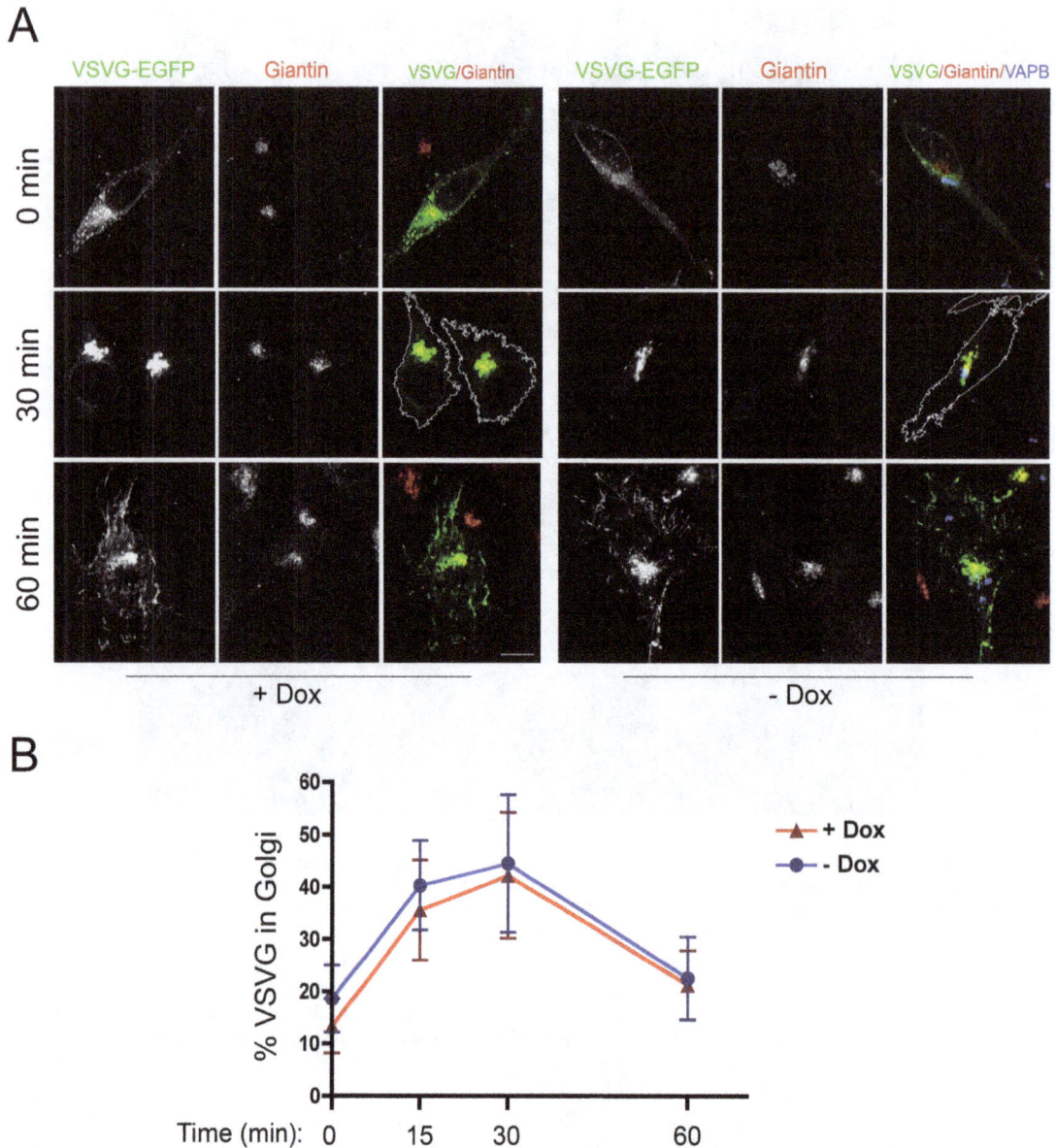

Figure 5. Transport of VSVG to the Golgi Complex occurs normally in cells expressing P56S-VAPB inclusions. A: HeLa-TetOff cells, induced (−Dox, right) or not induced (+Dox, left) to express *myc*-P56S-VAPB, were transfected with VSVG-EGFP at 39.3°C. After 24 h, one coverslip of each sample was fixed (0 min), while the others were shifted to 32°C and fixed after incubation for the indicated times. Cells were stained with anti-Giantin (red) and anti-*myc* (blue) antibodies. Maximum intensity projections of z-stacks are shown. The cell boundaries at the 30 min time point are indicated by the white line in the merge panel. Acquisition parameters were the same in all images. Scale bar, 10 μm. **B:** Time course (means ± SD) of VSVG transport through the Golgi. Significant differences between induced or non-induced samples were not detected by Student's t-test.

secretory membrane cargo was detected by Teuling et al. in primary hippocampal neuronal cultures [16]. In our system, we found that neither transport from the ER to the Golgi nor export to the cell surface were altered by the presence of P56S-VAPB inclusions. We conclude that cells can maintain secretory pathway function in the presence of P56S-VAPB inclusions, notwithstanding their close physical proximity to the Golgi apparatus demonstrated here.

The results reported in this study, showing a lack of interference of P56S-VAPB inclusions with basic cellular functions, are consistent with the outcome of analyses of transgenic animals. Restricting this discussion to mammals, four transgenic mouse lines have been reported so far [30,41,48,59]. Of these, only one,

in which the mutant protein was highly overexpressed (at seven fold higher levels than the endogenous protein), developed mild motor abnormalities and loss of cortical, but not spinal, motor neurons [41]. The other three strains, although presenting P56S-VAPB-containing inclusions in motor neurons, showed no motor abnormalities. These results suggest that the much lower levels of mutant protein expressed from a single allele in ALS8 patients may be devoid of pathogenic effect. Interestingly, in the study of Aliaga et al. [41], lower levels of mutant than of wild-type protein were detected in the brains of transgenic mouse strains that had comparable levels of mRNA expression. This observation demonstrates that the instability of the mutant protein first observed in cultured cells [25,31] is present also in animal tissues. P56S-VAPB

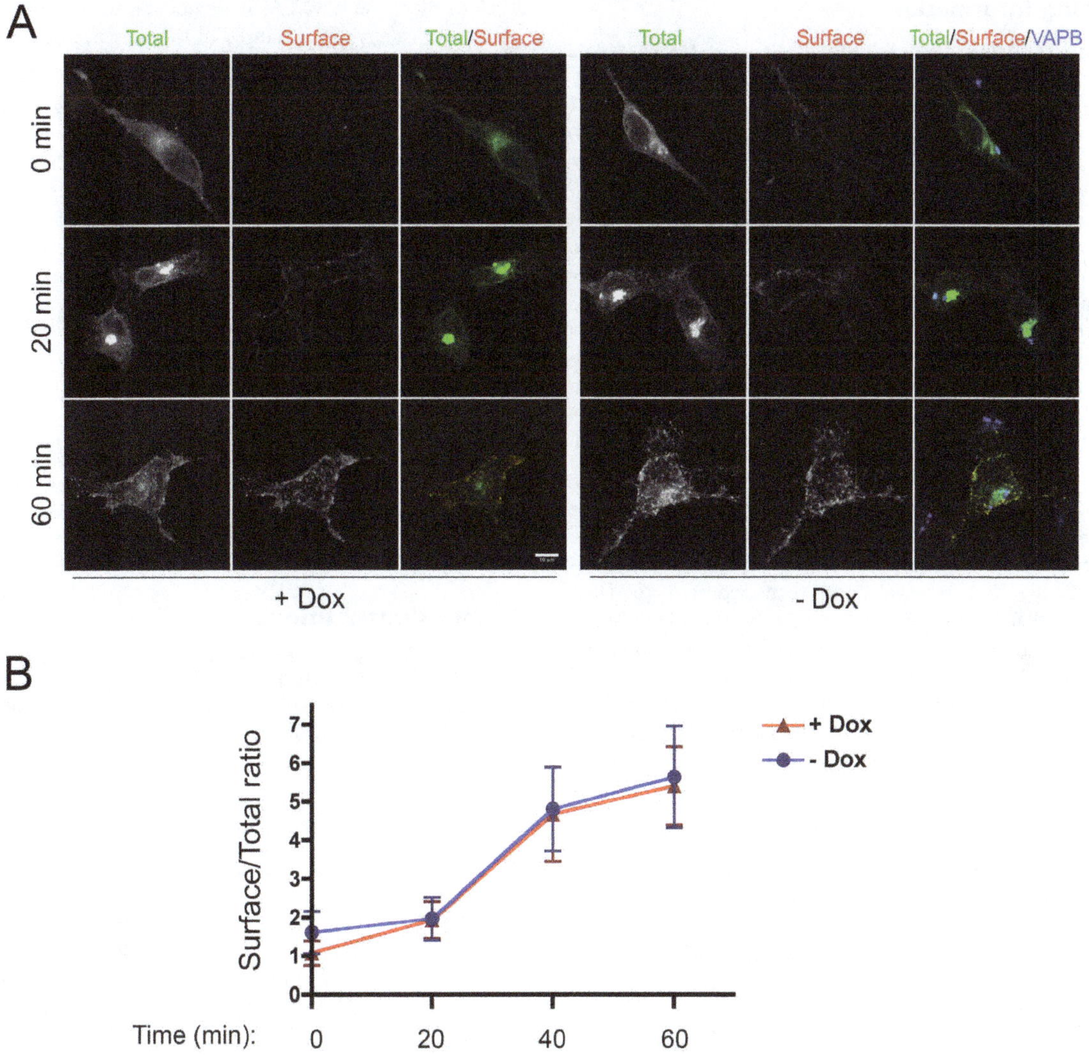

Figure 6. Transport of VSVG to the cell surface occurs normally in cells expressing P56S-VAPB inclusions. A: HeLa-TetOff cells, induced (−Dox) or not induced (+Dox) to express *myc*-P56S-VAPB, were transfected with VSVG-EGFP at 39.3°C. After 24 h, cells were shifted to 32°C. At the indicated times, the cells were chilled and incubated with anti-lumenal domain of VSVG under non-permeabilizing conditions (red). The cells were then permeabilized and stained with anti-VAPB antibodies (blue in merge panel - see Methods). Total VSVG (intracellular+surface) was revealed by GFP fluorescence (green). Maximum intensity projections of z-stacks are shown. The acquisition parameters were the same in all images. Scale bar, 10 μm. **B:** Time course (means ± SD) of VSVG surface labeling normalized to total EGFP fluorescence. Significant differences between induced or non-induced samples were not detected by Student's t-test.

instability most likely explains the lack of detectable VAPB inclusions in ALS8 (P56S-VAPB) patients' motor neurons generated from induced pluripotent stem cells (IPSC) [60].

In conclusion, our results provide an explanation for the discrepancy between the observations obtained in transiently transfected cells and transgenic mouse models, and support the hypothesis that haploinsufficiency alone underlies the dominant inheritance of VAPB mutations. In addition to the generally negative results obtained with the transgenic mice, this idea is supported also by the reduced levels of VAPB in iPSC-derived motor neurons of ALS8 patients [60] and in spinal motor neurons of sporadic ALS patients [16,53]. While strong effects of VAP deletion in cultured cells are obtained only when both homologues are silenced [3,5,16], the studies with mice specifically deleted for VAPB are in partial agreement with a pathogenic role of VAPB

haploinsufficiency in motor neuron disease. In one study, VAPB-deleted mice, although free from a full blown ALS phenotype, did develop mild, late onset defects in motor performance [22]; in another study, VAPB deletion was reported to cause alterations in muscle lipid metabolism [61]. To be noted, in the first of these studies [22], also the heterozygote mouse showed reduced motor performance in the Rotarod test, although the difference with respect to the controls was not statistically significant. This observation suggests that within the longer human lifespan, even a 50% reduction of the normal dosage of VAPB may affect motor neuron survival. Whether damage due to VAPB deficit is caused by the reduction of a unique VAPB function not carried out by VAPA, or whether long term motor neuron survival simply requires the full dosage of the sum of the two VAP homologues remains to be determined in appropriate cell and animal models.

Supporting Information

Figure S1 Purification of polyclonal anti-VAPB antibody. A: Western Blot analysis comparing the reactivity of anti-VAPB serum towards lysates from bacteria expressing GST-VAPA or GST-VAPB 1–225 (arrow) before and after adsorption to VAPA-coupled resin. Antibodies cross-reactive with VAPA are eliminated in this step of purification. The lower molecular weight bands recognized by the adsorbed antiserum in lysates from bacteria expressing the VAPB fusion protein are probably due to degradation products. **B:** Purification of adsorbed antiserum by affinity chromatography. Specificity of the antibodies was probed by western blotting against lysates from HeLa Tet-Off cells induced to express P56S-VAPB. Endogenous VAPB and P56S-VAPB induced by removal of Dox are indicated by the arrowhead and arrow, respectively. The asterisks indicate non-specific bands, of which the major ones are eliminated by the affinity purification. (PDF)

Figure S2 Comparison of the effect of proteasome inhibitors on endogenous wild-type VAPB and on the transfected mutant protein. Cells were induced to express P56S-VAPB by Dox removal, and then returned to Dox-containing media, as described in the legend to Figure 1. At the indicated times, cells were collected, and the lysates were analyzed by SDS-PAGE - immunoblotting, with the use of an anti-VAPB antibody. The endogenous wild-type protein is distinguished from the transfected *myc*-tagged mutant by its faster migration. The

levels of endogenous wt VAPB are not affected by drug treatments. Control cells (ctr) were cultured in the presence of Dox. (PDF)

Video S1 Maximum intensity projections of a field of P56S-VAPB- expressing HeLa Tet-Off cells doubly stained for VAPB with anti-*myc* antibodies (red) and for giantin (green). Shown are maximum intensity projections generated from rotation around the X-axis of a stack of 16 confocal sections acquired at 0.2 μm intervals. Each image is rotated by 5° with respect to the preceding one, for a total rotation of 180°. (AVI)

Acknowledgments

In addition to the people who kindly supplied reagents (listed in Materials and Methods), we acknowledge the Monzino Foundation (Milan, Italy), for its generous gift of the Zeiss LSM 510 Meta confocal microscope. We are grateful to Sara Francesca Colombo and Angelo Poletti for helpful discussion and to Cecilia Gotti and Milena Moretti for help with rabbit immunization.

Author Contributions

Conceived and designed the experiments: PG FN NB. Performed the experiments: PG GP AR LC. Analyzed the data: PG GP FN NB. Contributed reagents/materials/analysis tools: LC. Wrote the paper: PG FN NB. Critical revision of the manuscript: LC.

References

1. Lev S, Ben Halevy D, Peretti D, Dahan N (2008) The VAP protein family: from cellular functions to motor neuron disease. Trends Cell Biol 18: 282–290.
2. Yang Z, Huh SU, Drennan JM, Kathuria H, Martinez JS, et al. (2012) Drosophila Vap-33 is required for axonal localization of Dscam isoforms. J Neurosci 32: 17241–17250.
3. Kuijpers M, Yu KL, Teuling E, Akhmanova A, Jaarsma D, et al. (2013) The ALS8 protein VAPB interacts with the ER-Golgi recycling protein YIF1A and regulates membrane delivery into dendrites. EMBO J 32: 2056–2072.
4. Kawano M, Kumagai K, Nishijima M, Hanada K (2006) Efficient trafficking of ceramide from the endoplasmic reticulum to the Golgi apparatus requires a VAMP-associated protein-interacting FFAT motif of CERT. J Biol Chem 281: 30279–30288.
5. Peretti D, Dahan N, Shimoni E, Hirschberg K, Lev S (2008) Coordinated lipid transfer between the endoplasmic reticulum and the Golgi complex requires the VAP proteins and is essential for Golgi-mediated transport. Mol Biol Cell 19: 3871–3884.
6. Levine T, Loewen C (2006) Inter-organelle membrane contact sites: through a glass, darkly. Curr Opin Cell Biol 18: 371–378.
7. Rocha N, Kuijl C, van der Kant R, Janssen L, Houben D, et al. (2009) Cholesterol sensor ORP1L contacts the ER protein VAP to control Rab7-RILP-p150 Glued and late endosome positioning. J Cell Biol 185: 1209–1225.
8. Stefan CJ, Manford AG, Baird D, Yamada-Hanff J, Mao Y, et al. (2011) Osh proteins regulate phosphoinositide metabolism at ER-plasma membrane contact sites. Cell 144: 389–401.
9. De Vos KJ, Morotz GM, Stoica R, Tudor EL, Lau KF, et al. (2012) VAPB interacts with the mitochondrial protein PTPIP51 to regulate calcium homeostasis. Hum Mol Genet 21: 1299–1311.
10. Alpy F, Rousseau A, Schwab Y, Legueux F, Stoll I, et al. (2013) STARD3 or STARD3NL and VAP form a novel molecular tether between late endosomes and the ER. J Cell Sci 126: 5500–5512.
11. Amarilio R, Ramachandran S, Sabanay H, Lev S (2005) Differential regulation of endoplasmic reticulum structure through VAP-Nir protein interaction. J Biol Chem 280: 5934–5944.
12. Kanekura K, Nishimoto I, Aiso S, Matsuoka M (2006) Characterization of amyotrophic lateral sclerosis-linked P56S mutation of vesicle-associated membrane protein-associated protein B (VAPB/ALS8). J Biol Chem 281: 30223–30233.
13. Saita S, Shirane M, Natume T, Iemura S, Nakayama KI (2009) Promotion of neurite extension by protrudin requires its interaction with vesicle-associated membrane protein-associated protein. J Biol Chem 284: 13766–13777.
14. Ohnishi T, Shirane M, Hashimoto Y, Saita S, Nakayama KI (2014) Identification and characterization of a neuron-specific isoform of protrudin. Genes Cells 19: 97–111.
15. Nishimura AL, Mitne-Neto M, Silva HC, Oliveira JR, Vainzof M, et al. (2004) A novel locus for late onset amyotrophic lateral sclerosis/motor neurone disease variant at 20q13. J Med Genet 41: 315–320.
16. Teuling E, Ahmed S, Haasdijk E, Demmers J, Steinmetz MO, et al. (2007) Motor neuron disease-associated mutant vesicle-associated membrane protein-associated protein (VAP) B recruits wild-type VAPs into endoplasmic reticulum-derived tubular aggregates. J Neurosci 27: 9801–9815.
17. Kim S, Leal SS, Ben Halevy D, Gomes CM, Lev S (2010) Structural requirements for VAP-B oligomerization and their implication in amyotrophic lateral sclerosis-associated VAP-B(P56S) neurotoxicity. J Biol Chem 285: 13839–13849.
18. Shi J, Lua S, Tong JS, Song J (2010) Elimination of the native structure and solubility of the hVAPB MSP domain by the Pro56Ser mutation that causes amyotrophic lateral sclerosis. Biochemistry 49: 3887–3897.
19. Nishimura AL, Al-Chalabi A, Zatz M (2005) A common founder for amyotrophic lateral sclerosis type 8 (ALS8) in the Brazilian population. Hum Genet 118: 499–500.
20. Funke AD, Esser M, Kruttgen A, Weis J, Mitne-Neto M, et al. (2010) The p.P56S mutation in the VAPB gene is not due to a single founder: the first European case. Clin Genet 77: 302–303.
21. Chen HJ, Anagnostou G, Chai A, Withers J, Morris A, et al. (2010) Characterization of the properties of a novel mutation in VAPB in familial amyotrophic lateral sclerosis. J Biol Chem 285: 40266–40281.
22. Kabashi E, El Oussini H, Bercier V, Gros-Louis F, Valdmanis PN, et al. (2013) Investigating the contribution of VAPB/ALS8 loss of function in amyotrophic lateral sclerosis. Hum Mol Genet 22: 2350–2360.
23. van Blitterswijk M, van Es MA, Koppers M, van Rheenen W, Medic J, et al. (2012) VAPB and C9orf72 mutations in 1 familial amyotrophic lateral sclerosis patient. Neurobiol Aging 33: 2950 e2951–2954.
24. Fasana E, Fossati M, Ruggiano A, Brambillasca S, Hoogenraad CC, et al. (2010) A VAPB mutant linked to amyotrophic lateral sclerosis generates a novel form of organized smooth endoplasmic reticulum. Faseb J 24: 1419–1430.
25. Papiani G, Ruggiano A, Fossati M, Raimondi A, Bertoni G, et al. (2012) Restructured endoplasmic reticulum generated by mutant amyotrophic lateral sclerosis-linked VAPB is cleared by the proteasome. J Cell Sci 125: 3601–3611.
26. Chai A, Withers J, Koh YH, Parry K, Bao H, et al. (2008) hVAPB, the causative gene of a heterogeneous group of motor neuron diseases in humans, is functionally interchangeable with its Drosophila homologue DVAP-33A at the neuromuscular junction. Hum Mol Genet 17: 266–280.
27. Ratnaparkhi A, Lawless GM, Schweizer FE, Golshani P, Jackson GR (2008) A Drosophila model of ALS: human ALS-associated mutation in VAP33A suggests a dominant negative mechanism. PLoS One 3: e2334.
28. Suzuki H, Kanekura K, Levine TP, Kohno K, Olkkonen VM, et al. (2009) ALS-linked P56S-VAPB, an aggregated loss-of-function mutant of VAPB, predisposes

motor neurons to ER stress-related death by inducing aggregation of co-expressed wild-type VAPB. J Neurochem 108: 973–985.

29. Forrest S, Chai A, Sanhueza M, Marescotti M, Parry K, et al. (2013) Increased levels of phosphoinositides cause neurodegeneration in a Drosophila model of amyotrophic lateral sclerosis. Hum Mol Genet 22: 2689–2704.

30. Tudor EL, Galtrey CM, Perkinton MS, Lau KF, De Vos KJ, et al. (2010) Amyotrophic lateral sclerosis mutant vesicle-associated membrane protein-associated protein-B transgenic mice develop TAR-DNA-binding protein-43 pathology. Neuroscience 167: 774–785.

31. Moumen A, Virard I, Raoul C (2011) Accumulation of wildtype and ALS-linked mutated VAPB impairs activity of the proteasome. PLoS One 6: e26066.

32. Chen S, Zhang X, Song L, Le W (2012) Autophagy dysregulation in amyotrophic lateral sclerosis. Brain Pathol 22: 110–116.

33. Robberecht W, Philips T (2013) The changing scene of amyotrophic lateral sclerosis. Nat Rev Neurosci 14: 248–264.

34. Blokhuis AM, Groen EJ, Koppers M, van den Berg LH, Pasterkamp RJ (2013) Protein aggregation in amyotrophic lateral sclerosis. Acta Neuropathol 125: 777–794.

35. Tan CC, Yu JT, Tan MS, Jiang T, Zhu XC, et al. (2014) Autophagy in aging and neurodegenerative diseases: implications for pathogenesis and therapy. Neurobiol Aging 35: 941–957.

36. Kaiser SE, Brickner JH, Reilein AR, Fenn TD, Walter P, et al. (2005) Structural basis of FFAT motif-mediated ER targeting. Structure 13: 1035–1045.

37. Seelig HP, Schranz P, Schröter H, Wiemann C, Renz M (1994) Macrogolgin - a new 376 kD Golgi Complex outer membrane protein as target of antibodies in patients with rheumatic disease and HIV infections. J Autoimmun 7: 67–91.

38. Marra P, Salvatore L, Mironov A Jr, Di Campli A, Di Tullio G, et al. (2007) The biogenesis of the Golgi ribbon: the roles of membrane input from the ER and of GM130. Mol Biol Cell 18: 1595–1608.

39. Babetto E, Mangolini A, Rizzardini M, Lupi M, Conforti L, et al. (2005) Tetracycline-regulated gene expression in the NSC-34-tTA cell line for investigation of motor neuron diseases. Brain Res Mol Brain Res 140: 63–72.

40. Powers ET, Morimoto RI, Dillin A, Kelly JW, Balch WE (2009) Biological and chemical approaches to diseases of proteostasis deficiency. Annu Rev Biochem 78: 959–991.

41. Aliaga L, Lai C, Yu J, Chub N, Shim H, et al. (2013) Amyotrophic lateral sclerosis-related VAPB P56S mutation differentially affects the function and survival of corticospinal and spinal motor neurons. Hum Mol Genet 22: 4293–4305.

42. Bernasconi R, Molinari M (2011) ERAD and ERAD tuning: disposal of cargo and of ERAD regulators from the mammalian ER. Curr Opin Cell Biol 23: 176–183.

43. Yang M, Omura S, Bonifacino JS, Weissman AM (1998) Novel aspects of degradation of T cell receptor subunits from the endoplasmic reticulum (ER) in T cells: importance of oligosaccharide processing, ubiquitination, and proteasome-dependent removal from ER membranes. J Exp Med 187: 835–846.

44. Klionsky DJ, Abdalla FC, Abeliovich H, Abraham RT, Acevedo-Arozena A, et al. (2012) Guidelines for the use and interpretation of assays for monitoring autophagy. Autophagy 8: 445–544.

45. Kabeya Y, Mizushima N, Ueno T, Yamamoto A, Kirisako T, et al. (2000) LC3, a mammalian homologue of yeast Apg8p, is localized in autophagosome membranes after processing. Embo J 19: 5720–5728.

46. Cashman NR, Durham HD, Blusztajn JK, Oda K, Tabira T, et al. (1992) Neuroblastoma x spinal cord (NSC) hybrid cell lines resemble developing motor neurons. Dev Dyn 194: 209–221.

47. Gonatas NK, Stieber A, Gonatas JO (2006) Fragmentation of the Golgi apparatus in neurodegenerative diseases and cell death. J Neurol Sci 246: 21–30.

48. van Dis V, Kuijpers M, Haasdijk ED, Teuling E, Oakes SA, et al. (2014) Golgi fragmentation precedes neuromuscular denervation and is associated with endosome abnormalities in SOD1-ALS mouse motor neurons. Acta Neuropathol Commun 2: 38.

49. Prosser DC, Tran D, Gougeon PY, Verly C, Ngsee JK (2008) FFAT rescues VAPA-mediated inhibition of ER-to-Golgi transport and VAPB-mediated ER aggregation. J Cell Sci 121: 3052–3061.

50. Bergmann JE (1989) Using temperature-sensitive mutants of VSV to study membrane protein biogenesis. Methods Cell Biol 32: 85–110.

51. Andersen PM, Al-Chalabi A (2011) Clinical genetics of amyotrophic lateral sclerosis: what do we really know? Nat Rev Neurol 7: 603–615.

52. Ferraiuolo L, Kirby J, Grierson AJ, Sendtner M, Shaw PJ (2011) Molecular pathways of motor neuron injury in amyotrophic lateral sclerosis. Nat Rev Neurol 7: 616–630.

53. Anagnostou G, Akbar MT, Paul P, Angelinetta C, Steiner TJ, et al. (2010) Vesicle associated membrane protein B (VAPB) is decreased in ALS spinal cord. Neurobiol Aging 31: 969–985.

54. Gkogkas C, Middleton S, Kremer AM, Wardrope C, Hannah M, et al. (2008) VAPB interacts with and modulates the activity of ATF6. Hum Mol Genet 17: 1517–1526.

55. Tsuda H, Han SM, Yang Y, Tong C, Lin YQ, et al. (2008) The amyotrophic lateral sclerosis 8 protein VAPB is cleaved, secreted, and acts as a ligand for Eph receptors. Cell 133: 963–977.

56. Moustaqim-Barrette A, Lin YQ, Pradhan S, Neely GG, Bellen HJ, et al. (2014) The amyotrophic lateral sclerosis 8 protein, VAP, is required for ER protein quality control. Hum Mol Genet 23: 1975–1989.

57. Langou K, Moumen A, Pellegrino C, Aebischer J, Medina I, et al. (2010) AAV-mediated expression of wild-type and ALS-linked mutant VAPB selectively triggers death of motoneurons through a Ca2+-dependent ER-associated pathway. J Neurochem 114: 795–809.

58. Chattopadhyay D, Sengupta S (2014) First evidence of pathogenicity of V234I mutation of hVAPB found in Amyotrophic Lateral Sclerosis. Biochem Biophys Res Commun 448: 108–113.

59. Qiu L, Qiao T, Beers M, Tan W, Wang H, et al. (2013) Widespread aggregation of mutant VAPB associated with ALS does not cause motor neuron degeneration or modulate mutant SOD1 aggregation and toxicity in mice. Mol Neurodegener 8: 1.

60. Mitne-Neto M, Machado-Costa M, Marchetto MC, Bengtson MH, Joazeiro CA, et al. (2011) Downregulation of VAPB expression in motor neurons derived from induced pluripotent stem cells of ALS8 patients. Hum Mol Genet 20: 3642–3652.

61. Han SM, El Oussini H, Scekic-Zahirovic J, Vibbert J, Cottee P, et al. (2013) VAPB/ALS8 MSP ligands regulate striated muscle energy metabolism critical for adult survival in caenorhabditis elegans. PLoS Genet 9: e1003738.

15

Progressive Degradation of Crude Oil *n*-Alkanes Coupled to Methane Production under Mesophilic and Thermophilic Conditions

Lei Cheng[1,2], Shengbao Shi[3], Qiang Li[2], Jianfa Chen[3], Hui Zhang[2], Yahai Lu[1]*

1 College of Resources and Environmental Sciences, China Agricultural University, Beijing, 100193, China, 2 Key Laboratory of Development and Application of Rural Renewable Energy, Biogas Institute of Ministry of Agriculture, Chengdu, 610041, China, 3 State Key Laboratory of Petroleum Resources and Prospecting, China University of Petroleum (Beijing), Beijing, 102200, China

Abstract

Although methanogenic degradation of hydrocarbons has become a well-known process, little is known about which crude oil tend to be degraded at different temperatures and how the microbial community is responded. In this study, we assessed the methanogenic crude oil degradation capacity of oily sludge microbes enriched from the Shengli oilfield under mesophilic and thermophilic conditions. The microbial communities were investigated by terminal restriction fragment length polymorphism (T-RFLP) analysis of 16S rRNA genes combined with cloning and sequencing. Enrichment incubation demonstrated the microbial oxidation of crude oil coupled to methane production at 35 and 55°C, which generated 3.7 ± 0.3 and 2.8 ± 0.3 mmol of methane per gram oil, respectively. Gas chromatography-mass spectrometry (GC-MS) analysis revealed that crude oil *n*-alkanes were obviously degraded, and high molecular weight *n*-alkanes were preferentially removed over relatively shorter-chain *n*-alkanes. Phylogenetic analysis revealed the concurrence of acetoclastic *Methanosaeta* and hydrogenotrophic methanogens but different methanogenic community structures under the two temperature conditions. Candidate divisions of JS1 and WWE 1, *Proteobacteria* (mainly consisting of *Syntrophaceae*, *Desulfobacteraceae* and *Syntrophorhabdus*) and *Firmicutes* (mainly consisting of *Desulfotomaculum*) were supposed to be involved with *n*-alkane degradation in the mesophilic conditions. By contrast, the different bacterial phylotypes affiliated with *Caldisericales*, "Shengli Cluster" and *Synergistetes* dominated the thermophilic consortium, which was most likely to be associated with thermophilic crude oil degradation. This study revealed that the oily sludge in Shengli oilfield harbors diverse uncultured microbes with great potential in methanogenic crude oil degradation over a wide temperature range, which extend our previous understanding of methanogenic degradation of crude oil alkanes.

Editor: Stephen J. Johnson, University of Kansas, United States of America

Funding: This work was financially supported by National Nature Science Foundation of Science (40973059, 30900049), National High Technology Research and Development Program of China (2013aa064401), Sichuan Projects of International Cooperation and Exchanges (2013HH0018) and Basic Foundation for Scientific Research of State-level Public Welfare Institutes of China (2013ZL001). The funders had no role in study design, data collection and analysis, decision to publish, or preparation of the manuscript.

Competing Interests: The authors have declared that no competing interests exist.

* Email: yhlu@cau.edu.cn

Introduction

Crude oil is a complex mixture containing many thousands of different hydrocarbon compounds, which can be divided into four classes (saturated hydrocarbons, aromatic hydrocarbons, asphaltene and non-hydrocarbons). The biodegradation of crude oil by natural populations of microorganisms was reported over a century ago. Under aerobic conditions, general progressive biodegradation of petroleum hydrocarbon proceeds first with loss of *n*-alkanes, then isoprenoids, cyclic alkanes, and lower molecular weight aromatics, followed by the remainder of the more complex, higher molecular weight constituents [1,2]. Over the past two decades, it has consistently been shown that crude oil hydrocarbons could be degraded under nitrate-reducing [3], ferric iron-reducing [4], sulfate-reducing [5,6] and methanogenic conditions [7–9]. Multiple research teams have reported that the entire *n*-alkane fraction of crude oil can be consumed under sulfate reducing and/or methanogenic conditions [6,9–13]. Rueter *et al.*

[5] reported for the first time that C_8 to C_{11} *n*-alkanes of crude oil were completely degraded, and C_{12} to C_{16} *n*-alkanes were partially consumed by a thermophilic sulfate-reducing bacterium, but degradation of alkanes above hexadecane was not observed. Rabus *et al.* [14] enriched a denitrifying culture capable of degrading C_1–C_3 alkylbenzenes in the first generation incubation and growing on C_5–C_{20} *n*-alkanes and alkylbenzenes in the second subculture. Siddique *et al.* [15,16] reported the preferential use of *n*-alkanes $nC_{10} > nC_8 > nC_7 > nC_6$, but no preferential degradation of longer-chain *n*-alkanes (C_{14}, C_{16} and C_{18}) occurred. Additionally, the initial loss of the longer-chain *n*-alkanes of crude oil under sulphate reducing and methanogenic conditions has also been reported [17–19].

The anaerobic degradation of non-methane hydrocarbons is different from the aerobic process. Anaerobic hydrocarbon degradation is activated by fumarate addition, carboxylation, methylation, hydroxylation [20–22] and potentially other unknown mechanisms [23]. Over twenty culturable bacteria have been isolated and characterized that are capable of alkane degradation with nitrate and sulfate as electron acceptors (Widdel *et al.* 2010 and references therein). The methanogenic conversion of crude oil hydrocarbons requires syntrophic communities of acetogenic bacteria and methanogenic archaea from the thermodynamics point of view [24]. To date, no pure syntrophic hydrocarbon degraders have been isolated with the exception of a sulfate reducer capable of syntrophic hexadecane degradation in coculture with hydrogenotrophic methanogens [25]. The development of culture-independent approaches has revealed the vast majority of as-yet-uncultured syntrophic bacterial species present in methanogenic hydrocarbon degrading consortia [8]. For example, uncultured members of the *Syntrophaceae* family have been implicated in syntrophic alkane degradation under mesophilic conditions using qPCR and DNA-SIP [26,27]. *Thermotogae-* and *Firmicutes*-related members were the dominant phylotypes in thermophilic, methanogenic alkane degrading cultures [11,28,29]. Acetoclastic and hydrogenotrophic methanogens were always observed in the methanogenic communities, but research on the relative contribution of these two methanogenic pathways to total methane production during the anaerobic degradation of *n*-alkanes is limited [13,27].

Temperature influences petroleum biodegradation by affecting the chemical composition of the oil, the rate of hydrocarbon metabolism by microorganisms, and the composition of the microbial community [30]. Previous studies reported that microbial oxidation of *n*-alkanes coupled to methane production could occur in the temperature range 20–55°C [7,11,13,15,28,31]. Little is known about the potential for crude oil degradation and methane production under different temperature conditions, and how indigenous microbes respond to the crude oil and temperature shift. In this study, oily sludge microbes originating from Shengli oilfield were incubated at 35 and 55°C. The microbial activity of crude oil degradation was evaluated by detecting methane production and crude oil degradation over time. The microbial community structures were characterized using terminal restriction fragment length polymorphism (T-RFLP) fingerprinting and sequencing of 16S rRNA gene fragments.

Materials and Methods

Ethics statement

No specific permits were required for the described field studies. No specific permissions were required for these locations/ activities. The sampled locations are not privately owned or protected in any way, and the studies did not involve endangered or protected species.

Medium and incubation

Sample and medium. Oily sludge was sampled from a disposal field of Shengli oilfield, China, where oily sludge from oil tanks and pipelines in block Gudao of Shengli oil field was treated. The sludge samples were collected in February 2009, and stored at 4°C before experiments commenced. The dehydrated crude oil sampled from Block L801 in Shengli oilfield, with an average density of 0.926 g.mL^{-1}, was autoclaved at 121°C for 30 min, and repeated three times within a week. Fresh medium without sulfate and nitrate was prepared according to a previous report using the Hungate anaerobic technique [32]. Aliquots of medium were distributed into glass vials sealed with isobutyl rubber stoppers (Bellco, USA) and aluminum caps under a gas atmosphere of 80% N_2 and 20% CO_2, in which resazurin (1 mg.L^{-1}) was added as a redox indicator. Vials were autoclaved at 121°C for 30 min. Solutions of sterile $Na_2S.9H_2O$ (0.03%), $NaHCO_3$ (0.25%),

vitamin solution (2 ml.L^{-1}), trace elements solution SL-7 (2 ml. L^{-1}), vitamin B_{12} (2 ml.L^{-1}) and vitamin B_1 (2 ml.L^{-1}) [32] were injected into fresh medium before inoculation and the pH was adjusted to 7.0–7.2.

Methanogenic enrichment incubation. Approximately 50 g of oily sludge was dispersed into 600 mL vials (Fuxin, China) amended with 300 mL of fresh medium and *ca.* 3.2 g of sterile crude oil. The vials were incubated statically in the dark at 35 and 55°C to obtain pre-enrichment cultures. Three sets of enrichments were carried out in 120 mL glass vials containing 50 mL of freshwater medium: (1) the experiment group: 7.5 mL of pre-enrichment culture and 1 g of sterile crude oil, (2) the abiotic control group: 7.5 mL of pre-enrichment cultures; (3) the crude oil-free control group: 1 g of sterile crude oil. Numerous parallel cultures were prepared in each set, and incubated without shaking at 35 and 55°C in the dark. Two to three vials in each set were sacrificed for DNA extraction and/or crude oil determination at different time points.

Chemical Analyses

Methane determination. The gas sample (0.2 mL) was sampled by a gas-tight syringe with a pressure lock (Vici, USA), and injected into a gas chromatograph equipped with a thermal conductivity detector (Shimadzu GC 2010, Kyoto, Japan) for methane determination [33]. The gas pressure in the culture vials was determined with a barometer (Aukis, Shanghai, China), and the amount of methane production was calculated based on Avogadro's law after calibration with a gas mixture of N_2 (29.96%), CH_4 (39.99%), and CO_2 (30.05%).

Crude Oil Analysis. Crude oil (30–50 mg) was loaded into a silica gel column with neutral aluminum oxide (100–200 mesh). The column was subsequently eluted with *n*-hexane, methylene chloride: *n*-hexane (2:1) and chloroform: ethanol (98:2), to collect saturated hydrocarbons, aromatic hydrocarbons and non-hydrocarbons successively. The residues remaining in the column after elution contained the asphaltene fraction. The saturated hydrocarbons and internal standard (d_{50}-*n*-tetracosane) were analyzed by a gas chromatograph (Agilent 7890A, Santa Clara, CA) using an HP-5 MS fused silica capillary column (60 m×250 μm×25 μm film thickness). The carrier gas was Helium (99.999%) at a flow rate of 1 ml.min^{-1}. The temperature program was run from 50°C (1 min isotherm) to 120°C at 20°C per min, and further increased to 310°C at 3°C per min with a hold at 310°C for 25 min. Mass spectral data were generated by a mass spectrometer (Agilent 5975i) at an energy of 70 eV in SCAN/SIM mode.

Microbial analysis

DNA Extraction and PCR Amplification. The liquid cultures (2–4 mL) were centrifuged for 5 min at 14000 rpm at 4°C, and the pellets were maintained at −80°C. Genomic DNA was extracted using a bead-beating method [34]. DNA fragments were purified with a Wizard DNA clean-up system (Promega, Madison, WI) and checked using 1% agarose gel electrophoresis. PCR amplifications of the archaeal and bacterial 16S rRNA gene fragments for terminal restriction fragment length polymorphism (T-RFLP) and 16S rRNA clone library were accomplished using primers A109f/A934r [35,36] and B27f/B907r [37], respectively. The PCR amplifications of archaeal and bacterial 16S rRNA genes were performed as previously described [38].

T-RFLP Analysis. PCR amplifications for T-RFLP analysis used the same mixtures and programs as described above, but the 5′ end of primers A934r and B27f were labeled with 6-carboxyfluorescein (FAM) [39]. The FAM-labeled PCR products were purified with TIANquick Midi Purification Kit (TIANGEN,

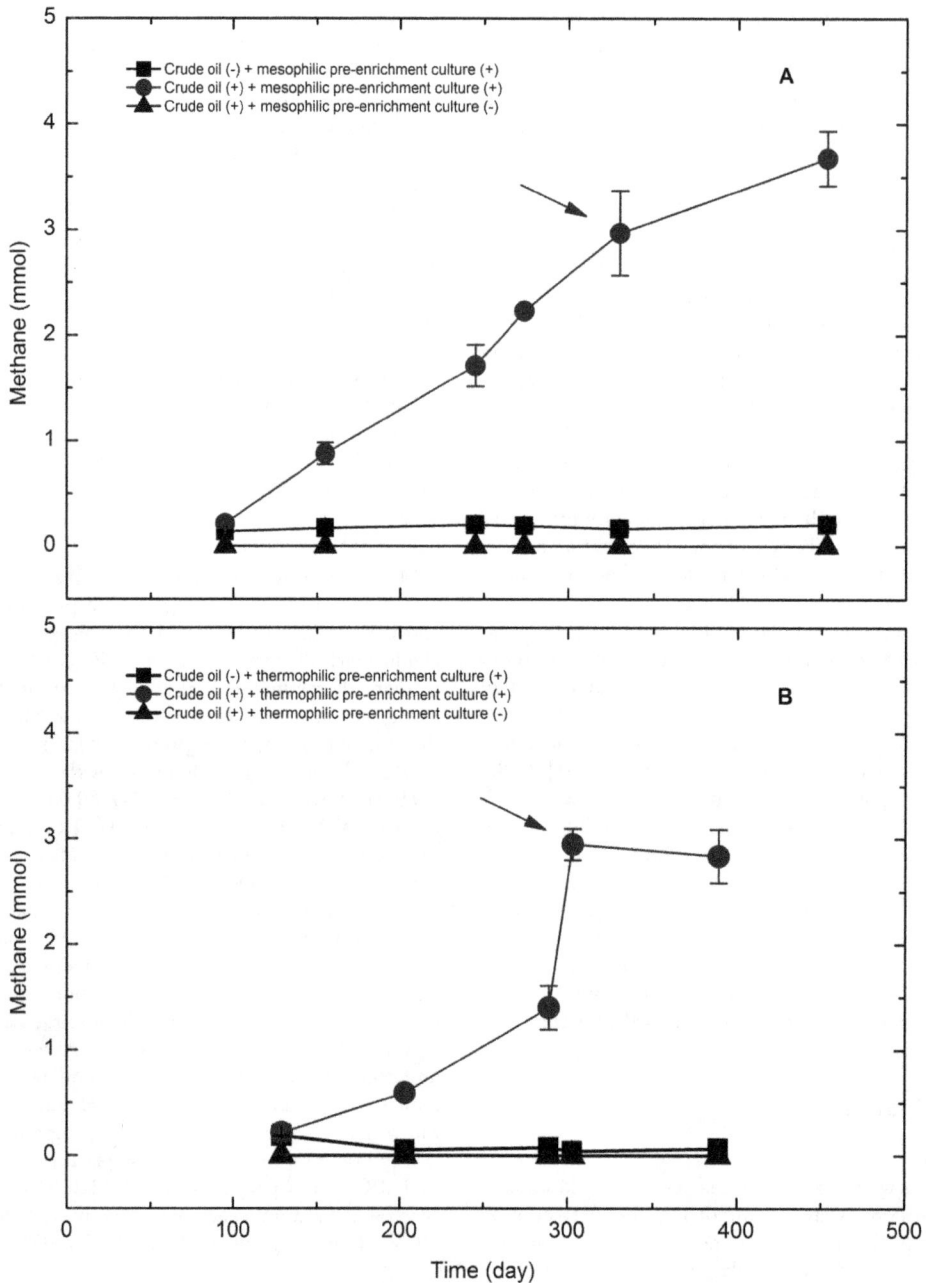

Figure 1. Time course of methane production at different temperatures. A: 35°C; B: 55°C. Arrows indicate sampling points for construction of clone libraries. ● and ▲: Cultures were grown in triplicate, error bars represent the standard deviation. ■: Cultures were grown in duplicate, error bars represent the standard deviation.

Beijing, China), then digested at 65°C for archaeal DNA using *Taq* I and 37°C for bacterial DNA using *Msp* I (TakaRa, Otsu, Japan) based on the manufacturer's instructions. The digestion products were further purified using the ethanol precipitation method [38]. Dried DNA samples were resuspended in 10 μL ddH₂O and a portion of each sample was mixed with deionized formamide containing 2% (v/v) internal standard ROX 30–1000 (Bioventure, Murfreesboro, TN). The mixtures were denatured at 95°C for 4 min and chilled on ice for 10 min. The DNA fragments were separated by capillary electrophoresis on a Genetic Analyzer 3130*xl* (Applied Biosystems, Foster City, CA). The relative terminal restriction fragment (T-RF) abundances of representative

phylotypes were analyzed with GeneMapper 4.0 (ABI). T-RFs with a peak height of less than 100 fluorescence units were excluded from analyses. Relative T-RFs abundance were determined as relative signal intensities of T-RFs with peak height analysis integration, and the relative abundance of T-RFs less than 2% were grouped together.

The experimental T-RFs were identified by comparison with *in silico* digested T-RFs using the clone library sequences generated from the same methanogenic consortia, or discriminated by T-RFLP analysis of 16S rRNA genes isolated from corresponding single colonies as templates. The relationship between methane production and bacterial T-RFs was assessed by redundancy

Table 1. Changes in group composition and biomarker ratios of crude oil during methanogenic degradation.

Incubation temperature	Time (d)	Group composition of crude oil (%)				Recovery efficiency (%)	Biomarker of crude oil
		Non-hydrocarbons	Saturated hydrocarbons	Aromatic hydrocarbons	Asphaltene		Pr/nC$_{17}$
	0	14.9	33.4	22.5	29.6	97.4	0.3
35°C	155	15.4±3.1	24.9±0.5	30.7±1.0	17.0±1.5	88.0	0.317±0.009
	245	18.7±1.3	23.1±0.5	30.8±2.8	16.9±2.3	89.6	0.390±0.046
	330	36.2±0.6	10.1±1.0	24.8±2.1	16.0±1.3	87.2	1.55[a]
	450	31.8±4.3	8.6±0.9	21.5±2.5	30.7±0.9	92.7	b
	Sterilized-330	28.9±3.0	25.2±1.2	22.5±2.0	12.4±1.9	89.0	0.347±0.005
55°C	203	18.0±3.4	23.1±0.7	30.0±3.4	17.4±1.6	88.6	0.301±0.008
	303	32.3±5.1	10.8±0.7	26.8±2.0	18.8±3.7	88.7	0.777±0.130
	Sterilized-303	31.3±10.6	21.6±2.8	21.8±2.8	14.5±4.9	89.2	0.3448 + 0.0008

Pr/nC$_{17}$: pristane/n-heptadecane; three replicates in each time points with exception of original day 0. a n-heptadecane was only detected in one of the three replicates; b: n-heptadecane was not detected above the detection limit; Sterilized-330: crude oil sampled from the sterilized control group after 330 days of incubation at 35°C, Sterilized-303: crude oil sampled from the sterilized control group after 303 days of incubation at 55°C.

analysis (RDA) using Canoco for Windows Version 4.5 [40]. T-RFs were assigned as species and methane production was the environmental variable. Significance of the factors was tested using Monte Carlo permutations (1499 permutations).

Construction of 16S rRNA Gene Clone Libraries. The PCR products were gel-purified with TIANgen Universal DNA Purification Kit (TIANGEN, China). The purified fragments were cloned into pMD 19-T vector (Takara, Otsu, Japan) and inserted into *E. coli* Competent Cells JM109 (Takara, Japan) according to the manufacturer's instructions. Single colonies were prepared and sequenced using a Genetic Analyzer 3730*xl* (ABI) as previously described [28].

Phylogenetic Analysis. The 16S rRNA gene sequences were checked with the "Chimera check with Bellerophon" program of the Greengene database [41]. The checked sequences were grouped into operational taxonomic units (OTUs) with a 97% threshold [42]. All sequences were submitted into the RDP database and classified into different taxonomic levels using RDP naïve Bayesian rRNA Classifier with an 80% confidence threshold [43], and representative sequences from each OTU were used to search for the most similar type strains using the Seqmatch program [44]. The diversity coverage of the 16S rRNA gene clone library was calculated using Good's formula as $C = [1-(n_1/N)] \times 100$, where n_1 is the number of unique OTUs and N is the total number of clones in the library [45]. The rarefaction curve was further generated using PAST version 2.00 [46]. Phylogenetic trees of the archaeal and bacterial 16S rRNA genes were created using the neighbor-joining method of Mega 5.1 [47]. Bootstrap values were calculated after 1,000 replications. Sequences were deposited in GenBank database with accession numbers JF946838–JF947178 and JN860886–JN860927.

Results

Methane Production

The pre-enrichment cultures accumulated methane consecutively at 35 and 55°C (Fig. S1 in File S1), indicating that the oily sludge microbes could grow at both temperatures. To further confirm methanogenic degradation of crude oil, the pre-enrichment cultures were subcultured and incubated at the same pre-enrichment temperature. The mesophilic culture (35°C), amended with crude oil, produced 3.7±0.3 mmol of methane with a maximum specific methane production rate of 0.01 mmol.d^{-1} after 453 days of incubation (Fig. 1A). The thermophilic consortium (55°C), amended with the same amount of crude oil, accumulated 2.8±0.3 mmol of methane after 389 days of incubation with maximum specific methane production rate of 0.02 mmol.d^{-1} (Fig. 1B). Less than 0.2 mmol methane was produced by the crude oil-free cultures at either temperature, and no methane was detected in the abiotic control.

Biodegradation of Petroleum Hydrocarbon

The continually accumulating methane generated in the crude oil-amended cultures relative to the oil-free controls indicated the biological degradation of crude oil via methanogenesis. The percentage of saturated hydrocarbons decreased from 33.4% to 8.6±0.9% at day 453 in the mesophilic consortium, and to 10.8±0.7% after 303 days of incubation in the thermophilic consortium (Table 1). From analysis of the saturated hydrocarbon profiles, we determined that *n*-alkanes were totally degraded after 453 days of incubation under mesophilic conditions (Fig. 2A). All of *n*-alkanes with chain length greater than 23 were totally degraded at 55°C after 303 days of incubation, and those less than 23 were incompletely degraded (Fig. 2B). The extent of degrada-

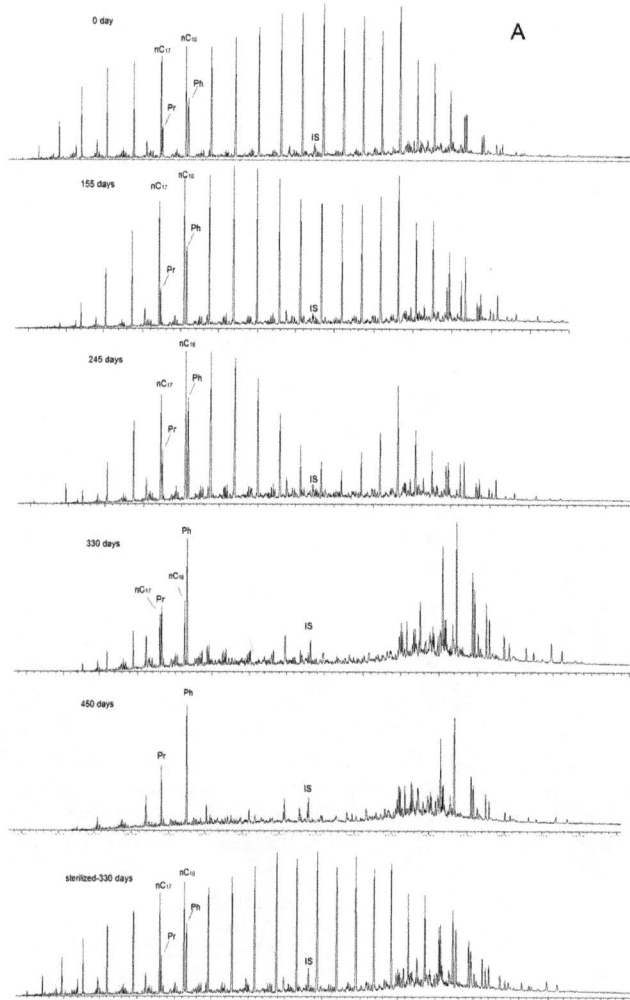

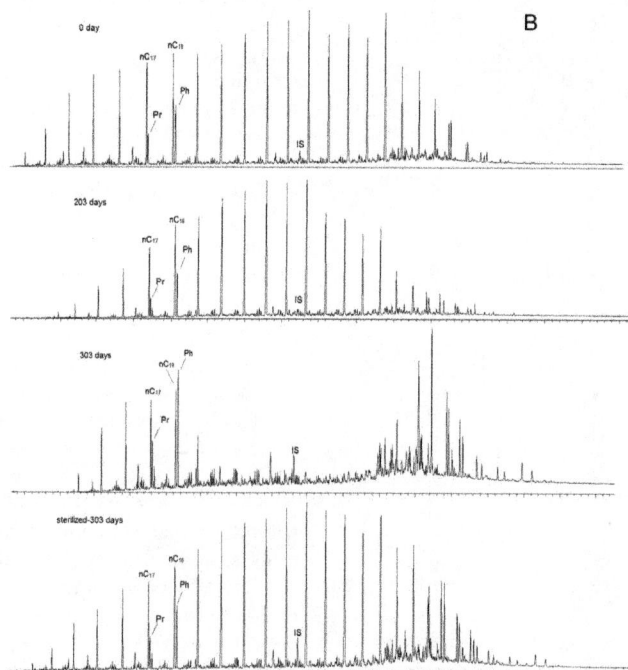

Figure 2. Gas chromatograms of the saturated hydrocarbon factions of crude oil. A: 35°C; B: 55°C. Pr: pristane; Ph: phytane; nC_{17}: *n*-heptadecane; nC_{18}: *n*-octadecane; IS: d_{50}-*n*-tetracosane. The crude oil samples were collected from the crude oil-degrading cultures at different time points, with the exception that "sterilized-330 days" crude oil was sampled from the second control after 330 days of incubation at 35°C, and "sterilized-303 days" crude oil was sampled from the second control after 303 days of incubation at 55°C.

tion was also assessed by the pristane/heptadecane (Pr/nC_{17}) ratio, which increased from 0.317 ± 0.009 at day 155 to 1.557 at day 330 under mesophilic conditions, and finally approached infinity with the complete degradation of *n*-heptadecane after 450 days of incubation. The value of the Pr/nC_{17} ratio increased to 0.706 ± 0.035 at day 303 in the thermophilic oil-degrading consortium. On the contrary, the ratio remained constant in the controls without microbial incubation at both temperatures (Table 1). In addition, the preferential degradation of longer-chain *n*-alkanes was observed at both temperatures (Fig. 2 and Table 1). GC analysis showed that the chromatograms of

saturated hydrocarbons shifted from a unimodal type to bimodal patterns after 245 days of incubation at 35°C, the bimodal pattern was well-established after 330 days of incubation (Fig. 2A).

Microbial community structure and dynamics

Archaeal domain. T-RFLP analysis of archaeal 16S rRNA genes revealed that the methanogenic crude oil-degrading cultures incubated at 35°C mainly consisted of T-RFs of 284-and 393-bp (78–94%). The community structure did not fluctuate much during 450 days of incubation, which was similar to the crude oil-free control (Fig. 3A). Under thermophilic conditions, the 228-

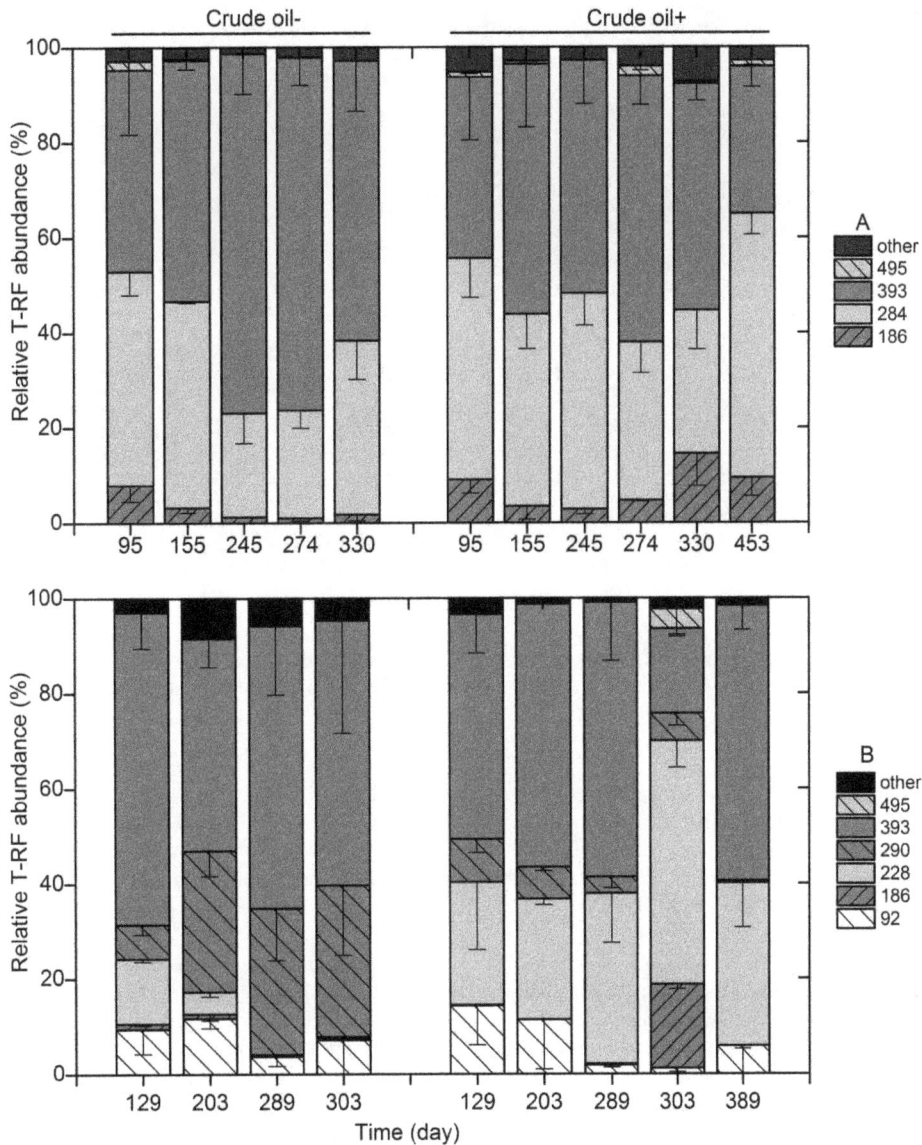

Figure 3. The archaeal T-RFLP profiles at 35°C (A) and 55°C (B). Crude oil -: consortium without crude oil addition; crude oil +: consortium with crude oil addition. Error bars represent the standard deviation, three replicates in crude oil-amended consortium and duplicates in crude oil-free consortium.

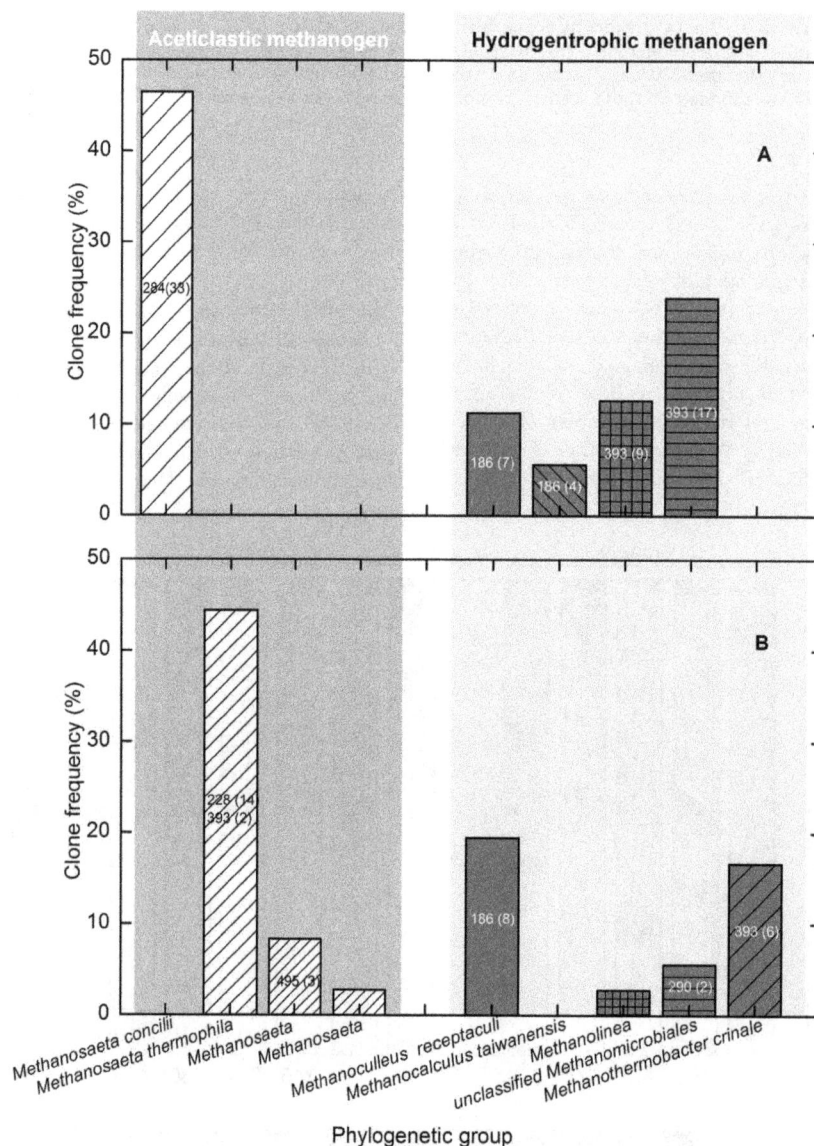

Figure 4. Phylogenetic affiliation of archaeal 16S rRNA gene sequences of the methanogenic crude oil-degrading consortia. A: 35°C; B: 55°C. The number in the column indicates the major OTUs represented by specific T-RFs, followed by the number in parentheses indicating the clone numbers of each OTU.

and 393-bp T-RFs dominated in the crude oil-degrading cultures, which accounted for 47 ± 16 and $35\pm21\%$ of the population, respectively. During 389 days of incubation; the 290-bp T-RF became the second most dominant phylotype ($24\pm19\%$) in the crude oil-free control after the most dominant T-RF of 393-bp ($56\pm29\%$) (Fig. 3B).

Two archaeal 16S rRNA gene clone libraries (L35A: total 71 clones retrieved from the mesophilic consortium; L55A: total 36 clones from the thermophilic consortium) were generated from the methanogenic crude oil-degrading consortia at 35 and 55°C, respectively, as shown in Table S1 in File S1. The analyses of rarefaction curves and Good coverage indexes revealed that saturation was reached (Fig. S2 and Table S1 in File S1). The predominant 284-bp T-RF retrieved from the mesophilic consortium was related to *Methanosaeta concilii* (>99% sequence similarity). The 186- and 393-bp T-RFs were related to the hydrogenotrophic *Methanomicrobiales* (Fig. 4A, Fig. S3 and Table

S1 in File S1). Similarly, the 228- and 495-bp T-RFs in the thermophilic consortium were mainly related to *Methanosaeta thermophila* (91–99% sequence similarity). The 393- and 186-bp T-RFs were mainly related to H$_2$-using *Methanothermobacter* and *Methanoculleus*, respectively (Fig. 4B, Fig. S3 and Table S1 in File S1).

Bacterial domain. The bacterial community structure was more diverse than the archaeal community structure, with 10–11 T-RFs routinely observed (Fig. 5). In the mesophilic cultures, The 161- and 164-bp T-RFs increased in relative abundance over time in the crude oil-amended cultures, compared with the crude oil-free control (Fig. 5A). Redundancy analysis revealed that the 154-, 161-, 164-, 207- and 218-bp T-RFs were positively associated with methane production in the mesophilic consortium (Fig. 6A). Redundancy analysis also revealed that the 63-, 193-, 281-, 290-, 484- and 568-bp T-RFs were highly correlated with methane production in the thermophilic consortium (Fig. 6B).

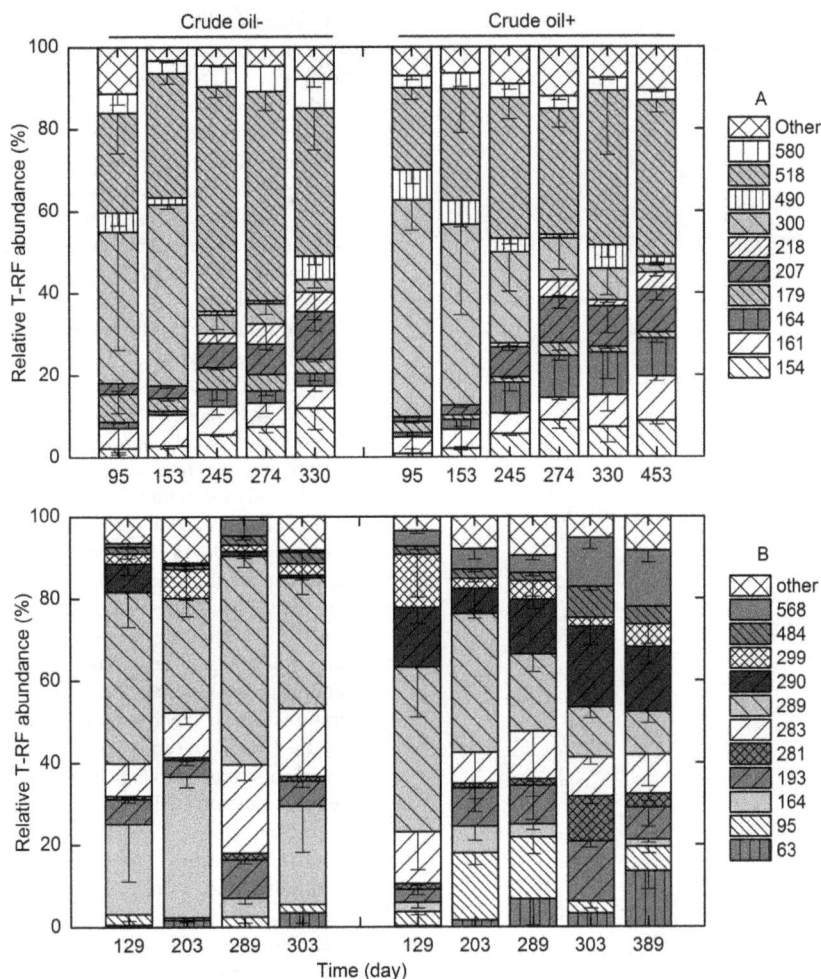

Figure 5. Bacterial T-RFLP profiles at 35°C (A) and 55°C (B). Crude oil -: consortium without crude oil addition; Crude oil +: crude oil amended consortium. Error bars represent the standard deviation, three replicates in crude oil-amended consortium and duplicates in crude oil-free consortium.

Two bacterial clone libraries (L35B: 156 clones from the mesophilic consortium; L55B: 119 clones from the thermophilic consortium) were also constructed from the same time points as for archaea. The rarefaction curves tended towards the saturation plateau (Fig. S2 in File S1), and coverage analysis suggested that 78.8 and 89.1% of predicted phylotypes were sampled from mesophilic and thermophilic consortia, respectively (Table S2 in File S1). The mesophilic bacterial community was mainly composed of unclassified *Bacteria* (39.1%), *Proteobacteria* (19.9%), *Chloroflexi* (18.6%) and *Firmicutes* (10.9%) (Fig. 7A, Fig. S4 and Table S2 in File S1). The bacterial community at 55°C was significantly different from that at 35°C, which mostly consisted of unclassified *Bacteria* (34.5%), *Caldisericales* (21.8%), *Firmicutes* (16.8%), *Synergistetes* (14.3%) and *Bacteroidetes* (7.5%) (Fig. 7B, Fig. S4 and Table S3 in File S1).

The combined analysis of T-RFLP profiles and clone library data revealed that the 161-bp T-RF presumably represented *in silico* T-RFs of 160, 161 and 162 bp as single clones, which could not be discriminated through capillary electrophoresis. The 160-bp T-RF represented uncultured bacteria with 86% sequence similarity to *Syntrophomonas zehnderi* [48], and the 162-bp T-RF represented uncultured members with 94–95% sequence similarity to syntrophic phenol degrading bacterium *Syntrophorhabdus*

aromaticivorans (Fig. S4 and Table S2 in File S1) [49]. The 161-bp T-RF represented members of uncultured candidate division WWE1 bacteria, with 94 – 100% sequence similarity to Candidatus *Cloacamonas acidaminovorans* (Fig. S4 and Table S2 in File S1) [50]. The 154- and 207-bp T-RFs represented uncultured candidate division JS1 bacterium. The 164-bp T-RF was most closely related to uncultured candidate division JS1, *Syntrophaceae* or *Desulfovibrionaceae* (Fig. S4 and Table S2 in File S1). In addition, the dominant 518-bp T-RF mainly represented uncultured members of *Chloroflexi* (Fig. S4 and Table S2 in File S1). The 300-bp T-RF representing *Soehngenia saccharolytica*-related microorganisms dominated at the early stage of incubation but became less abundant during the last 200 days of incubation (Fig. 5A and Table S2 in File S1). The 185-bp T-RF, the third most dominant OTU, was distantly related to *Smithella propionica* (92% sequence similarity) [51], and the 166-bp T-RF, the fifth most dominant OTU, shared 94% sequence similarity with *Desulfatibacillum alkenivorans* (Fig. S4 and Table S2 in File S1) [52], but these two fragments were not detected in the T-RFLP profile at high abundance (>2%).

In the thermophilic cultures, the 63-bp T-RF represented members of *Firmicutes*, which exhibited 91% sequence similarity to the thermophilic acetogen *Moorella thermoacetica* (Fig. S4 and

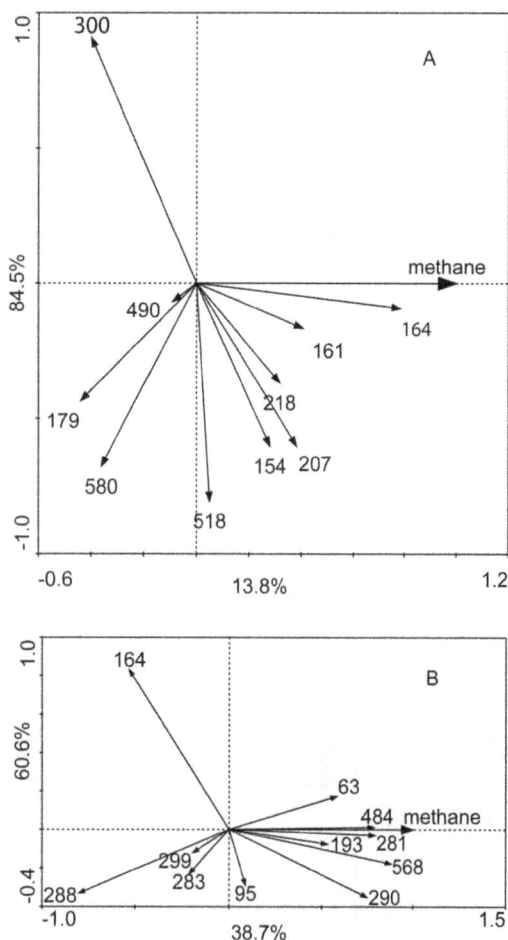

Figure 6. Redundancy analysis of bacterial TRFLP profiles versus methane production in methanogenic oil-degrading consortia at 35°C (A) and 55°C (B). Bold arrow denotes the explanatory variable of methane production. Values on the axes indicate the percentages of total variation explained by each axis.

from the Shengli oilfield possessed the ability to degrade crude oil n-alkanes and generate methane at both mesophilic (35°C) and thermophilic (55°C) temperatures. The mesophilic consortium, amended with the same amount of crude oil, could produce more methane than that at 55°C, but with a slower methane production rate. These results indicate that the oily sludge microcosms could degrade crude oil via methanogenesis over a wide temperature range. The difference in methane production between the mesophilic and thermophilic cultures may be attributed to the degree of crude oil degradation, and is probably correlated with the loss of n-alkanes. All n-alkanes completely disappeared at 35°C, while n-alkanes with chain lengths from C_{11} to C_{23} were incompletely degraded at 55°C. Crude oil is a complex mixture consisting of various n-alkanes, of which shorter chain n-alkanes are generally degraded faster than longer ones in previous reports [2,5]. Siddique et al. [15,16] reported that methanogenic microcosms enriched from the same mature fine tailings possessed different degradation patterns for short and long chain n-alkanes. The preferential degradation of mid- and high-range alkanes has also been reported [17–19], which differed from the expected pattern under aerobic conditions. Surprisingly, a relative enrichment of shorter n-alkanes was also detected [18,19], which was questioned by Galperin *et al.* [60] because of the unusual progression of n-alkane degradation. In this study, the preferential degradation of longer-chain n-alkanes was also observed at both temperatures tested, which is similar to results obtained in previous studies [16–19]. This result confirms the general progression of crude oil degradation under methanogenic conditions, however, a relative accumulation of shorter-chain n-alkanes was not observed.

The methanogenic degradation of hydrocarbons requires syntrophic cooperation of bacterial and archaeal communities, and thermodynamics analysis revealed five possible pathways for the conversion of hydrocarbons into methane [24]. The concurrence of acetoclastic and hydrogenotrophic methanogens, but with different methanogenic community structures, was observed at both temperatures in this study, which is similar to our previously reported consortia using hexadecane as substrate [38]. The dominant archaeal community during consecutive transfer and incubation with hexadecane shifted from acetoclastic *Methanosaeta* to hydrogenotrophic *Methanoculleus* [38]. *Methanoculleus* spp. were further reported as the dominant methane producers through DNA-SIP [27]. Methane production from crude oil was proposed to be generated through syntrophic acetate oxidation coupled to hydrogenotrophic *Methanothermobacter* under thermophilic conditions [11]. Similarly, the hydrogenotrophic *Methanothermobacter* has also been revealed as the dominant archaeal population in the hexadecane degrading consortium [28,33]. *Methanosaeta* spp., represented by the 228 bp T-RF, dominated in the thermophilic crude oil-degrading consortium, which suggests their potential role in methane production through acetate fermentation. Interestingly, it has been reported that archaeal populations in crude oil enrichment cultures were mainly composed of non-methanogenic archaea [29,61]. These results suggest that further research is needed for characterization of the relative contribution of archaea to methane production during crude oil degradation.

The complex components of crude oil provide a vast range of substrates for development of a complex microbial community. Knowledge about the key players initiating methanogenic hydrocarbon degradation is limited, as bacteria that degrade and grow on hydrocarbons via methanogenesis have to deal with the unfavorable energetics of the conversion processes [8,24,62]. The uncultured JS1 lineage and members of the Chloroflexi have been identified as the major clades of bacteria in the mesophilic

Table S3 in File S1) [53]. The 193- and 281-bp T-RFs represented members of *Synergistetes* (two taxa), which shared 88 - 98% sequence similarity with *Anaerobaculum thermoterrenum* [54]. The second most dominant OTU (22 clones) represented by the 290-bp T-RF, shared 81% sequence similarity with *Thermotoga maritima* [55], which could be clustered into "Shengli cluster" [28]. The other OTU (5 clones) represented by the 290-bp T-RF was closely related to protein-degrading bacterium *Coprothermobacter proteolyticus* (99% sequence similarity) [56,57]. The 484-bp T-RF represented unclassified bacteria, which shared 83.8% sequence similarity with *Carboxydothermus hydrogenoformans* (Fig. S4 and Table S3 in File S1) [58]. The 568-bp T-RF represented the dominant OTU (84% sequence similarity to *Caldisericum exile*) (Fig. S4 and Table S3 in File S1) [59].

Discussion

Research studies on the microbial oxidation of crude oil coupled to methane production have been well documented. The methane yield varied from 0.6 to 10.0 mmol per gram of oil from various oilfields [6,9,11–13], However, research on the effect of temperature on the methane potential and kinetics of crude oil degradation is scarce. In this study, oily sludge microbes collected

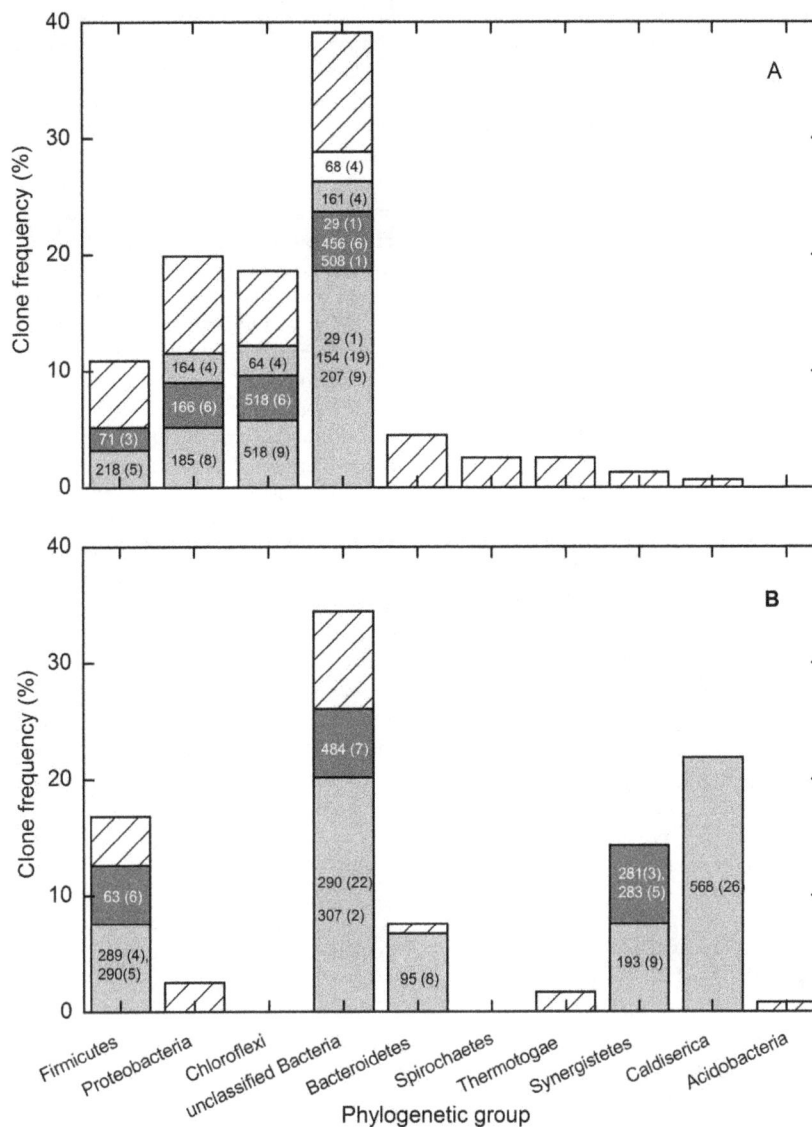

Figure 7. Phylogenetic affiliation of bacterial 16S rRNA gene sequences of the methanogenic crude oil-degrading consortia. A: 35°C; B: 55°C. The number in the column indicates the major OTUs represented by specific T-RFs, followed by the number in parentheses indicating the clone numbers of each OTU.

consortium, which have been widely detected in oil-impacted environments and proposed to play a role in hydrocarbon degradation [63,64]. Direct evidence for methanogenic degradation of *n*-alkanes by these two phylotypes has not been reported. The third major bacterial phylotype was related to *Proteobacteria* and was mainly composed of members of *Desulfobacteraceae*, *Syntrophaceae* and *Syntrophorhabdus*. The *Desulfobacteraceae*-affiliated member was most closely related to *D. alkenivorans*, which could oxidize alkanes with sulfate as an electron acceptor, or coupled to methane production [25,65]. The dominant *Syntrophaceae*-related OTU (type clone L35B_12) in this consortium shared 92% sequence similarity to syntrophic propionate oxidation bacterium *S. propionica* [51,66], and may represent an uncultured alkane degrader species. The *Syntrophaceae*-affiliated members have been identified as key players associated with alkane degradation under methanogenic conditions using culture-independent methods [26,27]. The *assA* related gene observed in the other *Syntrophaceae*-affiliated members may indicate fumarate

addition to alkanes [67,68]. Members of this family have also been detected in a number of methanogenic alkane-degrading cultures [7,12,13,15,26,27,31,69,70], suggesting their ecophysiological role in the anaerobic alkane degradation process. *S. aromaticivorans* species could use phenol, p-cresol, isophthalate, benzoate, and 4-hydroxybenzoate in syntrophic association with a hydrogeno-trophic methanogen with an optimum growth temperature range of 35–37°C [49], which indicates that the *Syntrophorhabdus*-related members may use aromatic hydrocarbons in crude oil to grow, and enhance methane production under mesophilic conditions. Further characterization of aromatic degradation by this consortium is beyond the scope of the study.

The thermophilic bacterial community structure is different from that at 35°C, which was mainly grouped into *Caldisericales* and "Shengli cluster" [28]. The *Caldisericales*-related members (type clone L55B_6) shared only 84% sequence similarity with *C. exile*, which grows anaerobically with yeast extract and sulfur compounds with thiosulfate, sulfite and elemental sulfur as electron acceptors [59]. The relatively few environmental clones related to *C. exile* were detected in hot spring sediment (GenBank *No.*: FJ638586), an anaerobic bioreactor (GenBank *No.*: EF515667) and terephthalate-degrading sludge [71]. The other dominant phylotype (type clone: L55B_1) exhibited 81% sequence similarity with extreme thermophile *T. maritima* [55], and shared 95% sequence similarity with representative clone BSK_100 of "Shengli Cluster" retrieved from a thermophilic hexadecane-degrading consortium, which was proposed to be involved in thermophilic hexadecane degradation [28]. The role of other observed less dominant bacterial phylotypes in both consortia is not clear. However, their persistent presence suggests that they are probably associated with the process of hydrocarbon degradation, possibly via multiple syntrophic interactions [72].

In summary, this present study shows that oily sludge microbes from Shengli oilfield possess the ability to degrade crude oil *n*-alkanes coupled to methane production at 35 and 55°C. The preferential degradation of long chain alkanes by oily sludge microbes was observed at both temperatures. Both acetoclastic and hydrogenotrophic methanogens are likely to play a role in methane production during oil degradation under mesophilic and thermophilic conditions. The dominant bacterial community structure is different between the two consortia, suggesting the adaptation of oily sludge microbes to the shift in temperature by selective growth of specific groups of microbes. [62] [14] [60].

Supporting Information

File S1 Figure S1, Time course of methane production of the pre-enrichment cultures incubated at 35°C (A) and 55°C (B). The square and rhombus represent the pre-enrichment cultures incubated at 35°C, The triangles represent the pre-enrichment cultures incubated at 55°C. The arrows indicates the sampling points for transfer incubation of the pre-enrichment cultures. **Figure S2,** Rarefaction curves constructed from archaeal and bacterial 16S rRNA gene libraries based on OTU cutoff of equal or higher 97%. **Figure S3,** Phylogenetic tree based on archaeal 16S rRNA gene sequences from representative clones of each OTUs, related type strains and environmental clones using neighbor-joining analysis of 779-nt alignment. Representative clones from the mesophilic consortium are indicated in red, followed by *in silico* T-RFs and clone numbers, and the representative clones from the thermophilic consortium are indicated in blue, followed by *in silico* T-RFs and clone numbers. **Figure S4,** Phylogenetic tree based on bacterial 16S rRNA gene sequences from representative clones of each OTUs and related strains and environmental clones using neighbor-joining analysis of 698-nt alignment. Representative clones from the mesophilic consortium are indicated in red, followed by *in silico* T-RFs and clone numbers, and the representative clones from the thermophilic consortium are indicated in blue, followed by *in silico* T-RFs and clone numbers, the OTUs less than 3 clones in the mesophilic consortium and 2 in the thermophilic consortium were not shown in the phylogenetic tree. *Methanothermobacter crinale* (EF554596) was used as outgroup. The scale bar represents 2% sequence divergence. **Table S1,** Phylogenetic affiliation of archaeal 16S rRNA genes and corresponding theoretical T-RFs retrieved from methanogenic oil-degrading consortia at 35 and 55°C, respectively. **Table S2,** Phylogenetic affiliation of bacterial 16S rRNA genes and corresponding theoretical T-RFs retrieved from mesophilic methanogenic oil-degrading consortium. **Table S3,** Phylogenetic affiliation of bacterial 16S rRNA gene and corresponding theoretical T-RFs retrieved from thermophilic methanogenic oil-degrading consortium.

(DOCX)

Author Contributions

Conceived and designed the experiments: LC HZ YHL. Performed the experiments: LC SBS QL JFC. Analyzed the data: LC SBS HZ YHL. Contributed reagents/materials/analysis tools: LC SBS YHL. Wrote the paper: LC HZ YHL.

References

1. Head IM, Jones DM, Larter SR (2003) Biological activity in the deep subsurface and the origin of heavy oil. Nature 426: 344–352.
2. Peters KE, Moldowan JM (1993) The biomarker guide: interpreting molecular fossils in petroleum and ancient sediments: Prentice Hall.
3. Zedelius J, Rabus R, Grundmann O, Werner J, Brodkorb D, et al. (2011) Alkane degradation under anoxic conditions by a nitrate-reducing bacterium with possible involvement of the electron acceptor in substrate activation. Env Microbiol Rep 3: 125–135.
4. Kunapuli U, Lueders T, Meckenstock RU (2007) The use of stable isotope probing to identify key iron-reducing microorganisms involved in anaerobic benzene degradation. ISME J 1: 643–653.
5. Rueter P, Rabus R, Wilkes H, Aeckersberg F, Rainey AF, et al. (1994) Anaerobic oxidation of hydrocarbons in crude oil by new types of sulphate-reducing bacteria. Nature 372: 455–458.
6. Townsend GT, Prince RC, Suflita JM (2003) Anaerobic oxidation of crude oil hydrocarbons by the resident microorganisms of a contaminated anoxic aquifer. Environ Sci Technol 37: 5213–5218.
7. Zengler K, Richnow HH, Rossello-Mora R, Michaelis W, Widdel F (1999) Methane formation from long-chain alkanes by anaerobic microorganisms. Nature 401: 266–269.
8. Gieg LM, Fowler SJ, Berdugo-Clavijo C (2014) Syntrophic biodegradation of hydrocarbon contaminants. Current Opinion in Biotechnology 27: 21–29.
9. Berdugo-Clavijo C, Gieg L (2014) Conversion of Crude Oil to Methane by a Microbial Consortium Enriched From Oil Reservoir Production Waters (10.3389/fmicb.2014.00197). Frontiers in Microbiology 5.
10. Caldwell ME, Garrett RM, Prince RC, Suflita JM (1998) Anaerobic biodegradation of long-chain n-alkanes under sulfate-reducing conditions. Environmental Science & Technology 32: 2191–2195.
11. Gieg LM, Davidova IA, Duncan KE, Suflita JM (2010) Methanogenesis, sulfate reduction and crude oil biodegradation in hot Alaskan oilfields. Environmental Microbiology 12: 3074–3086.
12. Gieg LM, Duncan KE, Suflita JM (2008) Bioenergy production via microbial conversion of residual oil to natural gas. Appl Environ Microbiol 74: 3022–3029.
13. Jones DM, Head IM, Gray ND, Adams JJ, Rowan AK, et al. (2008) Crude-oil biodegradation via methanogenesis in subsurface petroleum reservoirs. Nature 451: 176–180.
14. Rabus R, Wilkes H, Schramm A, Harms G, Behrends A, et al. (1999) Anaerobic utilization of alkylbenzenes and n-alkanes from crude oil in an enrichment culture of denitrifying bacteria affiliating with the β-subclass of *Proteobacteria*. Environmental Microbiology 1: 145–157.
15. Siddique T, Penner T, Semple K, Foght JM (2011) Anaerobic biodegradation of longer-chain *n*-alkanes coupled to methane production in oil sands tailings. Environmental Science & Technology 45: 5892–5899.
16. Siddique T, Fedorak PM, Foght JM (2006) Biodegradation of short-chain *n*-alkanes in oil sands tailings under methanogenic conditions. Environmental Science & Technology 40: 5459–5464.
17. Hasinger M, Scherr KE, Lundaa T, Bräuer L, Zach C, et al. (2011) Changes in iso- and n-alkane distribution during biodegradation of crude oil under nitrate and sulphate reducing conditions. Journal of Biotechnology 157: 490–498.
18. Bekins BA, Hostettler FD, Herkelrath WN, Delin GN, Warren E, et al. (2005) Progression of methanogenic degradation of crude oil in the subsurface Environmental Geosciences 12: 139–152.
19. Hostettler FD, Wang Y, Huang Y, Cao W, Bekins BA, et al. (2007) Forensic fingerprinting of oil-spill hydrocarbons in a methanogenic environment–Mandan, ND and Bemidji, MN. Environ Forensics 8: 139–153.
20. Meckenstock RU, Mouttaki H (2011) Anaerobic degradation of non-substituted aromatic hydrocarbons. Current Opinion in Biotechnology 22: 406–414.
21. Callaghan AV (2013) Enzymes involved in the anaerobic oxidation of n-alkanes: from methane to long-chain paraffins. Frontiers in Microbiology 4.
22. Callaghan AV (2012) Metabolomic investigations of anaerobic hydrocarbon-impacted environments. Current Opinion in Biotechnology.
23. Aitken CM, Jones DM, Maguire MJ, Gray ND, Sherry A, et al. (2013) Evidence that crude oil alkane activation proceeds by different mechanisms under sulfate-reducing and methanogenic conditions. Geochimica et Cosmochimica Acta 109: 162–174.
24. Dolfing J, Larter SR, Head IM (2008) Thermodynamic constraints on methanogenic crude oil biodegradation. ISME J 2: 442–452.

25. Callaghan AV, Morris BEL, Pereira IAC, McInerney MJ, Austin RN, et al. (2012) The genome sequence of *Desulfatibacillum alkenivorans* AK-01: a blueprint for anaerobic alkane oxidation. Environmental Microbiology 14: 101–113.

26. Gray ND, Sherry A, Grant RJ, Rowan AK, Hubert CRJ, et al. (2011) The quantitative significance of *Syntrophaceae* and syntrophic partnerships in methanogenic degradation of crude oil alkanes. Environmental Microbiology 13: 2957–2975.

27. Cheng L, Ding C, Li Q, He Q, Dai L-r, et al. (2013) DNA-SIP reveals that *Syntrophaceae* play an important role in methanogenic hexadecane degradation. PLoS ONE 8: e66784.

28. Cheng L, He Q, Ding C, Dai L-r, Li Q, et al. (2013) Novel bacterial groups dominate in a thermophilic methanogenic hexadecane-degrading consortium FEMS Microbiology Ecology 85: 568–577.

29. Mbadinga S, Li K-P, Zhou L, Wang L-Y, Yang S-Z, et al. (2012) Analysis of alkane-dependent methanogenic community derived from production water of a high-temperature petroleum reservoir. Applied Microbiology and Biotechnology 96: 531–542.

30. Atlas RM (1975) Effects of Temperature and Crude Oil Composition on Petroleum Biodegradation. Appl Environ Microbiol 30: 396–403.

31. Wang L-Y, Gao C-X, Mbadinga SM, Zhou L, Liu J-F, et al. (2011) Characterization of an alkane-degrading methanogenic enrichment culture from production water of an oil reservoir after 274 days of incubation. International Biodeterioration & Biodegradation 65: 444–450.

32. Widdel F, Boetius A, Rabus R (2006) Anaerobic biodegradation of hydrocarbons including methane. In: Dworkin M, Falkow S, Rosenberg E, Schleifer K-H, Stackebrandt E, editors. The Prokaryotes: Ecophysiology and Biochemistry. New York: Springer. pp.1028–1049.

33. Cheng L, Dai L, Li X, Zhang H, Lu Y (2011) Isolation and characterization of *Methanothermobacter crinale* sp. nov, a novel hydrogenotrophic methanogen from Shengli Oilfields. Applied and Environmental Microbiology 77: 5212–5219.

34. Peng J, Lü Z, Rui J, Lu Y (2008) Dynamics of the methanogenic archaeal community during plant residue decomposition in an anoxic rice field soil. Appl Environ Microbiol 74: 2894–2901.

35. Grosskopf R, Janssen PH, Liesack W (1998) Diversity and structure of the methanogenic community in anoxic rice paddy soil microcosms as examined by cultivation and direct 16S rRNA gene sequence retrieval. Applied and Environmental Microbiology 64: 960–969.

36. Stahl DA, Amann R (1991) Nucleic acid techniques in bacterial systematics. In: Stackebrandt E, Goodfellow M, Development and application of nucleic acid probes.editors. New York: John Wiley & Son Ltd. pp. 205–248.

37. Lane D (1991) 16S/23S rRNA Sequencing. In: Stackebrandt E, Goodfellow M, editors. Development and application of nucleic acid probes. New York: John Wiley & Son Ltd. pp. 115–175.

38. Cheng L, Rui J, Li Q, Zhang H, Lu Y (2013) Enrichment and dynamics of novel syntrophs in a methanogenic hexadecane-degrading culture from a Chinese oilfield. FEMS Microbiology Ecology 83: 756–766.

39. Chin K-J, Lukow T, Conrad R (1999) Effect of temperature on structure and function of the methanogenic archaeal community in an anoxic rice field soil. Applied and Environmental Microbiology 65: 2341–2349.

40. ter Braak CJF, Šmilauer P (2002) Canoco reference manual and canodraw for windows user's guide: software for canonical community ordination (version 4.5). Ithaca.

41. DeSantis TZ, Hugenholtz P, Larsen N, Rojas M, Brodie EL, et al. (2006) Greengenes, a chimera-checked 16S rRNA gene database and workbench compatible with ARB. Appl Environ Microbiol 72: 5069–5072.

42. Schloss PD, Westcott SL, Ryabin T, Hall JR, Hartmann M, et al. (2009) Introducing mothur: open-source, platform-independent, community-supported software for describing and comparing microbial communities. Appl Environ Microbiol 75: 7537–7541.

43. Wang Q, Garrity GM, Tiedje JM, Cole JR (2007) Naïve bayesian classifier for rapid assignment of rRNA sequences into the new bacterial taxonomy. Applied and Environmental Microbiology 73: 5261–5267.

44. Cole JR, Wang Q, Cardenas E, Fish J, Chai B, et al. (2009) The ribosomal database project: improved alignments and new tools for rRNA analysis. Nucleic Acids Research 37: 141–145.

45. Good I (1953) The population frequencies of species and the estimation of population parameters. Biometrika 40: 237–264.

46. Hammer Ø, Harper D, Ryan P (2001) PAST: Paleontological Statistics Software Package for education and data analysis. Palaeontologia Electronica 4: 9.

47. Tamura K, Peterson D, Peterson N, Stecher G, Nei M, et al. (2011) MEGA5: molecular evolutionary genetics analysis using maximum likelihood, evolutionary distance, and maximum parsimony methods. Molecular Biology and Evolution 28: 2731–2739.

48. Sousa DZ, Smidt H, Alves MM, Stams AJM (2007) *Syntrophomonas zehnderi* sp. nov., an anaerobe that degrades long-chain fatty acids in co-culture with *Methanobacterium formicicum*. International Journal of Systematic and Evolutionary Microbiology 57: 609–615.

49. Qiu Y-L, Hanada S, Ohashi A, Harada H, Kamagata Y, et al. (2008) *Syntrophorhabdus aromaticivorans* gen. nov., sp. nov., the first cultured anaerobe capable of degrading phenol into acetate in obligate syntrophic associations with a hydrogenotrophic methanogen. Appl Environ Microbiol 74: 2051–2058.

50. Pelletier E, Kreimeyer A, Bocs S, Rouy Z, Gyapay G, et al. (2008) "Candidatus *Cloacamonas acidaminovorans*": genome sequence reconstruction provides a first glimpse of a new bacterial division. Journal of Bacteriology 190: 2572–2579.

51. Liu Y, Balkwill DL, Aldrich HC, Drake GR, Boone DR (1999) Characterization of the anaerobic propionate-degrading syntrophs *Smithella propionica* gen. nov., sp. nov. and *Syntrophobacter wolinii*. Int J Syst Bacteriol 49: 545–556.

52. Cravo-Laureau C, Matheron R, Joulian C, Cayol J-L, Hirschler-Rea A (2004) *Desulfatibacillum alkenivorans* sp. nov., a novel n-alkene-degrading, sulfate-reducing bacterium, and emended description of the genus *Desulfatibacillum*. Int J Syst Evol Microbiol 54: 1639–1642.

53. Drake HL, Daniel SL (2004) Physiology of the thermophilic acetogen *Moorella thermoacetica*. Research in Microbiology 155: 422–436.

54. Rees GN, Patel BKC, Grassia GS, Sheehy AJ (1997) *Anaerobaculum thermoterrenum* gen. nov., sp. nov., a Novel, Thermophilic Bacterium Which Ferments Citrate. International Journal of Systematic Bacteriology 47: 150–154.

55. Huber R, Langworthy TA, König H, Thomm M, Woese CR, et al. (1986) *Thermotoga maritima* sp. nov. represents a new genus of unique extremely thermophilic eubacteria growing up to 90°C. Archives of Microbiology 144: 324–333.

56. Rainey FA, Stackebrandt E (1993) Transfer of the Type Species of the Genus *Thermobacteroides* to the Genus *Thermoanaerobacter* as *Thermoanaerobacter acetoethylicus*(Ben-Bassat and Zeikus 1981) comb. nov., Description of *Coprothermobacter* gen. nov., and Reclassification of *Thermobacteroides proteolyticus* as *Coprothermobacter proteolyticus*(Ollivier et al. 1985) comb. nov. International Journal of Systematic Bacteriology 43: 857–859.

57. Sasaki K, Morita M, Sasaki D, Nagaoka J, Matsumoto N, et al. (2011) Syntrophic degradation of proteinaceous materials by the thermophilic strains *Coprothermobacter proteolyticus* and *Methanothermobacter thermautotrophicus*. Journal of Bioscience and Bioengineering In Press, Corrected Proof.

58. Svetlichny VA, Sokolova TG, Gerhardt M, Ringpfeil M, Kostrikina NA, et al. (1991) *Carboxydothermus hydrogenoformans* gen. nov., sp. nov., a CO-utilizing thermophilic anaerobic bacterium from hydrothermal environments of Kunashir Island. Systematic and Applied Microbiology 14: 254–260.

59. Mori K, Yamaguchi K, Sakiyama Y, Urabe T, Suzuki K-i (2009) *Caldisericum exile* gen. nov., sp. nov., an anaerobic, thermophilic, filamentous bacterium of a novel bacterial phylum, *Caldiserica* phyl. nov., originally called the candidate phylum OP5, and description of *Caldisericaceae* fam. nov., *Caldisericales* ord. nov. and *Caldisericia* classis nov. International Journal of Systematic and Evolutionary Microbiology 59: 2894–2898.

60. Galperin Y, Kaplan IR (2008) Comments on the reported unusual progression of petroleum hydrocarbon distribution patterns during environmental weathering. Environ Forensics 9: 117–120.

61. Zhou L, Li K-P, Mbadinga S, Yang S-Z, Gu J-D, et al. (2012) Analyses of n-alkanes degrading community dynamics of a high-temperature methanogenic consortium enriched from production water of a petroleum reservoir by a combination of molecular techniques. Ecotoxicology 21: 1680–1691.

62. Widdel F, Knittel K, Galushko A (2010) Anaerobic hydrocarbon-degrading microorganisms: an overview. In: Timmis KN, editor. Handbook of Hydrocarbon and Lipid Microbiology: Springer Berlin Heidelberg. pp.1997–2021.

63. Pham VD, Hnatow LL, Zhang S, Fallon RD, Jackson SC, et al. (2009) Characterizing microbial diversity in production water from an Alaskan mesothermic petroleum reservoir with two independent molecular methods. Environmental Microbiology 11: 176–187.

64. Ficker M, Krastel K, Orlicky S, Edwards E (1999) Molecular characterization of a toluene-degrading methanogenic consortium. Appl Environ Microbiol 65: 5576–5585.

65. Cravo-Laureau C, Matheron R, Cayol J-L, Joulian C, Hirschler-Rea A (2004) *Desulfatibacillum aliphaticivorans* gen. nov., sp. nov., an n-alkane- and n-alkene-degrading, sulfate-reducing bacterium. Int J Syst Evol Microbiol 54: 77–83.

66. de Bok FAM, Stams AJM, Dijkema C, Boone DR (2001) Pathway of propionate oxidation by a syntrophic culture of *Smithella propionica* and *Methanospirillum hungatei*. Appl Environ Microbiol 67: 1800–1804.

67. Tan B, Nesbo C, Foght J (2014) Re-analysis of omics data indicates Smithella may degrade alkanes by addition to fumarate under methanogenic conditions. ISME J.

68. Embree M, Nagarajan H, Movahedi N, Chitsaz H, Zengler K (2013) Single-cell genome and metatranscriptome sequencing reveal metabolic interactions of an alkane-degrading methanogenic community. ISME J 8: 757–767.

69. Siddique T, Penner T, Klassen J, Nesbø C, Foght J (2012) Microbial communities involved in methane production from hydrocarbons in oil sands tailings. Environmental Science & Technology 46: 9802–9810.

70. Callaghan AV, Davidova IA, Savage-Ashlock K, Parisi VA, Gieg LM, et al. (2010) Diversity of benzyl- and alkylsuccinate synthase genes in hydrocarbon-impacted environments and enrichment cultures Environmental Science & Technology 44: 7287–7294.

71. Chen C-L, Macarie H, Ramirez I, Olmos A, Ong SL, et al. (2004) Microbial community structure in a thermophilic anaerobic hybrid reactor degrading terephthalate. Microbiology 150: 3429–3440.

72. Lykidis A, Chen C-L, Tringe SG, McHardy AC, Copeland A, et al. (2011) Multiple syntrophic interactions in a terephthalate-degrading methanogenic consortium. ISME J 5: 122–130.

16

Modulation of K_{Ca}3.1 Channels by Eicosanoids, Omega-3 Fatty Acids, and Molecular Determinants

Michael Kacik[1], Aida Oliván-Viguera[2], Ralf Köhler[2,3]*

1 Faculty of Medicine, Philipps-University Marburg & Medical Center I, Clemenshospital/University Hospital of University Münster, 48153 Münster, Germany, **2** Aragon Institute of Health Sciences I+CS/IIS, 50009 Zaragoza, Spain, **3** Fundación Agencia Aragonesa para la Investigación y Desarrollo (ARAID), 50018 Zaragoza, Spain

Abstract

Background: Cytochrome P450- and ω-hydrolase products (epoxyeicosatrienoic acids (EETs), hydroxyeicosatetraenoic acid (20-HETE)), natural omega-3 fatty acids (ω3), and pentacyclic triterpenes have been proposed to contribute to a wide range of vaso-protective and anti-fibrotic/anti-cancer signaling pathways including the modula-tion of membrane ion channels. Here we studied the modulation of intermediate-conductance Ca^{2+}/calmodulin-regulated K^+ channels (K_{Ca}3.1) by EETs, 20-HETE, ω3, and pentacyclic triterpenes and the structural requirements of these fatty acids to exert channel blockade.

Methodology/Principal Findings: We studied modulation of cloned human hK_{Ca}3.1 and the mutant hK_{Ca}3.1^{V275A} in HEK-293 cells, of rK_{Ca}3.1 in aortic endothelial cells, and of mK_{Ca}3.1 in 3T3-fibroblasts by inside-out and whole-cell patch-clamp experiments, respectively. In inside-out patches, Ca^{2+}-activated hK_{Ca}3.1 were inhibited by the ω3, DHA and α-LA, and the ω6, AA, in the lower μmolar range and with similar potencies. 5,6-EET, 8,9-EET, 5,6-DiHETE, and saturated arachidic acid, had no appreciable effects. In contrast, 14,15-EET, its stable derivative, 14,15-EEZE, and 20-HETE produced channel inhibition. 11,12-EET displayed less inhibitory activity. The K_{Ca}3.1^{V275A} mutant channel was insensitive to any of the blocking EETs. Non-blocking 5,6-EET antagonized the inhibition caused by AA and augmented cloned hK_{Ca}3.1 and rK_{Ca}3.1 whole-cell currents. Pentacyclic triterpenes did not modulate K_{Ca}3.1 currents.

Conclusions/Significance: Inhibition of K_{Ca}3.1 by EETs (14,15-EET), 20-HETE, and ω3 critically depended on the presence of electron double bonds and hydrophobicity within the 10 carbons preceding the carboxyl-head of the molecules. From the physiological perspective, metabolism of AA to non-blocking 5,6,- and 8,9-EET may cause AA-de-blockade and contribute to cellular signal transduction processes influenced by these fatty acids.

Editor: Alexander G. Obukhov, Indiana University School of Medicine, United States of America

Funding: The study was supported by grants of the Deutsche Forschungsgemeinschaft (DFG): KO1899/10-1 and 11-1 to RK (www.dfg.de). European Community (FP7-PEOPLE Project 321721), and the Fondo de Investigación Sanitaria (Red HERACLES RD12/0042/0014). MK and RK were supported by intramural funding of the University Hospital Marburg & Giessen. The funders had no role in study design, data collection and analysis, decision to publish, or preparation of the manuscript.

Competing Interests: The authors have declared that no competing interests exist.

* Email: rkohler.iacs@aragon.es

Introduction

The intermediate-conductance Ca^{2+}/calmodulin-activated K^+ channel, K_{Ca}3.1 (encoded by the *KCNN4* gene) produces K^+-efflux and cell membrane hyperpolarization to mobilization of intracellular Ca^{2+} [1,2,3]. The channel is mainly expressed in red and white blood cells [4,5,6], secretory epithelia of salivary glands [7], intestine [8], bronchioles [9], vascular endothelium [10], proliferating smooth muscle [11,12,13,14] and fibroblasts [15,16], and malignant brain cancers ([17,18], for review see [19,20]). In these tissues, the channel contributes to the regulation of cell volume [4], anion and water secretion [8], cytokine production [21], endothelial vasodilator responses [10], Ca^{2+}-dependent cell cycle progression, cell migration, and mitogenesis [14,22,23], respectively.

At the molecular level, the most important determinant of channel activation is an increase of intracellular Ca^{2+} that causes conformational changes of constitutively bound calmodulin [1,2], leading to channel gating. Besides this principal mechanism, c-terminal phosphorylation of the channel by cAMP/PKA-dependent mechanisms [24] has been proposed to cause endogenous positive-regulation of channel activity. The omega-6 fatty acid (ω6), arachidonic acid (AA), was identified by Dan Devor and coworkers as the first negative endogenous regulator of K_{Ca}3.1 [25]. Moreover, their seminal work revealed also major mechanisms of membrane trafficking and internalization/recycling/degradation of hK_{Ca}3.1 [26,27]. AA-inhibition of the channel is presumably caused by AA-interaction with lipophilic residues (T250/V275) lining the channel cavity below the selectivity filter and presumed gate of K_{Ca}3.1 [25]. Yet, the structural requirements of the fatty acid itself for K_{Ca}3.1-blockade are unknown.

Here, we hypothesized that structurally related omega-3 fatty acids (ω3), docosahexaenoic acid (DHA) and α-linolenic acid (α-LA), the cytochrome-P450-epoxygenase (CYP)-generated metabolites of AA, epoxyeicotrienoic acids (5,6-EET, 8,9-EET, 11,12-EET, and 14,15-EET) as well as the ω-hydroxylase product, 20-

hydroxyeicosatetraeonic acid (20-HETE), are additional lipid modulators of $K_{Ca}3.1$. Moreover, epoxygenation of AA to 5,6-EET, 8,9-EET, 11,12-EET, or 14,15-EET may shed light on the structural requirements for channel modulation. In addition, a potential $K_{Ca}3.1$-regulation by EETs, 20-HETE, and ω3 could be of help to understand the physiological actions of these fatty acids in physiological systems like the vascular endothelium and arteries, in which they have been shown to exert vasodilator or vasoconstrictor actions, respectively (for review see [28,29,30]). Moreover, EETs and ω3 have been proposed to have anti-inflammatory and anti-atherosclerotic activity and to modulate angiogenesis, cardiac fibrosis and cancer growth [31,32,33,34,35]. In this respect, EETs and $K_{Ca}3.1$-functions have overlapping impacts and may be mechanistically linked as components of the same signal transduction pathway(s). Today, several downstream targets such as G-protein-coupled receptors have been proposed to mediate EET-actions but specific receptors for EETs, HETEs, as well as for ω3 are still elusive (for review see [30,31]). So far it is unknown whether these fatty acids modulate $hK_{Ca}3.1$-functions.

In addition to these fatty acids, we tested whether lipids of the pentacyclic triterpene class, uvaol, erythrodiol, oleanolic acid, and maslinic acid, exert $K_{Ca}3.1$-modulatory actions. These natural triterpenes are found in virgin olive oil and have been suggested having antioxidant, antifibrotic, anti-atherosclerotic, and, both, pro- as well as anti-inflammatory activities [35,36,37,38]. However, whether these actions are related to - at least in part - $K_{Ca}3.1$-modulation has not been studied before.

We therefore conducted an electrophysiological study on cloned $hK_{Ca}3.1$ and endothelial $rK_{Ca}3.1$ and studied channel modulation by selected ω3, the four EETs, and 20-HETE, synthetic stable analogues, and other related fatty acids with structural differences or similarities (for structures see Figure 1). To further study potential binding/interaction sites within the $K_{Ca}3.1$ channel, we investigated blocking efficacy of the fatty acids on the AA-insensitive $K_{Ca}3.1$-mutant V275A [25]. Moreover, we studied the interactivity of EETs with its precursor, AA. In murine fibroblast, we tested the modulation of $mK_{Ca}3.1$ by DHA and by pentacyclic triterpenes.

Our major findings were that the 14,15-EET, 20-HETE, DHA, and α α-LA, were negative modulators of $K_{Ca}3.1$ while non-blocking 5,6-EET antagonized AA-mediated inactivation. KCa3.1 blockade critically depended on hydrophobicity of the 10 carbons preceding the carboxyl head and the presence of at least one electron double bond in this part of the carbon chain.

Materials and Methods

Cells, channel clones, and cell culture

HEK-293 cells stably expressing $hK_{Ca}3.1$ were a kind gift from Dr. Khaled Houamed, University of Chicago and Dr. Heike Wulff, Department of Pharmacology, University of California, Davis. Stably expressing cells were selected with puromycin (1 µg/ml; Sigma, Deisenhofen, Germany). The $hK_{Ca}3.1^{V275A}$, $hK_{Ca}3.1^{T250S}$, and $hK_{Ca}3.1^{T250S/V275A}$ mutants were kind gifts from Dr. Dan Devor, University of Pittsburgh, Department of Cell Biology. The clones were stably expressed in HEK-293 using FuGENE 6 Transfection kit (Roche, Basel, Switzerland) and manufacturer's protocols. Stably expressing HEK-293 cells selected using geneticin (G-418, 100 µl/10 ml; Sigma, Deisenhofen, Germany). Rat aortic endothelial cells with endogenous $rK_{Ca}3.1$ were provided by the BMFZ of the Philipps-University Marburg [39]. Murine 3T3 fibroblasts were obtained from ATCC (3T3-L1, ref# CL-173, ATCC, Rockville, MD). As usual cell culture medium, we used Dulbecco's Modified Eagle Medium

(DMEM) supplemented with 10% calf serum and 1% penicillin/streptomycin (all from Biochrom KG, Berlin, Germany). Before patch-clamp, cells were trypsinized and seeded on cover slips for 4–24 hrs.

Patch-clamp electrophysiology

Membrane currents in excised inside-out patches and whole-cell currents were recorded with an EPC-9 patch-clamp amplifier (HEKA, Lambrecht Pfalz, Germany) using borosilicate glass pipettes with a tip resistance of 2–3 MOhm. Seal resistance was above 1 GOhm. In inside-out experiments, we continuously monitored outward currents at a holding potential of 0 mV prior to patch excision and thereafter. Activation of $K_{Ca}3.1$-mediated currents occurred immediately after excision of the patch and exposure of the intracellular side of the patch to the Ca^{2+}-containing bath solution ("intracellular" solution see below). For conventional whole-cell current measurements, we used voltage ramps (voltage range for recording: -120 mV to $+100$ mV; duration, 1 sec; applied every 3 sec; voltage range evaluated: -110 to $+30$ mV). Series resistance was between 7–15 MegaOhms and membrane resistance was >1 GigaOhm. In such experiments, the "intracellular" Ca^{2+}-containing solution was "infused" into the cell via the patch-pipette after seal rupture activating K_{Ca}-currents usually within 2–10 sec. Current amplitudes remained stable thereafter over 5 min and longer in some. The solution was composed of (mM): 140 KCl, 1 $MgCl_2$, 1 Na_2ATP, 2 EGTA, 1.92 $CaCl_2$ (3 µM $[Ca^{2+}]_{free}$) and 5 HEPES (adjusted to pH 7.2 with KOH). In a subset of experiments, $[Ca^{2+}]_{free}$ was buffered to 0.01, 0.3, 0.5 µM $[Ca^{2+}]_{free}$ (0,07, 0.72, 1.25, and 1.48 mM $CaCl_2$, each combined with 2 mM EGTA). The "extracellular" solution was composed of (mM): 137 NaCl, 4.5 Na_2HPO_4, 5 KCl, 1.5 KH_2PO_4, 1 $MgCl_2$, 1 $CaCl_2$, 10 EGTA (10 nM $[Ca^{2+}]_{free}$), 10 glucose and 10 HEPES (adjusted to pH 7.4 with NaOH). For additional details, see [16]. In inside-out experiments, the high Na^+ solution served as pipette solution and the high K^+ solution as bath solution; in whole-cell experiments, vice versa. For measurements of $rK_{Ca}3.1$ currents in RAEC, we performed the experiments in the presence of the $K_{Ca}2$ blocker UCL-1684 (250 nM) [40] to eliminate $rK_{Ca}2.3$ currents in these cells.

Chemicals and drugs

Standard chemicals were obtained from Sigma-Aldrich (Deisenhofen, Germany). 5,6-EET (4-{3-[(2Z,5Z,8Z)-tetradeca-2,5,8-trien-1-yl]oxiran-2-yl}butanoic acid), 8,9-EET ((5Z)-7-{3-[(2Z,5Z)-undeca-2,5-dien-1-yl]oxiran-2-yl}hept-5-enoic acid), 11,12-EET ((5E,8Z)-10-{3-[(2E)-oct-2-en-1-yl]oxiran-2-yl}deca-5,8-dienoic acid), 14,15-EET ((5Z,8Z,11Z)-13-(3-pentyloxiran-2-yl)trideca-5,8,11-trienoic acid), 5,6-DiHETE ((8Z,11Z,14Z)-5,6-dihydroxy-8,11,14-icosatrienoic acid), 14,15-EEZE ((5Z)-13-[(2S,3R)-3-pentyl-2-oxiranyl]-5-tridecenoic acid), and 20-HETE ((5Z,8Z,11Z,14Z)-20-hydroxy-5,8,11,14-icosatetraenoic acid) were purchased from Cayman Chemicals (Michigan, IL, USA). Arachidonic acid ((5Z,8Z,11Z,14Z)-5,8,11,14-icosatetraenoic acid), arachidonyl glycerol (1,3-dihydroxy-2-propanyl (5Z,8Z,11Z,14Z)-5,8,11,14-icosatetraenoate), arachidic acid (icosanoic acid), charybdotoxin, docosahexaenoic acid ((4Z,7Z,10Z,13Z,16Z,19Z)-4,7,10,13,16,19-docosahexaenoic acid), α-linolenic acid ((9Z,12Z,15Z)-9,12,15-octadecatrienoic acid), dimethyl sulfoxide (DMSO) and acetonitrile were obtained from Sigma-Aldrich. Arachidonyl trifluoromethyl ketone (AACOCF$_3$; (6Z,9Z,12Z,15Z)-1,1,1-trifluoro-6,9,12,15-henicosatetraen-2-one), anandamide ((5Z,8Z,11Z,14Z)-N-(2-hydroxyethyl)-5,8,11,14-icosatetraenamide), and UCL-1684 (17,24-diaza-1,9-diazoniaheptacyclo[23.6.2.29,16.219,22.13,7.010,15.026,-31]octatriaconta-1(31),3(38),4,6,9,11,13,15,19,21,25,27,29,32,34,36-hexadecaene) were obtained from TOCRIS

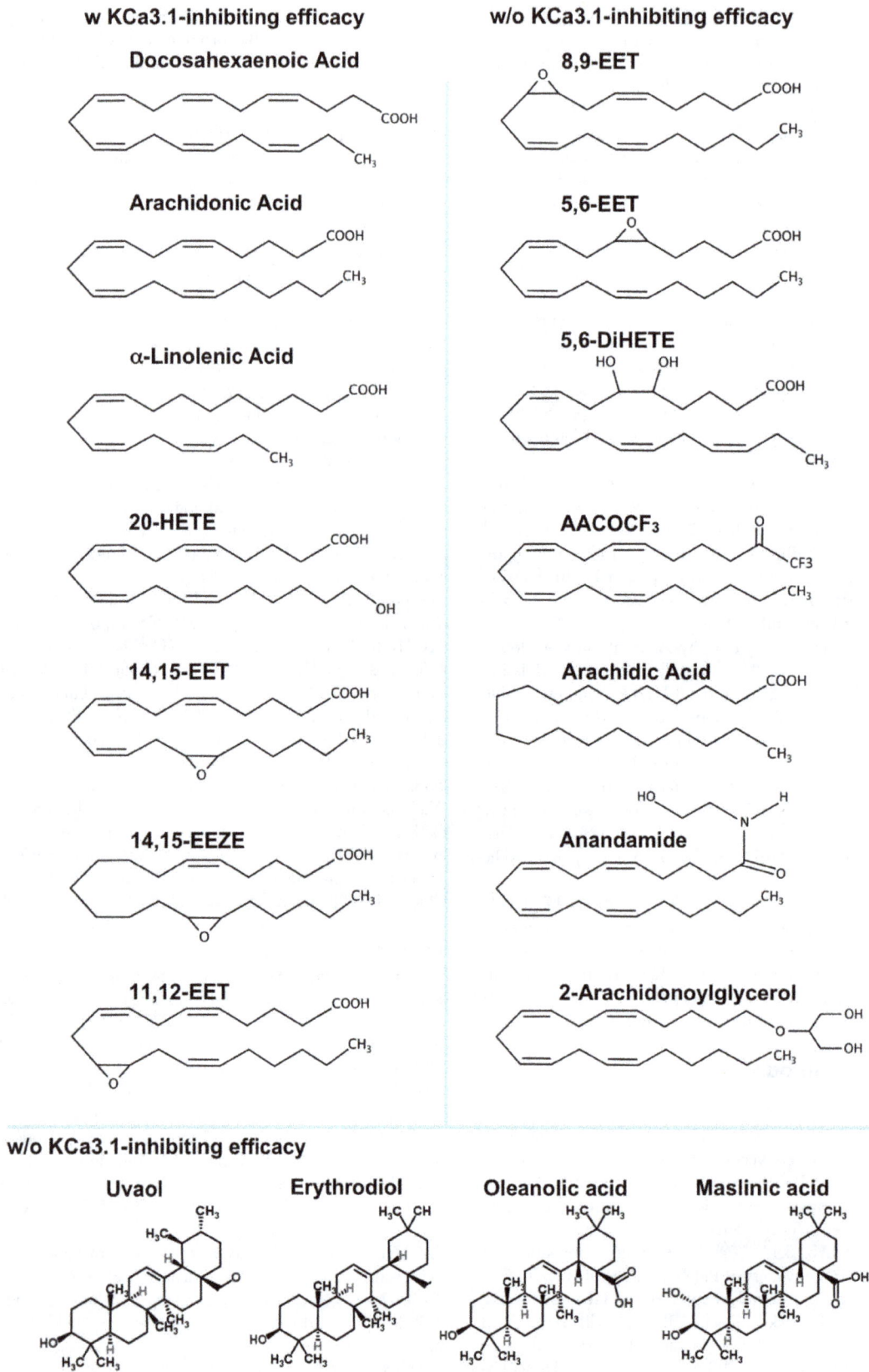

Figure 1. Chemical structures of eicosanoids, ω3, and pentacyclic triterpenes and schematic overview of blocking efficacy (decreasing from top to bottom) or non-blocking efficacy.

(Germany). Uvaol ((3β)-Urs-12-ene-3,28-diol), erythrodiol ((3β)-Olean-12-ene-3,28-diol), oleanolic acid ((3β)-3-Hydroxyolean-12-en-28-oic acid), and maslinic acid ((2α,3β)-2,3-Dihydroxyolean-12-en-28-oic acid) were kind gifts from Dr. Jesús Osada, Department of Biochemistry and Molecular and Cellular Biology, Veterinary School, Health Research Institute of Aragon, CIBEROBN, Zaragoza, Spain. EETs were delivered as ethanol stock solutions. Ethanol was evaporated under nitrogen stream and the EETs were reconstituted in DMSO at a concentration of 10 mM. Stocks were stored at $-20°C$ until use. Stock solutions of the other fatty acids (10 mM) were also prepared with DMSO. Ahead of use stock solutions were diluted 1:10 with the bath buffer and the final DMSO concentration did not exceed 0.2%. Since unsaturated fatty acids are sensitive to oxidative degradation, we minimized exposure times in aqueous solutions and to air and prepared the aqueous pre-dilutions of the compounds immediately before starting the experiments. Bath solutions were not gassed with oxygen.

Statistics

Data are given as mean $\pm$ SEM. For statistical comparison of multiple data sets we used one-way ANOVA and the Tukey *post hoc* and p-values of <0.05 were considered significant.

Results

In inside-out experiments on HEK-293 expressing cloned $hK_{Ca}3.1$, excision of the patch into the 3 µM Ca^{2+}-containing bath solution caused immediate activation of K^+-outward currents that were stable over several minutes (Figure 2A). Non-transfected cells did not display these currents. In $hK_{Ca}3.1$-HEK-293, K^+-outward currents were virtually absent in the continuing presence of the classical $K_{Ca}3.1$-blocking toxin, charybdotoxin, in the "extracellular" pipette solution (Figure 2A). Likewise, in the continuing presence of the selective small molecule blocker of $K_{Ca}3.1$, TRAM-34 [6], in the bath solution prevented K^+-outward currents, although we observed an initial spike-like outward current (Figure 2A) after excision of the patch.

In the continuing presence of 1 or 10 µM of the ω3, docosahexaenoic acid (DHA), arachidonic acid (AA), and α-linolenic acid (α-LA), $hK_{Ca}3.1$ currents could still be activated by patch-excision but the currents did not last and were inhibited after 30 sec (Figure 2B and C). The inhibition by 1 µM was less pronounced than inhibition by 10 µM for all ω3 tested here (Figure 2C). However, potencies and kinetics of current inhibition differed between the ω3 with the following order of potency and time to full inhibition: DHA≥AA>α-LA (Figure 2D). In contrast, the saturated fatty acid, arachidic acid (ArA), did not produce channel inhibition (Figure 2B and C).

With respect to the four EETs, 5,6-EET, 8,9-EET, 11,12-EET, 14,15-EET (Figure 3A for current traces and B for summary data), we found that only 14,15-EET displayed substantial inhibition with potency and kinetics similar to those observed with α-LA. 11,12-EET produced less inhibition. 5,6-EET, 8,9-EET, and 5,6-DiHETE produced virtually no inhibition. The stable analogue of 14,15-EET, 14,15-EEZE, produced an inhibition similar to that caused by 14,15-EET. The ω-hydroxylase product, 20-HETE that was hydroxylated at C20 (end) of the carbon chain, inhibited the current with kinetics and potency similar to other blocking fatty acids (Figure 3A and B). In contrast, molecules that differed from EETs and ω3 because of a major modification of the carboxyl group to hydroxyethylamide like in arachidonoyl ethanolamide (AEA), also known as anandamide, to a 1,3-dihydroxy-2-propanyl as in 2-arachidonoylglycerol (2-AG), and to trifluoromethyl ketone

as in arachidonyl trifluoromethyl ketone (AACOCF₃) did not produce inhibition (Figure 3C).

The single mutants, $hK_{Ca}3.1^{V275A}$ and $hK_{Ca}3.1^{T250S}$, and the double mutant, $hK_{Ca}3.1^{V275A/T250S}$, were largely insensitive to AA and TRAM-34 (data shown for $hK_{Ca}3.1^{V275A}$), although the $hK_{Ca}3.1^{T250S}$ mutant appeared to have a smaller impact compared to the virtually complete insensitivity of the $hK_{Ca}3.1^{V275A}$ mutant to AA (Figure 4A and B). With respect to the other $hK_{Ca}3.1$-blocking fatty acids, $hK_{Ca}3.1^{V275A}$ mutant was also insensitive to 11,12 EET, 14,15-EEZE, and 20-HETE as examples of fully (14,15-EEZE, 20-HETE) or partially (11,12-EET) $hK_{Ca}3.1$-blocking fatty acids (Figure 4A and B).

We next tested the idea whether the 5,6-EET as a non-blocking EET antagonizes AA-mediated channel blockade. These experiments showed that in the presence of both fatty acids, 1 µM 5,6-EET did not significantly prevent channel inhibition by 10 µM AA although the time period to achieve channel inhibition appeared to be increased (Figure 5A and B). At 1 µM AA we observed a significant antagonism of channel blockade by 1 µM 5,6-EET at a later time point (Figure 5A and B).

An increase of intracellular Ca^{2+} stimulates Ca^{2+}-dependent PLA₂ activity and AA-release. In our fast-whole cell experiments using a pipette solution with 0.3 µM Ca^{2+}_{free}, we expected Ca^{2+}-dependent activation of $hK_{Ca}3.1$ and also Ca^{2+}-dependent PLA-2-mediated AA-release. In keeping with the idea that 5,6-EET antagonizes endogenous AA effects, we hypothesized that 5,6-EET augments total $hK_{Ca}3.1$-currents in the HEK-293 cells and tested this in a small series of fast-whole cell experiments (Figure 6). We found that 5,6-EET (at 1 µM) produced significant potentiation by ≈twofold of the $K_{Ca}3.1$ current that was pre-activated by 0.3 µM intracellular Ca^{2+} (Figure 6A). A high concentration of AA (10 µM) abolished these 5,6-EET-potentiated currents. Whole-cell currents produced by the $hK_{Ca}3.1^{T250S/V275A}$ mutant did not show potentiation by 5,6-EET (Figure 6A, right panel).

We performed another series of whole-cell experiments on rat aortic endothelial cells (RAEC) as an established and physiologically relevant cell system involving Ca^{2+}-dependent AA and CYP/EETs signaling as well as $K_{Ca}3.1$-dependent hyperpolarization as two mechanisms for endothelium-dependent vasodilation besides the nitric oxide pathway [29]. We tested specifically whether 1) AA and 14,15-EET produced a similar inhibition of endogenous $rK_{Ca}3.1$ channels in RAEC, 2) $rK_{Ca}3.1$ currents displayed a similar sensitivity to inhibition by AA, and 3) 5,6-EET produced potentiation of the current. As shown in figure 6B, these experiments revealed that 14,15-EET at 1 µM abolished calcium-activated $rK_{Ca}3.1$ currents in these RAEC, in this regard similar to the findings in $hK_{Ca}3.1$-overexpressing HEK-293. With respect to 5,6-EET-potentiation we found that 5,6-EET at 1 µM potentiated by ≈2.5-fold these endothelial calcium-activated $rK_{Ca}3.1$ currents being pre-activated by 0.5 µM and 3 µM intracellular Ca^{2+} but not at 0.1 µM, a Ca^{2+}-concentration that did not allow channel pre-activation (Figure 6B). AA at a concentration of 10 µM substantially blocked this 5,6-EET-potentiated current. Similar to the inside-out experiments, we did not see appreciable antagonistic effects at this lower concentration (1 µM) of 5,6-EET in these whole-cell experiments.

The ω3, DHA, and pentacyclic triterpenes as e.g. uvaol have been demonstrated experimentally to protect against cardiac fibrosis [35,36], in addition to their documented vaso-protective and anti-inflammatory actions [37,38]. Recently, we reported membrane expression of $K_{Ca}3.1$ channels in proliferating murine 3T3-fibroblasts [16]. In the present study, we performed a series of whole-cell experiments and tested whether DHA and pentacyclic triterpenes inhibited $mK_{Ca}3.1$ in murine fibroblasts. We found

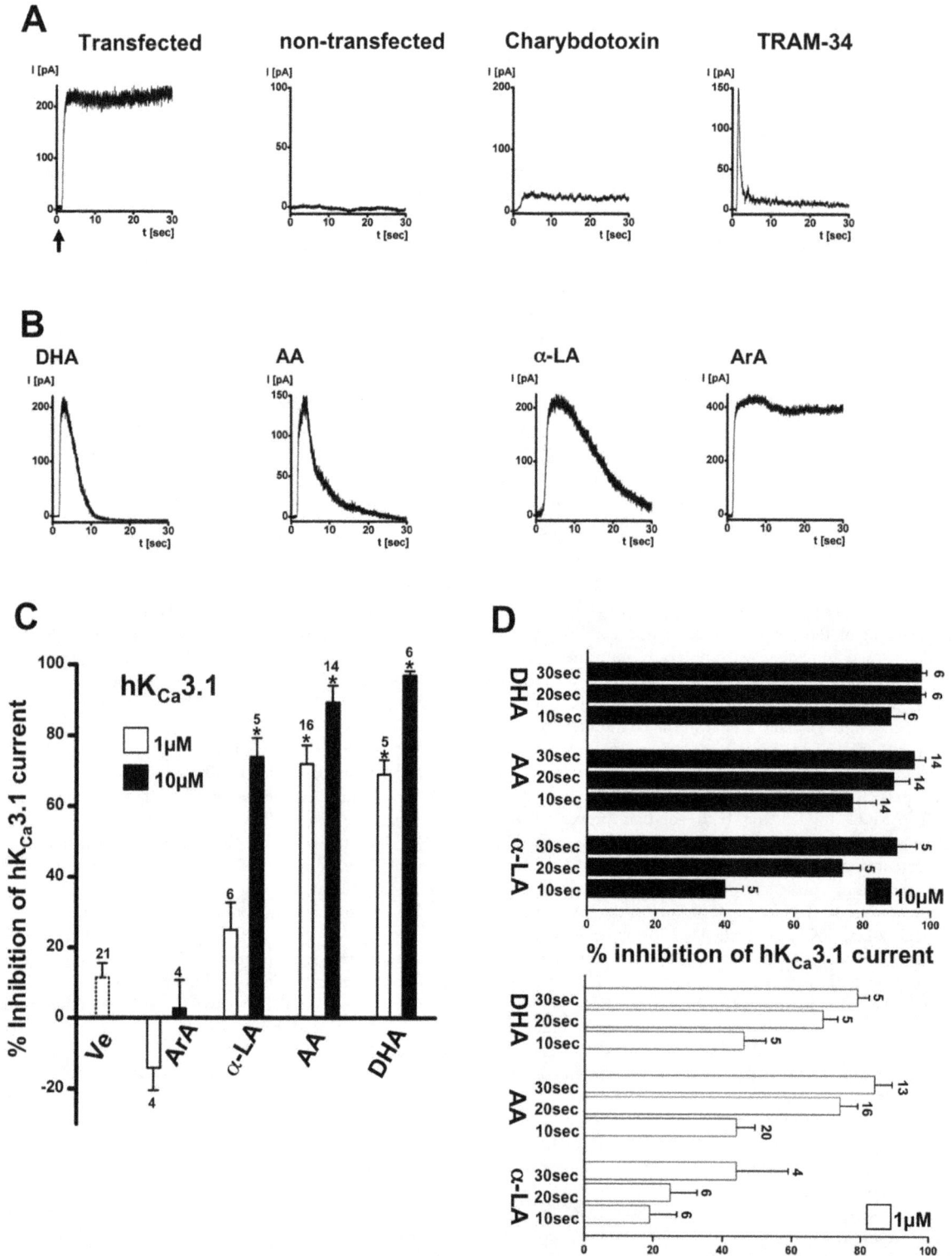

Figure 2. Membrane expression of cloned human $K_{Ca}3.1$ in HEK-293 in inside-out patches and basic pharmacological characterization. A) From left to right: Exemplary traces of immediate activation of $hK_{Ca}3.1$-outward currents upon excision of the patch into 3 μM Ca^{2+}-containing bath solution (as indicated by arrow). K_{Ca}-outward currents are absent in non-transfected HEK-293. Inhibition of $hK_{Ca}3.1$-outward currents by charybdotoxin (100 nM, in the pipette solution) and TRAM-34 (1 μM, in the bath solution). B) Inhibition of $hK_{Ca}3.1$ by ω3 and arachidonic acid. From left to right: Time course of inactivation of $hK_{Ca}3.1$ by docosahexaenoic acid (DHA, 10 μM), arachidonic acid (AA, 10 μM), α-linolenic acid (α-LA, μM) over time. Saturated arachidic acid (ArA, 10 μM) did not affect channel activity. C) Concentration-dependence of inhibition. Note that half of the current was inhibited by AA, DHA, and α-LA at approx. 1 μM. D) Time course of channel inactivation by two concentrations of AA, DHA, and α-LA over time. Data are means ± SEM (% inhibition of $K_{Ca}3.1$-current normalized to initial peak amplitude after patch-excision); numbers in the graphs indicate the number of inside-out experiments; *$P < 0.05$ vs. vehicle (Ve); One-way ANOVA and Tukey *post hoc* test.

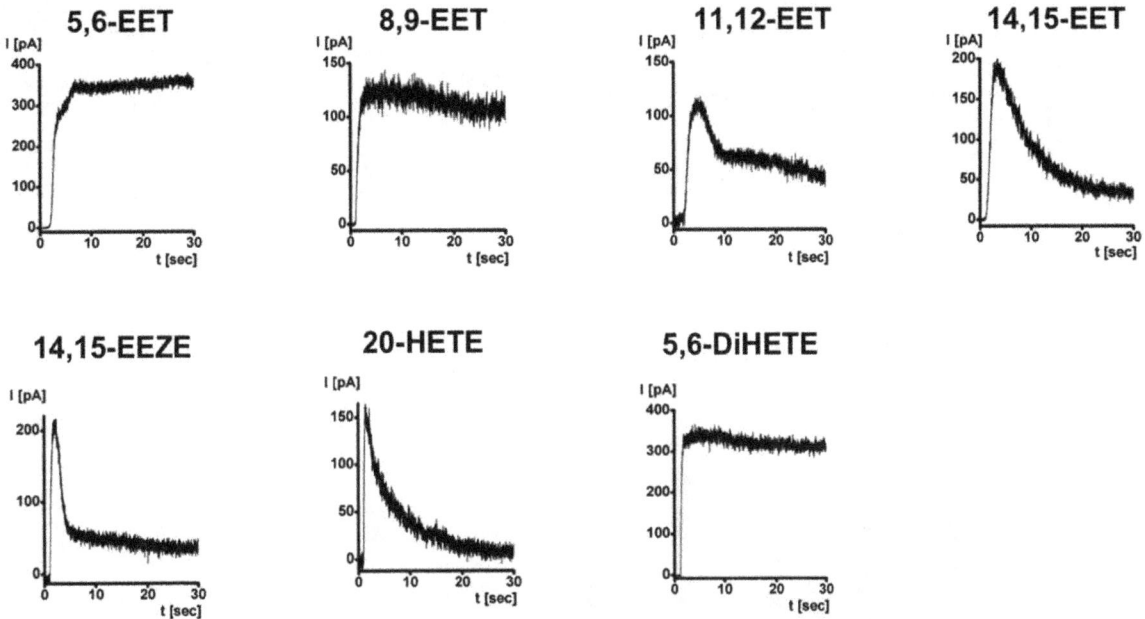

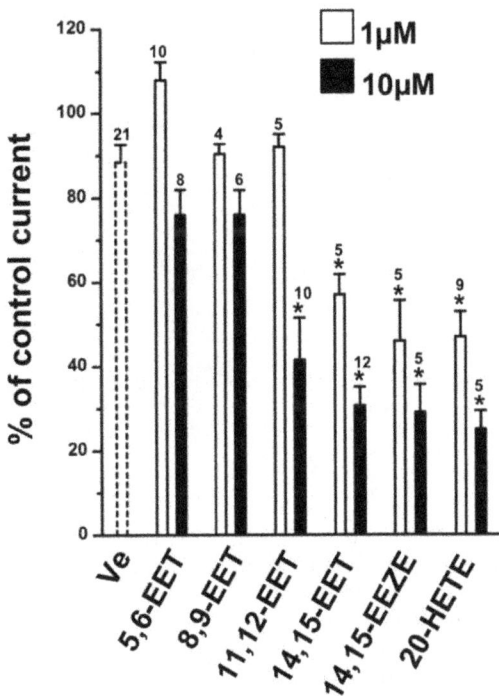

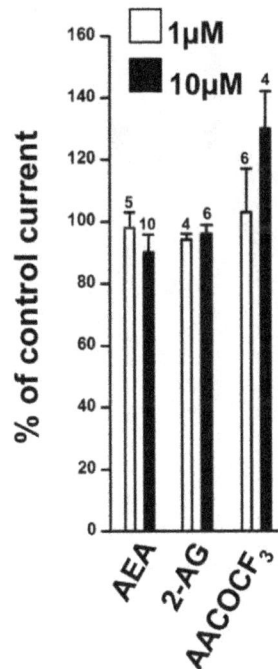

Figure 3. Heterogeneous sensitivity of hK$_{Ca}$3.1 to the four EETs, stable 14,15-EEZE, 20-HETE and 5,6-DiHETE. A) Representative traces of hK$_{Ca}$3.1 outward-currents in inside-out patches overtime in the continuing presence of the fatty acids at 10 μM. B) Summary data of maximal change of current (% of control) at two concentrations (1 and 10 μM). 5,6 DiHETE was tested at 10 μM (0±10%, n = 4). C) No K$_{Ca}$3.1-blockade in the presence of anandamide (AEA; 10 μM), arachidonoylglycerol (2-AG; 10 μM), arachidonyl trifluoromethyl ketone (AACOCF$_3$; 10 μM). Numbers in the graphs indicate the number of inside-out experiments. Data are means ± SEM (% inhibition of K$_{Ca}$3.1-current normalized to initial peak amplitude after patch-excision); *P<0.05 vs. vehicle (Ve); One-way ANOVA and Tukey *post hoc* test.

$$K_{Ca}3.1^{V275A}$$

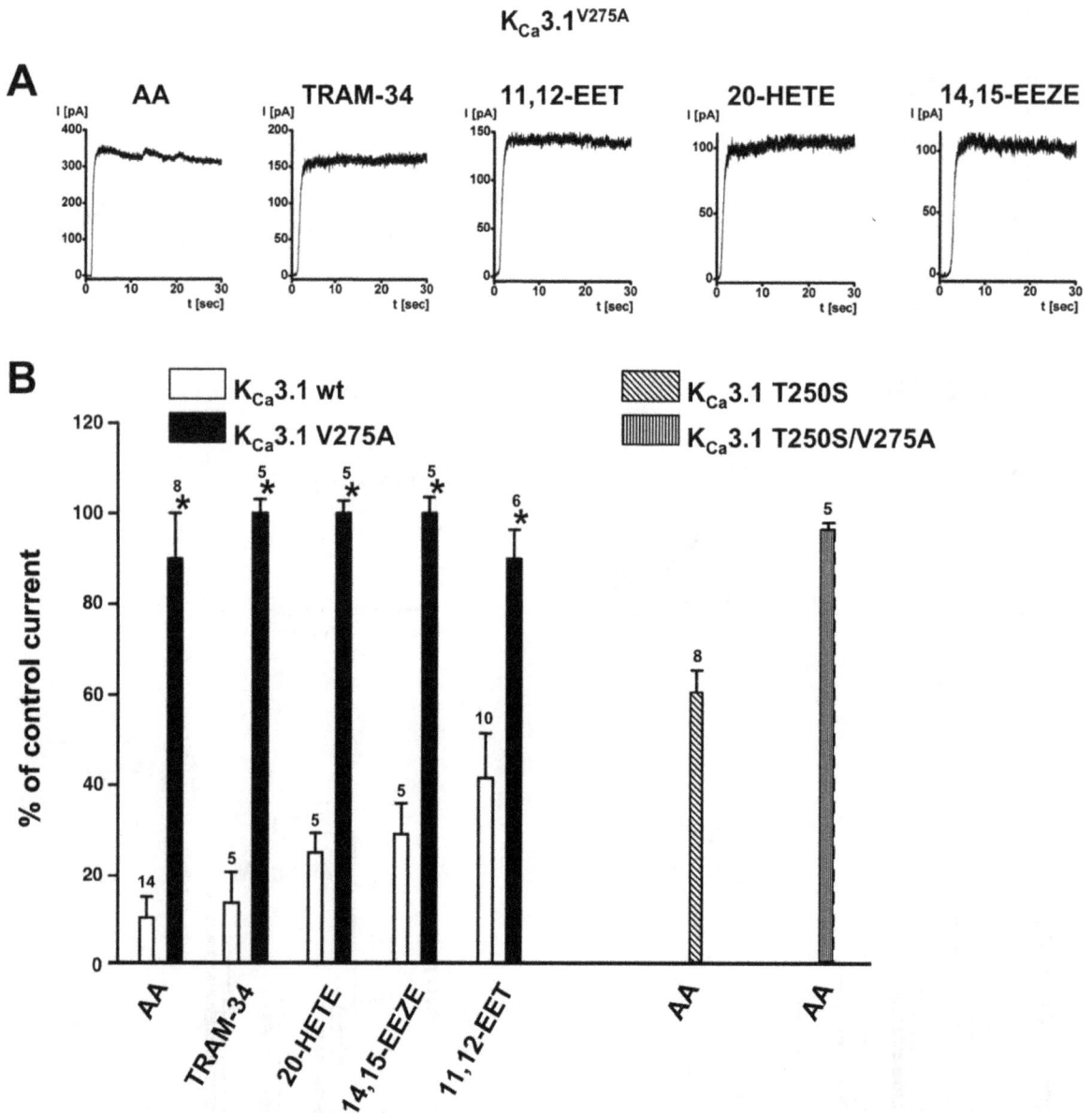

Figure 4. Insensitivity of hK_{Ca}3.1 mutants. A) Representative current traces obtained from inside-out recordings using HEK-293 expressing the hK_{Ca}3.1^{V275A} mutant. B) Summary data from experiments using the three different hK_{Ca}3.1 mutants and wt hK_{Ca}3.1. Concentration of all compounds was 10 μM. Data are means ± SEM; numbers in the graphs indicate the number of inside-out experiments. *$P<0.05$ vs. wt; One-way ANOVA and Tukey *post hoc* test.

that DHA at 1–10 μM abolished virtually mK_{Ca}3.1 (Figure 6C). In contrast, the pentacyclic triterpenes, uvaol, erythrodiol, maslinic acid, and oelanic acid, did not modulate mK_{Ca}3.1-currents at 1 μM (Figure 6D).

Discussion

Here we studied modulation of K_{Ca}3.1 channel by CYP-products, 5,6-EET, 8,9-EET, 11,12-EET, and 14,15-EET, the ω-hydrolase product, 20-HETE, and the ω3, DHA, and α-LA, and identified structural requirements of these fatty acids for K_{Ca}3.1-modulation. Our major findings were that 14,15-EET and 20-HETE as well as DHA and α-LA produced K_{Ca}3.1 inhibition with potencies in the lower μmolar range. 11,12-EET was less potent and 5,6-EET and 8,9-EET did not cause inhibition. However, 5,6-

EET was able to antagonize AA-induced inhibition. The observation that 14,15-EET and 20-HETE were efficient inhibitors while 5,6 and 8,9-EET not, identified the hydrophobic carbon stretch from C1–10 of the carboxyl head of the molecule as structural requirement for channel inhibition (for schematic overview of structural features of K_{Ca}3.1-blocking and non-blocking fatty acids see Figure 1).

Several down-stream targets and receptors for propagation of intracellular or paracrine actions of EETs and ω3 have been proposed and, particularly, ion channel modulation by these fatty acid emerged as an additional mechanistic step. Yet, a plethora of channels have been shown to be directly activated by EET or to be a downstream target of EETs [41,42,43,44,45,46,47]. For instance, the TRPV4 channel, a member of the transient receptor potential gene family of cation channels, have been proposed to be

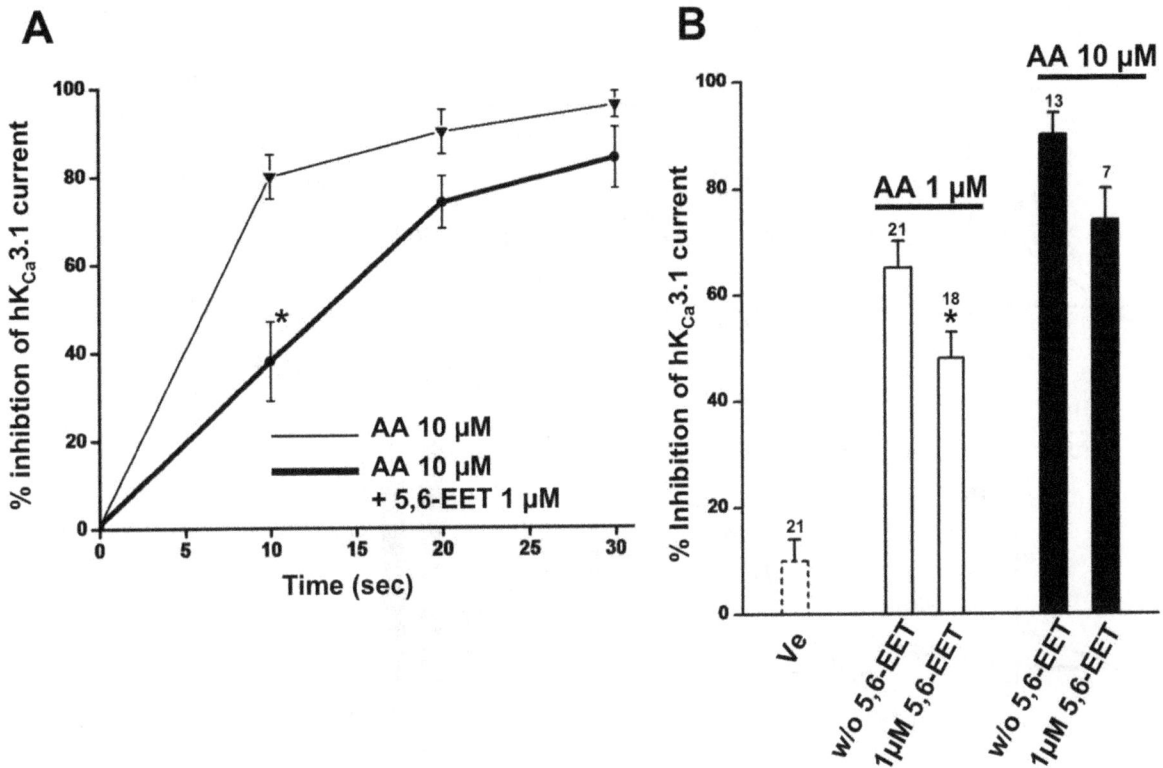

Figure 5. Moderate antagonism of AA-mediated hK_{Ca}3.1-inhibtion by 5,6-EET. A) Time course of channel inhibition by 10 μM of AA in the presence of 1 μM 5,6-EET. B) Summary data of channel inhibition at 20 s after seal excision and with two concentrations (1 and 10 μM) of 5,6-EET and AA. Data are means ± SEM; numbers in the graphs indicate the number of inside-out experiments; *$P<0.05$ vs. AA alone, One-way ANOVA and Tukey *post hoc* test.

activated by 5,6-EET and 8,9-EET and the resulting Ca^{2+}-influx into the vascular endothelium caused vasorelaxation [43,44]. TRPA1 channels in afferent neurons were activated by 5,6-EET leading to an increase in nociception in mice [48]. Yet, another TRP channel, the TRPC6 channel, has been shown to be translocated in a PKA-dependent manner to the cell membrane that required 11,12-EET binding to Gs-receptors in endothelial cells [49]. Moreover, 11,12-EET has been proposed to induce hypoxic vasoconstriction in the lung involving TRPC6 mechanism [50]. Other studies showed that 14,15-EET mediates phosphorylation of epithelial sodium channel (ENAC) activity in an ERK1/2 dependent mechanism [51].

With respect to K^+ channels, 8,9-EET, 11,12-EET, and 14,15-EET have been reported to activate ATP-sensitive K^+- channels by allosteric interaction with the ATP-binding site of the channel [52]. Two-pore tandem K^+ channels (K2P) and large-conductance $K_{Ca}1.1$ channels were known since long to be activated by ω3 and ω6 [45,46,53,54,55]. Moreover, 11,12-EET activation of $K_{Ca}1.1$ channels was considered a main mechanism in smooth muscle, by which EET produced vasorelaxation [56]. In contrast, 20-HETE has been shown recently to enhance angiotensin-II-induced vasoconstriction by inactivating $K_{Ca}1.1$ channels [57]. Interestingly, AA has also been shown to inhibit voltage-gated K^+ channels such as the T lymphocyte KV1.3 channel [58] and the endogenous KV channels in HEK-293 (unpublished observation by our group). To our knowledge there were no data on direct or downstream modulation of $K_{Ca}3.1$ channels by EETs that were not simply linked to EET-mediated increase in intracellular Ca^{2+}. Hence, it was well established that $K_{Ca}3.1$ channels were inhibited by the ω6, AA, that required mechanistically interaction with the

lipophilic residues, V275 and T250, lining the channel cavity [25]. The structural requirements of the AA molecule to produce this inhibition remained however unclear. Our present study confirmed AA-mediated inhibition and the requirements of residue V275 and to some extent of T250 (Figure 4). Moreover, we provided additional insight by showing that the ω3, DHA and α-LA produced similar inhibition of the cloned human channel. Moreover, we showed here that AA abolished endogenous $rK_{Ca}3.1$ (Figure 6) that suggested that AA could be an endogenous negative regulator of $K_{Ca}3.1$ in the endothelium and could thereby influence the $K_{Ca}3.1$-dependent endothelium-derived hyperpolarization (EDH)-mediated type of arterial vasodilation [28,59]. However, this has not been further clarified by the present study. Interestingly, our inside-out experiments showed that $K_{Ca}3.1$ could still be activated in the continuous presence of the AA but inactivated rapidly following Ca^{2+}-dependent activation (Figure 2). This suggested a major impact of AA on $K_{Ca}3.1$-gating unlike charybdotoxin (Figure 2) that obstructs simply the pore and ion flow by binding to the outer vestibule of the channel, independently of gating. However, we cannot exclude that this transient activation seen in the presence of AA reflected a delay of inhibition caused by diffusion of AA and the other compounds from the bath solution towards the excised membrane patch in the patch pipette.

With respect to eicosanoid-modulation of $K_{Ca}3.1$, our study demonstrated that 14,15-EET, the stable analogue, 14,15-EEZE, and 20-HETE were $K_{Ca}3.1$-inhibitors with potencies slightly below that of AA. Structurally, this inhibition required apparently hydrophobicity and 2 double electron bonds within the first 10 carbons of the carboxyl head of the molecules. This was concluded

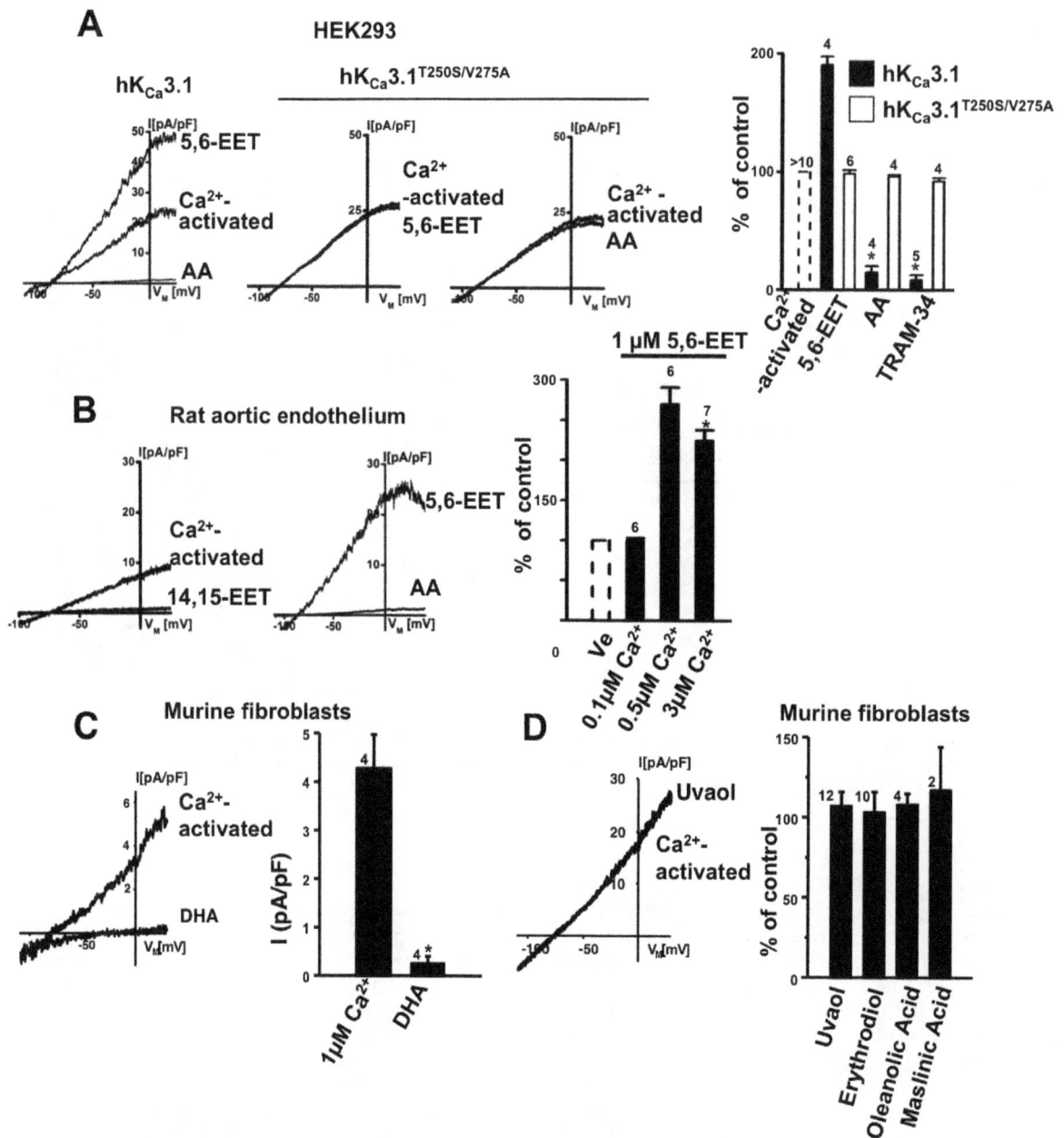

Figure 6. 5,6-EET-potentiation of $K_{Ca}3.1$ currents. A) Whole-cell current traces; from left to right: potentiation of Ca^{2+}-pre-activated $hK_{Ca}3.1$ by 5,6-EET (1 μM) followed by inhibition of the current by AA (10 μM), insensitivity of the $hK_{Ca}3.1^{T250S/V275A}$ mutant to 5,6-EET, and insensitivity of the $hK_{Ca}3.1^{T250S/V275A}$ mutant to AA (10 μM) and TRAM-34 (1 μM). The $hK_{Ca}3.1$ currents were pre-activated by 250 nM Ca^{2+}. Panel on the right: summary data. B) From left to right: Ca^{2+}-pre-activation of rat endothelial $rK_{Ca}3.1$ by 3 μM Ca^{2+} and current inhibition by 14,15-EET (1 μM), larger currents in the presence of 5,6-EET (1 μM) and inhibition by AA (10 μM). Panel on right: Summary data: dependence of 5,6-EET-potentiation on the intracellular Ca^{2+}. Note that at a low intracellular Ca^{2+} (0.1 μM) that is below/near the threshold for $K_{Ca}3.1$ activation, 5,6-EET did not potentiate the current. In contrast, potentiation occurred at an intracellular Ca^{2+} concentration that is near the EC_{50} for Ca^{2+}-activation of $K_{Ca}3.1$ as well as at a saturating Ca^{2+} concentration. C) DHA (1 μM) blocked Ca^{2+}-pre-activated $mK_{Ca}3.1$ in murine fibroblasts. D) Pentacyclic triterpenes did not modulate murine fibroblast $mK_{Ca}3.1$ at a concentration of 1 μM. Data are means ± SEM (% inhibition of $K_{Ca}3.1$-current normalized to initial peak amplitude after establishing electrical access (by seal rupture) and stable Ca^{2+}-activation of $K_{Ca}3.1$-outward currents); Numbers in the graphs indicate the number of whole-cell experiments; *$P<0.05$ vs. control (peak amplitude of the $K_{Ca}3.1$-current in the respective cell); One-way ANOVA and Tukey *post hoc* test.

from the lack of inhibitory activity of 5,6-EET and 8,9-EET, in which this part of the fatty acid chain was epoxygenated. The partial inhibition caused by 11,12-EET could be explained by the conserved hydrophobicity within carbons 1–10 although 11,12-epoxygenation appeared to have efficacy-reducing impact. In respect to channel-eicosanoids interactions, it was likely that

epoxygenation as in 5,6,-EET and 8,9-EET did not allow the proper interactions of these molecules with hydrophobic residues of the cavity below the selectivity filter as they have been postulated for AA [25]. The intactness of carboxyl head of the molecule was another structural need since major alterations as in anandamide and 2-arachidonoylglycerol let to a loss of inhibitory efficacy (see figure 1 for structures and scheme of blocking efficacy of the fatty acids). However, detailed structural analysis on yet not available crystal structures of the open and closed $K_{Ca}3.1$ channel and mapping of AA and EETs interaction/binding will be needed to provide more definite insight into this lipid modulation of $K_{Ca}3.1$ channels. In contrast to eicosanoids and $\omega 3$, the pentacyclic triterpenes studied here did not modulate $mK_{Ca}3.1$ channel, which might be explained by their more "rigid" and larger structures that may not fit into the internal cavity of the channel.

From the physiological and pharmacological perspective, mircomolar EETs, stable EET-analogues, and 20-HETE have been used to study mechanisms of vasodilation or vasoconstriction. Since $K_{Ca}3.1$ has been demonstrated a major component in the EDH-mediated type of endothelium-dependent vasodilation [59] and considering that this channel modulates also functions in several other tissues [1,2], interactions of the different EET and of 20-HETE with $K_{Ca}3.1$ channels as described in the present study needs to be taken into account.

An additional interesting observation was that 5,6-EET was capable to antagonize AA-inhibition of $K_{Ca}3.1$-activity in isolated patches (Figure 4). Moreover, at the whole-cell level, 5,6-EET potentiated Ca^{2+}-pre-activated $K_{Ca}3.1$-currents. While 5,6-EET did not have a direct effect on channel-gating per se as concluded from the inside-out experiments (Figure 3), it was tempting to speculate that 5,6-EET antagonized the -at least partial - channel inhibition caused by endogenous Ca^{2+}-dependent PLA_2-mediated AA-release. This view was fostered by the insensitivity of the $hK_{Ca}3.1^{V275A}$ to 5,6-EET-potentiation (Figure 6). Such a mechanism may represent a novel mechanism of endogenous $K_{Ca}3.1$-modulation beyond Ca^{2+}-regulation of the channel. Moreover, the 5,6-EET-mediated de-blockade of $K_{Ca}3.1$ could be a thus far

unrecognized mechanism underlying EDH-mediated vasodilation, in which both EETs and $K_{Ca}3.1$ have been implicated to play major roles.

It is worth to mention that $K_{Ca}3.1$ channels contribute to a variety of pathologies such as acute and chronic inflammation [60,61], vasculo-occlusive disease (neointima formation) [12], atherosclerosis [62], angiogenesis [22], polycystic kidney disease [63], ulcerative colitis [21,64], tumor growth and metastasis (e.g. glioblastoma [17]), transplant vasculopathy [65,66], and organ fibrosis [67]. EETs, $\omega 3$, and pentacyclic triterpenes have also been reported to mechanistically contribute to/influence such disease states [31,32,33,34,35,36,37,38]. In this respect, some of the reported anti-inflammatory, vaso-protective, and anti-cancerogenic actions of EETs and $\omega 3$ as well as anti-hypotensive actions of 20-HETE, but possibly not that of pentacyclic triterpenes, could be explained by inhibition of pro-proliferative $K_{Ca}3.1$ functions. This also raised the possibility to use stable 14,15-EET or 20-HETE mimetics [68] to target $K_{Ca}3.1$ in disease states, to which this channel adds patho-mechanistically.

In conclusion, the present electrophysiological and structure-activity-relationship study demonstrated modulation of cloned and endogenous $K_{Ca}3.1$ channels by selective EETs, 20-HETE, and $\omega 3$ and revealed major structural determinants of the molecules for channel interaction.

Author Contributions

Conceived and designed the experiments: RK MK AOV. Performed the experiments: RK MK AOV. Analyzed the data: RK MK AOV. Contributed reagents/materials/analysis tools: RK MK AOV. Contributed to the writing of the manuscript: RK MK AOV.

References

1. Ishii TM, Silvia C, Hirschberg B, Bond CT, Adelman JP, et al. (1997) A human intermediate conductance calcium-activated potassium channel. Proc Natl Acad Sci U S A 94: 11651–11656.
2. Wei AD, Gutman GA, Aldrich R, Chandy KG, Grissmer S, et al. (2005) International Union of Pharmacology. LII. Nomenclature and molecular relationships of calcium-activated potassium channels. Pharmacol Rev 57: 463–472.
3. Logsdon NJ, Kang J, Togo JA, Christian EP, Aiyar J (1997) A novel gene, hKCa4, encodes the calcium-activated potassium channel in human T lymphocytes. J Biol Chem 272: 32723–32726.
4. Grgic I, Kaistha BP, Paschen S, Kaistha A, Busch C, et al. (2009) Disruption of the Gardos channel (KCa3.1) in mice causes subtle erythrocyte macrocytosis and progressive splenomegaly. Pflugers Arch 458: 291–302.
5. Vandorpe DH, Shmukler BE, Jiang L, Lim B, Maylie J, et al. (1998) cDNA cloning and functional characterization of the mouse Ca^{2+}-gated K^+ channel, mIK1. Roles in regulatory volume decrease and erythroid differentiation. J Biol Chem 273: 21542–21553.
6. Wulff H, Miller MJ, Hansel W, Grissmer S, Cahalan MD, et al. (2000) Design of a potent and selective inhibitor of the intermediate-conductance Ca^{2+}-activated K^+ channel, IKCa1: a potential immunosuppressant. Proc Natl Acad Sci U S A 97: 8151–8156.
7. Begenisich T, Nakamoto T, Ovitt CE, Nehrke K, Brugnara C, et al. (2004) Physiological roles of the intermediate conductance, Ca^{2+}-activated potassium channel Kcnn4. J Biol Chem 279: 47681–47687.
8. Devor DC, Singh AK, Frizzell RA, Bridges RJ (1996) Modulation of Cl-secretion by benzimidazolones. I. Direct activation of a Ca(2+)-dependent K^+ channel. Am J Physiol 271: L775–784.
9. Kroigaard C, Dalsgaard T, Nielsen G, Laursen BE, Pilegaard H, et al. (2012) Activation of endothelial and epithelial K(Ca) 2.3 calcium-activated potassium channels by NS309 relaxes human small pulmonary arteries and bronchioles. Br J Pharmacol 167: 37–47.
10. Köhler R, Ruth P (2010) Endothelial dysfunction and blood pressure alterations in K^+-channel transgenic mice. Pflugers Arch 459: 969–976.
11. Neylon CB, Lang RJ, Fu Y, Bobik A, Reinhart PH (1999) Molecular cloning and characterization of the intermediate-conductance Ca(2+)-activated K(+) channel in vascular smooth muscle: relationship between K(Ca) channel diversity and smooth muscle cell function. Circ Res 85: e33–43.

12. Köhler R, Wulff H, Eichler I, Kneifel M, Neumann D, et al. (2003) Blockade of the intermediate-conductance calcium-activated potassium channel as a new therapeutic strategy for restenosis. Circulation 108: 1119–1125.
13. Tharp DL, Wamhoff BR, Wulff H, Raman G, Cheong A, et al. (2008) Local delivery of the KCa3.1 blocker, TRAM-34, prevents acute angioplasty-induced coronary smooth muscle phenotypic modulation and limits stenosis. Arterioscler Thromb Vasc Biol 28: 1084–1089.
14. Bi D, Toyama K, Lemaitre V, Takai J, Fan F, et al. (2013) The intermediate conductance calcium-activated potassium channel KCa3.1 regulates vascular smooth muscle cell proliferation via controlling calcium-dependent signaling. J Biol Chem 288: 15843–15853.
15. Pena TL, Rane SG (1999) The fibroblast intermediate conductance K(Ca) channel, FIK, as a prototype for the cell growth regulatory function of the IK channel family. J Membr Biol 172: 249–257.
16. Olivan-Viguera A, Valero MS, Murillo MD, Wulff H, Garcia-Otin AL, et al. (2013) Novel phenolic inhibitors of small/intermediate-conductance Ca(2)(+)-activated K(+) channels, KCa3.1 and KCa2.3. PLoS One 8: e58614.
17. D'Alessandro G, Catalano M, Sciaccaluga M, Chece G, Cipriani R, et al. (2013) KCa3.1 channels are involved in the infiltrative behavior of glioblastoma in vivo. Cell Death Dis 4: e773.
18. Lambertsen KL, Gramsbergen JB, Sivasaravanaparan M, Ditzel N, Sevelsted-Moller LM, et al. (2012) Genetic KCa3.1-Deficiency Produces Locomotor Hyperactivity and Alterations in Cerebral Monoamine Levels. PLoS One 7: e47744.
19. Pardo LA, Stuhmer W (2014) The roles of K(+) channels in cancer. Nat Rev Cancer 14: 39–48.
20. Wulff H, Kolski-Andreaco A, Sankaranarayanan A, Sabatier JM, Shakkottai V (2007) Modulators of small- and intermediate-conductance calcium-activated potassium channels and their therapeutic indications. Curr Med Chem 14: 1437–1457.
21. Di L, Srivastava S, Zhdanova O, Ding Y, Li Z, et al. (2010) Inhibition of the K^+ channel KCa3.1 ameliorates T cell-mediated colitis. Proc Natl Acad Sci U S A 107: 1541–1546.
22. Grgic I, Eichler I, Heinau P, Si H, Brakemeier S, et al. (2005) Selective blockade of the intermediate-conductance Ca^{2+}-activated K^+ channel suppresses proliferation of microvascular and macrovascular endothelial cells and angiogenesis in vivo. Arterioscler Thromb Vasc Biol 25: 704–709.

23. Si H, Grgic I, Heyken WT, Maier T, Hoyer J, et al. (2006) Mitogenic modulation of Ca^{2+}-activated K$^+$ channels in proliferating A7r5 vascular smooth muscle cells. Br J Pharmacol 148: 909–917.

24. Gerlach AC, Gangopadhyay NN, Devor DC (2000) Kinase-dependent regulation of the intermediate conductance, calcium-dependent potassium channel, hIK1. J Biol Chem 275: 585–598.

25. Hamilton KL, Syme CA, Devor DC (2003) Molecular localization of the inhibitory arachidonic acid binding site to the pore of hIK1. J Biol Chem 278: 16690–16697.

26. Bertuccio CA, Lee SL, Wu G, Butterworth MB, Hamilton KL, et al. (2014) Anterograde trafficking of KCa3.1 in polarized epithelia is Rab1- and Rab8-dependent and recycling endosome-independent. PLoS One 9: e92013.

27. Balut CM, Hamilton KL, Devor DC (2012) Trafficking of intermediate (KCa3.1) and small (KCa2.x) conductance, Ca(2+)-activated K(+) channels: a novel target for medicinal chemistry efforts? ChemMedChem 7: 1741–1755.

28. Feletou M, Kohler R, Vanhoutte PM (2010) Endothelium-derived vasoactive factors and hypertension: possible roles in pathogenesis and as treatment targets. Curr Hypertens Rep 12: 267–275.

29. Feletou M, Kohler R, Vanhoutte PM (2012) Nitric oxide: orchestrator of endothelium-dependent responses. Ann Med 44: 694–716.

30. Campbell WB, Fleming I (2010) Epoxyeicosatrienoic acids and endothelium-dependent responses. Pflugers Arch 459: 881–895.

31. Pfister SL, Gauthier KM, Campbell WB (2010) Vascular pharmacology of epoxyeicosatrienoic acids. Adv Pharmacol 60: 27–59.

32. Fleming I (2011) The cytochrome P450 pathway in angiogenesis and endothelial cell biology. Cancer Metastasis Rev 30: 541–555.

33. Zhang G, Kodani S, Hammock BD (2014) Stabilized epoxygenated fatty acids regulate inflammation, pain, angiogenesis and cancer. Prog Lipid Res 53: 108–123.

34. Chen C, Wei X, Rao X, Wu J, Yang S, et al. (2011) Cytochrome P450 2J2 is highly expressed in hematologic malignant diseases and promotes tumor cell growth. J Pharmacol Exp Ther 336: 344–355.

35. Siddesha JM, Valente AJ, Yoshida T, Sakamuri SS, Delafontaine P, et al. (2014) Docosahexaenoic acid reverses angiotensin II-induced RECK suppression and cardiac fibroblast migration. Cell Signal 26: 933–941.

36. Martin R, Miana M, Jurado-Lopez R, Martinez-Martinez E, Gomez-Hurtado N, et al. (2012) DIOL triterpenes block profibrotic effects of angiotensin II and protect from cardiac hypertrophy. PLoS One 7: e41545.

37. Lou-Bonafonte JM, Arnal C, Navarro MA, Osada J (2012) Efficacy of bioactive compounds from extra virgin olive oil to modulate atherosclerosis development. Mol Nutr Food Res 56: 1043–1057.

38. Marquez-Martin A, De La Puerta R, Fernandez-Arche A, Ruiz-Gutierrez V, Yaqoob P (2006) Modulation of cytokine secretion by pentacyclic triterpenes from olive pomace oil in human mononuclear cells. Cytokine 36: 211–217.

39. Köhler R, Eichler I, Schonfelder H, Grgic I, Heinau P, et al. (2005) Impaired EDHF-mediated vasodilation and function of endothelial Ca-activated K channels in uremic rats. Kidney Int 67: 2280–2287.

40. Rosa JC, Galanakis D, Ganellin CR, Dunn PM, Jenkinson DH (1998) Bis-quinolinium cyclophanes: 6,10-diaza-3(1,3),8(1,4)-dibenzena-1,5(1,4)- diquinoli-nacyclodecaphane (UCL 1684), the first nanomolar, non-peptidic blocker of the apamin-sensitive Ca^{2+}-activated K$^+$ channel. J Med Chem 41: 2–5.

41. Xiao YF (2007) Cyclic AMP-dependent modulation of cardiac L-type Ca^{2+} and transient outward K$^+$ channel activities by epoxyeicosatrienoic acids. Prostaglandins Other Lipid Mediat 82: 11–18.

42. Earley S (2011) Endothelium-dependent cerebral artery dilation mediated by transient receptor potential and Ca^{2+}-activated K$^+$ channels. J Cardiovasc Pharmacol 57: 148–153.

43. Watanabe H, Vriens J, Prenen J, Droogmans G, Voets T, et al. (2003) Anandamide and arachidonic acid use epoxyeicosatrienoic acids to activate TRPV4 channels. Nature 424: 434–438.

44. Vriens J, Owsianik G, Fisslthaler B, Suzuki M, Janssens A, et al. (2005) Modulation of the Ca2 permeable cation channel TRPV4 by cytochrome P450 epoxygenases in vascular endothelium. Circ Res 97: 908–915.

45. Wang RX, Chai Q, Lu T, Lee HC (2011) Activation of vascular BK channels by docosahexaenoic acid is dependent on cytochrome P450 epoxygenase activity. Cardiovasc Res 90: 344–352.

46. Nielsen G, Wandall-Frostholm C, Sadda V, Olivan-Viguera A, Lloyd EE, et al. (2013) Alterations of N-3 polyunsaturated fatty acid-activated K2P channels in hypoxia-induced pulmonary hypertension. Basic Clin Pharmacol Toxicol 113: 250–258.

47. Fernandes J, Lorenzo IM, Andrade YN, Garcia-Elias A, Serra SA, et al. (2008) IP3 sensitizes TRPV4 channel to the mechano- and osmotransducing messenger 5′-6′-epoxyeicosatrienoic acid. J Cell Biol 181: 143–155.

48. Sisignano M, Park CK, Angioni C, Zhang DD, von Hehn C, et al. (2012) 5,6-EET is released upon neuronal activity and induces mechanical pain hypersensitivity via TRPA1 on central afferent terminals. J Neurosci 32: 6364–6372.

49. Ding Y, Fromel T, Popp R, Falck JR, Schunck WH, et al. (2014) The biological actions of 11,12-epoxyeicosatrienoic acid in endothelial cells are specific to the R/S enantiomer and require the Gs protein. J Pharmacol Exp Ther.

50. Keseru B, Barbosa-Sicard E, Popp R, Fisslthaler B, Dietrich A, et al. (2008) Epoxyeicosatrienoic acids and the soluble epoxide hydrolase are determinants of pulmonary artery pressure and the acute hypoxic pulmonary vasoconstrictor response. FASEB J 22: 4306–4315.

51. Pidkovka N, Rao R, Mei S, Gong Y, Harris RC, et al. (2013) Epoxyeicosa-trienoic acids (EETs) regulate epithelial sodium channel activity by extracellular signal-regulated kinase 1/2 (ERK1/2)-mediated phosphorylation. J Biol Chem 288: 5223–5231.

52. Lu T, Hong MP, Lee HC (2005) Molecular determinants of cardiac K(ATP) channel activation by epoxyeicosatrienoic acids. J Biol Chem 280: 19097–19104.

53. Kirber MT, Ordway RW, Clapp LH, Walsh JV, Jr., Singer JJ (1992) Both membrane stretch and fatty acids directly activate large conductance Ca^{2+}-activated K$^+$ channels in vascular smooth muscle cells. FEBS Lett 297: 24–28.

54. Blondeau N, Petrault O, Manta S, Giordanengo V, Gounon P, et al. (2007) Polyunsaturated fatty acids are cerebral vasodilators via the TREK-1 potassium channel. Circ Res 101: 176–184.

55. Maingret F, Patel AJ, Lesage F, Lazdunski M, Honore E (1999) Mechano- or acid stimulation, two interactive modes of activation of the TREK-1 potassium channel. J Biol Chem 274: 26691–26696.

56. Zou AP, Fleming JT, Falck JR, Jacobs ER, Gebremedhin D, et al. (1996) Stereospecific effects of epoxyeicosatrienoic acids on renal vascular tone and K(+)-channel activity. Am J Physiol 270: F822–832.

57. Fan F, Sun CW, Maier KG, Williams JM, Pabbidi MR, et al. (2013) 20-Hydroxyeicosatetraenoic acid contributes to the inhibition of K$^+$ channel activity and vasoconstrictor response to angiotensin II in rat renal microvessels. PLoS One 8: e82482.

58. Szekely A, Kitajka K, Panyi G, Marian T, Gaspar R, et al. (2007) Nutrition and immune system: certain fatty acids differently modify membrane composition and consequently kinetics of KV1.3 channels of human peripheral lymphocytes. Immunobiology 212: 213–227.

59. Wulff H, Kohler R (2013) Endothelial small-conductance and intermediate-conductance KCa channels: an update on their pharmacology and usefulness as cardiovascular targets. J Cardiovasc Pharmacol 61: 102–112.

60. Wulff H, Castle NA (2010) Therapeutic potential of KCa3.1 blockers: recent advances and promising trends. Expert Rev Clin Pharmacol 3: 385–396.

61. Grgic I, Wulff H, Eichler I, Flothmann C, Kohler R, et al. (2009) Blockade of T-lymphocyte KCa3.1 and Kv1.3 channels as novel immunosuppression strategy to prevent kidney allograft rejection. Transplant Proc 41: 2601–2606.

62. Toyama K, Wulff H, Chandy KG, Azam P, Raman G, et al. (2008) The intermediate-conductance calcium-activated potassium channel KCa3.1 con-tributes to atherogenesis in mice and humans. J Clin Invest 118: 3025–3037.

63. Albaqumi M, Srivastava S, Li Z, Zhdnova O, Wulff H, et al. (2008) KCa3.1 potassium channels are critical for cAMP-dependent chloride secretion and cyst growth in autosomal-dominant polycystic kidney disease. Kidney Int 74: 740–749.

64. Koch Hansen L, Sevelsted-Moller L, Rabjerg M, Larsen D, Hansen TP, et al. (2014) Expression of T-cell K1.3 potassium channel correlates with pro-inflammatory cytokines and disease activity in ulcerative colitis. J Crohns Colitis.

65. Hua X, Deuse T, Chen YJ, Wulff H, Stubbendorff M, et al. (2013) The potassium channel KCa3.1 as new therapeutic target for the prevention of obliterative airway disease. Transplantation 95: 285–292.

66. Chen YJ, Lam J, Gregory CR, Schrepfer S, Wulff H (2013) The Ca(2)(+)-activated K(+) channel KCa3.1 as a potential new target for the prevention of allograft vasculopathy. PLoS One 8: e81006.

67. Grgic I, Kiss E, Kaistha BP, Busch C, Kloss M, et al. (2009) Renal fibrosis is attenuated by targeted disruption of KCa3.1 potassium channels. Proc Natl Acad Sci U S A 106: 14518–14523.

68. Tunctan B, Korkmaz B, Sari AN, Kacan M, Unsal D, et al. (2013) Contribution of iNOS/sGC/PKG pathway, COX-2, CYP4A1, and gp91(phox) to the protective effect of 5,14-HEDGE, a 20-HETE mimetic, against vasodilation, hypotension, tachycardia, and inflammation in a rat model of septic shock. Nitric Oxide 33: 18–41.

17

In Vitro Evolution and Affinity-Maturation with Coliphage Qβ Display

Claudia Skamel¹, Stephen G. Aller², Alain Bopda Waffo³*

1 Campus Technologies Freiburg (CTF) GmbH, Agency for Technology Transfer at the University and University Medical Center Freiburg, Freiburg, Germany, **2** Department of Pharmacology and Toxicology and Center for Structural Biology, University of Alabama at Birmingham, Birmingham, Alabama, United States of America, **3** Department of Biological Sciences, Alabama State University, Montgomery, Alabama, United States of America

Abstract

The *Escherichia coli* bacteriophage, Qβ (Coliphage Qβ), offers a favorable alternative to M13 for *in vitro* evolution of displayed peptides and proteins due to high mutagenesis rates in Qβ RNA replication that better simulate the affinity maturation processes of the immune response. We describe a benchtop *in vitro* evolution system using Qβ display of the VP1 G-H loop peptide of foot-and-mouth disease virus (FMDV). DNA encoding the G-H loop was fused to the A1 minor coat protein of Qβ resulting in a replication-competent hybrid phage that efficiently displayed the FMDV peptide. The surface-localized FMDV VP1 G-H loop cross-reacted with the anti-FMDV monoclonal antibody (mAb) SD6 and was found to decorate the corners of the Qβ icosahedral shell by electron microscopy. Evolution of Qβ-displayed peptides, starting from fully degenerate coding sequences corresponding to the immunodominant region of VP1, allowed rapid *in vitro* affinity maturation to SD6 mAb. Qβ selected under evolutionary pressure revealed a non-canonical, but essential epitope for mAb SD6 recognition consisting of an Arg-Gly tandem pair. Finally, the selected hybrid phages induced polyclonal antibodies in guinea pigs with good affinity to both FMDV and hybrid Qβ-G-H loop, validating the requirement of the tandem pair epitope. Qβ-display emerges as a novel framework for rapid *in vitro* evolution with affinity-maturation to molecular targets.

Editor: Mark Isalan, Imperial College London, United Kingdom

Funding: The authors acknowledge the Deutsche Forschungsgemeinschaft for grant # DFG-BI 521/2-3 and the CNBR of ASU for grant # NSF-CREST (HRD-1241701). The funders had no role in study design, data collection and analysis, decision to publish, or preparation of the manuscript.

Competing Interests: Dr. Claudia Skamel, a co-author is currently working with a commercial company (Campus Technologies Freiburg GmbH).

* Email: abopdawaffo@alasu.edu

Introduction

Following its discovery by George Smith in the early 1980's, phage display technologies have been built predominantly from DNA phage platforms, particularly that of M13 [1–5]. M13 is DNA-filamentous bacteriophage with a genome size of 6.4 kb [6] and have very low mutation rates that limit their use in *in vitro* evolution processes. On the contrary, RNA-based replication systems possess attractive features, including high mutation rates, high population size and short replication times, that can be exploited for rapid *in vitro* evolution [7]. Additionally, RNA-replication systems lack recombination processes that can further complicate DNA-based replication systems and technologies. Early efforts to generate recombinant RNA had limited success due to limitations in technology and RNA instability. However, with the improvement of recombinant DNA technology, and the existence of reverse transcription techniques, the generation of recombinant RNA is now straightforward. Recent advancements have led to the generation and cloning of Qβ cDNA into several stable plasmids that are able to liberate phage upon bacterial transformation [8]. The cDNA of Qβ coliphage RNA has become amenable for use in displaying random peptide libraries *in vitro* followed by *in vivo* translation and phage production.

Qβ belongs to the family of *Leviviridae* and is found throughout the world in bacteria isolates associated with sewage [9]. Of the four groups of RNA coliphages, the genome and proteins of Qβ phages have been the most extensively characterized [10]. Some representatives of these groups are: group I (f2, MS2, R17, fr) group II (GA) group III (Qβ) and group IV (SP) [11]. In this report we present a framework of peptide display and affinity maturation using Qβ phage and the integrin receptor of Foot-and-Mouth-Disease-Virus (FMDV) as a proof-of-concept for acquiring binders to a highly infectious agent with many different serotypes. FMDV, the causative agent of the most economically important infectious diseases in farm animals, has seven serotypes (O, A, C, SAT₁, SAT₂, SAT₃ and Asia 1, [12]). The varied nature of the serotypes compromises the ability to control this disease using present vaccination strategies. Furthermore, the instability of currently available vaccines leaves farmers with no practical option but to slaughter, emphasizing the urgent need for new vaccines [13]. FMDV is a single stranded positive-sense RNA virus of ~8 kilobases (kb). FMDV particles consist of four major polypeptides, three outer capsid proteins (VP1, VP2 and VP3) and a fourth smaller capsid protein (VP4). The G-H loop of VP1 is of particular interest due to its major antigenic site at the carboxyl terminal [14–16].

Both FMDV and Qβ have icosahedral shells of 30 nm and 25 nm in diameter, respectively [17,18]. The Qβ genome is ~4.2 kb surrounded by a shell of 180 coat protein molecules

[17,18]. Of these proteins, A2, A1 (known as readthrough) and the replicase are encoded by the phage genome and are important for the formation of infectious phage [19]. Due to its copy number and position [20], we hypothesize that A1 can be utilized for phage display. Phage display, previously called phage exposition, consists of an insertion of a foreign DNA fragment into the minor structural phage A1 gene to create a fusion protein, which is then incorporated into a virion that retains its infectivity and exposes the foreign peptides in an accessible form at the surface [1].

We constructed hybrid phages bearing FMDV VP1 G-H loop C-terminus that efficiently binds monoclonal antibodies directed against the antigenic loop. Furthermore, display of randomized peptides allowed *in vitro* Qβ phage selection, evolution and convergence on a displayed peptide containing a tandem amino acid sequence required for anti-FMDV monoclonal antibody recognition. The specificity, productivity, affinity and efficiency of the hybrid phage were characterized. Additionally, our data provides an insight into FMDV antigen motif representing candidates for development of vaccines for livestock.

Materials and Methods

Reagents

All media for bacteria culture and phages were purchased from Fisher Scientific (Pittsburgh, PA). Restriction enzymes and *T4 DNA ligase* were purchased from New England BioLabs (Ipswich, MA). Unless otherwise indicated, chemical reagents (ie. RbCl and CaCl$_2$) were purchased from Sigma-Aldrich (St. Louis, MO).

Microorganisms

Escherichia coli MC1016 (Invitrogen, Grand Island, NY) was used to grow and maintain plasmids. *E. coli* HB101 was used to grow and maintain pBRT7Qβ, pQβ8 plasmids and all their recombinant derivatives. Three different indicator bacteria were used for phage production and titration: K12 (*E. coli* ATCC 23725), HfrH (*E. coli* ATCC23631) and Q13 (*E. coli* ATCC 29079) purchased from ATCC. The *E. coli* bacteriophage Q-β *ATCC 23631-B1* was used as a positive wild type (wt) control in experiments.

Antibodies

The FMDV VP1 G-H loop specific antibody, SD6, was obtained from Professor Esteban Domingo's laboratory from the Department of Virology and Microbiology of the University of Madrid, Spain. Anti-green fluorescent protein (GFP) polyclonal antibody was from Biofuture Group in Goettingen, Germany. Anti-protein tHisF and HisJ polyclonal antibodies were obtained from Professor Hans-Joachim Fritz'laboratory from the Institute for Microbiology and Genetics of the University of Goettingen, Germany.

Hybrid phage construction

Plasmids pBRT7Qβ and pQβ8 were obtained from Professor Weber [21] and from Professor Kaesberg groups [8] respectively. These plasmids pBRT7Qβ having 7489 bp (from 1 to 7489 when restricted with *Sma*I [21]) and pQβ8 having 7393 bp (from 1 to 7393 when restricted with *Sma*I endonuclease [8]) were used for this work since they both contain the entire cDNA of the Qβ phage with different orientation. For the cloning procedure into the pBRT7Qβ plasmid, *Afl*II and *Nsi*I restriction sites were used. All the primers used to amplify foreign functional protein genes were flanked with *Afl*II or Bpu10I (forward) and *Esp*I or *Nsi*I (reverse) restriction sites (Table 1). These primers were designed to confer some important features to the foreign gene after cloning: to

maintain the reading frame of the vector and to maintain the important secondary structure of phage RNA for replication transcription, translation, regulation and assembly. The *Esp*I site is absent at the end sequence of the A1 protein gene of the pBRT7Qβ plasmid. To introduce this site, DNA fragments were transiently cloned into some intermediate plasmids. After PCR, the foreign gene insert was cloned into the pCR2.1 Topo vector. The pCR2.1 vector is a linearized vector ready for direct ligation of unmodified, unpurified PCR products. This vector has a single overhanging "T" which facilitates the cloning of a *Taq* and *Phusion* amplified fragment. This vector does not contain the *Afl*II and *Esp*I restriction sites and the PCR products are cloned between two *Eco*RI restriction sites. The recombinant pCR2.1 plasmid enables amplification with higher fidelity and sequencing of the PCR fragments for further cloning.

The correct insert was later cloned into a pUC-cassette vector, which allows the introduction of new restriction sites like *Esp*I. The pUC-cassette is a recombinant pUC18, containing the C-terminal of the cDNA of the A1 protein gene (from 2129–2402), which introduces the *Esp*I restriction site prior to cloning, into pBRT7Qβ. This part of the A1 gene is cloned between the *Hind*III and *Kpn*I restriction sites. Modifications aiming to add or subtract part of the foreign gene fragment to be cloned into the pBRT7Qβ plasmid were performed on the recombinant pUC18. When these manipulations were successfully performed in small size plasmids, the foreign protein gene inserts were cloned back into the pBRT7Qβ/pQβ8 plasmid as presented in Fig. 1. The recombinant pBRT7Qβ was used to generate RNA phage displaying the exogenous functional peptide. To further explore this new display technology, other functional proteins with different specific motifs that are larger than, but related to the FMDV GH-loop in the structure were also studied. These functional proteins were: the green fluorescent protein (GFP), the imidazole glycerol phosphate subunit of the synthase thermostable subunit (tHisF) and the periplasmic histidine-binding protein (HisJ).

Deletion of A1 protein construction in pBRT7Qβ and pQβ8

To insert larger DNA fragments at the end of the A1 gene, we deleted the last 162 nucleotides of this gene keeping the interregional A1 gene and the replicase gene. These plasmids are called pBRT7QβΔA1 or pQβ8ΔA1 derived from pBRT7Qβ or pQβ8 respectively. To construct recombinant QβΔA1 plasmid, either a short portion of the cDNA of Qβ of about 420 bp was amplified with PCR or a gene part of 162 bp (between *Afl*II and *Nsi*I restriction sites) was removed and replaced by a short adaptor gene sequence. In the case of PCR, the forward primer (Table 1) used was flanked by *Bpu*10I and the reverse primer was flanked by the tag and the *Nhe*I and *Nsi*I sequences. *Nhe*I was added to monitor the cloning process. The PCR product was cloned into the pBRT7Qβ or pQβ8 plasmids using *Bpu*10I and *Nsi*I. The adaptor oligos were annealed and ligated into the *Esp*I enzyme restriction site.

Strategy for phage production

Positive clones were transformed into *E. coli* HB 101 after sequencing using the method of Taniguchi and collaborators [22]. The supernatants of overnight clones were checked by agar overlay method for the presence of phages. The phages were amplified using an indicator bacteria cell, *E. coli* HfrH, Q13, 1101 or K12. Fresh overnight cultures (on standard nutrient agar I plate) were amplified at 37°C to reach an OD_{600} of 0.6–0.8 (after 2–3 h) and inoculated with phage suspension at a multiplicity of 3.

Table 1. Oligonucleotides.

Name	Sequence	Functions
ABW1	atgcatttcatccttagGCTAGCttactacgacttaagatagatgaattgttcgatgttaccg	For A1 deletion with NheI and NsiI restriction site reverse
ABW2	cagctgaacccagcgtatTGAacgttgctcattgccggtggtggctc	For A1 deletion forward before Bpu10I site used
CB191	TTACACCGCCAGTGCACGCGCGGGGATCTTGCTCACCTAACGACGAC	FMD-loop adaptor
CB192	TAAGTCGTCGTTAGGTGAGCAAGATCCCCGCGTGCACTGGCGGTG	Complementary of CB191
CB193	TTACACCGCCAGNGCANNNNNNNNNNNNNNNTCACCTAACGACGA	Randomize-FMD-loop adaptor
CB194	CTGGCGGTG	Complementary of CB193 first
CB195	TAAGTCGTCGTTAGGTG	Complementary of CB193 second
CB197	ACCTTCAACCTCAATTCTTGTGTTC	For sequencing Qβ cDNA for reverse from G 2410
CB198	TGCGTGATCAGAAGTATGATATTCG	For sequencing of Qβ cDNA of end of A1 gene and insert forward from T 2083

The infected cells were incubated at 37°C by shaking (150 rpm) for 5 h. After this incubation period, the phage titer was checked and a second round of amplification was performed to scale up the phage titer according to the same procedure. At the end of amplification, indicator cells were allowed to complete the lysis by adding few drops of chloroform to the culture suspension.

Agar overlays for spot test

This test was done according to Adam [23]. A bacteria culture was grown to log phase (OD$_{600}$ of 0.6–0.8). A volume of 100 µl of this culture was added to 3 ml of YT-Top-agar and the mixture was poured on the surface of nutrient agar plates. The plates were left to solidify at 37°C for a few minutes. Thereafter, 4–7 µl of the phage suspension was dropped on the solidified plates. The plates were incubated at 37°C for 24 h and examined for lysis of the *E. coli* lawn where the droplet of phage suspension was placed.

Measurement of phage yield and plaque quality

Indicator bacteria (100 µl) were infected with 100 µl of the appropriate serial dilution of phage-containing tryptone glucose yeast (TGY) solution. After 10 minutes of incubation at room temperature, 3 ml of soft-agar (TGY with 0.6% agar) was added and the mixture was poured onto plain agar plates. The plates were allowed to solidify and incubate for 16 h at 37°C. The plaque count was done following the method of Pace & Spiegelman [24]. We observed both quality and size of plaques.

Reverse transcription (RT) PCR from plaques and/or purified phages

Phages from the clear zone of specific plaques were extracted by excising the soft media to a tube, adding 10–15 µl of H$_2$O and centrifuged at 3000 rpm for 5 min to remove the media. 10 µl of the supernatant was used for RT. For the purified phage, following RNA extraction, 2–5 µg of RNA was incubated at 99.6°C with 50 pmol of reverse primers for 2 min in a total volume of 11 µl (filled with RNase-free water). This was followed by incubation on ice to allow the annealing of primers to the template. To the mixture, the following components were added: 10 µl 5×RT-buffer, 2 µl MgSO$_4$ (25 mM), 1 µl 10 mM dNTP mix, 1.5 µl AMV-RT (5 U/µl) and 24.5 µl of H$_2$O RNase-free.

To allow reverse transcription, the mixture was incubated at 42°C for 1 h followed by AMV denaturation at 94°C for 2 min. For the PCR reaction, 2 µl of the reverse transcriptase was used with the following protocol. The cycling protocol consisted of 25 cycles of three temperatures: 94°C, 30 s (strand denaturation), 50–57°C, 1 min (primer annealing), 68°C, 2 min (primer extension), followed by a final extension at 68°C for 7 min.

Selection with biopanning

The antibodies were adsorbed to Xenobind™ microtiter plates (Dunn in Asbach, Germany) for biopanning. The middle wells of the plates were covered with 150 µl of the antibody solution (2.5 µg/ml in carbonate buffer: 15 mM Na$_2$CO$_3$ and 35 mM NaHCO$_3$ pH 9.6) and incubated at room temperature for overnight. To cover the surface of the wells, a solution of 5% bovine serum albumin (BSA) in the antibody solution was added and then incubated for 1 hour at room temperature. The excess of unbound BSA was removed by washing 3 times with wash buffer (137 mM NaCl, 2.7 mM KCl, 8.3 mM Na$_2$HPO$_4$ 2H$_2$O, 1.5 mM NaH$_2$PO$_4$ at pH 7.2 and 0.05% Triton X-100). A phage solution of 150 µl was then added to experimental wells of the plate and the plate was incubated at room temperature for 4 h and washed twice with wash buffer and 2 additional times with phage buffer. The experimental wells were then covered with 200 µl of *E coli* Q13 or HfrH culture (grown to OD$_{600}$ of 0.7) and incubated at 37°C for 20 min. The bacteria culture from experimental wells was transferred into tubes as aliquots under sterile conditions after incubation. One aliquot was plated for phage titration and the rest incubated at 37°C overnight. For the next round of biopanning, 150 µl of the previous overnight bacteria and panning were used as phage solution. To further characterize phages from rounds of panning, 50 µl of phages from each round were used to extract RNA. This RNA was subjected to RT-PCR and sequencing reactions.

Phage purification and analysis

Phage was collected using polyethylene glycol (PEG) precipitation as described in [25] with minor modifications (using PEG$_{8000}$). Phage suspension (cell debris and phage) was incubated with 8% PEG and 0.5 M NaCl (final concentration, respectively) overnight at 4°C. The phages were pelleted by centrifugation at 3000 rpm for 20 min at 4°C (Sorvall GSA). The pellet was resuspended in phage buffer (10 mM Tris HCl pH 7.5, 1 mM MgCl2, 100 mM NaCl, 10 mg/l gelatine with 1/5 of the volume phage suspension after amplification) at 4°C for 20 min and pelleted by centrifugation at 10000 rpm for 20 min at 4°C (Sorvall RC5B, SS34). The procedure was repeated. After overnight incubation the phages were collected by centrifugation and the pellet was suspended in a

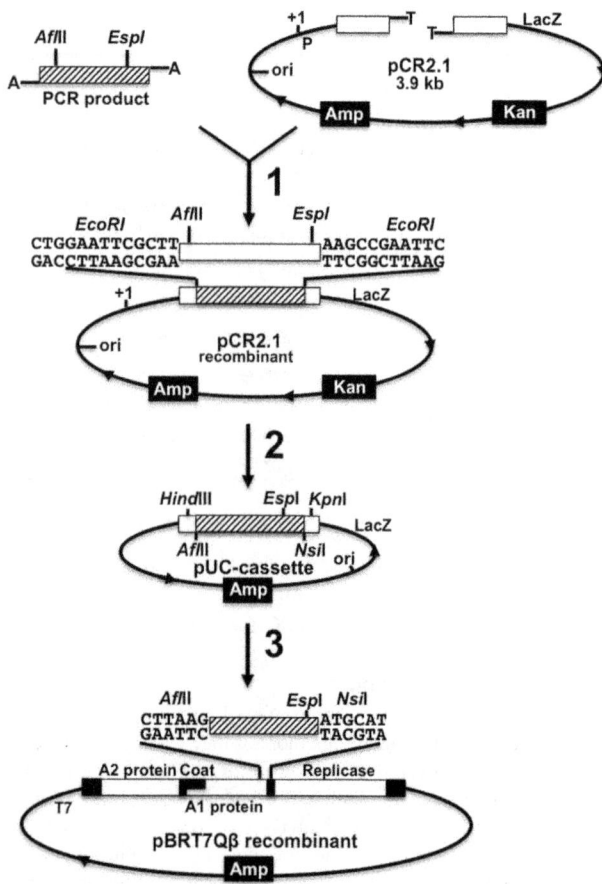

Figure 1. Schematic representation of the RNA phage display vector construction. General cloning procedure from PCR fragments to pBRT7Qβ with transient cloning in the pUC18-cassette working plasmid. Step 1: cloning of PCR fragment into pCR2.1 vector; Step 2: cloning of the foreign gene from PCR into the pUC-cassette (with *Nsi*I) using *Afl*II and *Esp*I sites; Step 3: Cloning of the foreign gene into pBRT7Qβ using *Afl*II and *Nsi*I. P: promoter; *Amp*: ampicillinase gene; *Kan*: kanamycin resistance gene; ori: origin of replication.

small amount of phage buffer (50 μl) without gelatine. The suspension was centrifuged at 15000 rpm for 20 min at 37°C (Sorvall RC5B, SS34) and the supernatant containing phages were collected and subjected to DEAE sepharose or CsCl-gradient purification. For long-term storage the phage phase was stored in 50% glycerol at −80°C.

Another phage amplification procedure was done based on Gschwender and Hofschneider [26]. In this procedure at the log phase infected cells were incubated in high Mg^{2+} (200 mM) to inhibit cell lysis after infection. Phages were extracted from bacterial sedimentation after lysis were induced with 50 mM EDTA on ice. The suspension was adjusted to pH 9.5 by the addition of 1 M NaOH under vigorous stirring. The cellular debris was removed by low-speed centrifugation.

Immuno-precipitation of hybrid phages against respective displayed peptide-antibody

Agarose double diffusion of Ouchterlony and Nilsson [27] was used to test the presence of foreign protein on phage surface with modifications. 1% agarose gel solution in assay buffer (50 mM Tris-HCl pH 7.5; 0.1 M NaCl) was poured into 10 cm petri dish, and allowed to solidify. Six wells were punched at equal distance

from the center well. To each well 50 μl of the appropriate concentration of phages or antibody was added and incubated for 24 h at room temperature.

Electron microscopy

A carbon-coated Formvar grid was filled with 5 μl of purified phage solution diluted to the titer of 10^9 plaque-forming units per ml (p.f.u/ml). The solution was left on the grid for a short while and then a few drops of aqueous uranyl acetate were added. Slides were then observed under a JEOL 1200EX electron microscope.

Results

Tolerance of Qβ A1 gene to manipulation

Initially, two variants of the pBRT7Qβ plasmid were constructed: pBRT7QβESPI and pBRT7QβNOTI. In these plasmids, additional nucleotides were added to the 3′end of A1 gene to introduce multiple cloning sites (Fig. 1). For pBRT7QβESPI, 6 nucleotides were added to introduce an *Esp*I site, and 9 nucleotides were added to pBRT7QβNOTI to introduce a *Not* I site. We tested if these extensions allowed proper DNA packing and the production of infectious-competent phage. Indeed, we show that 3 different gene fusions with A1 placed in front of the natural opal and ochre stop codons (TGA and TAA), produced phage plaques in bacterial lawns (Fig. 2). These results suggest that the 3′- end of A1 can accept minor extensions without disturbing the function of phage infectivity. We next explored the lengths of extensions and their effect on infectivity. Various DNA lengths (15–850 bp) were successfully fused with the A1 gene (Figs. 2 & 3), but only recombinant plasmids containing foreign inserted DNA with lengths between 15–300 bp produced phage plaques. These results show that the length of the inserted DNA is critically important for this novel system. Next we tested whether the 3′ end of the A1 gene is critical and important. To accomplish this, we constructed the plasmid pBRT7QβΔA1, in which non-essential sequences of the cDNA of the Qβ genome were deleted from the 3′-terminus of the A1 protein gene. Specifically, we deleted a 162 bp part of the 3′ terminus of the A1 gene (between nucleotides 2271 and 2333) and replaced it with a short adaptor gene sequence of 33 bp leaving the original intercistronic region between A1 gene and the replicase gene intact. Interestingly, these recombinant plasmids with 3′ truncations of A1, still produced phage plaques. However, further deletion of the A1 protein gene beyond nucleotides 2271 at 5′ end or 2333 at 3′ end abolished phage production. Furthermore, we tested whether the orientation of Qβ cDNA within the plasmid is critical. We created identical constructs using pBRT7Qβ and pQβ8, both of which contain the entire cDNA of phage Qβ albeit in opposite orientations. These plasmids yielded phages with similar titers to the wt, suggesting that the orientation of phage cDNA does not influence the phage production. Positive recombinant pBRT7Qβ or pQβ8 plasmids were identified via restriction enzyme (Fig. 3) prior to sequencing and transforming into Qβ for characterizing the display of foreign peptides and proteins.

Phage production and resulting titers

To produce wild type and recombinant phages, all plasmid vectors and variant constructs were transformed into *E. coli* HB101. *E. coli* HB101 bacteria were selected because they lack the pili appendage (F⁻) necessary for Qβ absorption and infection. This insures exclusive usage of high-fidelity DNA polymerase-mediated replication of Qβ genes and prevents premature evolutionary events. Similar phage titers for both wt and recombinant phages (~10^8 to 10^9 p.f.u/ml) were obtained with

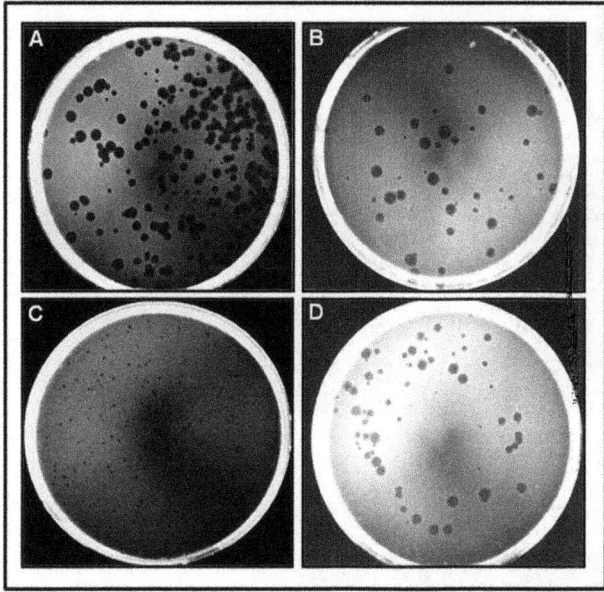

Figure 2. Morphology of wild type vs. hybrid Qβ phage plaques. Panel A) wild type Qβ phages; Panel B) Qβ-FMDV VP1 G-H loop phages; Panel C) Qβ-GFP rescued phages from *E. coli* SURE (expression host with F⁺) over-expressing A1-GFP protein infected with wild type Qβ. Panel D) QβΔA1 phages. All at very low multiplicity of infection (MOI), and all plates are exactly 1 day (24 hours) old when photographed.

plaque sizes ranging between 1 mm and 3 mm in diameter in both wt and variants. The plate of phage harboring the 3′-truncation of A1 minor coat protein gene was dominated by smaller (1 mm and 70%) than larger (3 mm and 30%) sized plaques as shown in Fig. 2. However, some minor differences in plaque size were also observed. We interpret these differences as due to either the nature of the quasispecies within the phage population and/or effects associated with the insert size (Fig. 2). Next, we analyzed the phages with different sized plaques using RT-PCR and wt Qβ was used as a standard (Fig. 4). Finally, all cDNAs were sequenced to confirm the presence of the appropriate foreign gene within the hybrid phage. Results show that sequences encoding foot-and-mouth disease virus (FMDV), HisJ and HisF, appended onto Qβ-A1 allow assembly of plaque-forming phage particles containing the gene fusions.

Efficient Qβ-FMDV phage library display and biopanning

Rapid fitness gains are the main goal of molecular evolution and are directly proportional to the population size and the selection pressure. To mimic this process *in vitro*, we synthesized a randomized VP1 G-H loop library (YTAXA**XXXX**XHLTT) that corresponded to the immunodominant region of VP1 including the canonical RGD epitope (YTASA**RGD**LAHLTT) using three oligonucleotides CB193, CB194 and CB195 as depicted in (Fig. 5). We then cloned the randomized library into Qβ plasmids (pBRT7Qβ and derivatives) and used monoclonal antibody (mAb) SD6 as the constant selective target in a biopanning assay. Additionally, the original sequence of the VP1 G-H loop of FMDV serotype C clone C-S8cl was cloned into pBRT7Qβ using the annealed oligonucleotides: CB191 and CB192 (Fig. 5) as a control. Recombinant plasmid (pBRT7Qβ-FMDV), derived from the previous vector harboring the VP1 G-H loop fused with A1 within the phage cDNA was used to transform *E. coli* HB101.

Only 50–55% of these clones produced phage plaques (not shown). Positive VP1 G-H loop clones were validated through RT-PCR and sequencing. We further confirmed the presence of the G-H loop of the hybrid phages through dot blot using mAb SD6 (Fig. 6). We further randomized the G-H loop to form a library, which was directly ligated with A1 of cDNA of Qβ. The Qβ-FMDV phages were found to produce clear plaques in all wt Qβ natural hosts namely: *E. coli* K12, Q13, and HfrH. As before, we recapitulated our data in HB101, showing that, as with wt phage, Qβ-FMDV phage can be propagated in the other *E. coli* strains. There were no significant differences in the yield of phage particles between Qβ-FMDV and Qβ. Finally, these Qβ-FMDV phages were amplified in Q13 cells (chosen among other Qβ hosts) and a high titer (10^9 p.f.u./ml) was obtained and purified by ultracentrifugation on CsCl gradient for guinea pig immunization.

Non-canonical FMDV epitope

To gain fitness, the synthesized library of the G-H loop was selected using a modified biopanning protocol [1,2,28] with mAb SD6. This modified protocol selects and amplifies phage while avoiding acidic elution of phages selected (Fig. 7). We reasoned that removal of an acidic elution step would enhance phage viability and the overall efficiency of *in vitro* evolution. Additionally, media containing the indicator bacteria *E. coli* Q13 grown to the log phase was added directly to the plate to further enhance survival of hybrid phages. After each round, an aliquot of phage was used for RT-PCR and the resulting DNA sequence was compared to the wild type sequence (Fig. 8). The sequence comparison of the randomized VP1 G-H loop after six rounds of biopanning revealed a shift of mAb SD6 binding motif from *Arg-Gly-Asp* to *Xxx-Arg-Gly*. Due to the preservation of the *Arg* and *Gly* in all three rounds of biopanning, we conclude that this tandem pair is essential for mAb SD6 binding. The third amino acid was substituted without disturbing the binding capacity of the peptide to the mAb SD6. Over 80% of glycine exposed by the phage was in contact with the mAb SD6. This clearly shows that arginine and glycine were not just only together in the antibody binding motif but were representing optimized amino acid from a randomized pool.

Immunization characterization of Qβ-FMDV phages

To assess the immunogenicity of the hybrid phages, guinea pigs were immunized with purified Qβ-FMDV phages. The serum obtained after immunization tested positive for FMDV antibody (Professor Esteban Domingo, personal communication). We validated this finding with a qualitative Ouchterlony assay (27) using serum obtained from immunization with Qβ-FMDV phages against the same phages on one hand, then Qβ wt. In this assay, antibody and antigen solution are placed in nearby wells cut out of a thin layer of agarose and allowed to diffuse toward each other forming a visible line of precipitation where they meet. The lines produced by the two adjacent wells containing Qβ-FMDV and Qβ wt join together in a pattern of partial identity (Fig. 9A). More specifically, at least two epitopes on Qβ-FMDV were recognized by the serum antibody. A similar result was obtained with immunoglobulin purified from the serum with protein A affinity column (results not shown). The fractionated Igs from the column did not react with phage displaying the C-terminal deletion of A1 (pQβ8ΔA1; Fig. 9C). Furthermore, the thickness of the precipitation line in the double diffusion was reduced with the reduction in the serum amount shifting towards the phages wells (Fig. 9B). The second line of precipitation close to the phage wells was reduced with half of the phages titer (Fig. 9C). The additional line of precipitation on Ouchterlony assay plate showed the presence

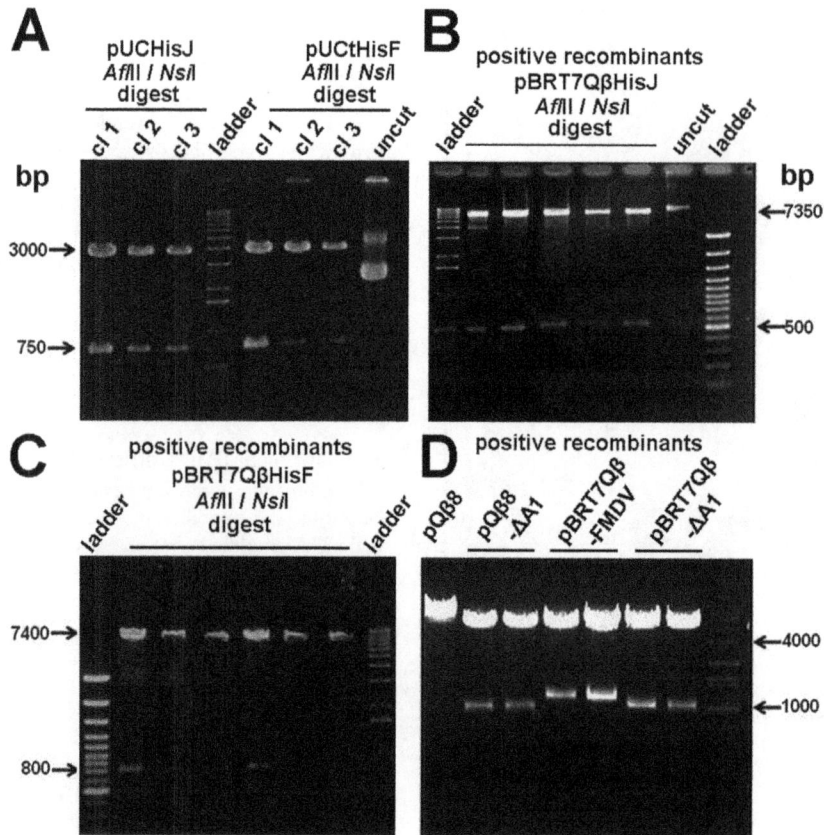

Figure 3. Agarose gel electrophoresis of the RNA display system vector construction. Panel A) Lanes 1–3: positive recombinant pUCHisJ plasmid clone (cl) restricted with *Afl*II and *Nsi*I; Lanes 5–7: positive pUCtHisF and Lane 8: negative clone. Panel B) Lanes 2–6: positive recombinants pBRT7QβHisJ restricted with *Afl*II and *Nsi*I; Lane 7: negative clone. Panel C) Lanes 2–7: positive recombinants pBRT7QβtHisF restricted with *Afl*II and *Nsi*I. Panel D): Lane 1: pQβ8 negative control; Lanes 2 and 3: positive recombinants pQβ8ΔA1; Lanes 4 and 5: positive recombinants pBRT7Qβ-FMDV; Lanes 6 and 7: positive recombinants pBRT7QβΔA1 all restricted with *Nhe*I. Lanes "ladder" were loaded with the 100 bp or 1 kb DNA ladder.

of an epitope on Qβ-FMDV that is absent on both wt Qβ and pQβ8ΔA1. We conclude that this additional line of precipitation can only be the VP1 G-H loop exposed on the exterior surface of Qβ. This result illustrated and clarified the heterogeneity and specificity of the serum obtained from immunized guinea pigs.

Finally, Qβ-FMDV presence and display was validated using negative stain electron microscopy that shows the presence of antigen-antibody interaction. The wt phage particle was found with a clear zone around its particles (Fig. 10A), while Qβ-FMDV phages displaying the G-H loop of FMDV showed dots decorating its surface by mAb SD6 (Fig. 10C). We theorize that, these dots

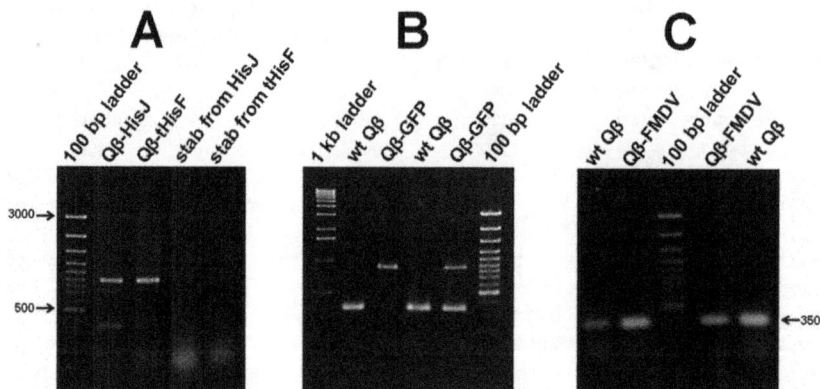

Figure 4. RT-PCR of RNA purified from Qβ-phage plaques. Panel A) Lane 2: Qβ-HisJ; lane 3: Qβ-tHisF; lane 4: soft agar stab from HisJ plate; lane 4: soft agar stab from tHisF. Panel B) Lanes 2 and 4: wild type Qβ; Lanes 3 and 5: Qβ-GFP. Panel C) Lanes 2 and 4: Qβ-FMDV; Lanes 1 and 5: wild type Qβ (positive control). The 100 bp and 1 kb DNA ladder were used.

CB 191
5' - <u>TTA</u>CACCGCCAGTGCACGCGGGGATCTTGCTCACCTAACGACGAC___ - 3'
3' - GTGGCGGTCACGTGCGCCCCTAGAACGAGTGGATTGCTGCTGAAT - 5'
CB 192

CB 193
5' - <u>TTA</u>CACCGCCAGNGCNNNNNNNNNNNNNNNNNTCACCTAACGACGAC___ - 3'
3' - GTGGCGGTC   GTGGATTGCTGCTGAAT - 5'
CB 194   **CB 195**

Figure 5. Design and schematic of FMDV VP1 G-H loop (serotype C-S8cl) oligonucleotide sequences. Panel A) Original sequence used for cloning (with *Esp*I site at both ends) into Qβ and production of phages Qβ-FMDV for guinea pig immunization. Panel B) Randomized sequence synthesis (with ends similar to Panel A) for library generation and phage population production and selection against mAb SD6.

represent the position of A1 fused with the G-H loop of FMDV on the exterior surface of phage Qβ (arrow). The combined results of animal immunization, EM and serological assays, suggested that, on the exterior surface of Qβ phages, there can be exposed epitopes which may be used to induce the production of specific antibody.

Discussion

Current phage display technologies have been exclusively designed using DNA-based phage platforms such as M13 [1–5,28]. Use of such systems for library screening purposes requires highly diverse starting libraries encoding the displayed proteins since *in vitro* evolution is difficult, due to the relatively high-fidelity of proof-reading bacterial DNA polymerases. The work of Drake [29] showed a surprisingly consistent overall genomic mutation rate for DNA-based replication in plants, yeast, bacteria and bacteriophage. When the mutation rate per base was multiplied by the genome size, the mutation rate per genome for these DNA-based replicating organisms fell within an astonishingly narrow range (0.0022–0.0046; Table 2). Genomic replication error rates for RNA-based replicating MNV11RNA and bacteriophage Qβ were 7× and ~500× greater, respectively, revealing a high degree of inherent genomic evolution potential. Compared to M13-based mutation rates, Qβ has a 420× greater mutation rate per base, indicating a major advantage in utilization of Qβ for *in vitro* evolution. Moreover, relatively harsh acidic elution procedures common to DNA phage systems add another obstacle in

developing a practical system for *in vitro* evolution. The RNA phage Qβ possesses key features that can be exploited for *in vitro* evolutionary display that overcomes both obstacles [30]. We describe here a framework for Qβ display that allows a robust *in vitro* evolution at the lab benchtop.

Developing Qβ phage for peptide display included four steps. Firstly, the DNA sequence coding for the displayed peptides or proteins was fused to the end of the A1 minor coat protein gene in the DNA plasmids utilized. A1 is essential for infection, but specific roles in the Qβ virion cycle have not yet been elucidated. Our results indicated that the C-terminus of A1 (nucleotides 2271 to 2333), plays only a minor role in function and is consistent with a previous report describing the importance of the N-terminus [31]. Secondly, recombinant plasmids were sequenced and transformed in bacteria cells to produce a population of phages. Thirdly, hybrid

Figure 6. Dot blotting analysis of the FMDV VP1 G-H loop displayed on Qβ phages with SD6 mAb. Spot A: Supernatant of the Qβ wild-type culture infection (negative control); Spot B: Qβ-G-H loop phages from the supernatant of culture after 5 h of infection; Spot C: Qβ-G-H loop phages from the supernatant after overnight infection (higher concentration); Spot D: the phage buffer only as a negative control. Phages from spots A, B and C were purified by PEG₈₀₀₀/NaCl precipitation and CsCl gradient. The same pattern was obtained with the spotted corresponding crud lysat.

Figure 7. Schematic of biopanning assay with Qβ phage derivatives without the usual acidic elution. A) A population of phages displaying the library of interest (here randomized VP1 G-H loop) was added to the well of a plate pre-coated with the desired target (in this case, mAb SD6 covalently immobilized with F$_c$ region). B) Indicator *E coli* are added to the well after phages having low-affinity to the target are removed. C) High-affinity phage bound to target can infect *E. coli* Q13 by adsorbing and injecting its RNA via the F⁺ pilus. D) Phages newly obtained after indicator *E. coli* infection were transferred to new wells containing the immobilized target for the next round of biopanning.

AGTGCACGCGGGGATCTTGCT Ser Ala Arg Gly Asp Leu Ala	FMDV strain S8C
AGnGCAnnnnnnnnnnnnnnnT	Randomized sequence
AGgGCAng t nggngnn t nn t T	after transformation
AGgCARnTASGGTGcMSATGT	1 biopan round
AGaGCRnTASKGKSYCCaYGT	2 biopan rounds
AGAGCaRTASgGGG t CCRYGT	3 biopan rounds
Arg Ala V/I R/G Gly Pro R/C	3 biopan rounds

Figure 8. Sequences comparison of the randomized FMDV G-H loop displayed after three rounds of Biopanning. The first line up is the original loop motif *Arg-Gly-Asp* of VP1 G-H loop of FMDV strain S8C. On the left hand side under the original sequence are the sequences obtained after a round of biopanning. The low case letters show the different between the sequences. The last line is the expression of the evolution of the original motif sequence on the 3rd round shown the maintanance of *Arg-Gly* and change of the last amino acid.

phages were subjected to several rounds of selection and amplification to optimize interactions with the target. Fourthly, selected phages were analyzed for the presence of the correct recombinant RNA and the ability to display the appropriate protein. We examined the limits of the sizes of displayed proteins by attempting display of HisJ (726 bp), HisF (753 bp) and GFP (714 bp). Unfortunately, all three proved too large for Qβ, as evidenced by poor infectivity and/or plaque formation, even when fused to full-length A1 or A1 3′ truncation. The transformation efficiency of these constructs dropped with the increasing size of the insert but could be somewhat improved by using the rubidium chloride method with heat shock at 43.5°C, in contrast to conventional methods [32]. The Qβ phage remains functional (able to absorb and infect) with up to 60 nucleotides inserted into the A1 gene. The phage can also function with 162 nucleotides removed from the 3′ end of the A1 gene. These observations together give the A1 gene a total loading capacity of ~222 nucleotides (74 amino acids protein). When more than 300 nucleotides of foreign gene were inserted, no viable, stable, infective replicable phages were obtained. Although spontaneous Qβ particles have been previously obtained with the A1 protein extended to 195 amino acids [33], neither replication competency

nor recombinant RNA-genomic packaging limits were determined. An enormous drop in phage viability was found using MS2 phage to display five amino acids (Ala-Ser-Ile-Ser-Ile) on the exterior surface [34] revealing potential limitations on the secondary structure of the recombinant RNA on its influence on the replication, regulation and assembly to form viable phage [35].

Successful transformation and subsequent amplification was achieved with the A1 3′ truncation plasmid and the 14 amino acid G-H loop of FMDV plasmid. FMDV is a particular threat for animal livestock worldwide and the large number of serotypes and diversity of strains make the development of a universal vaccine very challenging. Through its RNA replication system, FMDV has a very high mutation rate which allows the virus to escape drug suppression [7]. Considering the highly mutable character of RNA viruses, we reasoned that a vaccine system that can also exploit this feature of RNA viruses would be highly valuable. The displayed G-H loop was found to occupy the corners of the icosahedron of Qβ as visualized by negative-stain electron microscopy, which corresponds to its natural positions in the FMDV structure [36]. The recombinant A1-G-H loop was found to decorate the phage at 12 copies per virion, indicating that the wt A1 protein must also have 12 copies per Qβ phage. Prior to these results, the exact copy number of the A1 protein was not well known but was estimated to be between 3–7% of total phage protein [20,37,38]. An octapeptide of β-tubulin motif was also found to decorate the 12 corners of the Qβ icosahedron (unpublished data). This is in contrast to the commonly used phage M13 where any fusion to the minor coat protein gene, pIII, would display the foreign peptide on only one side of the phage [39].

We next exploited the high error-rate of the RNA-dependent RNA polymerase of Qβ to test the feasibility of using this system for *in vitro* evolution. Transforming a completely degenerate DNA that appended seven randomly encoded amino acids to the A1 protein, with a total theoretical diversity of 2×10^{14}, has resulted in a six amino acid peptide library with an actual diversity of 2×10^{8}. In only six rounds of biopanning and selective pressure in the presence of immobilized SD6 antibody, we isolated the Qβ phage that contained a conserved tandem pair, *(Arg-Gly)*, that is essential for mAb SD6 recognition. The main motif of the G-H loop was known to be Arg-Gly-Asp that recognizes and binds to mAb SD6 [14,15]. Upon randomization and selection, the Arg-Gly motif was found to be enough for binding to the same mAb SD6, amongst the spectrum of variants generated. Panning with Qβ-FMDV phages has a double advantage of binding to antibody (selection)

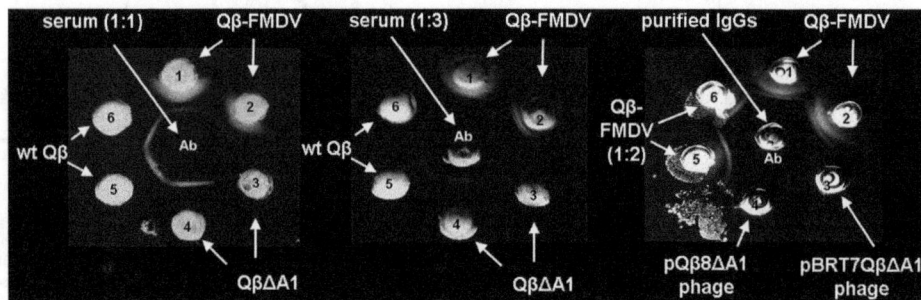

Figure 9. Ouchterlony double diffusion assay. A) Wells 1 and 2 represent Qβ-FMDV phages; wells 3 and 4 represent QβΔA1 phages; wells 5 and 6 represent wild-type Qβ; center well contains polyclonal serum from immunized guinea pig (labeled "Ab"). B) Same as panel A but with 1/3 of the serum concentration. C) Wells 1 and 2 Qβ-FMDV are the same as panel A; wells 5 and 6 contain half the phage titer of wells 1 and 2; well 3 represents phages from pBRT7QβΔA1 and well 4 represents phages from pQβ8ΔA1; center well contains IgGs purified from serum panel A and B (labeled "Ab"). The line of precipitation is visible as a white haze forming a half-circle around some of the wells in the experiments.

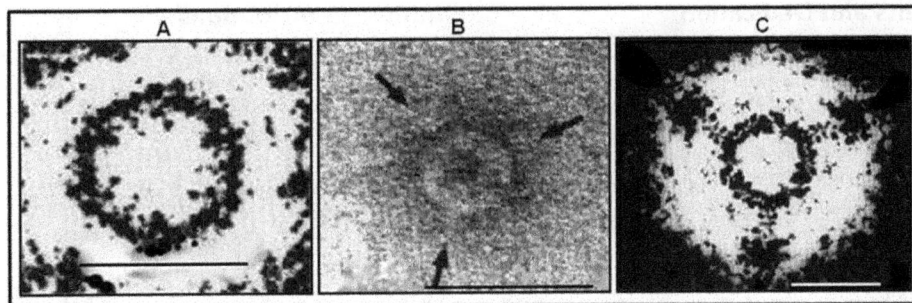

Figure 10. Field light micrograph of modified Qβ. A) Unlabelled Qβ virion: the original particle projection obtained with conventional transmission electron microscopy of a negatively stained sample was treated by Marham rotation 3 times 120\grad\intervals and printed at reversed contrast; Magnification Bar: 50 nm. B) Negatively stained Qβ, modified at A1 gene products by additional of FMDV G-H loop motif decorating the corners by IgG of mAb SD6 against VP1 G-H loop motif at 120\grad\intervals (arrows); Magnification Bar 25 nm. C) Phage particles projection depicted in B was treated by Makham rotation and printed at reversed contrast; arrows showing antibody against integrin motif attached to the corners of the virus particle; Magnification 25 nm.

and being amplified without conventional acidic elution of the phage. The phages were selected by the A1 protein extension which does not hinder adsorption via the pilus, allowing the phage to inject RNA. To mimic the natural infection of Qβ during panning, prevents acidic elution and neutralization steps previously needed before further enrichment and amplification [40–42]. This newly developed panning method can be exploited in new antibody selection from a pool of mRNA since it is very similar to the natural one without intervention of rough chemicals like in the case of most DNA phage display technologies.

Serum from guinea pigs immunized with Qβ-FMDV exhibited cross reactivity with intact FMDV as well as the hybrid Qβ phage displaying the G-H loop in a qualitative Ouchterlony assay. The fact that Qβ exist as quasi species with very high mutation rate [43–48] is a double advantage for Qβ-FMDV hybrid phages which contain a pool of antigens (quasispecies with mutant spectra) important for immunization and potential vaccine. A pool of different antigen FMDV G-H loop strains can be exposed on Qβ surface and used as vaccine.

In conclusion, we have developed a peptide library display system using the Qβ RNA-coliphage that efficiently mimics evolutionary adaptation and affinity maturation. A randomized G-H loop of the FMDV VP1 protein was exposed on the exterior surface of Qβ and selected against the G-H loop mAb SD6. Guinea pig serum immunized with the hybrid phages (Qβ-FMDV) contained immunoglobulin specific to FMDV and the Qβ-FMDV hybrid phages. These hybrid phages could principally serve as good candidates for FMDV vaccine development. Robust viability and infectivity was achieved with a C-terminal A1 deletion that maintained 12 copies per virion. Current size limitations of display are ~74 amino acid peptide/protein domain since larger domains (e.g. GFP) could not be displayed. With further optimization of A1-appended sequences, Qβ display of larger protein domains may eventually prove possible. The Qβ *in vitro* evolution platform we describe here may be readily adapted for the development of nanotechnology including novel biosensors, therapeutics, immunization reagents or crystallization scaffolds.

Table 2. Table adapted from Drake [29].

Species name	Genome size (bases)	Target	Mutation rate (per base)[a]	Mutation rate (per genome)
N. crassa	4.19×10^7	*ad-3AB, mtr*	7.2×10^{-11}	0.0030
S. cerevisiae	1.38×10^7	*URA3, CAN1*	2.2×10^{-10}	0.0031
E. coli	4.70×10^6	*lacI, hisGDCBHAFE*	4.6×10^{-10}	0.0022
bacteriophage M13	6.41×10^3	*lacZα*	7.2×10^{-7}	0.0046
bacteriophage λ	4.85×10^4	*cI*	7.7×10^{-8}	0.0038
bacteriophage T2	1.60×10^5	*rII*	2.7×10^{-8}	0.0043
bacteriophage T4	1.66×10^5	*rII*	2.0×10^{-8}	0.0033
MNV11RNA[b]	86	itself	3×10^{-4}	0.026
bacteriophage Qβ[b]	4.2×10^3	replicase	3×10^{-4}	1.9
Taq polymerase	n/a	n/a	2×10^{-5}	n/a

[a]In the cases where multiple targets were measured, the average is presented.
[b]Taken from Domingo [48].

Acknowledgments and Dedication

This paper is dedicated to the memory of Prof. Dr. Christof K Biebricher of the Max-Planck-Institute of Biophysical Chemistry of Gottingen, Germany, who was an initial sponsor of this research project. Our thanks and sincere appreciation to Professor Esteban Domingo, for the SD6 mAb, neutralization assay, and the serum production against Qβ-FMDV. The authors are grateful to Professor Weber for pBRT7Qβ and to Professor Alexander Chetverin for pQβ7 and pQβ8 plasmids.

Author Contributions

Conceived and designed the experiments: ABW CS. Performed the experiments: ABW CS. Analyzed the data: ABW CS SGA. Contributed reagents/materials/analysis tools: ABW SGA. Wrote the paper: ABW SGA.

References

1. Smith GP (1985) Filamentous fusion phage: novel expression vectors that display cloned antigens on the virion surface. Science 234: 211–212.

2. Smith GP, Scott JK (1993) Libraries of peptides and proteins displayed on filamentous phage. Meth Enzymol 217: 228–257.

3. Smith GP, Petrenko VA (1997) Phage display. Chem Rev 97: 391–410

4. Petrenko VA, Smith GP (2000) Phages from landscape libraries as substitute antibodies. Protein Eng 13: 589–592.

5. Rakonjac J, Bennett NJ, Spagnuolo J, Gagic D, Russel M (2011) Filamentous bacteriophage: biology, phage display and nanotechnology application. Curr Issues Mol Biol 13: 51–76.

6. Marvin DA (1998) Filamentous phage structure, infection and assembly. Curr Opin Struct Biol 8: 150–158.

7. Domingo E, Holland JJ (1997) RNA virus mutations and fitness for suvival. Annu Rev Microbiol 51: 151–178.

8. Shaklee PN, Miglietta JJ, Palmenberg AC, Kaesberg P (1988) Infectious positive- and negative-strand transcript RNAs from bacteriophage Qβ cDNA clones. Virology 163: 209–213.

9. Furuse K, Osawa K, Kawashiro J, Tanaka R, Ozawa A et al. (1983) Bacteriophage distribution in human faeces: continuous survey of healthy subjects and patients with internal and leukaemic diseases. J Gen Virol 64: 2039–2043.

10. Furuse K (1987) Distribution of coliphage in the general environment: general considerations. In Phage Ecology, pp. 87–124. Edited by S. M. . Goyal, C. P. Gerba & G. . Bitton. New York: Wiley.

11. Bollback JP, Huelsenbeck JP (2001) Single-stranded RNA bacteriophage (family Leviviridae). J Mol Evol 52: 117–128.

12. Brown F (1999) Foot-and-mouth disease and beyond: vaccine design, past, present and future. Arch Virol 15: 179–188.

13. Kahn S, Geale DW, Kitching PR, Bouffard A, Allard DG et al. (2002) Vaccination against foot-and-mouth disease: the implications for Canada. Can Vet J 43: 349–354.

14. Logan D, Abu-Ghazaleh R, Blakemor W, Curry S, Jackson T (1993) Structure of a major immunogenic site on foot-and-mouth disease virus. Nature 362: 566–568.

15. Domingo E, Verdaguer N, Ochoa WF, Ruiz-Jarabo CM, Sevilla N et al. (1999) Biochemical and structural studies with neutralizing antibodies raised against foot-and-mouth disease virus. Virus Res 62: 169–175.

16. Verdaguer N, Sevilla N, Valero ML, Stuart D, Brocchi E et al. (1998) A similar pattern of interaction for different antibodies with a major antigenic site of foot and mouth disease virus: implications for intratypic antigenic variation. J Virol 72: 739–748.

17. Brown F, Cartwright B (1961) Dissociation of foot-and-mouth disease virus into its nucleic acid and protein components. Nature 192: 1163–1164.

18. Blumenthal T, Carmichael GC (1979) RNA replication: function and structure of Qβ replicase. Annu Rev Biochm 48: 525–548.

19. Weber K, Konigsberg W (1975) Proteins of RNA phages. In RNA Phages, pp 51–84. Edited by N. D. . Zinder. Cold Spring Harbor, New York: Cold Spring Harbor Laboratory.

20. Hofstetter H, Monstein HJ, Weissmann C (1974) The read-through protein A$_1$ is essentiel for the formation of viable Qβ particles. Biochim Biophs Acta 374: 238–251.

21. Barrera I, Schuppli D, Sogo JM, Weber H (1993) Different mechanismus of recognition of bacteriophage Qβ plus and minus-strand RNAs by Qβ replicase. J Mol Biol 232: 512–521.

22. Taniguchi T, Palmiri M, Weissmann C (1978) Qβ DNA-containing hybrid plasmids giving rise to Qβ phage formation in the bacteria host. Nature 274: 223–228.

23. Adam MH (1959) In: Bacteriophages, pp. 473–490. Inter-Science Publishers: New York.

24. Pace NR, Spiegelman S (1966) In vitro synthesis of an infectious mutant RNA with a normal RNA replicase. Science 153: 64–67.

25. Yamamoto KR, Alberts BM, Benzinger R, Lawhorne L, Treiber G (1970) Rapid bacteriophage sedimentation in the presence of polyethylene glycol and its application to large-scale virus purification. Virology 40: 734–744.

26. Gschwender HH, Hofschneider PH (1969) Lysis inhibition of φX174-, M12 and Q β–infected Escherichia coli bacteria by magnesium ions. Biochim Biophys Acta 190: 454–459.

27. Ouchterlony O, Nilsson LA (1978) Immunodiffusion and immunoelectrophoresis. In: Handbook of experimental immunology. Edited by, Weir, D. M., 3rd Edition, Oxford: Blackwell Scientific Publication.

28. Azzazy HME, Highsmith Jr EW (2002) Phage display technology: clinical applications and recent innovations. Clininical Biochemistry 35: 425–445.

29. Drake JW (1991) A constant rate of spontaneous mutation in DNA-based microbes. Proceedings of the National Academy of Sciences 88(16): 7160–7164.

30. Ferrer-Orta C, Arias A, Escarmis C, Verdaguer N (2006) A comparison of viral RNA-dependent RNA polymerases. Curr Opin Struct Biol 16: 27–34.

31. Vasiljeva I, Kolzlovska T, Cielens I, Strelnnikova A, Kazaks A et al. (1998) Mosaic Qβ coats as a new presentation model. FEBS Letters 431: 7–11.

32. Kerri M, Titball RW (1996) Transformation of Burkholderia pseudomallei by electroporation. Anal Biochem 242: 73–76.

33. Kozlovska TM, Cielens I, Vasiljeva I, Strelnikova A, Kazaks A et al. (1996) RNA phage Qβ coat protein as a carrier for foreign epitopes. Intervirology 39(1–2): 9–15.

34. Van Meerten D, Olsthoorn RCL, van Duin J, Verhaert RM D (2001) Peptide display on live MS2 phage: restriction at the RNA genome level. J Gen Virol 82: 1797–1805.

35. Beekwilder MJ, Nieuwenhuizen R, van Duin J (1995) Secondary structure model for the last two domains of single-stranded RNA phage Qβ. J Mol Biol 247: 903–917.

36. Long D, Abu-Ghazaleh R, Blakemore W, Curry S, Jackson T et al. (1993) Structure of a major immunogenic site on foot-and-mouth disease virus. Nature 362: 566–568.

37. Rumnieks J, Kaspars T (2011) Crystal structure of the read-through domain from bacteriophage Qβ A1 protein. Protein Science 20: 1707–1712.

38. Weiner AM, Weber K (1971) Natural read-through at the UGA termination signal of Qβ coat protein cistron. Nat New Biol 234: 206–209.

39. van Rooy I, Hennink WE, Storm G, Schiffelers RM, Mastrobattista E (2012) Attaching the phage display-selected GLA peptide to liposomes: factors influencing target binding. Eur J Pharm Sci 45: 330–335.

40. Jenkins GM, Rambaut A, Pybus OG, Holmes E (2002) Rates of molecular evolution in RNA viruses: a quantitative phylogenetic analysis. J Mol Evol 54: 156–165.

41. Scott JK, Smith GP (1990) Searching for peptides ligands with an epitope library. Science 349: 386–390.

42. Beer M, Liu C-Q (2012) Panning of a phage display library against a synthetic capsule for peptide ligands that bind to the native capsule of Bacillus anthracis. Plos One 7: e45472. doi:10.1371/journal.pone.0045472

43. Eigen M (1971) Selforganization of matter and evolution of biological macromolecules. Naturwissenschaften 58: 465–523.

44. Eigen M, Schuster P (1977) The hypercycle – a principal of natural selforganisation. Naturwissenschaften 64: 541–565.

45. Eigen M, McCaskill J, Schuster P (1989) The molecular quasispecies. In Prigogine, I. & Rice, S. A. Edited by, Adv Chem Phys 75: 149–263 John Wiley & Sons, Inc.

46. Steinhauer D, Domingo E, Holland JJ (1992) Lack of evidence for proofreading mechanisms associated with an RNA virus polymarase. Gene 122: 281–288.

47. Schuster P, Stadler PF (1994) Landscapes: complex optimization problems and biopolymer structures. Comput Chem 18: 295–324.

48. Domingo E, Biebricher CK, Eigen M, Holland JJ (2001) Quasispecies and RNA virus evolution: principles and consequences (p. 173). Austin: Landes Bioscience.

18

Carcinoembryonic Antigen-Related Cell Adhesion Molecules (CEACAM) 1, 5 and 6 as Biomarkers in Pancreatic Cancer

Florian Gebauer[1,2*,∮]**, Daniel Wicklein**[2∮]**, Jennifer Horst**[2]**, Philipp Sundermann**[1]**, Hanna Maar**[2]**,
Thomas Streichert[3]**, Michael Tachezy**[1]**, Jakob R. Izbicki**[1]**, Maximilian Bockhorn**[1]**, Udo Schumacher**[2]

1 Department of General, Visceral and Thoracic Surgery, University Medical Center Hamburg-Eppendorf, University of Hamburg, Hamburg, Germany, **2** Institute of Anatomy and Experimental Morphology and University Cancer Center Hamburg (UCCH), University Medical-Center Hamburg-Eppendorf, Hamburg, Germany, **3** Institute of Clinical Chemistry, University Medical-Center Hamburg-Eppendorf, Hamburg, Germany

Abstract

Background: Aim of this study was to assess the biological function in tumor progression and metastatic process carcinoembryonic antigen-related cell adhesion molecules (CEACAM) 1, 5 and 6 in pancreatic adenocarcinoma (PDAC).

Experimental Design: CEACAM knock down cells were established and assessed in vitro and in a subcutaneous and intraperitoneal mouse xenograft model. Tissue and serum expression of patients with PDAC were assessed by immunohistochemistry (IHC) and by enzyme linked immunosorbent assays.

Results: Presence of lymph node metastasis was correlated with CEACAM 5 and 6 expression (determined by IHC) and tumor recurrence exclusively with CEACAM 6. Patients with CEACAM 5 and 6 expression showed a significantly shortened OS in Kaplan-Meier survival analyses. Elevated CEACAM6 serum values showed a correlation with distant metastasis and. Survival analysis revealed a prolonged OS for patients with low serum CEACAM 1 values. In vitro proliferation and migration capacity was increased in CEACAM knock down PDAC cells, however, mice inoculated with CEACAM knock down cells showed a prolonged overall-survival (OS). The number of spontaneous pulmonary metastasis was increased in the CEACAM knock down group.

Conclusion: The effects mediated by CEACAM expression in PDAC are complex, though overexpression is correlated with loco-regional aggressive tumor growth. However, loss of CEACAM can be considered as a part of epithelial-mesenchymal transition and is therefore of rather importance in the process of distant metastasis.

Editor: Aamir Ahmad, Wayne State University School of Medicine, United States of America

Funding: The authors have no support or funding to report.

Competing Interests: The authors declare that no competing interests exist.

* Email: fgebauer@uke.uni-hamburg.de

∮ These authors contributed equally to this work.

Introduction

In recent years, the prognosis of patients suffering from pancreatic ductal adenocarcinoma (PDAC) has not improved significantly [1,2]. Most patients present with advanced tumor stages making curative treatment impossible. Complete surgical resection, followed by adjuvant chemotherapy is nowadays the gold-standard in therapy of PDAC. However, even in patients after complete surgical tumor resection, the overall survival remains poor with a median survival time of 20–24 months [3–5]. Aggressive loco-regional tumor growth in addition to early distant and peritoneal metastasis and a high degree of chemoresistance make PDAC one of the most lethal gastrointestinal tumors [6].

Today, neither reliable serum nor tissue markers predicting the clinical course of patients after diagnosis of PDAC are available. Furthermore, the molecular interactions of the tumor with the host and the local factors that allow PDAC to display such an aggressive progression are poorly understood. Therefore, there is an imperative need for a better understanding of the tumor biology with respect to the mechanisms of local tumor invasion and recurrence.

Carcioembryonic antigen-related cell adhesion molecules (CEACAMs) are members of the glycosylphosphatidylinositol (GPI)-linked immunoglobulin (Ig) superfamily [7]. There are more than 17 genes that belong to this family, with their gene products primarily integrated into the cell membrane. Within the CEACAM family the CEACAM subtypes are structurally similar and physiologically expressed on the apical surface of numerous cell types, e.g. endothelial and hematopoietic cells as well as epithelial cells of different organs. Depending on the cell type and CEACAM subtype, the transmitted effect after binding a certain partner varies, including regulation of cell adhesion, tumor suppression, angiogenesis, activation of leukocytes and other immuno-reactive cells, and regulation of the cell cycle [8–11]. The CEACAM 5 gene, its product also known as CD66e codes for the carcioembryonic antigen (CEA) and has become one of the best-known members of the Ig-superfamily since it has a significant role in the clinical routine as a tumor marker for several tumor entities including gastrointestinal and respiratory malignancies [12]. However, due to lacking sensitivity and specificity its predictive value alone is still unsatisfying [13–16]. The CEACAM subtypes 1 and 6 are described to be under- or overexpressed in several tumor entities like lung cancer, colon cancer and melanoma [17–23]. Though overexpression is widely observed, some studies even report a decreased expression in certain tumor entities at different tumor stages.

Interestingly, recent studies found CEACAM 1 and 6 expression in primary PDAC correlated with a shortened overall patient survival [24,25]. The biological principles why CEACAM expression mediates tumor progression are not fully understood. We therefore analyzed the effects of CEACAM 1, 5 and 6 in vitro and established a xenograft mouse model to investigate the functional role of CEACAM expression in PDAC. To assess whether CEACAM expression has an impact on the tumor progression in patients, we combined serum and immunohisto-chemical analysis for CEACAM molecules 1, 5 and 6 to analyze whether there is a correlation of CECACM expression with clinico-pathologcial data of patients with PDAC.

Material and Methods

Cell line and CEACAM knock down

The human pancreatic adenocarcinoma cell line PaCa 5061 was established from a primary tumor. Detailed characteristics of establishment and culture of the cell line have been described previously [26]. Briefly, cells were obtained from a patient with a PDAC who underwent surgical resection in the Department of General, Visceral and Thoracic Surgery at the University Medical Center Hamburg-Eppendorf. The final tumor classification according to the UICC 7th ed. revealed a pT3, N1, L1, V1, R0, G2 PDAC of the pancreatic head.

CEACAM knock down variants were conducted by shRNA interference as described previously [27]. Oligonucleotides were cloned into a pSIREN-RetroQ vector and transfected by FuGENE transfection agents (Roche Diagnostics, Hilden, Germany) with the tumor cells. Selection of knock down clones was performed by expression of a Puromycin chemoresistance (Clontech, Saint-Germain-en-Laye, France) and flow-cytometer sorting (FACS-LSR Fortessa, BD Bioscience, San Jose, USA). Only cells with a knock down of >90%, as determined by flow cytometry, were used for in vitro and in vivo experiments. Unconjugated antibodies against CEACAM 1, 5 and 6 and the corresponding mouse biotinylated IgM or rat IgM isotype control (Dako, Glostrup, Denmark) were detected with goat anti-mouse Ig-APC (BD Biosciences). Cells were analyzed using a CyFlow cytometer (Partek, Münster, Germany) with a subsequent addition of AlexaFluor488 conjugated streptavidin (Invitrogen) before staining.

In vitro characterization of CEACAM knock down cells

Cell proliferation was assessed using colorimetric XTT assay (Roche Diagnostics, Basel, Switzerland) according to manufactur-er's instructions. Cells were plated in 96-well plates at 3000 cells/well. After 48 h to allow for cell adherence, cells were incubated with colorimetric substrates. Colorimetric changes were measured in a multi-well spectrophotometer (MR5000 Multiplate Reader, Dynatech, Denkendorf, Germany).

Differences in cell migration were assessed using FluoroBlok Migration Assay (BD Bioscience, San Jose, USA) with 24-well 8 micron pore size inserts. Cells were trypsinized and re-suspended in serum-free RPMI1640 (Invitrogen, Darmstadt, Germany) in a concentration of 300.000 cells/mL. 400 µl cell suspension was added to the apical chamber and 800 µl RPMI1640 with 10% fetal calf serum (FCS, Invitrogen) was added to the bottom chamber. The assay was incubated for 24 h under standard cell culture conditions.

After removal of the chemo attractant of the bottom chamber, visualization of migrated cells was performed by adding 500 µl/well HBSS buffer with Calcein AM (Invitrogen) 4 µg/mL in the bottom well and incubation for 1 h. Readout was conducted at 494/517 nm (Ex/Em) on a Genios bottom-reading fluorescence plate reader (Tecan, Männedorf, Swizerland).

Laminar flow experiments were performed using IBIDI microslides VI (IBIDI, Munich, Germany) connected to a syringe pump (Model 100 Series; kdScientific, Holliston, MA) and cell movement was observed with an inverted microscope (Zeiss, Jena, Germany; Axiovert 200). Tumor cells were suspended in cell culture medium (20 ml, 200 000 cells/ml) and microslides were coated with Human Pulmonary Microvascular Endothelial Cells (HPMEC) (PromoCell, Heidelberg, Germany). HPMEC were suspended in cell culture medium, seeded in microslides at a concentration of 5×10^5 cells/ml and 20 µl medium with cells was pipetted into each flow channel. Cell grew confluent over night under standard conditions. Applied shear rates ranged from 0.05 dyn/cm^2 to 10.0 dyn/cm^2. Cell movement was recorded and analyzed with regard to the quality of movement (adhesion, rolling and tethering) and rolling velocity using CapImage 8.5 program (Dr. Heinrich Zeintl, Heidelberg, Germany).

Animal experiments

Animal experiments were conducted according to the UKCCR guidelines for the welfare of animals in experimental neoplasia [28], the locals Ethics committee for animal experiments (Behörde für Soziales, Familie, Gesundheit, Verbraucherschutz; Amt für Gesundheit und Verbraucherschutz; Billstr. 80, D-20539 Hamburg, Germany, project No. G58/09) and as well the institutional animal welfare officer of the University Medical-Center recommended and approved the study.

For the subcutaneous tumor model, one million PaCa 5061 tumor cells were injected in the right scapula region in 8–12 week old C57BL/6N pfp$^{-/-}$/rag2$^{-/-}$ double-knockout mice. Animals were sacrificed when primary tumors exceeded 2 cm^3 or ulcerated the mouse skin, the mice were terminally narcotized and sacrificed by cardiocentesis.

For assessment of the influence of CEACAM expression on peritoneal dissemination, we established an intraperitoneal tumor model as previously described [29]. Briefly, one million tumor cells were injected in the lower left abdominal quadrant intraperitoneally in suspension volume of 200 µL. Assessment and time points of termination of the experiment was conducted according to a previous established scoring system. Assessment of the extend of the intraperitoneal tumor growth was done with a modified peritoneal carcinomatosis index (PCI) as described previously [30]. Briefly, the peritoneal cavity is divided in 9 abdomino-pelvic regions, depending on the extend of tumor growth, scores between 0 and 3 points are assigned (0 points: no tumor present; 1 point:

tumor <1 mm^3; 2 points: tumor >1 and <3 mm^3; 3 points: tumor >3 mm^3) that lead to a PCI score from 0 to 27.

Quantification of pulmonary metastasis, disseminated tumor cells (DTC) and CTC by Alu-PCR

The left lungs were homogenized in a sample disruptor (TissueLyser II, Qiagen, Hilden, Germany) and subjected to DNA-isolation (QIAamp DNA Mini Kit, Qiagen). Bone marrow was collected by flushing the left femora with 1 ml NaCl 0.9%. 200 µl blood and the bone marrow suspensions were subjected to DNA- isolation using the QIAamp DNA Blood Mini Kit.

DNA concentrations of all samples were quantified using a NanoDrop spectrophotometer (Peqlab, Erlangen, Germany). As the content of detectable Alu-sequences in the following qPCR would have been affected simply by varying DNA-concentrations, all lung- and bone marrow-DNA samples were normalized to 30 ng/µl using AE buffer (Qiagen). The concentrations of blood-DNA were quite similar in all samples (approx. 10 ng/µl) and were therefore not normalized. qPCR was performed with established human-specific Alu-primers [31]. 2 µl total DNA (i.e., 60 ng lung/bone marrow-DNA; 20 ng blood-DNA) were used for each qPCR. Numerical data were determined against a standard curve as described [32]. The detection limit for specific human Alu-sequence signals was determined for each tissue type by testing DNA from five healthy (non-injected) pfp$^{-/-}$/rag2$^{-/-}$ mice of similar sex and age. For each sample, analyses were performed in duplicates and as independent experiments at least twice.

Patients and surgical procedures

Between 1992 and 2009, all patients who underwent major resectional pancreatic surgery at the Department of General, Visceral and Thoracic Surgery at the University Medical Centre Hamburg-Eppendorf were included in a prospective, pancreatic database. The study was approved by the Ethics Committee of the Chamber of Physicians in Hamburg, Germany. Written consent for using the samples for research purposes was obtained from all patients prior to surgery or blood drawing.

Patients with PDAC of the pancreatic head region routinely underwent either partial pancreatoduodenectomy (PD) or pylorus-preserving duodenopancreatectomy (PPDP) and organ-preserving resection methods in cases of chronic pancreatitis (CP). Only patients with macroscopic complete tumor resection were included in the final analysis. In-hospital mortality was defined as death at any time during the entire period of hospitalization. Follow-up information was obtained from our institution's outpatient clinic, from the appropriate general practitioners' offices, or from the regional cancer registry. When the date of death was not recorded, patients were censored at the last recorded contact.

Tissue micro array construction and immounohistochemistry

Tissue cores were obtained from formalin-fixed paraffin-embedded (FFPE) tissue blocks from patients with pathologically proven PDAC. Representative areas of the tumor were selected based on hematoxylin-eosin staining.

TMA construction was performed as previously described [33–35]. Briefly, 252 tissue cylinders with a diameter of 0.6 mm were punched from the "donor" tissue blocks using a custom-made semiautomatic robotic precision instrument and placed into one paraffin block that contained the 252 individual samples. Within these samples, there were 142 PDACs, 40 neuroendocrine pancreatic tumors (NET), 33 intraductal papillary mucinous

neoplasm (IPMN), and 37 samples of healthy tissue as a negative control. The resulting TMA blocks were used to produce 4-µm sections that were transferred to an adhesive-coated slide system (Instrumedics Inc., Hackensack, New Jersey, USA).

The immunohistochemical staining protocols were optimized on various benign and malignant tissues in an extensive multistep procedure that modified the staining protocol until the required selective staining was achieved with the lowest possible background signal (according to [36]).

Sections were deparaffinized and dried overnight at 37°C. Antigen retrieval was performed by microwave oven treatment in citrate buffer (pH 6.0) for 1 min, sections were then rehydrated in Tris-buffered saline (TBS; 0.05 M Tris-HCl at pH 7.6 and 0.15 M NaCl) and blocked with rabbit AB serum (Biotest Diagnostics, Dreirach, Germany) diluted 1:10 in TBS for 60 minutes. CEACAM staining was performed using a specific CEACAM monoclonal antibody (CEACAM1 clone 4D1/C2 IgG2a, in-house clone (previously described by [37]) at a dilution of 1:200, CEACAM5 clone #2383 IgG1 at a dilution of 1:50 (Cell Signaling, Beverly, USA); CEACAM6 clone IgG1 (9A6), at a dilution of 1:40 [Sigma Aldrich, Hamburg, Germany]) overnight at 4°C. Biotinylated secondary polyclonal rabbit anti-mouse antibodies (Dako, Hamburg, Germany) were used for binding the CEACAM primary antibody. Epithelial-mesenchymal transition (EMT) markers were studies by immunohistochemistry as well. ZEB 1 (IgG at a dilution of 1:100, polyclonal rabbit anti-human, Atlas Antibodies, Stockholm, Sweden), ZEB 2 (IgG at a dilution of 1:100, polyclonal rabbit anti-human, Atlas Antibodies, Stockholm, Sweden), E-Cadherin (), Pan-Cytokeratin ().

The binding sites were detected using the ABC-AP-Kit (Vector Laboratories Inc., Burlingame, USA). Alkaline phosphatase activity of a biotin-streptavidin–alkaline phosphatase complex was visualized using naphthol-AS bisphosphate as substrate and hexatozised New Fuchsin was used for simultaneous coupling. The sections were counterstained with Mayer's hemalum (Merck, Darmstadt, Germany).

The staining intensity (0, 1+, 2+, 3+) and the fraction of positive tumor cells were scored for each tissue spot as recently published [34]. Spots without staining and with a staining intensity of 1+ in $<70\%$ and 2+ in $<30\%$ of the tumor cells were scored as CEACAM low, medium scores were given for a staining intensity of 1+ in $\geq70\%$, 2+ in $\geq30\%$ or 3+ in $<30\%$ of the tumor cells, and high scores were given for a staining intensity of 2+ in $\geq70\%$ or 3+ in $\geq30\%$ of the tumor cells. Immunohistochemical analysis of the sections was performed without knowledge of the patients' identity or clinical status. Immunohistochemical analysis and scoring were performed by two independent investigators who were unaware of the patient outcome or other clinical findings. In 95% of the samples, the evaluations of the two observers were identical, the remaining slides were re-evaluated, and consensus decisions were made.

The staining protocol was as well used for mice grown tumors and showed similar sensitivity and specificity compared to the TMA staining.

Enzyme linked immunosorbent assay (ELISA)

For quantification CEACAM subtypes in peripheral blood, serum samples of 46 Caucasian patients with PDAC and 47 Caucasian patients with CP, who were indicated for surgical treatment, were analyzed with an enzyme linked immunoassay (ELISA). All blood samples were obtained directly before surgery. As healthy controls, 50 Caucasian blood-bank donors, obtained from the institute for transfusion medicine (University Medical Centre Hamburg-Eppendorf), were included in the study.

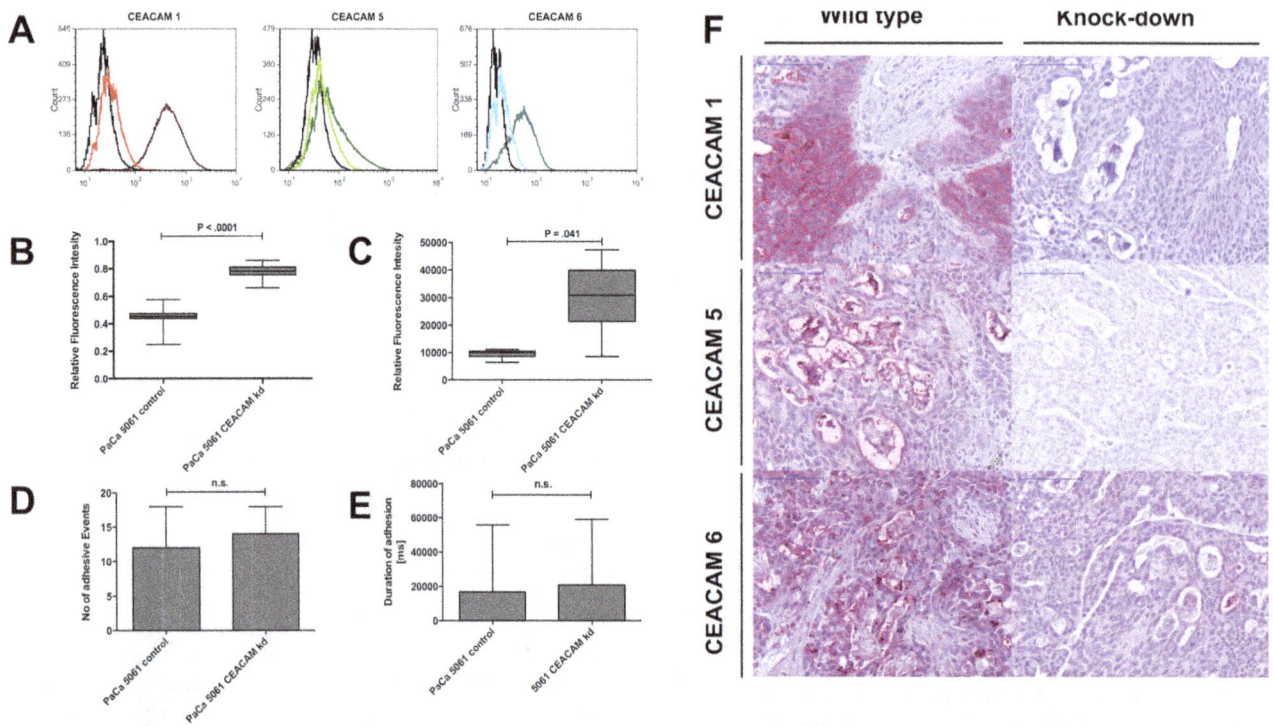

Figure 1. CEACAM levels decreased to <5% for CEACAM 1, 5 and 6 after shRNA knock down as shown in flow cytometry. (A). Basal CEACAM expression varied depending on the CEACAM subtype, highest levels were found for CEACAM 1, intermediate for CEACAM 6 and lowest levels for CEACAM 5. As seen in proliferation (B) and migration assays (C), CEACAM knock down cells showed a higher proliferative and migrative potential compared to control cells. Differences in adhesion on stimulated (TNFα) endothelial cells were not detectable between CEACAM kd and control cells. Neither the average number (D) nor the average adhesion time to the endothelial cells was different (E). Murine xenograft tumors were highly positive for CEACAM 1, 5 and 6. Immunohistochemical staining in control cells showed complete absence of CEACAM expression in immunohistochemistry in knock down cells (F).

Preparation of serum samples were conducted according to a standardized protocol [38]. Median age was 62.4 years at time of diagnosis (range 36.3–90.4 years, 24 male [52.2%], 21 female [47.8%]). Serum values of 47 patients who underwent surgery due to chronic pancreatitis (median age at time of diagnosis 47.0 years, range 31.1–76.1 years) and 50 samples of healthy blood donors were used as controls (25 male [50%], 25 female [50%]).

For the detection of CEACAM 1 and CEACAM 5 in the serum, 96-well flexible microtiter plates (Costar 9019, USA) were coated with 50 µl per well of 2 µg/ml of monoclonal mouse capture antibody (Clone 283324 and 843130, mouse IgG1, R&D systems, USA) overnight at 4°C. Wells were blocked with 3% w/v bovine serum albumin (BSA; Fraction V, 98% purity, Sigma Aldrich, Germany) in PBS/T (phosphate buffered saline, pH 7.3, containing 0.05% v/v Tween) for 45 min at room temperature and then incubated for 1 h with human sera diluted 1:50 in PBS at room temperature. After five washes with PBS/T, bound protein was detected with biotin-conjugated polyclonal antibody anti-CEA-CAM 1 or 5 (CEACAM 1 Goat IgG, clone 842284; CEACAM 5 sheep IgG, clone 843131Goat IgG, R&D systems, USA), respectively, followed by streptavidin-conjugated peroxidase using TMB (3,3′, 5,5″-tetramethylbenzidine) as substrate. The color reaction was stopped by addition of 10 mM H_2SO_4 and analyzed at 450 nm using an ELISA reader (DynaTech MR 5000, USA). Human recombinant CEACAM 1 and 5 proteins (R&D systems) served as an internal standard for the assay.

CEACAM 6 was determined by a direct ELISA coating 50 µl patient serum in a 1:50 dilution over night a 4°C in the 96-well plate. Wells were blocked with 3% bovine serum albumin (BSA;

Fraction V, 98% purity, Sigma Aldrich) in PBS/T (phosphate buffered saline, pH 7.3, containing 0.05% v/v Tween) for 45 min at room temperature and then incubated for 1 h with the detection antibody anti-CEACAM 6 (mouse IgG1, clone 843158, R&D systems). The color reaction was stopped by addition of 10 mM H_2SO_4 and analyzed at 450 nm. To ensure that the immunoassay was suitable for measuring clinical serum samples, reproducibility and linearity were examined (according to [39]).

Statistical analysis

For explorative statistical analysis of the individual patient groups, either a two-sided chi-square test or a Fisher exact test was used. Quantitative variables were either tested by means of the Student *t-test* or by medians of the Wilcoxon test. Test for normal distribution of the quantitative variables was performed by Kolmogorov-Smirnov-Test. Kaplan–Meier analysis (log-rank test) was used for disease free- and overall-survival analysis excluding in-hospital mortality. All variables achieving a P value ≤ 0.05 were included in a multivariate cox-regression model. The cut-off level of serum CEACAM quantification was determined by using the Youden-index and as described previously [33,40].

Results

In vitro characteristics of CEACAM knock down PDAC cells

As analyzed by flow cytometry, basal CEACAM expression was present on the tumor cells. Surface levels of the CEACAM

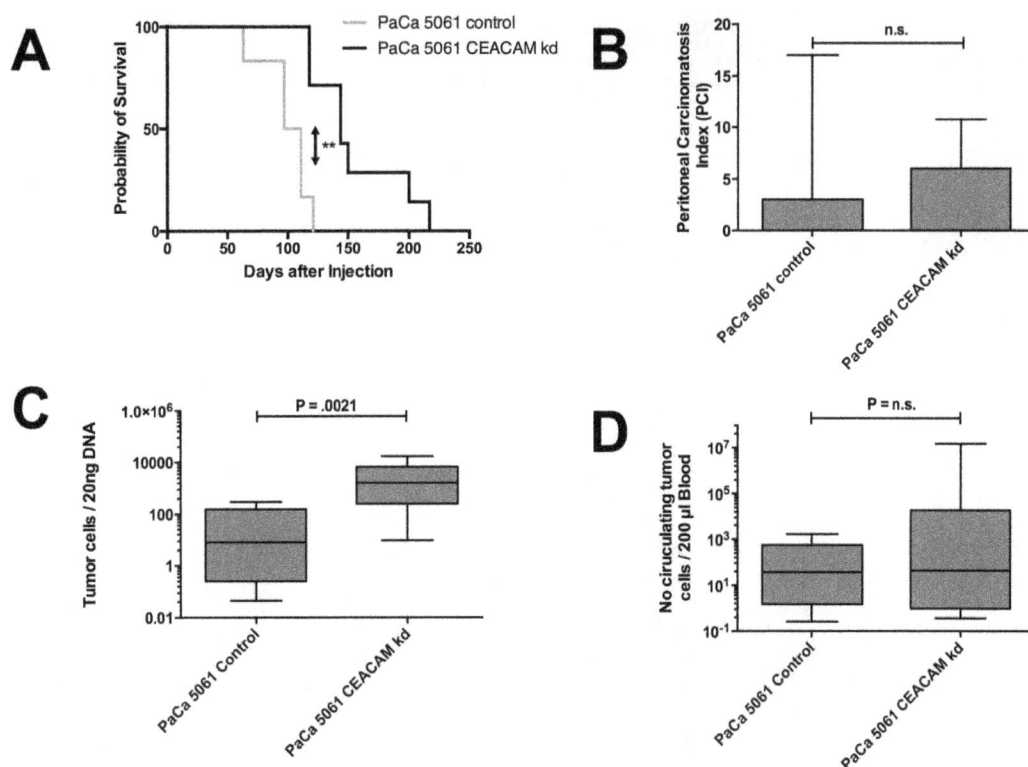

Figure 2. Subcutaneous murine xenograft model showed prolonged survival in CEACAM knock down group compared to the control group (A). Intraperitoneally, no differences in the peritoneal carcinomatosis index (PCI) were observed at time of scarification (80 days after injection). CEACAM knock down showed more DNA copies of human DNA in the left lung (P = 0.021) (C) which is correlated with a higher metastatic load in the lung, however, no difference was observed for circulating tumor cells in the blood (D).

subtypes 1, 5 and 6 were reduced to <5% compared with CEACAM expression on control cells (Figure 1A). Proliferation and migration increased in the CEACAM knock down cells compared with the control cells (Figure 1B and C), but CEACAM knockdown did not affect the adherence on stimulated endothelium (HPMEC) (Figure 1D and E).

For control of presence of the CEACAM knock down in the murine grown tumors, IHC staining of CEACAM 1, 5 and 6 was performed and, as depicted in Figure 1F, the knock down levels were stable and still present in the tumors grown in the mice.

Xenograft model for functional analysis of CEACAMs in PDAC

To answer the question whether CEACAM family members have a functional effect on tumor formation, growth and metastasis, a tumor xenograft experiment with human PDAC cells (with and without CEACAM knockdown) in pfp⁻/rag2⁻ mice was performed. After subcutaneous tumor cell injection, mice inoculated with CEACAM knock down cells showed a significantly prolonged overall survival until reaching the termination criteria (median survival 144 d) when compared with PaCa 5061 control (median survival 104 d; P = 0.014) and wild-type PACA 5061 (median survival 79 d; P = 0.01) (Figure 2A). The tumor size and weight at time of death did not differ significantly between the groups (data not shown).

While the subcutaneous xenograft model showed a prolonged OS for mice with CEACAM knockdown, the influence of the CEACAM knock down in the intraperitoneally xenograft model was not present. The peritoneal carcinomatosis index did not

differ between the groups, which was also represented in missing differences in the dissemination as there were no differences between the CEACAM knock down and the control group in circulating tumor cells in peripheral blood (Figure 2B and D). However, in the CEACAM knock down group, we found higher amounts of human DNA (meaning significantly more PDAC cells) in the lungs compared with the control group (Figure 2C). PDAC cells showed no affinity for dissemination to the bone marrow, human tumor cell DNA was not detected in any of the groups. Markers for EMT showed no significant differences between the wild type and CEACAM knock down tumors as determined by immunohistochemistry (Fig S1).

Immunohistochemical CEACAM 1, 5 and 6 expression in patients samples

Out of the 142 tumor spots, 5 (3.5%) specimen were not evaluable on the TMA due to missing or unrepresentative tumor tissue. Tissue specimens of 137 patients were finally evaluable and correlated with clinic-pathological data. Patients were aged between 33.1–85.0 years (median 63.5 years). There were 82 male patients (59.9%) and 55 female (40.1%).

CEACAM 1 expression was found in 62.8% of all tumor specimens (n = 86), CEACAM 5 in 87 patients (63.5%), CEACAM 6 expression was observed in 99 patients (72.3%). The majority of the tumors showed a homogeneous staining within each specimen, though in a minority of tumor specimens, we observed an inhomogeneous IHC staining pattern within the tumor area. The expression pattern of all three CEACAMs was membranous and cytoplasmic as well (Figure 3). Expression of CEACAM 5 was

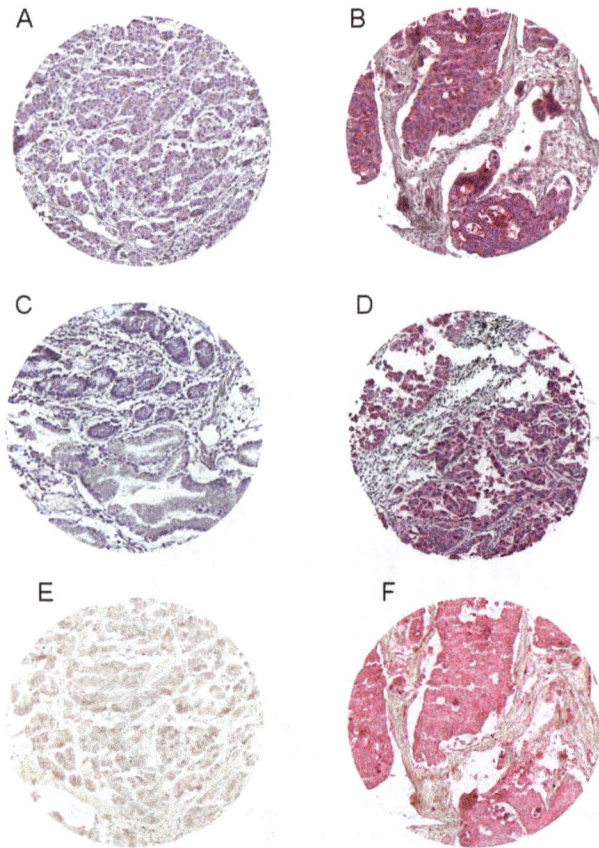

Figure 3. TMA immunohistochemical staining of CEACAM 1 negative (A) and positive (B), CEACAM 5 negative (C) and positive (D), CEACAM 6 negative (E) and positive (F).

associated with presence of CEACAM 6 expression (P<.001). A correlation between CEACAM 1 and CEACAM 5 expression (P = 0.113) or CEACAM 6 was not observed (P = 0.09).

A correlation with clinico-pathological data revealed no significant association with any parameter for CEACAM 1 except a correlation with distant metastasis (P = 0.008). CEACAM 5 and 6 expression was correlated with a positive lymph node status (P = 0.017 and P = 0.046, respectively) and distant metastasis (P< 0.001) (Table 1).

The Kaplan-Meier survival analysis showed no correlation between the CEACAM 1 expression and the overall (OS) or disease-free survival (DFS), respectively. Patients with a positive CEACAM 5 and/or 6 expression had a shortened OS and DFS (P = 0.025 and P = 0.007, P = 0.010 and P = 0.030, respectively) (Figure 4A–C, Table 2). In patients with positive expression for all three CEACAM proteins no significant differences in DFS or OS compared with those patients that were negative for all CEACAM subtypes (P = 0.144 and P = 0.742, data not shown) was observed.

CEACAM 1, 5 and 6 in serum

CEACAM 1 values in PDAC were not elevated compared with patients with CP, but were higher compared to BD (PDAC median 33.0 μg/l, range 3.3–136.7 μg/l; CP median 23.1 μg/l, range 1.8–110.1; BD median 16.1 μg/l, range 7.8–36.5 μg/l; P = 0.059 and P<0.001, respectively). Similar results were found for CEACAM 5: serum values were higher in the PDAC group compared to BD, but not for CP (PDAC median 8.5 μg/l, range

0.7–75.2 ng/ml; CP median 4.8 μg/l, range 0.7–24.0; BD median 1,9 μg/l, range 0.2–9.2 μg/l; P = 0.122 and P = 0.002, respectively). Patients with PDAC showed elevated CEACAM 6 serum expression compared to both, CP and BD (PDAC median 2.90 μg/l, range 1.34–5.46 μg/l; CP median 2.25 μg/l, range 0.77–5.15; BD median 2.34 μg/l, range 1.25–6.99 μg/l; P = 0.06 and P = 0.029, respectively). In none of the performed analyses, a significant difference between CP and BD was detectable (Figure 2G–I).

Receiver operating characteristic curves were used to establish the sensitivity-specificity relationship for CEACAM 1, 5 and 6. The optimal cut-off values were determined by Youdens-Index calculation. The area-under-the-curve (AUC) for CEACAM 1 was 0.711 (cut-off value 184.1 μg/l), for CEACAM 5 0.689 (cut-off value 1.95 μg/l) and for CEACAM 6 0.664 (cut-off value 3.58 μg/ l) (Figure 2J) The sensitivity of CEACAM 1 in detecting PDAC was 53.5% with a corresponding specificity of 54.7%. For CEACAM 5, the sensitivity is 79.1% with a specificity of 44.2%. Sensitivity of CEACAM 6 was 47.0% with a specificity of 82.6%. The AUC for a combination for all three CEACAMs showed no improvement compared with the determination of each of the CEACAMs alone (AUC 0.680, data not shown).

For a correlation between CEACAM serum values with clinico-pathological data, the serum values were divided into a low level (<75th percentile) and a high level group (≥75th percentile). High CEACAM 6 values with presence of distant metastasis (P = 0.009) and grading (P = 0.019) (Table 1). A Kaplan-Meier survival analysis calculated with the previous mentioned cut-off values showed a significantly prolonged overall-survival in patients with a low serum expression of CEACAM 1 (P = 0.022) (Figure 2D–F, Table 2). Comparing those patients that showed elevated serum levels for all three CEACAM subtypes to those who showed normal CEACAM serum values revealed no significant differences in DFS and OS (data not shown). Elevated serum levels of CEACAM were not found to be correlated with increased tissue expression in any CEACAM subtype (data not shown).

Multivariate analysis

In the multivariate cox regression analysis, grading, lymph node status and distant metastasis were found to be independent prognosticators for overall survival in the TMA (Table 3). None of the analyzed CEACAM subtypes reached statistical significance in the multivariate analysis, neither in the IHC nor in the ELISA analysis.

Discussion

The function of CEACAM expression in malignant tumors is still under debate. Several studies that focused either on immunohistochemical or serum expression of one of the numerous CEACAM molecules were previously published [25,41–46]. Here, we assessed the clinical relevance of CEACAM 1, 5 and 6 expression in both, immunohistochemical and serum analysis in patients with PDAC. In addition, we implemented a CEACAM knock down xenograft model for assessment of the potential functional role of CEACAM expression in PDAC.

In our immunohistochemical TMA analysis we found that the majority of tumors expressed CEACAM proteins. About 70% of all analyzed tumor spots showed an expression for either CEACAM 1, 5 or 6, or a combination of all of them. In univariate analysis, CEACAM 5 and 6 expression were correlated with lymph node metastasis. The survival analysis revealed both a shortened overall and disease free survival in patients with a high CEACAM 5 or 6 expression.

Table 1. Immunohistochemical expression of CEACAM 1, 5 and 6, serum expression (ELISA) and correlation with patients' demographic characteristics and clinico-pathological data.

| | | Tissue Micro Array (IHC) | | | | | | | | | Serum Expression (ELISA) | | | | | | | | | |
| | | CEACAM 1 | | | CEACAM 5 | | | CEACAM 6 | | | | CEACAM 1 | | | CEACAM 5 | | | CEACAM 6 | | |
Variables	N	neg. / pos.	P*		neg. / pos.	P*		neg. / pos.	P*	n	neg. / pos.	pP		neg. / pos.	P*		neg. / pos.	P	
Total	137	51 (37.2%) / 86 (62.8%)			50 (36.4%) / 87 (63.5%)			38 (27.7%) / 99 (72.3%)		46	35 (76.1%) / 11 (23.9%)			31 (67.4%) / 15 (32.6%)			35 (76.1%) / 11 (23.9%)		
Sex			.074			.140			0.222			.718			.781			.306	
Male	82	26 (31.7%) / 56 (68.3%)			26 (31.9%) / 56 (68.1%)			20 (24.3%) / 62 (75.7%)		21	17 (81.0%) / 4 (23.5%)			17 (81.0%) / 4 (19.0%)			14 (66.7%) / 7 (33.3%)		
Female	55	25 (45.5%) / 30 (54.5%)			24 (43.1%) / 31 (56.9%)			18 (32.1%) / 37 (67.9%)		25	18 (72.0%) / 7 (38.9%)			18 (72.0%) / 7 (28.0%)			21 (84.0%) / 4 (16.0%)		
Age. Years			.245			.129			.392			.264			.351			.624	
≤60	74	30 (40.5%) / 44 (59.5%)			23 (31.3%) / 51 (68.7%)			19 (25.8%) / 55 (74.2%)		15	14 (93.3%) / 1 (7.1%)			10 (66.7%) / 5 (33.3%)			11 (73.3%) / 4 (26.7%)		
>60	63	21 (33.3%) / 42 (66.7%)			27 (42.9%) / 36 (57.1%)			19 (29.5%) / 44 (70.5%)		31	21 (67.7%) / 10 (47.6%)			24 (77.4%) / 7 (22.6%)			24 (77.4%) / 7 (22.6%)		
Tumor stage			.777			.886			.326			.222			.158			.430	
T1	7	2 (28.6%) / 5 (71.4%)			2 (33.3%) / 5 (66.7%)			2 (28.6%) / 5 (71.4%)		1	0 (0.0%) / 1 (100.0%)			1 (100.0%) / 0 (0.0%)			1 (100.0%) / 0 (0.0%)		
T2	37	12 (32.4%) / 25 (67.6%)			15 (41.7%) / 22 (58.3%)			14 (37.8%) / 23 (62.2%)		6	5 (83.3%) / 1 (20.0%)			6 (100.0%) / 0 (0.0%)			6 (100.0%) / 0 (0.0%)		
T3	85	33 (38.8%) / 52 (61.2%)			31 (36.0%) / 54 (64.0%)			21 (24.4%) / 64 (75.6%)		30	21 (70.0%) / 9 (42.9%)			22 (73.3%) / 8 (26.7%)			22 (73.3%) / 8 (26.7%)		
T4	8	4 (50.0%) / 4 (50.0%)			3 (33.3%) / 6 (66.6%)			0 (0.0%) / 7 (100.0%)		9	8 (88.9%) / 1 (12.5%)			5 (55.6%) / 4 (44.4%)			6 (66.7%) / 3 (33.3%)		
Nodal status			.732			.017			.046			.230			.846			.863	
Negative	55	21 (38.5%) / 34 (61.5%)			26 (47.8%) / 29 (52.2%)			23 (38.4%) / 32 (61.6%)		24	14 (58.3%) / 10 (71.4%)			19 (79.2%) / 5 (20.8%)			19 (79.2%) / 5 (20.8%)		
Positive	82	29 (35.9%) / 53 (64.1%)			22 (26.8%) / 60 (73.2%)			21 (25.5%) / 61 (74.5%)		22	20 (90.9%) / 2 (10.0%)			14 (63.6%) / 8 (36.4%)			15 (68.2%) / 7 (31.8%)		
Grading			.555			.331			.325			.827			.497			**.016**	
G1	7	4 (57.1%) / 3 (42.9%)			4 (57.1%) / 3 (42.9%)			4 (50.0%) / 4 (50.0%)		3	2 (66.7%) / 1 (50.0%)			3 (100.0%) / 0 (0.0%)			3 (100.0%) / 0 (0.0%)		
G2	58	21 (36.2%) / 37 (63.8%)			18 (30.6%) / 40 (69.4%)			16 (26.9%) / 42 (73.1%)		28	21 (75.0%) / 7 (33.3%)			20 (71.4%) / 8 (28.6%)			25 (89.3%) / 3 (10.7%)		
G3	72	27 / 45			28 / 44			18 / 54		15	10 / 5			12 / 3			8 / 7		

Table 1. Cont.

Variables	N	Tissue Micro Array (IHC) CEACAM 1 neg.	pos.	P*	CEACAM 5 neg.	pos.	P*	CEACAM 6 neg.	pos.	P*	n	Serum Expression (ELISA) CEACAM 1 neg.	pos.	pP	CEACAM 5 neg.	pos.	P*	CEACAM 6 neg.	pos.	P
M		37.5%	62.5%	.008	38.5%	61.5%	<.001	25.4%	74.6%	<.001		66.7%	50.0%	.729	80.0%	20.0%	.056	53.3%	46.7%	.002
negative (N0) (n)	119	39	80		35	84		25	94		36	26	10		28	8		31	5	
%		32.8%	67.2%		29,40%	70.6%		21%	79%			72.2%	38.5%		77.8%	22.2%		86.1%	13.9%	
positive (N1) (n)	18	12	6		15	3		13	5		10	8	2		5	5		3	7	
%		66.7%	33.3%		83,30%	16,70%		72,20%	27,80%			80.0%	25.0%		50.0%	50.0%		30.0%	70.0%	

Additionally, we quantified CEACAM 1, 5 and 6 in serum of patients with PDAC. Compared to serum samples of healthy blood donors, CEACAM values of all subtypes were elevated. CEACAM 5 and 6 values were higher in PDAC than in patients with CP. Similar to the IHC results, we found that an elevated CEACAM 5 expression correlated with lymph node metastasis and increased CEACAM 6 values were correlated with the presence of distant metastasis and tumor grading. CEACAM 1 was detectable in IHC as well as in blood serum, but a correlation with clinic-pathological data was not evident in our analysis. Interestingly, when serum concentrations were evaluated as predictors for OS, only CEACAM 1 was associated with a shortened OS in the Kaplan-Meier survival analysis.

We were not able to find a correlation between tissue expression and elevated levels in the blood serum of these patients. This might have several reasons: For example, the mere expression of the proteins does not have to result in an increased shedding of them. Furthermore, flushing of the shedded molecule into the blood stream might be a consequence of the disruption of anatomical barriers surrounding host tissues and endothelial cells. Taken together, the mechanisms regulating the shedding of CEACAMs and their dissemination into the surrounding tissue and their entry into the blood system are barely understood and further investigations are needed.

CEACAM 5 is widely used as a serum marker in patients with PDAC in the clinical routine work-up. Our analysis showed a significant correlation between CEACAM 5 and the patients' lymph node status alone but not with survival or any additional clinico-pathological parameter. Therefore, the question arises, whether CEACAM 5 determination preoperatively is of use in the clinical routine work-up. As shown in the serum analysis, the predictive value for primary diagnosis of PDAC with CEACAM determination in the serum alone is unsatisfying, according to our analysis, the sensitivity and specifity is poor.

CEACAMs were shown to be overexpressed in numerous other tumor entities, e.g. colon, breast and lung cancer [17,20,21]. CEACAM 1 expression, for example, was found to be associated with a poor clinical outcome in patients with non-small cell lung cancer (NSCLC) and neuroendocrine pancreatic tumors which suggests a pro-tumorgenic effect when the protein is up-regulated [19,47].

The biological role of CEACAM expression in PDAC is still widely unknown. Since CEACAM expression in normal tissue has pro-angiogenetic effects, regulation of the cell adhesion and may play a part in regulating apoptosis, these effects could also be of importance in tumor progression when CEACAM expression is up-regulated within the tumor cells [8,9,48–50]. However, increased CEACAM expression is not only seen in malignant, but also observed in inflammatory tissue. One of the physiological roles of the molecule might be pro-inflammatory, which could explain the missing difference between PDAC and CP in our serum analysis [51].

Expression of CEACAM 6 was not only seen in invasive PDAC but also in pancreatic epithelial neoplasia (PanIN) lesions which are considered as tumor precursors, but without an invasive growth [43]. Unfortunately, so far no data exist, whether PanIN lesions with a high CEACAM 6 expression show a higher or faster conversion rate into malignant tumors than PanINs without CEACAM 6 expression. This could rather be of interest whether CEACAM 6 itself promotes a transformation of benign tumor lesions into an aggressive invasive carcinoma. Moreover, an analysis of further CEACAM subtypes would be interesting, since our analysis suggests protumorgenic effects of CEACAM 1 and 5

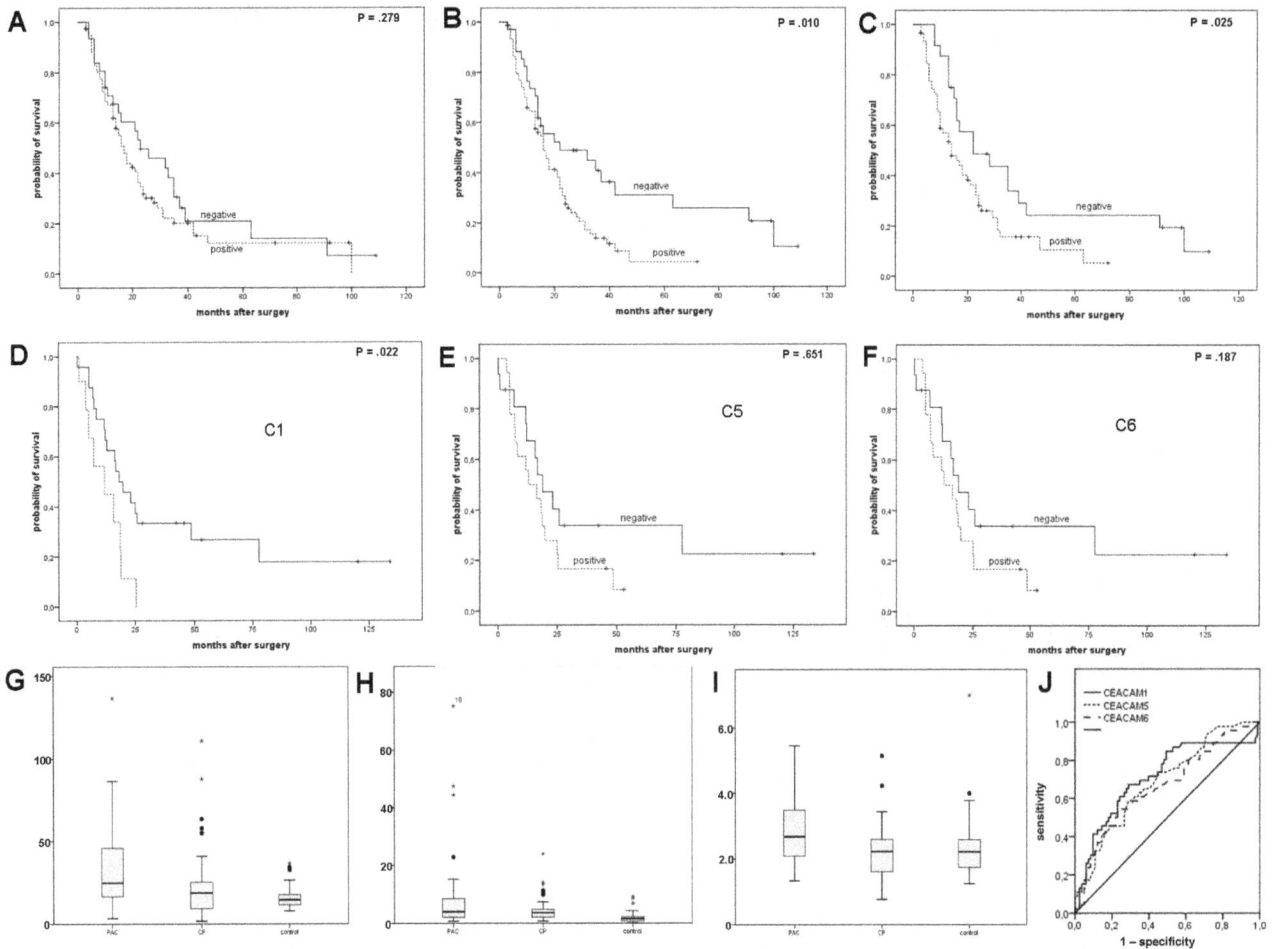

Figure 4. Kaplan-Meier survival analysis (log-rank test) for correlation between overall-survival (OS) and CEACAM expression on the TMA. (A–C). CEACAM 1 positive patients showed a median OS of 17.0 months (14.0–20.1 months), CEACAM 1 negative 23.0 months (9.5–36.5 months, P =.279). OS of CEACAM 5 positive patients was 16.0 months (12.8–19.2 months) vs. 22.0 months (4.1–47.9 months) (p =.025). CEACAM 6 positive patients showed an OS of 14.0 months (7.9–20.0) vs. 22.0 months (5.6–38.4 months, P =.010). Survival curves for correlation between overall-survival and CEACAM serum expression (D–F). CEACAM 1 positive patients showed an OS of 11.8 months (2.4–25.2) vs. 18.3 months (10.0–26.2 months, P =.022). Median OS for patients positive for CEACAM 5 was 15.7 months (2.4–32.3 months) vs. 18.6 months (8.7–28.6 months, P =.651), and median OS for patients with CEACAM 6 positivity was 18.8 months (9.5–28.1 months) vs. 12.8 months (3.3–22.3 months, P =.187). Boxplots for CEACAM 1 serum expression (G), CEACAM 5 (H) and CEACAM 6 (I) compared pancreatic ductal adenocarcinoma (PDAC), chronic pancreatic (CP) and healthy blood donors (BD). Receiver operating curve (ROC) for CEACAM serum expression (J).

as well. So far data of CEACAM 1 and 5 expressions in PanIN do not exist.

Even with a prospective database, due to the study design, the obtained data is of retrospective character, which may result in impairment of the significance of those results. With IHC and ELISA studies we are not able to determine, whether CEACAM proteins have a direct effect on local and distant tumor progression or if the revealed statistical correlations are an epiphenomenon of a different process being active in tumor-host interactions. We therefore established a xenograft mouse model with a CEACAM knock down variant of the previously established cell line PACA 5061. Flow cytometry analysis and immunohistochemistry before and after tumor cell injection into the mice showed a stable CEACAM knock down of >90% of all observed CEACAM subtypes. The overall survival in the mice with the CEACAM knock down cell line was significantly prolonged compared to the wild type cell line. This suggests a direct influence of CEACAM-meditated functions in tumor progression. Previous studies already showed anti-tumor effects when CEACAM 6 targeted therapies

were used [44,52]. Binding of Fab antibody fragments against CEACAM 5 and/or CEACAM 6 led to reduced tumor growth in xenograft mouse models and was found to be associated with increased chemoresistance against gemcitabine [41,53]. In our experiments, we focused on a whole knock down, not only of CEACAM 6. Tumor growth was slower in the knock down cell line and led to an increased overall survival in the knock down group. The interactions between human and murine CEACAM molecules were studied previously [16]. Mice itself do not express CEACAM 5 and 6 subtypes but homo- and heterophilic interactions between different CEACAM subtypes were described extensively [54,55]. Thus, it is not surprising that CEACAM knock down effects could be observed in the murine xenograft model even without CEACAM 5 and 6 expressions in the mice.

The observed effect of a more aggressive tumor growth in those cells with CEACAM expression is in concordance with the observed effects in the clinical data. As previously shown, tumors with an increased CEACAM expression have to be generally considered as more aggressive and seem to be correlated with a

Table 2. Overall and disease-free survival.

			Overall survival			Disease free survival		
			Negative	positive	p	Negative	positive	p
IHC	CEACAM 1	Median (95%CI)	23.0 (9.5–36.5)	17.0 (14.0–20.1)	.279	9.0 (4.6–13.4)	7.0 (5.6–8.2)	.308
	CEACAM 5	Median (95%CI)	22.0 (4.1–47.9)	16.0 (12.8–19.2)	**.025**	13.0 (11.3–14.7)	6.0 (5.1–6.7)	**.007**
	CEACAM 6	Median (95%CI)	22.0 (5.6–38.4)	14.0 (7.9–20.0)	**.010**	13.0 (6.8–19.2)	7.0 (5.8–8.2)	**.030**
Serum samples (ELISA)	CEACAM 1	Median (95%CI)	18.3 (10.0–26.2)	11.8 (2.4–25.2)	**.022**	12.0 (4.6–18.4)	12.5 (5.0–19.1)	.467
	CEACAM 5	Median (95%CI)	18.6 (8.7–28.6)	15.7 (2.4–32.3)	.651	13.4 (4.5–22.4)	12.0 (3.9–21.4)	.192
	CEACAM 6	Median (95%CI)	12.8 (3.3–22.3)	18.8 (9.5–28.1)	.187	11.9 (5.6–21.0)	13.1 (4.5–20.4)	.091

Kaplan-Meier survival analysis (log-rank test) depending on CEACAM 1, 5 and 6 expression in immunohistochemistry (IHC) and serum samples (ELISA) for patients with pancreatic ductal adenocarcinoma.

Table 3. Multivariate cox-regression analysis for immunohistochemical analysis (IHC) and serum analysis (ELISA).

	IHC				Serum samples (ELISA)			
	significance	HR	95%CI Min	95%CI Max	significance	HR	95%CI Min	95%CI Max
sex	.480	.824	.481	1.410	.654	.752	.216	2.618
age (<65 yrs vs. >65 yrs)	.876	.958	.562	1.633	.591	.731	.233	2.294
pT group (T1/2 vs. T3/4)	.626	1.150	.655	2.020	.753	1.227	.343	4.388
pN	**.016**	2.064	1.146	3.717	.402	.563	.147	2.161
M	**.005**	3.157	1.546	6.705	.401	2.447	.304	19.722
Grading (G1 vs. G2/3)	**.001**	2.410	1.409	4.123	.392	1.659	.521	5.283
CEACAM 1	.423	1.300	.684	2.471	.059	3.971	.950	16.595
CEACAM 5	.086	1.860	.915	3.780	.952	1.040	.296	3.649
CEACAM 6	.071	1.797	.952	3.391	.147	.435	.142	1.339

HR = Hazard ratio. T = tumor stage. N = lymph node stage. M = distant metastasis.

poor prognosis for the individual patient. In most adenocarcinoma (e.g. colon carcinoma), main clinical complication (and ultimately cause of death) is distant metastasis (to lung, liver, bone marrow, etc.) whereas in pancreatic carcinoma locally recurrent tumors and intraperitoneal carcinomatosis is of major importance. Interestingly, the xenograft model used in this study seems to model this situation: The CEACAM knockdown primary tumors showed a slower, less aggressive growth prolonging the animals OS, whereas distant metastasis to the lung increased. These findings are in concordance to our in vitro findings showing the CEACAM knock down cells a higher proliferative and migratory potential than the control cells. Obviously, CEACAMs have no clear protumorigentic or tumorsupressive function in PDAC moreover the mediated effects are of higher complexity. On the one hand, the local tumor growth is significantly impaired in the subcutaneous compartment while at the same time the number of distant metastases increases. Similar effects could be observed for different types of cell surface proteins, like EpCAM whose role of tumor progression is of comparable complexity. Probably, the effect of epithelial-mesenchymal-transition is of some importance in this context. However, we were not able to detect any increased expression in the canonical EMT drivers ZEB1 and ZEB2 in the CEACAM knock down tumors or decreased expression of E-Cadherin or cytokeratin. Obviously, CEACAM expression itself is not correlated with the expression of the canonical EMT drivers as mentioned above. However, as our results suggest, CEACAM expression has a direct influence on the tumor progression and metastatic behavior but without affecting the expression of EMT markers.

So far, the exact functional role of the CEACAM molecules in PDAC is still not fully understood, though we were able to show distinct functional aspects of CEACAM interactions in vitro and in vivo. However, these finding may help understand the inconclusive results that were revealed, not only in our study, with respect to CEACAM expression and the individual patients' prognosis.

Supporting Information

Figure S1 Immunohistchemical staining of murine xenograft tumors for markers of epithelial-mesenchymal transtition (EMT).
(TIF)

Author Contributions

Conceived and designed the experiments: FG JRI MB MT US DW. Performed the experiments: FG DW HM JH PS TS. Analyzed the data: FG DW MT JRI US MB. Contributed reagents/materials/analysis tools: FG HM DW JH US JRI. Wrote the paper: FG DW US JRI MB.

References

1. Loos M, Kleeff J, Friess H, Buchler MW (2008) Surgical treatment of pancreatic cancer. Ann N Y Acad Sci 1138: 169–180.
2. Alexakis N, Halloran C, Raraty M, Ghaneh P, Sutton R, et al. (2004) Current standards of surgery for pancreatic cancer. Br J Surg 91: 1410–1427.
3. Neoptolemos JP, Stocken DD, Bassi C, Ghaneh P, Cunningham D, et al. (2010) Adjuvant chemotherapy with fluorouracil plus folinic acid vs gemcitabine following pancreatic cancer resection: a randomized controlled trial. JAMA 304: 1073–1081.
4. Neoptolemos JP, Stocken DD, Tudur Smith C, Bassi C, Ghaneh P, et al. (2009) Adjuvant 5-fluorouracil and folinic acid vs observation for pancreatic cancer: composite data from the ESPAC-1 and -3(v1) trials. Br J Cancer 100: 246–250.
5. Oettle H, Post S, Neuhaus P, Gellert K, Langrehr J, et al. (2007) Adjuvant chemotherapy with gemcitabine vs observation in patients undergoing curative-intent resection of pancreatic cancer: a randomized controlled trial. JAMA 297: 267–277.
6. Sperti C, Pasquali C, Piccoli A, Pedrazzoli S (1997) Recurrence after resection for ductal adenocarcinoma of the pancreas. World J Surg 21: 195–200.
7. Thompson JA, Eades-Perner AM, Ditter M, Muller WJ, Zimmermann W (1997) Expression of transgenic carcinoembryonic antigen (CEA) in tumor-prone mice: an animal model for CEA-directed tumor immunotherapy. Int J Cancer 72: 197–202.
8. Obrink B (1997) CEA adhesion molecules: multifunctional proteins with signal-regulatory properties. Curr Opin Cell Biol 9: 616–626.
9. Beauchemin N, Draber P, Dveksler G, Gold P, Gray-Owen S, et al. (1999) Redefined nomenclature for members of the carcinoembryonic antigen family. Exp Cell Res 252: 243–249.
10. Kuespert K, Pils S, Hauck CR (2006) CEACAMs: their role in physiology and pathophysiology. Curr Opin Cell Biol 18: 565–571.
11. Horst AK, Ito WD, Dabelstein J, Schumacher U, Sander H, et al. (2006) Carcinoembryonic antigen-related cell adhesion molecule 1 modulates vascular remodeling in vitro and in vivo. J Clin Invest 116: 1596–1605.
12. Gold P, Freedman SO (1965) Specific carcinoembryonic antigens of the human digestive system. J Exp Med 122: 467–481.
13. Ona FV, Zamcheck N, Dhar P, Moore T, Kupchik HZ (1973) Carcinoembryonic antigen (CEA) in the diagnosis of pancreatic cancer. Cancer 31: 324–327.
14. Hockey MS, Stokes HJ, Thompson H, Woodhouse CS, Macdonald F, et al. (1984) Carcinoembryonic antigen (CEA) expression and heterogeneity in primary and autologous metastatic gastric tumours demonstrated by a monoclonal antibody. Br J Cancer 49: 129–133.
15. Thompson JA, Grunert F, Zimmermann W (1991) Carcinoembryonic antigen gene family: molecular biology and clinical perspectives. J Clin Lab Anal 5: 344–366.
16. Horst AK, Wagener C (2004) CEA-Related CAMs. Handb Exp Pharmacol: 283–341.
17. Arabzadeh A, Chan C, Nouvion AL, Breton V, Benlolo S, et al. (2012) Host-related carcinoembryonic antigen cell adhesion molecule 1 promotes metastasis of colorectal cancer. Oncogene.
18. Scholzel S, Zimmermann W, Schwarzkopf G, Grunert F, Rogaczewski B, et al. (2000) Carcinoembryonic antigen family members CEACAM6 and CEACAM7 are differentially expressed in normal tissues and oppositely deregulated in hyperplastic colorectal polyps and early adenomas. Am J Pathol 156: 595–605.
19. Thom I, Schult-Kronefeld O, Burkholder I, Schuch G, Andritzky B, et al. (2009) Expression of CEACAM-1 in pulmonary adenocarcinomas and their metastases. Anticancer Res 29: 249–254.
20. Thies A, Moll I, Berger J, Wagener C, Brummer J, et al. (2002) CEACAM1 expression in cutaneous malignant melanoma predicts the development of metastatic disease. J Clin Oncol 20: 2530–2536.
21. Xie S, Luca M, Huang S, Gutman M, Reich R, et al. (1997) Expression of MCAM/MUC18 by human melanoma cells leads to increased tumor growth and metastasis. Cancer Res 57: 2295–2303.
22. Dango S, Sienel W, Schreiber M, Stremmel C, Kirschbaum A, et al. (2008) Elevated expression of carcinoembryonic antigen-related cell adhesion molecule 1 (CEACAM-1) is associated with increased angiogenic potential in non-small-cell lung cancer. Lung Cancer 60: 426–433.
23. Huang S, Xie K, Bucana CD, Ullrich SE, Bar-Eli M (1996) Interleukin 10 suppresses tumor growth and metastasis of human melanoma cells: potential inhibition of angiogenesis. Clin Cancer Res 2: 1969–1979.
24. Duxbury MS, Ito H, Benoit E, Ashley SW, Whang EE (2004) CEACAM6 is a determinant of pancreatic adenocarcinoma cellular invasiveness. Br J Cancer 91: 1384–1390.
25. Simeone DM, Ji B, Banerjee M, Arumugam T, Li D, et al. (2007) CEACAM1, a novel serum biomarker for pancreatic cancer. Pancreas 34: 436–443.
26. Kalinina T, Gungor C, Thieltges S, Moller-Krull M, Murga Penas EM, et al. (2010) Establishment and characterization of a new human pancreatic adenocarcinoma cell line with high metastatic potential to the lung. BMC Cancer 10: 295.
27. Wicklein D (2012) RNAi technology to block the expression of molecules relevant to metastasis: the cell adhesion molecule CEACAM1 as an instructive example. Methods Mol Biol 878: 241–250.
28. Workman P, Aboagye EO, Balkwill F, Balmain A, Bruder G, et al. (2010) Guidelines for the welfare and use of animals in cancer research. Br J Cancer 102: 1555–1577.
29. Gebauer F, Wicklein D, Stubke K, Nehmann N, Schmidt A, et al. (2013) Selectin binding is essential for peritoneal carcinomatosis in a xenograft model of human pancreatic adenocarcinoma in pfp-/rag2- mice. Gut 62: 741–750.
30. Portilla AG, Sugarbaker PH, Chang D (1999) Second-look surgery after cytoreduction and intraperitoneal chemotherapy for peritoneal carcinomatosis from colorectal cancer: analysis of prognostic features. World J Surg 23: 23–29.
31. Nehmann N, Wicklein D, Schumacher U, Muller R (2010) Comparison of two techniques for the screening of human tumor cells in mouse blood: quantitative real-time polymerase chain reaction (qRT-PCR) versus laser scanning cytometry (LSC). Acta Histochem 112: 489–496.
32. Lange T, Ullrich S, Muller I, Nentwich MF, Stubke K, et al. (2012) Human prostate cancer in a clinically relevant xenograft mouse model: identification of beta(1,6)-branched oligosaccharides as a marker of tumor progression. Clin Cancer Res 18: 1364–1373.

33. Tachezy M, Zander H, Marx AH, Stahl PR, Gebauer F, et al. (2012) ALCAM (CD166) expression and serum levels in pancreatic cancer. PLoS One 7: e39018.

34. Dancau AM, Simon R, Mirlacher M, Sauter G (2010) Tissue microarrays. Methods Mol Biol 576: 49–60.

35. Gebauer F, Tachezy M, Effenberger K, von Loga K, Zander H, et al. (2011) Prognostic impact of CXCR4 and CXCR7 expression in pancreatic adenocarcinoma. J Surg Oncol 104: 140–145.

36. Simon R, Mirlacher M, Sauter G (2010) Immunohistochemical analysis of tissue microarrays. Methods Mol Biol 664: 113–126.

37. Stoffel A, Neumaier M, Gaida FJ, Fenger U, Drzeniek Z, et al. (1993) Monoclonal, anti-domain and anti-peptide antibodies assign the molecular weight 160,000 granulocyte membrane antigen of the CD66 cluster to a mRNA species encoded by the biliary glycoprotein gene, a member of the carcinoembryonic antigen gene family. J Immunol 150: 4978–4984.

38. Tuck MK, Chan DW, Chia D, Godwin AK, Grizzle WE, et al. (2009) Standard operating procedures for serum and plasma collection: early detection research network consensus statement standard operating procedure integration working group. J Proteome Res 8: 113–117.

39. Wright PF, Nilsson E, Van Rooij EM, Lelenta M, Jeggo MH (1993) Standardisation and validation of enzyme-linked immunosorbent assay techniques for the detection of antibody in infectious disease diagnosis. Rev Sci Tech 12: 435–450.

40. Youden WJ (1950) Index for rating diagnostic tests. Cancer 3: 32–35.

41. Duxbury MS, Ito H, Benoit E, Waseem T, Ashley SW, et al. (2004) A novel role for carcinoembryonic antigen-related cell adhesion molecule 6 as a determinant of gemcitabine chemoresistance in pancreatic adenocarcinoma cells. Cancer Res 64: 3987–3993.

42. Duxbury MS, Ito H, Benoit E, Zinner MJ, Ashley SW, et al. (2004) Overexpression of CEACAM6 promotes insulin-like growth factor I-induced pancreatic adenocarcinoma cellular invasiveness. Oncogene 23: 5834–5842.

43. Duxbury MS, Matros E, Clancy T, Bailey G, Doff M, et al. (2005) CEACAM6 is a novel biomarker in pancreatic adenocarcinoma and PanIN lesions. Ann Surg 241: 491–496.

44. Strickland LA, Ross J, Williams S, Ross S, Romero M, et al. (2009) Preclinical evaluation of carcinoembryonic cell adhesion molecule (CEACAM) 6 as potential therapy target for pancreatic adenocarcinoma. J Pathol 218: 380–390.

45. Blumenthal RD, Leon E, Hansen HJ, Goldenberg DM (2007) Expression patterns of CEACAM5 and CEACAM6 in primary and metastatic cancers. BMC Cancer 7: 2.

46. Blumenthal RD, Hansen HJ, Goldenberg DM (2005) Inhibition of adhesion, invasion, and metastasis by antibodies targeting CEACAM6 (NCA-90) and CEACAM5 (Carcinoembryonic Antigen). Cancer Res 65: 8809–8817.

47. Serra S, Asa SL, Bamberger AM, Wagener C, Chetty R (2009) CEACAM1 expression in pancreatic endocrine tumors. Appl Immunohistochem Mol Morphol 17: 286–293.

48. Gerstel D, Wegwitz F, Jannasch K, Ludewig P, Scheike K, et al. (2011) CEACAM1 creates a pro-angiogenic tumor microenvironment that supports tumor vessel maturation. Oncogene 30: 4275–4288.

49. Chen Z, Chen L, Baker K, Olszak T, Zeissig S, et al. (2011) CEACAM1 dampens antitumor immunity by down-regulating NKG2D ligand expression on tumor cells. J Exp Med 208: 2633–2640.

50. Skubitz KM, Skubitz AP (2008) Interdependency of CEACAM-1, -3, -6, and -8 induced human neutrophil adhesion to endothelial cells. J Transl Med 6: 78.

51. Gray-Owen SD, Blumberg RS (2006) CEACAM1: contact-dependent control of immunity. Nat Rev Immunol 6: 433–446.

52. Duxbury MS, Ito H, Ashley SW, Whang EE (2004) CEACAM6 as a novel target for indirect type 1 immunotoxin-based therapy in pancreatic adenocarcinoma. Biochem Biophys Res Commun 317: 837–843.

53. Duxbury MS, Matros E, Ito H, Zinner MJ, Ashley SW, et al. (2004) Systemic siRNA-mediated gene silencing: a new approach to targeted therapy of cancer. Ann Surg 240: 667–674; discussion 675–666.

54. Heine M, Nollau P, Masslo C, Nielsen P, Freund B, et al. (2011) Investigations on the usefulness of CEACAMs as potential imaging targets for molecular imaging purposes. PLoS One 6: e28030.

55. Han E, Phan D, Lo P, Poy MN, Behringer R, et al. (2001) Differences in tissue-specific and embryonic expression of mouse Ceacam1 and Ceacam2 genes. Biochem J 355: 417–423.

19

A Simplified and Versatile System for the Simultaneous Expression of Multiple siRNAs in Mammalian Cells Using Gibson DNA Assembly

Fang Deng[1,2], Xiang Chen[2], Zhan Liao[2,3], Zhengjian Yan[2,4], Zhongliang Wang[2,4], Youlin Deng[2,4], Qian Zhang[2,4], Zhonglin Zhang[2,5], Jixing Ye[2,6], Min Qiao[2,4,3], Ruifang Li[2,5], Sahitya Denduluri[2], Jing Wang[2,4], Qiang Wei[2,4], Melissa Li[2], Nisha Geng[2], Lianggong Zhao[2,7], Guolin Zhou[2], Penghui Zhang[2,4], Hue H. Luu[2], Rex C. Haydon[2], Russell R. Reid[2,8], Tian Yang[1]*, Tong-Chuan He[2,3]*

1 Department of Cell Biology, Third Military Medical University, Chongqing, 400038, China, 2 Molecular Oncology Laboratory, Department of Orthopaedic Surgery and Rehabilitation Medicine, The University of Chicago Medical Center, Chicago, IL, 60637, United States of America, 3 Department of Orthopaedic Surgery, the Affiliated Xiang-Ya Hospital of Central South University, Changsha, 410008, China, 4 Ministry of Education Key Laboratory of Diagnostic Medicine, and the Affiliated Hospitals of Chongqing Medical University, Chongqing, 400016, China, 5 Department of Surgery, the Affiliated Zhongnan Hospital of Wuhan University, Wuhan, 430071, China, 6 School of Bioengineering, Chongqing University, Chongqing, 400044, China, 7 Department of Orthopaedic Surgery, the Second Affiliated Hospital of Lanzhou University, Lanzhou, Gansu, 730000, China, 8 The Laboratory of Craniofacial Biology, Department of Surgery, The University of Chicago Medical Center, Chicago, IL, 60637, United States of America

Abstract

RNA interference (RNAi) denotes sequence-specific mRNA degradation induced by short interfering double-stranded RNA (siRNA) and has become a revolutionary tool for functional annotation of mammalian genes, as well as for development of novel therapeutics. The practical applications of RNAi are usually achieved by expressing short hairpin RNAs (shRNAs) or siRNAs in cells. However, a major technical challenge is to simultaneously express multiple siRNAs to silence one or more genes. We previously developed pSOS system, in which siRNA duplexes are made from oligo templates driven by opposing U6 and H1 promoters. While effective, it is not equipped to express multiple siRNAs in a single vector. Gibson DNA Assembly (GDA) is an *in vitro* recombination system that has the capacity to assemble multiple overlapping DNA molecules in a single isothermal step. Here, we developed a GDA-based pSOK assembly system for constructing single vectors that express multiple siRNA sites. The assembly fragments were generated by PCR amplifications from the U6-H1 template vector pB2B. GDA assembly specificity was conferred by the overlapping unique siRNA sequences of insert fragments. To prove the technical feasibility, we constructed pSOK vectors that contain four siRNA sites and three siRNA sites targeting human and mouse β-catenin, respectively. The assembly reactions were efficient, and candidate clones were readily identified by PCR screening. Multiple β-catenin siRNAs effectively silenced endogenous β-catenin expression, inhibited Wnt3A-induced β-catenin/Tcf4 reporter activity and expression of Wnt/β-catenin downstream genes. Silencing β-catenin in mesenchymal stem cells inhibited Wnt3A-induced early osteogenic differentiation and significantly diminished synergistic osteogenic activity between BMP9 and Wnt3A *in vitro* and *in vivo*. These findings demonstrate that the GDA-based pSOK system has been proven simplistic, effective and versatile for simultaneous expression of multiple siRNAs. Thus, the reported pSOK system should be a valuable tool for gene function studies and development of novel therapeutics.

Editor: Jun Sun, Rush University Medical Center, United States of America

Funding: The reported work was supported in part by research grants from the National Institutes of Health (AT004418, AR50142, AR054381 to TCH, RCH and HHL), and the National Natural Science Foundation of China (NSFC grant #81271770 to TY). DF was a recipient of a doctorate fellowship from the China Scholarship Council. This work was also supported in part by The University of Chicago Core Facility Subsidy grant from the National Center for Advancing Translational Sciences (NCATS) of the National Institutes of Health through grant UL1 TR000430. The funders had no role in study design, data collection and analysis, decision to publish, or preparation of the manuscript.

Competing Interests: The authors have declared that no competing interests exist.

* Email: tche@uchicago.edu (TCH); tiany@163.net (TY)

Introduction

RNA interference (RNAi) was first discovered in *C. elegans* as a protecting mechanism against invasion by foreign genes and has subsequently been demonstrated in diverse eukaryotes, such as insects, plants, fungi and vertebrates [1–7]. RNAi is a cellular process of sequence-specific, post-transcriptional gene silencing initiated by double-stranded RNAs (dsRNA) homologous to the gene being suppressed. The dsRNAs are processed by Dicer to generate duplexes of approximately 21nt, so-called short interfering RNAs (siRNAs), which cause sequence-specific mRNA degradation. Dicer-produced siRNA duplexes comprise two 21 nucleotide strands, each bearing a 5′ phosphate and 3′ hydroxyl group, paired in a way that leaves two-nucleotide overhangs at the 3′ ends. Target regulation by siRNAs is mediated by the RNA-induced silencing complex (RISC). Since its discovery, RNAi has become a valuable and powerful tool to analyze loss-of-function phenotypes *in vitro* and *in vivo* [2–7]. Given its gene-specific targeting nature, RNAi also offers unprecedented opportunities for developing novel and effective therapeutics for human diseases [8–13].

The practical applications of siRNA duplexes to interfere with the expression of a given gene require target accessibility and effective delivery of siRNAs into target cells and for certain applications long-term siRNA expression [8–14]. While RNAi can be achieved by delivering synthetic short double-stranded RNA duplexes into cells, a more commonly-used approach is to express short hairpin RNAs (shRNAs) or siRNAs in cells [11,12,14]. In this case, the endogenous expression of siRNAs is achieved by using various Pol III promoter expression cassettes that allow transcription of functional siRNAs or their precursors [14]. However, one of the formidable technical challenges is to effectively construct these RNAi expression vectors, especially when gene silencing necessitates the use of multiple siRNA target sites for a gene of interest. We previously developed the pSOS system, in which the siRNA duplexes are made from an oligo template driven by opposing U6 and H1 promoters [15]. While effective, it usually requires to make multiple vectors and multiple-round infections to achieve effective knockdown when multiple siRNA sites are used. On the other hand, there are clear needs to simultaneously deliver multiple siRNAs that target more than one genes.

Gibson DNA Assembly (GDA), so named after the developer of the method [16], is one of commonly-used synthetic biology techniques that offer restriction enzyme-free, scarless, largely sequence-independent, and multi-fragment DNA assembly [17,18]. GDA is an *in vitro* recombination system that has the capacity to assemble and repair multiple overlapping DNA molecules in a single isothermal step [16,17]. The optimized GDA contains three essential components: an exonuclease (e.g., 5′-T5 exonuclease) that removes nucleotides from the ends of double-stranded (ds) DNA molecules so exposing complementary single-stranded (ss) DNA overhangs that are specifically annealed; a DNA polymerase (e.g., Phusion DNA polymerase) that fills in the ssDNA gaps of the joined molecules; and a DNA ligase (e.g., Taq ligase) that covalently seals the nicks [17]. Thus, this assembly method can be used to seamlessly construct synthetic and natural genes, genetic pathways, and entire genomes as useful molecular engineering tools [16–18].

Here, we sought to use the GDA technique to establish a simplified one-step assembly system for constructing a single vector that expresses multiple siRNA target sites. To achieve this, we have engineered the GDA destination retroviral vector pSOK, based on our previously reported pSOS vector [15], which can be linearized with SwaI for assembly reactions. The assembly fragments containing multiple siRNA sites are generated by PCR amplifications using the back-to-back U6-H1 promoter vector pB2B as a template. The first fragment overlaps with the 3′-end of U6 promoter while the last fragment overlaps with the 3′-end of H1 promoter. The ends of the middle fragments overlap the specific siRNA target sequences, which confers assembly specificity. After the GDA reactions, single vectors expressing multiple siRNA target sites are generated. To prove the feasibility of this pSOK system, we have developed the vectors that contain four siRNA sites and three siRNA sites that target human and mouse β-catenin, respectively. We demonstrate that the assembly reactions are

efficient, and that candidate clones are readily identified by PCR screening, although vectors containing three siRNAs are seemingly more favorably assembled under our assembly condition. Functional analyses demonstrate that the multiple β-catenin siRNA constructs can effectively silence endogenous β-catenin expression, inhibit Wnt3A-induced β-catenin/Tcf4 reporter activity and the expression of Wnt/β-catenin downstream target genes. In mesenchymal stem cells, silencing β-catenin inhibits Wnt3A-induced early osteogenic differentiation and significantly diminishes the synergistic osteogenic activity between BMP9 and Wnt3A both *in vitro* and *in vivo*. Taken together, our results have demonstrated that the GDA-based pSOK system is proven simplistic, effective and versatile for simultaneous expression of multiple siRNA target sites. Thus, the pSOK system should be a valuable tool for gene function studies and the development of therapeutics.

Materials and Methods

Cell culture and chemicals

HEK-293 and human colon cancer SW480 lines were purchased from ATCC (Manassas, VA) and maintained in complete Dulbecco's Modified Eagle's Medium (DMEM) containing 10% fetal bovine serum (FBS, Invitrogen, Carlsbad, CA), 100 units of penicillin and 100 μg of streptomycin at 37°C in 5% CO_2 [19–24]. The reversibly immortalized mouse embryonic fibroblasts (iMEFs) were previously characterized [25,26]. The recently engineered 293pTP line was used for adenovirus amplification [27]. Both 293pTP and iMEFs were maintained in complete DMEM. Unless indicated otherwise, all chemicals were purchased from Sigma-Aldrich (St. Louis, MO) or Fisher Scientific (Pittsburgh, PA).

Construction of the retroviral vector pSOK and PCR template vector pB2B for Gibson DNA Assembly reactions

As illustrated in **Figure 1A** and **Figure S1A**, the MSCV retroviral vector pSOK was constructed on the base of our previously reported pSOS vector [15], which contains the opposing U6 and H1 promoters to drive siRNA duplex expression. The linker sites of the pSOS vector were modified and a SwaI site was engineered for linearizing the vector for Gibson Assembly (**Figure 1A, panel a**). This vector also confers Blasticidin S resistance for generating stable mammalian cell lines. The pB2B vector was constructed on the base of our previously reported pMOLuc vector [28]. Briefly, the high-fidelity PCR amplified U6 and H1 promoter fragments were subcloned into the EcoRI/HindIII sites of pMOLuc in a back-to-back orientation, and ligated at MluI site (**Figure S1B**). Both U6 and H1 promoters contain a string of "AAAAA" immediately preceding their transcription start sites, which serves as transcription termination signal for the reverse strand. The full-length vector sequences and maps are available at: http://www.boneandcancer.org/MOLab%20Vectors%20after%20Nov%201%202005/pSOK.pdf and http://www.boneandcancer.org/MOLab%20Vectors%20after%20Nov%201%202005/pBOK%20vector%20map%20and%20sequence%202013-12-02.pdf.

Gibson DNA Assembly (GDA) reactions for generating pSOK vectors expressing siRNA sites targeting human and mouse β-catenin and the generation of stable cell lines

The GDA reactions were carried out by using the Gibson Assembly Master Mix from New England Biolabs (Ipswich, MA)

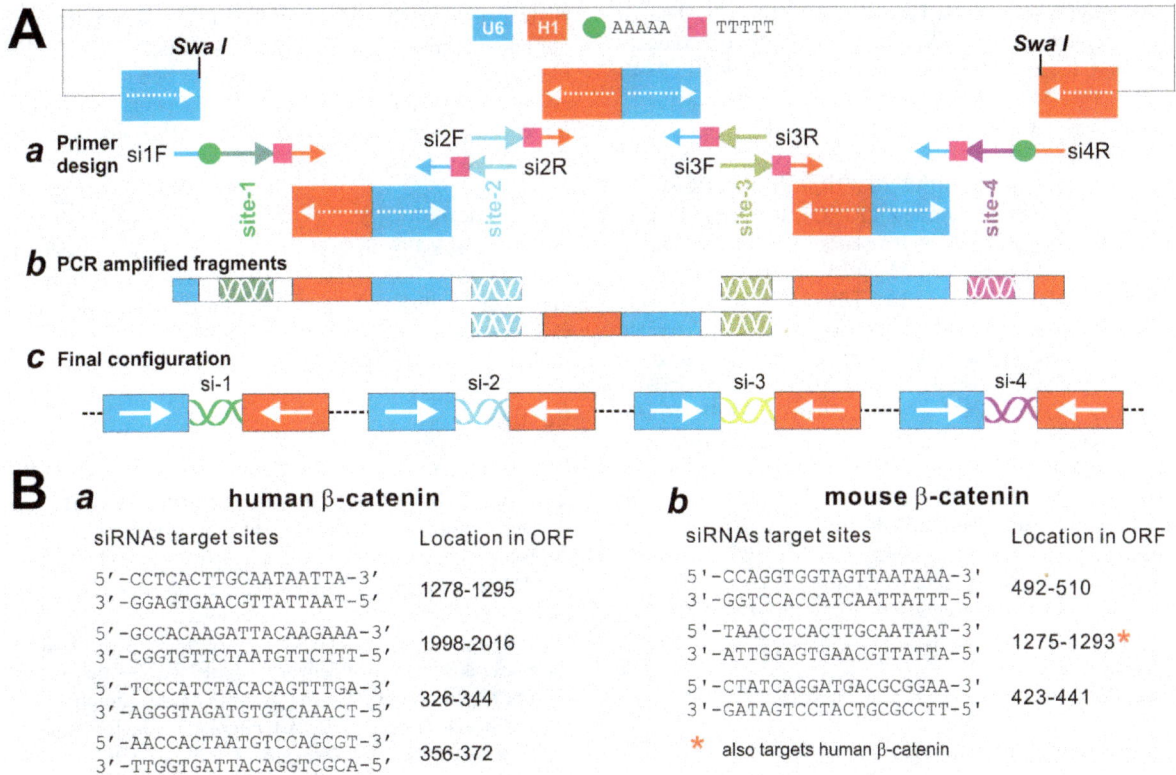

Figure 1. Schematic depiction of the one-step system pSOK for expressing multiple siRNAs. (A) Schematic representation of a tandem siRNA targeting configuration (4 sites listed as an example). The pSOK vector was constructed based on the previously reported pSOS vector, which contains opposing U6 and H1 promoters to drive siRNA duplex expression (**Figure S1A**) [15]. The linker sites of the pSOS vector were modified and a SwaI site was created for linearizing the vector for Gibson Assembly (**a**). The primers were designed according to the guidelines outlined in **Figure S2A**. Using the pB2B as a template vector (**Figure S1B**), the back-to-back U6-H1 promoter fragments with different siRNA target sites were generated. The first fragment overlaps with the 3′-end of the U6 promoter while the last fragment overlaps with the 3′-end of the H1 promoter (**b**). The ends of the middle fragments overlaps the specific siRNA target sequences (**b**). After the Gibson Assembly reaction, a single vector expressing 4 siRNA target sites is constructed (**c**). It is noteworthy that the siRNA sites may target the same or different genes. (B) The targeting sequences and locations of the designed siRNA sites on human (**a**) and mouse (**b**) β-catenin open reading frame (ORF). All of these sites have been validated in previous studies [15,40]. Note that one of the mouse siRNAs also targets human β-catenin coding sequence.

following the manufacturer's instructions. The overlapping inserts were prepared by PCR amplifications using the Phusion High-Fidelity PCR kit (New England Biolabs). Each assembly reaction contained approximately 100 ng of each insert and 50 ng of the SwaI-linearized pSOK vector and incubated at 50°C for 45 min. After the assembly reactions, the reaction mix was briefly digested with SwaI and transformed into electro-competent DH10B cells. Colony PCR screening was carried out using a forward and reverse primer pair of the two neighboring siRNA sites. Positive clones were sequencing verified. Regardless the compositions of the obtained clones, vectors containing one, two, three or four siRNA sites targeting human β-catenin were designated as pSOK-siBC1, pSOK-siBC2, pSOK-siBC3, and pSOK-siBC4, respectively. For the mouse β-catenin siRNAs, we only chose the vector that contains all three siRNA sites, namely pSOK-simBC3 for this study. A control vector containing three scrambled sites that do not target any human and mouse genes (5′-GCAAAGACGCAA-TAATACA-3′; 5′-GCACAAAGAACGACTATAA-3′; 5′-GAAACACGATTAACAGACA-3′) was also constructed, designated as pSOK-siControl.

The stable knockdown lines were generated using a retrovirus system as previously reported [15,29–31]. Briefly, the siRNA-containing pSOK vectors were co-transfected with the retrovirus packaging plasmid pCL-Ampho into HEK-293 cells. The packaged retrovirus supernatants were used to infect 293, SW480 (for siBC vectors) and iMEFs (for simBC3 vector). The infected cells were selected in Blasticidin S (4 μg/ml) for 5–7 days. The stable pools of cells were kept in LN2 for long-term storage. The resultant stable lines were designated such as 293-siBC4, 293-siControl, SW480-siBC4, SW480-siControl, iMEF-simBC3, and iMEF-siControl, to name a few.

Generation and amplification of recombinant adenoviruses expressing BMP9, Wnt3A, and GFP

Recombinant adenoviruses were generated using the AdEasy technology as described [30,32–34]. The coding regions of human BMP9 and mouse Wnt3A were PCR amplified and cloned into an adenoviral shuttle vector, and subsequently used to generate and amplify recombinant adenoviruses in HEK-293 or 293pTP cells [27]. The resulting adenoviruses were designated as AdBMP9 and AdWnt3A, both of which also express GFP [35–38]. Analogous adenovirus expressing only GFP (AdGFP) was used as controls [39–42]. For all adenoviral infections, polybrene (4–8 μg/ml) was added to enhance infection efficiency as previously reported [23].

Cell transfection and firefly luciferase reporter assay

Subconfluent cells were transfected with the Tcf/Lef reporter pTOP-Luc using Lipofectamine Reagent (Invitrogen) by following the manufacturer's instructions. For 293 and iMEF cells, the cells were co-transfected with pCMV-Wnt3A. At 48 h post transfection, cells were lysed for luciferase assays using Luciferase Assay System (Promega, Madison, WI) by following the manufacturer's instructions. Easy conditions were done in triplicate.

RNA isolation and quantitative real-time PCR (qPCR)

Total RNA was isolated by using TRIZOL Reagents (Invitrogen) and used to generate cDNA templates by reverse transcription reactions with hexamer and M-MuLV reverse transcriptase (New England Biolabs, Ipswich, MA). The cDNA products were used as PCR templates. The sqPCR were carried out as described [43–47]. PCR primers (**Table S1**) were designed by using the Primer3 program and used to amplify the genes of interest (approximately 150–250 bp). For qPCR analysis, SYBR Green-based qPCR analysis was carried out by using the thermocycler Opticon II DNA Engine (Bio-Rad, CA) with a standard pUC19 plasmid as described elsewhere [21,48–50]. The qPCR reactions were done in triplicate. The sqPCR was also carried out as described [15,24,25,27,29,39,47,51,52]. Briefly, sqPCR reactions were carried out by using a touchdown protocol: $94°C \times 20''$, $68°C \times 30''$, $70°C \times 20''$ for 12 cycles, with $1°C$ decrease per cycle, followed by 25–30 cycles at $94°C \times 20''$, $56°C \times 30''$, $70°C \times 20''$. PCR products were resolved on 1.5% agarose gels. All samples were normalized by the expression level of GAPDH.

Immunofluorescence staining

Immunofluorescence staining was performed as described [30,40,43,50,53,54]. Briefly, cells were infected with AdWnt3A or AdGFP for 48 h, fixed with methanol, permeabilized with 1% NP-40, and blocked with 10% BSA, followed by incubating with β-catenin antibody (Santa Cruz Biotechnology). After being washed, cells were incubated with Texas Red-labeled secondary antibody (Santa Cruz Biotechnology). Stains were examined under a fluorescence microscope. Stains without primary antibodies, or with control IgG, were used as negative controls.

Qualitative and quantitative assays of alkaline phosphatase (ALP) activity

ALP activity was assessed quantitatively with a modified assay using the Great Escape SEAP Chemiluminescence assay kit (BD Clontech, Mountain View, CA) and qualitatively with histochemical staining assay (using a mixture of 0.1 mg/ml napthol AS-MX phosphate and 0.6 mg/ml Fast Blue BB salt), as previously described [29,30,32,33,39,40,44,53]. Each assay condition was performed in triplicate and the results were repeated in at least three independent experiments.

iMEF cell implantation and ectopic bone formation

All animal studies were conducted by following the guidelines approved by the Institutional Animal Care and Use Committee (IACUC) of The University of Chicago (protocol #71108). Stem cell-mediated ectopic bone formation was performed as described [20,25,29,30,33,47,55–57]. Briefly, subconfluent iMEFsimBC3 and iMEF-siControl cells were infected with AdBMP9 and/or AdWnt3A, or AdGFP for 16 h, collected and resuspended in PBS for subcutaneous injection (5×10^6/injection) into the flanks of athymic nude (nu/nu) mice (5 animals per group, 4–6 wk old, female, Harlan Laboratories, Indianapolis, IN). At 4 weeks after implantation, animals were sacrificed, and the implantation sites were retrieved for histologic evaluation and Trichrome staining as described below.

Histological evaluation and Trichrome staining

Retrieved tissues were fixed, decalcified in 10% buffered formalin, and embedded in paraffin. Serial sections of the embedded specimens were stained with hematoxylin and eosin (H & E). Trichrome staining was carried out as previously described [20,25,26,44,47,52,55,56].

Statistical analysis

The quantitative assays were performed in triplicate and/or repeated three times. Data were expressed as mean ± SD. Statistical significances were determined by one-way analysis of variance and the student's t test. A value of $p < 0.05$ was considered statistically significant.

Results

Construction of the GDA vector pSOK for expressing multiple siRNA target sites in mammalian cells

We previously developed the pSOS system, in which the siRNA duplexes are made from an oligo template driven by opposing U6 and H1 promoters [15]. While effective, it usually requires to make multiple vectors and multiple-round infections to achieve effective knockdown if multiple siRNA target sites are used. In other cases, there are clear needs to deliver multiple siRNAs that target more than one genes. Here, we sought to establish a simplified one-step approach, based on the GDA technology, which will allow us to make a single vector that express multiple siRNA target sites against one gene or multiple genes.

As depicted in **Figure 1A**, the pSOK vector was constructed based on the previously reported pSOS vector [15], which contains the opposing U6 and H1 promoters to drive siRNA duplex expression (**Figure S1A**). The linker sites of the pSOS vector were modified and a SwaI site was engineered for linearizing the vector for Gibson Assembly (**Figure 1A, panel a**). This vector confers Blasticidin S resistance for generating stable mammalian cell lines. For examples, four siRNA target sites are exemplified to illustrate the primer design and construction process, the primers were designed according to the guidelines outlined in **Figure S2A**. Using the pB2B as a template vector (**Figures S1B, and S2B**), the back-to-back U6-H1 promoter fragments with different siRNA target sites were generated. The first fragment overlaps with the 3'-end of the U6 promoter while the last fragment overlaps with the 3'-end of the H1 promoter (**Figure 1A, panel b**). Thus, the ends of the middle fragments overlap the specific siRNA target sequences (**Figure 1A, panel b**). After the GDA reaction, a single vector expressing four siRNA target sites is constructed (**Figure 1A, panel c**).

To prove the principle and feasibility of the pSOK system, we designed four siRNA sites and three siRNA sites that target human and mouse β-catenin, respectively (**Figure 1B**). The targeting sequences and locations of the designed siRNA sites on human (**Figure 1B, panel a**) and mouse (**Figure 1B, panel b**) β-catenin open reading frame (ORF). These siRNA sites were previously demonstrated to effectively silence β-catenin expression [15,40]. As indicated, these siRNA sites target a broad region of β-catenin coding regions, and one of the mouse siRNAs also target human β-catenin (**Figure 1A, panel b**).

Construction and characterization of pSOK vectors that express multiple siRNAs targeting human β-catenin

We first chose to construct the pSOK vector expressing the four siRNAs that target human β-catenin. After performing PCR amplifications of the three inserts as depicted in **Figure 1A**, we carried out the GDA reactions using the three inserts and the SwaI-linearized pSOK vector. The potential recombinants were screened by colony PCR using the forward and reverse primer pairs of the neighboring siRNA sites (**Figure 2A**). Since there were multiple repetitive U6-H1 promoter units in the construct, we found the most robust and specific amplifications were obtained when the forward and reverse primer pairs of the neighboring siRNA sites were used. To further demonstrate this phenomenon, we used a representative pSOK-siBC4 clone as the template and tested PCR amplifications with different combinations of primer pairs. We found that the primer pairs 1F/2R, 2F/3R, and 3F/4R yielded robust and relatively specific products while primer pairs covering two or more siRNA sites produced multiple bands (**Figure 2A**).

Since the overlapping sequences for the inserts are only 19 bp, it is conceivable that a high exonuclease activity in the assembly reaction may over digest the overlapping sequences and cause mispairing of the siRNA sites. In fact, we found clones containing one to four siRNA sites as shown by the restriction digestion to release the inserts (in roughly 500 bp U6-H1 cassettes) (**Figure 2B**). More than candidate clones were verified by DNA sequencing and

blasting against the query sequence outlined in **Figure S2C**. Sequencing analysis of these clones revealed that different clones may contain different siRNA sites (**Figure 2C**). Statistically, clones containing three siRNA sites (i.e, siBC3) are the most abundant (accounted for about 50% of the clones), followed by clones containing two or four siRNA sites (i.e., siBC2 or siBC4) at about 20% each (**Figure 2D**). Surprisingly, clones containing one siRNA site only accounted for about 10% of the screened clones (**Figure 2D**). These results indicate that the assembly reactions are efficient although the assembly of three siRNA sites (e.g., two inserts) may be a more favorable event, at least under our reaction conditions. After sequencing verification, the clones containing one, two, three and four siRNA sites were pooled and designated as pSOK-siBC1, pSOK-siBC2, pSOK-siBC3 and pSOK-siBC4, respectively. An analogous vector containing three scrambled sites was also constructed as a control (e.g., pSOK-siControl). While we chose to construct multiple siRNAs to target the same gene (i.e., β-catenin), it is conceivable that one can assemble multiple siRNAs that target more than one genes.

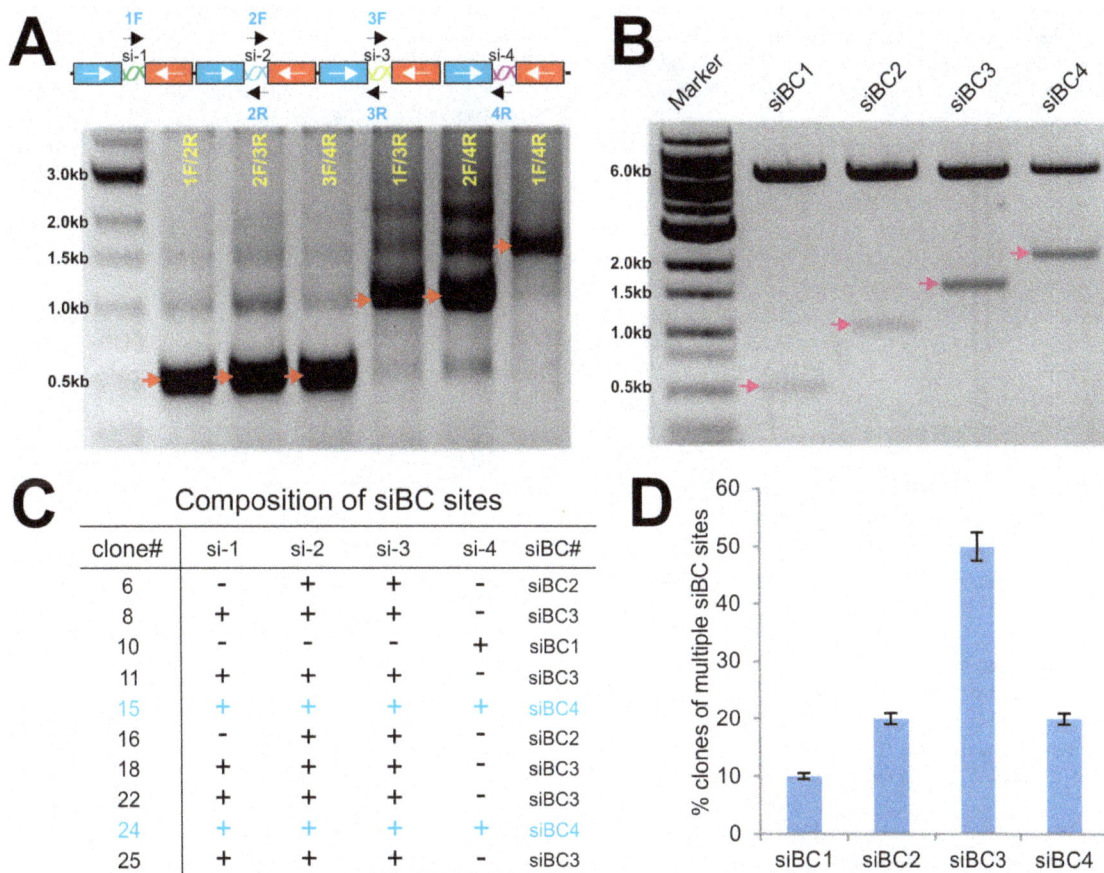

Figure 2. Construction and characterization of pSOK vectors that express multiple siRNAs targeting human β-catenin. (A) PCR screening strategy for candidate clones. A representative pSOK-siBC4 clone was used as a template and PCR amplified with different combinations of primer pairs, as depicted above the gel image. The PCR products were resolved on a 0.8% agarose gel. The arrows indicate the expected products. **(B)** Restriction digestion confirmation of the obtained clones. Representative clones containing 1 to 4 siRNA sites were digested with PmeI/HindIII to release the inserts. The digested products were resolved on a 0.8% agarose gel. The arrows indicate the expected products. **(C)** The frequency of the clones containing multiple copies of siRNA target expression units of human β-catenin. Plasmid DNA was isolated from approximately 80 individual SwaI-resistant clones and subjected to DNA sequencing. The presence of different copy numbers of siRNA sites was tabulated. **(D)** siRNA target site composition of 10 representative clones of human β-catenin. "+", site present, "−", site absent.

Functional validation of the pSOK-siBC4 vector that contains four siRNA sites targeting human β-catenin

Although most of these siRNA target sites have been tested for their silencing efficiency and effect on Wnt/β-catenin signaling activity, the knockdown efficiency may be compromised due to promoter competitions in the pSOK system because multiple U6-H1 expression cassettes are engineered in a single vector. To test this possibility, we established stable lines of HEK-293 and SW480 cells expressing the siBC sites or siControl using a retroviral system. We first assessed the knockdown efficiency of endogenous β-catenin in 293 cells and SW480 cells. Using qPCR analysis, we found that the endogenous β-catenin expression in 293-siBC4 was significantly lower than that of the 293-siControl's (**Figure 3A, panel a**). Similarly, using the human colon cancer line SW480 we found the β-catenin expression was drastically reduced in SW480-siBC4 cells compared with that in the SW480-siControl cells (**Figure 3A, panel b**). Overall, the siBC4-expressing 293 and SW480 cells exhibited marked decreases in the β-catenin expression, only about 32% and 4% of the control cells' (**Figure 3A, panel c**). It is noteworthy that we also found that endogenous β-catenin expression was significantly reduced in the 293 and SW480 cells that express siBC1, siBC2, and siBC3 (data not shown). When the 293-siBC4 (co-transfected with Wnt3A) and SW480-siBC4 cells were transfected with the β-catenin/Tcf reporter pTOP-Luc, the reporter activities were marked reduced at the tested time points in both 293 cells ($p<0.001$) (**Figure 3B, panel a**) and SW480 cells ($p<0.001$) (**Figure 3B, panel b**). Moreover, the Wnt3A-stimulated reporter activities in 293 cells stably expressing siBC1, siBC2, and siBC3 were also remarkably inhibited (**Figure S3A**), and similar results were obtained in SW480 cells, in which the Wnt/β-catenin signaling is constitutively active (**Figure S3B**). Furthermore, we examined the β-catenin knockdown efficiency in SW480 cells by immunofluorescence staining. We found that cytoplasmic/nuclear accumulation of β-catenin in SW480-siBC4 cells was significantly diminished, compared with that in the SW480-siControl cells (**Figure 3C**). Taken the above results together, the pSOK-siBC4 vector that expresses four human β-catenin siRNA sites can effectively silence β-catenin expression in human cells.

pSOK-simBC3 contains multiple siRNAs targeting mouse β-catenin and effectively inhibits canonical Wnt signaling activity in iMEFs

Our results in **Figure 2** indicate that three siRNA sites (two inserts) are seemingly more favorably assembled into pSOK vector. It is conceivable that in most cases three siRNA sites should be sufficiently effective in silencing a given gene. Here, we tested this possibility by constructing a vector, designated as pSOK-simBC, that expressed three siRNA sites targeting mouse β-catenin (**Figure 1B, panel b**). The construction and screening process were very efficient. After sequencing verification, the pSOK-simBC was packaged as retrovirus and used to generate the stable line iMEF-simBC3, along with a control line iMEF-siControl. The iMEFs were previously characterized multi-potent mesenchymal stem cells (MSCs) [20,25]. When the iMEF stable lines were infected with AdWnt3A or AdGFP and analyzed for β-catenin expression, we found that β-catenin expression was significantly reduced in iMEF-simBC3 cells, compared with that in iMEF-siControl cells ($p<0.001$) (**Figure 4A**).

Using the β-catenin/Tcf luciferase reporter, we found that iMEF-simBC3 cells exhibited significantly lower β-catenin/Tcf reporter activity upon Wnt3A stimulation ($p<0.001$) (**Figure 4B**). Accordingly, when the expression of two well-characterized Wnt/

β-catenin downstream target genes, Axin2 [58] and c-Myc [59], was examined, we found that Wnt3A was shown to significantly induce the expression of Axin2 and c-Myc in iMEF-siControl cells; but silencing β-catenin in iMEFs significantly diminished Wnt3A-induced expression of c-Myc (**Figure 4C, panel a**) and Axin2 (**Figure 4C, panel b**). Furthermore, immunofluorescence staining indicate that Wnt3A-induced cytoplasmic/nuclear accumulation of β-catenin protein was effectively reduced in iMEF-simBC3 cells, compared with that in iMEF-siControl cells (**Figure 4D**). Therefore, the above data demonstrate that the three-siRNA site-containing pSOK-simBC3 can effectively blunt the functional activities of Wnt3A/β-catenin in iMEF cells.

Silencing β-catenin diminishes the synergistic osteogenic activity between BMP9 and Wnt3A in iMEF cells

We further analyzed the functional consequences of β-catenin knockdown on MSC differentiation. We and others demonstrated that canonical Wnt signaling can induce osteogenic differentiation of mesenchymal stem cells [39,40,48,60]. We sought to determine if Wnt3A can induce early osteogenic marker alkaline phosphatase (ALP) activity in iMEFs, and if the induced ALP activity would be reduced when β-catenin is silenced in iMEFs. We found that Wnt3A effectively induced ALP activity in iMEF-siControl cells, which was significantly blunted in iMEF-simBC3 cells (**Figure 5A**). We previously demonstrated that BMP9 is one of the most potent osteogenic BMPs in mesenchymal cell stems [26,30,32,33,61]. We found that BMP9 stimulated robust ALP activity in iMEF-siControl cells while the BMP9-induced ALP activity was remarkably reduced in iMEF-simBC3 cells (**Figure 5A**).

We previously showed that Wnt3A and BMP9 act synergistically in regulating osteogenic differentiation of MSCs [40]. We found that the iMEF-siControl cells co-transduced with Wnt3A and BMP9 exhibited higher ALP activity than that of the cells transduced with either Wnt3A or BMP9 alone, which was remarkably blunted by β-catenin knockdown (**Figure 5A**). Quantitative ALP activity analysis revealed a similar trend, BMP9 and Wnt3A+BMP9 stimulated ALP activities were significantly inhibited by β-catenin knockdown $p<0.001$ (iMEF-simBC3 vs. iMEF-siControl) (**Figure 5B**). Thus, these results suggest that β-catenin may play an important role in this synergistic action between BMP9 and Wnt3A in osteogenic differentiation of MSCs.

BMP9-induced ectopic bone formation from iMEFs is potentiated by Wnt3A but attenuated by β-catenin knockdown

Using our previously established stem cell implantation assays [20,25,26,30,38,47,52,57], we tested the *in vivo* effect of β-catenin knockdown on BMP9 and Wnt3A-induced ectopic bone formation. Subconfluent iMEF-simBC3 and iMEF-siControl cells were transduced with AdBMP9, AdWnt3A, AdBMP9+AdWnt3A, or AdGFP, and injected subcutaneously into the flanks of athymic nude mice for 4 weeks. No recoverable masses were detected in the GFP or Wnt3A group. Robust bony masses were retrieved from both BMP9 and BMP9+Wnt3A transduced iMEF-siControl groups, while significantly smaller masses were recovered from the iMEF-simBC3 group (**Figure 6A, panels a & b vs. c**). BMP9+Wnt3A group formed a slightly larger bone masses (**Figure 6A, panels a vs. b**).

When the retrieved samples were subjected to H & E staining, we found that BMP9-transduced iMEF-siControl cells formed evident trabecular bone, which was even more robust in the

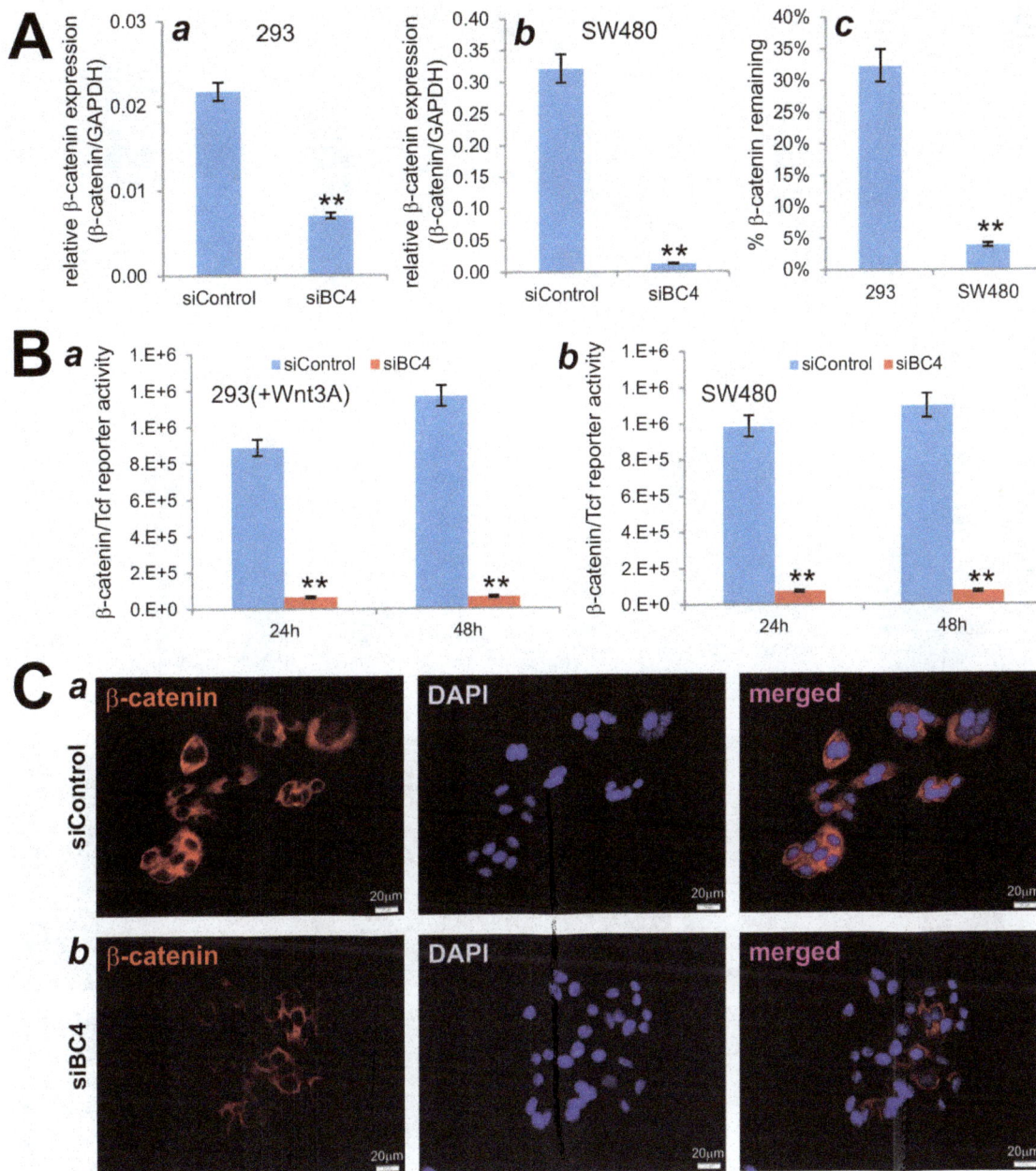

Figure 3. Functional validation of siRNAs targeting human β-catenin. (A) Efficient knockdown of endogenous β-catenin in HEK-293 and SW480 cells. Total RNA was isolated from subconfluent 293-siBC4, 293-siControl, SW480-siBC4, and SW480-siControl cells, and subjected to qPCR analysis using primers specific for human β-catenin. All samples were normalized with GAPDH. Each reaction was done in triplicate. Relative β-catenin expression was calculated by dividing β-catenin expression levels with respective GAPDH levels in 293 (*a*) and SW480 (*b*) cells. The % of remaining β-catenin expression was calculated by dividing the relative β-catenin expression in siBC4 with that of the respective siControl's (*c*). "**", $p < 0.001$. **(B)** β-Catenin/Tcf transcription activity is significantly reduced in siBC4 cells. Subconfluent 293-siBC4 and 293-siControl cells were co-transfected with TOP-Luc reporter and pCMV-Wnt3A plasmids using Lipofectamine reagent (*a*), while SW480-siBC4 and SW480-siControl cells were transfected with TOP-Luc reporter plasmid using Lipofectamine reagent (*b*). At 24 h and 48 h after transfection, the cells were lysed and subjected to firefly luciferase assay using the Luciferase Reporter Assay System (Promega). Each assay condition was done in triplicate. "**", $p < 0.001$. **(C)** siBCs can effectively block Wnt3a-induced β-catenin accumulation. Subconfluent SW480-siBC4 and SW480-siControl cells fixed and subjected to immunofluorescence staining with an anti-β-catenin antibody. The cell nuclei were counter stained with DAPI. Control IgG and minus primary antibody were used as negative controls (data not shown).

presence of both BMP9 and Wnt3A (**Figure 6B**). However, silencing β-catenin expression in iMEF-simBC3 cells significantly reduced trabecular bone formation induced by BMP9 or BMP9+Wnt3A, and formed cartilage-like small masses (**Figure 6B**). Trichrome staining confirmed that iMEF-siControl cells transduced with BMP9 formed apparently mature and mineralized bone matrices, while a combination of BMP9 and Wnt3A induced more mature and highly mineralized bone matrices (**Figure 6C**). However, the maturity and mineralization were significantly diminished in iMEF-simBC3 cells transduced with

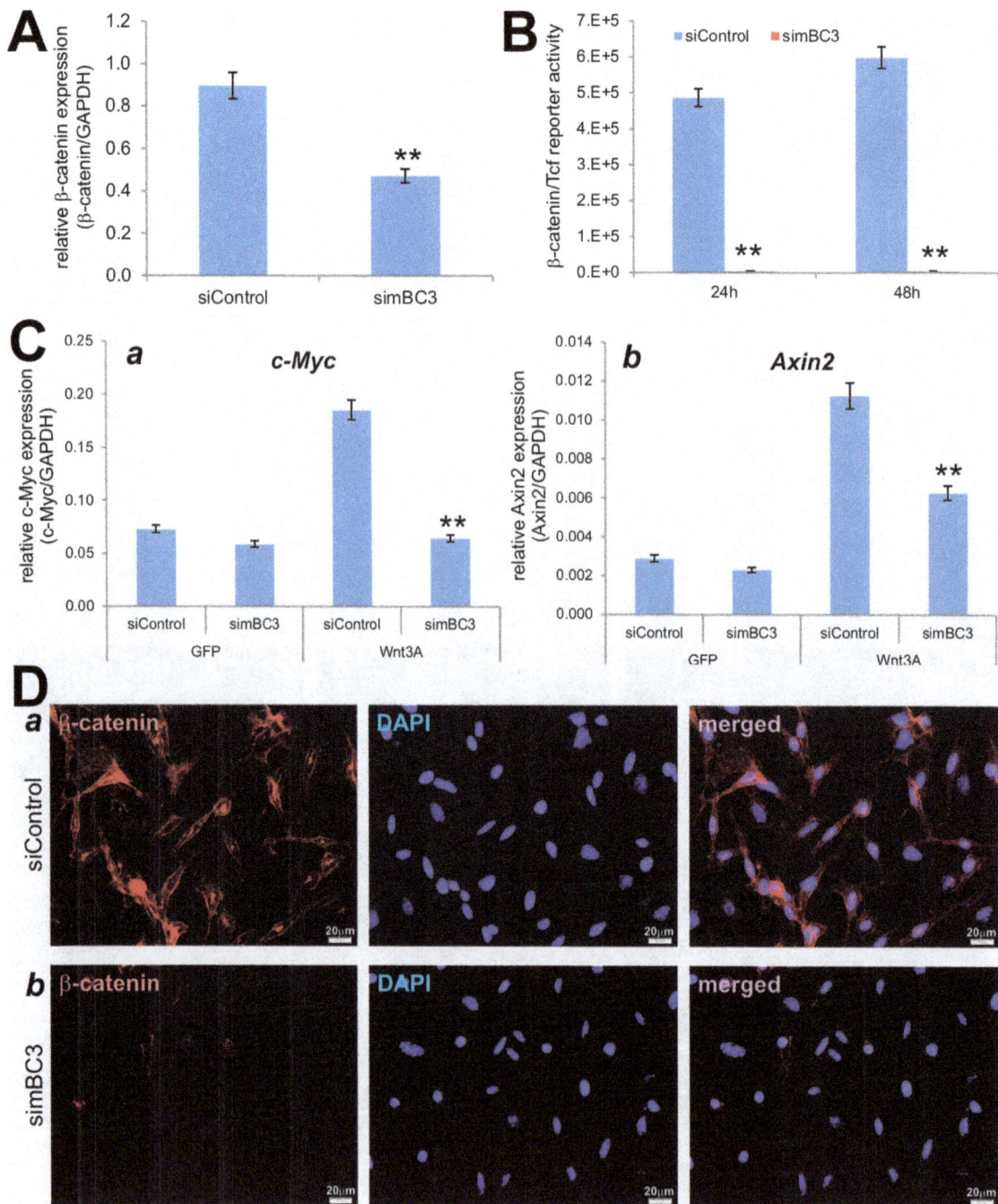

Figure 4. Multiple siRNAs targeting mouse β-catenin simBC3 effectively inhibit canonical Wnt signaling activity in iMEFs. (A) Reduced β-catenin expression in iMEF-simBC3 cells. Subconfluent iMEF-simBC3 and iMEF-siControl cells were infected with AdWnt3A or AdGFP. At 36 h after infection, total RNA was isolated and subjected to qPCR analysis using primers for mouse β-catenin and GAPDH. Relative expression was calculated by dividing the β-catenin expression levels with respective GAPDH expression. All samples were subjected to the subtraction of baseline (i.e., AdGFP infected cells) expression. Each assay was done in triplicate. "**", $p < 0.001$. (**B**) iMEF-simBC3 cells exhibit significantly lower β-catenin/Tcf reporter activity upon Wnt3A stimulation. Subconfluent iMEF-simBC3 and iMEF-siControl cells were transfected with TOP-Luc reporter plasmid and infected with AdWnt3A or AdGFP. At 24 h and 48 h post transfection/infection, cells were lysed for luciferase assays. Relative β-catenin/Tcf reporter activity was subjected to subtractions of basal activity (i.e., AdGFP groups). Easy conditions were done in triplicate. "**", $p < 0.001$. (**C**). Wnt3A-induced expression of Wnt/β-catenin target genes was significantly decreased in iMEF-simBC3 cells. Subconfluent iMEF-simBC3 and iMEF-siControl cells were infected with AdWnt3A or AdGFP for 36 h. Total RNA was isolated and subjected to reverse transcription. The resultant cDNAs were used as templates for qPCR analysis using primers specific for mouse Axin2 and c-Myc transcripts. All samples were normalized by GAPDH levels. Each assay condition was done in triplicate. "**", $p < 0.001$. (**D**) simBC3 can effectively block Wnt3a-induced β-catenin accumulation. Subconfluent iMEF-siControl (a) cells fixed and subjected to immunofluorescence staining with an anti-β-catenin antibody. The cell nuclei were counter stained with DAPI. Control IgG and minus primary antibody were used as negative controls (data not shown). Representative images are shown.

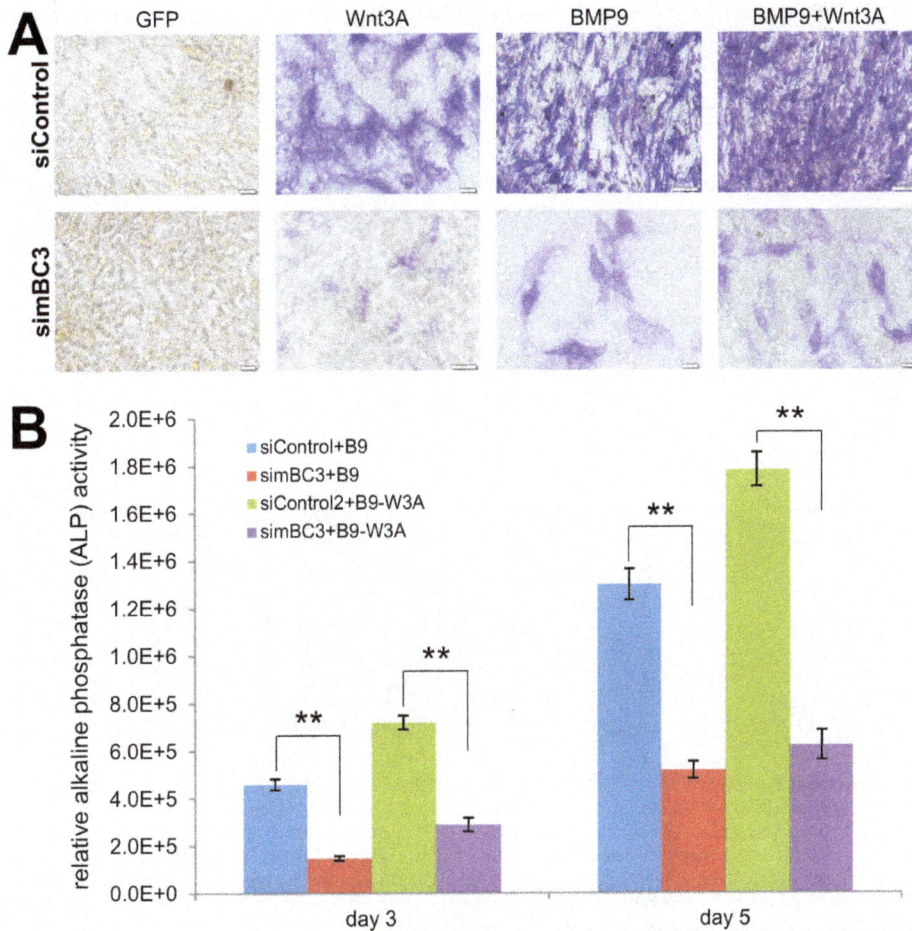

Figure 5. Silencing β-catenin diminishes the synergistic osteogenic activity between BMP9 and Wnt3A in iMEF cells. (**A**) Wnt3A and/or BMP9-induced early osteogenic marker alkaline phosphatase (ALP) activity is reduced in iMEF-simBC3 cells. Subconfluent iMEF-simBC3 and iMEF-siControl cells were infected with AdWnt3A, AdBMP9, AdGFP, or AdWnt3A+AdBMP9. At day 5 post infection, cells were fixed for ALP histochemical staining assay. Each assay conditions were done in triplicate. Representative results are shown. (**B**) Wnt3A and/or BMP9-induced ALP activity is decreased in the β-catenin silenced iMEFs. The experiments were set up in a similar fashion to that described in (**A**). At days 3 and 5, cells were lysed for quantitative ALP activity assays. Basal ALP activities (e.g., GFP groups) were subtracted from all BMP9, Wnt3A, and Wnt3A+BMP9 groups. Each assay conditions were done in triplicate. "**", $p < 0.001$ (iMEF-simBC3 vs. iMEF-siControl).

either BMP9 or BMP9+Wnt3A (**Figure 6C**). Taken together, these in vivo results strongly suggest that β-catenin may play an important role in mediating BMP9-induced bone formation, and the BMP9-Wnt3A may crosstalk in inducing osteoblastic differentiation of MSCs.

Discussion

To overcome the technical challenges in simultaneously expressing multiple siRNAs that silence one specific gene or different genes, here we sought to develop a simple, efficient and versatile method to express multiple siRNAs in a single vector by exploring the possible utility of the Gibson DNA Assembly. We take advantages of our previously established pSOS system, in which the siRNA duplexes are generated from oligo templates driven by opposing U6 and H1 promoters [15]. Since there are only a few Pol III promoters that are well characterized, we choose to use the same U6-H1 promoter cassette to drive the expression of multiple siRNA sites. However, the use of these repetitive U6-H1 expression units poses a technical challenge for choosing overlapping sequences for Gibson DNA Assembly. To overcome this hurdle we design an assembly scheme that takes advantages of the unique sequences of different siRNA sites (e.g., stretches of 19 nucleotides). The assembly fragments containing multiple siRNA sites are generated by PCR amplifications using the back-to-back U6-H1 promoter vector pB2B as a template while the vector is SwaI-linearized pSOK.

We carried out the proof-of-principle studies using multiple siRNAs targeting human and mouse β-catenin. We demonstrate that the assembly reactions are efficient, and that candidate clones are readily identified by PCR screening. Functional analyses demonstrate that multiple β-catenin siRNA constructs can effectively silence endogenous β-catenin expression, inhibit Wnt3A-induced β-catenin/Tcf reporter activity and the expression of Wnt/β-catenin downstream target genes. Furthermore, in mesenchymal stem cells we found that silencing β-catenin inhibits Wnt3A-induced early osteogenic differentiation and significantly diminishes the synergistic osteogenic activity between BMP9 and Wnt3A both in vitro and in vivo. Therefore, our results have demonstrated that the Gibson Assembly-based pSOK system is proven simplistic, effective and versatile for simultaneous expression of multiple siRNA target sites.

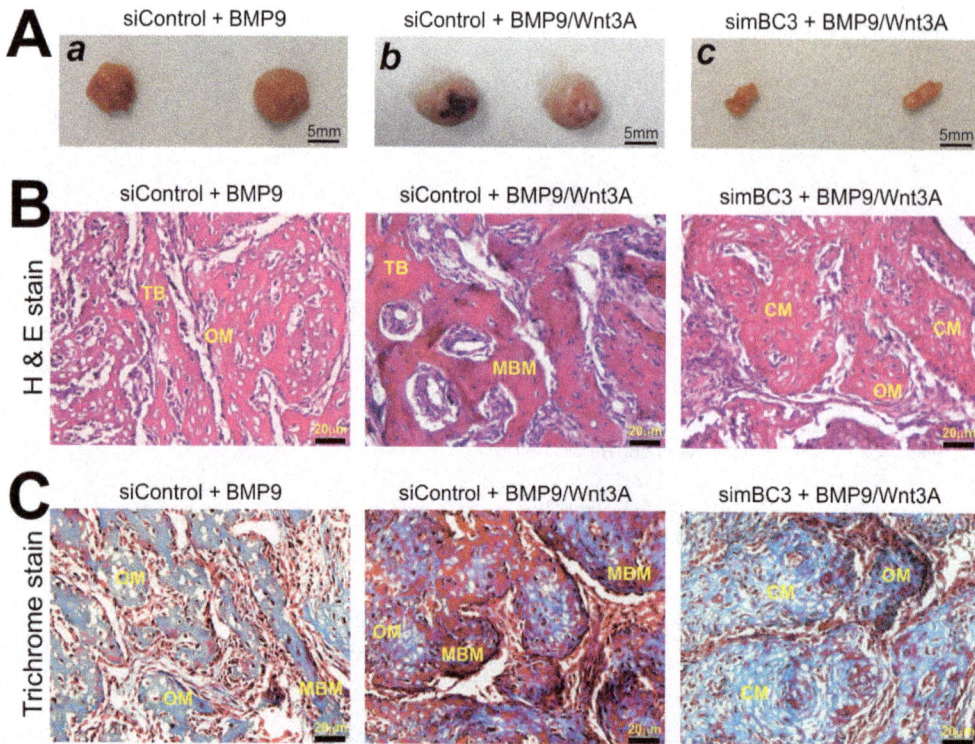

Figure 6. BMP9-induced ectopic bone formation from iMEFs is potentiated by Wnt3A but attenuated by β-catenin knockdown. (A) Gross images. Subconfluent iMEF-simBC3 and iMEF-siControl cells were infected with AdBMP9, AdWnt3A, AdBMP9+AdWnt3A, or AdGFP for 16 h. Cells were collected for subcutaneous injections (3×10^6 cells/site in 100 μl PBS) into the flanks of athymic nude mice (n = 5 each group). At 4 weeks after injection, the animals were sacrificed. Masses formed at the injection sites were retrieved from the groups injected with the iMEF-siControl cells transduced with BMP9 (*a*) or BMP9+Wnt3A (*b*), while very small masses were retrieved from the animals injected with the iMEFs transduced with BMP9+Wnt3A (c) or BMP9 (not shown). No masses were retrieved from the animals injected with Wnt3A- or GFP-transduced iMEF cells. Representative results are shown. (**B**) and (**C**) Histologic evaluation and Trichrome staining. The retrieved samples were decalcified and subjected to paraffin-embedded sectioning for histologic evaluation, including H & E staining (*B*) and Trichrome staining (*C*). Representative results are shown. TB, trabecular bone; MBM, mineralized bone matrix; OM, osteoid matrix; CM, chondroid matrix.

Our findings have addressed at least two technical and functional concerns over the pSOK system. First, our design for the overlapping ends of the inserts is only 19 nucleotides. It's conceivable the overlapping sequences are too short and may comprise the assembly reactions. Our results indicate that the assembly efficiency is reasonably high although, in our attempt to assemble four siRNA sites, we do obtain clones that contain one, two or three sites. In fact, vectors containing three siRNAs are seemingly more favorably assembled under our assembly condition. Second, the repeated U6-H1 promoter cassettes may compromise the expression of multiple siRNA sites due to possible promoter competition [62]. Our results using the clones containing different numbers of siRNA sites strongly suggest that the use of repetitive U6-H1 expression cassettes may pose little or insignificant impact on the efficient expression of multiple siRNA sites although we did not analyze the precise expression levels of these siRNA duplexes. Given the nature of the short 19-nt overlapping sequences, we have found two critical technical parameters should be taken into consideration for an efficient assembly: 1) using 3~5 times more inserts than conventional ligation reactions; and 2) using shorter assembly reaction time (e.g., 30–45 min at 50°C). Furthermore, it is conceivable that the same assembly system can be introduced into recombinant adenovirus, adenovirus-associated virus, and other gene delivery vector systems.

In this study, we examined the functional consequences of β-catenin knockdown on Wnt3A and/or BMP9-induced osteogenic differentiation of mesenchymal stem cells. Wnts are a family of secreted glycoproteins that regulate many developmental processes [63]. Wnt signaling plays an important role in skeletal development [60,64]. Wnt proteins bind to their cognate receptor frizzled (Fz) and LRP-5/6 co-receptors, and activate distinct signaling pathways, including the canonical β-catenin pathway. In the absence of Wnt signaling, β-catenin is degraded by the proteasome system after GSK3β dependent phosphorylation. In the presence of Wnt signaling, unphosphorylated β-catenin accumulates in the cytoplasm and translocates into the nucleus where it associates with Tcf/LEF transcription factors to regulate the expression of target genes [59,65–67]. However, the precise function of Wnt/β-catenin in osteoblastic differentiation remains to be fully elucidated. We previously found that BMP9 (aka, GDF2) is one of the most potent osteogenic BMPs [26,30,32,33,61,68]. Through gene expression profiling, we found that Wnt3A and BMP9 regulated the expression of overlapping but distinct sets of downstream target genes in MSCs [39,48], suggesting that there may be an important crosstalk between BMP and Wnt-induced osteogenic signaling. In this study, we used iMEFs and demonstrated that Wnt3A and BMP9 can potentiate each other's ability to induce osteogenic differentiation *in vitro* and *in vivo*. Furthermore, β-catenin knockdown significantly diminishes BMP9-induced osteogenic differentiation of iMEFs, indicating that BMP9-induced osteoblastic differentiation requires functional β-catenin signaling.

In summary, we provide a conceptual design of a simplified and versatile system for the simultaneous expression of multiple siRNAs that silence one or different genes. A series of proof-of-concept studies have validated the technical feasibility and functional efficiency of the pSOK system by silencing human and mouse β-catenin expression. Thus, our results have demonstrated that the GDA-based pSOK system should be a valuable tool for gene function studies and the development of therapeutics.

Supporting Information

Figure S1 Schematic representations of the pSOK and pB2B vectors developed in this study. (**A**) The pSOK vector is a Murine Stem Cell Virus (MSCV) retroviral vector. It was derived from the previously developed pSOS vector [15]. The pSOK is a destination vector used for the one-step Gibson Assembly after SwaI linearization. This vector confers Blasticidin S resistance for generating stable mammalian cell lines. (**B**) The pB2B vector is a common template for PCR amplifications to generate the fragments with distinct siRNA target sites, which are subsequently used for Gibson Assembly with the SwaI-linearized pSOK vector. The full-length sequences and maps of these vectors are available at: http://www.boneandcancer.org/MOLab%20Vectors%20after %20Nov%201%202005/pSOK.pdf and http://www.boneandcan cer.org/MOLab%20Vectors%20after%20Nov%201%202005/p BOK%20vector%20map%20and%20sequence%202013-12-02. pdf.
(TIF)

Figure S2 A Guide for primer design and essential sequences for assembly analysis. (**A**) Primer design guide. To make a construct containing four siRNA target sites driven by opposing U6-H1

promoters, three PCR fragments will be made for the assembly reaction. Please note the sense-strand (upper case; driven by U6 promoter) and reverse-complement strand (lower case) of the chosen siRNA sites. (**B**) The DNA sequence of the H1-U6 back-to-back promoters in pB2B is used to amplify the different siRNA fragments. Please note the template sequence contains the "TTTTT" and "AAAAA" sequences to terminate siRNA transcripts. (**C**) The assembled query sequence for BLAST analysis of sequenced candidadte clones. One can simply replace the designed "X", "Y" and "Z" target site sequences (red and underlined) and use the modified sequence as a template to perform BLAST2 analysis and verify colony authenticity.
(TIF)

Figure S3 Function validation of the silencing efficiency of four siRNA sites targeting human β-catenin. 293 and SW480 cells stably expressing one, two, three, four siRNA sites, or siControl were generated as described in Methods. Subconfluent 293 lines were co-transfected with TOP-Luc and pCMV-Wnt3A plasmids (**A**) while the SW480 lines were just transfected with TOP-Luc reporter plasmid (**B**). At 24 h and 48 h after transfection, cells were lysed and subjected to firefly luciferase activity assays as described in Methods. Each assay condition was done in triplicate.
(TIF)

Table S1 Primers used for PCR analysis.
(XLS)

Author Contributions

Conceived and designed the experiments: TCH TY RRR RCH HHL FD. Performed the experiments: FD XC ZL ZY ZW. Analyzed the data: FD XC ZL ZY ZW. Contributed reagents/materials/analysis tools: YD QZ ZZ JY MQ RL SD JW QW ML NG LZ GZ PZ. Wrote the paper: TCH TY RCH HHL FD.

References

1. Hammond SM, Bernstein E, Beach D, Hannon GJ (2000) An RNA-directed nuclease mediates post-transcriptional gene silencing in Drosophila cells. Nature 404: 293–296.
2. Castel SE, Martienssen RA (2013) RNA interference in the nucleus: roles for small RNAs in transcription, epigenetics and beyond. Nat Rev Genet 14: 100–112.
3. Dykxhoorn DM, Novina CD, Sharp PA (2003) Killing the messenger: short RNAs that silence gene expression. Nat Rev Mol Cell Biol 4: 457–467.
4. Ghildiyal M, Zamore PD (2009) Small silencing RNAs: an expanding universe. Nat Rev Genet 10: 94–108.
5. Hammond SM, Caudy AA, Hannon GJ (2001) Post-transcriptional gene silencing by double-stranded RNA. Nat Rev Genet 2: 110–119.
6. Okamura K, Lai EC (2008) Endogenous small interfering RNAs in animals. Nat Rev Mol Cell Biol 9: 673–678.
7. Sarkies P, Miska EA (2014) Small RNAs break out: the molecular cell biology of mobile small RNAs. Nat Rev Mol Cell Biol 15: 525–535.
8. Bumcrot D, Manoharan M, Koteliansky V, Sah DW (2006) RNAi therapeutics: a potential new class of pharmaceutical drugs. Nat Chem Biol 2: 711–719.
9. Czech MP, Aouadi M, Tesz GJ (2011) RNAi-based therapeutic strategies for metabolic disease. Nat Endocrinol 7: 473–484.
10. de Fougerolles A, Vornlocher HP, Maraganore J, Lieberman J (2007) Interfering with disease: a progress report on siRNA-based therapeutics. Nat Rev Drug Discov 6: 443–453.
11. Iorns E, Lord CJ, Turner N, Ashworth A (2007) Utilizing RNA interference to enhance cancer drug discovery. Nat Rev Drug Discov 6: 556–568.
12. Kim DH, Rossi JJ (2007) Strategies for silencing human disease using RNA interference. Nat Rev Genet 8: 173–184.
13. Pecot CV, Calin GA, Coleman RL, Lopez-Berestein G, Sood AK (2011) RNA interference in the clinic: challenges and future directions. Nat Rev Cancer 11: 59–67.
14. Fellmann C, Lowe SW (2014) Stable RNA interference rules for silencing. Nat Cell Biol 16: 10–18.
15. Luo Q, Kang Q, Song WX, Luu HH, Luo X, et al. (2007) Selection and validation of optimal siRNA target sites for RNAi-mediated gene silencing. Gene 395: 160–169.

16. Gibson DG, Young L, Chuang RY, Venter JC, Hutchison CA 3rd, et al. (2009) Enzymatic assembly of DNA molecules up to several hundred kilobases. Nat Methods 6: 343–345.
17. Gibson DG (2011) Enzymatic assembly of overlapping DNA fragments. Methods Enzymol 498: 349–361.
18. Lienert F, Lohmueller JJ, Garg A, Silver PA (2014) Synthetic biology in mammalian cells: next generation research tools and therapeutics. Nat Rev Mol Cell Biol 15: 95–107.
19. Wang N, Zhang H, Zhang BQ, Liu W, Zhang Z, et al. (2014) Adenovirus-mediated efficient gene transfer into cultured three-dimensional organoids. PLoS One 9: e93608.
20. Wang N, Zhang W, Cui J, Zhang H, Chen X, et al. (2014) The piggyBac Transposon-Mediated Expression of SV40 T Antigen Efficiently Immortalizes Mouse Embryonic Fibroblasts (MEFs). PLoS One 9: e97316.
21. Lamplot JD, Liu B, Yin L, Zhang W, Wang Z, et al. (2014) Reversibly Immortalized Mouse Articular Chondrocytes Acquire Long-Term Proliferative Capability while Retaining Chondrogenic Phenotype. Cell Transplant.
22. Wen S, Zhang H, Li Y, Wang N, Zhang W, et al. (2014) Characterization of constitutive promoters for piggyBac transposon-mediated stable transgene expression in mesenchymal stem cells (MSCs). PLoS One 9: e94397.
23. Zhao C, Wu N, Deng F, Zhang H, Wang N, et al. (2014) Adenovirus-mediated gene transfer in mesenchymal stem cells can be significantly enhanced by the cationic polymer polybrene. PLoS One 9: e92908.
24. Li R, Zhang W, Cui J, Shui W, Yin L, et al. (2014) Targeting BMP9-Promoted Human Osteosarcoma Growth by Inactivation of Notch Signaling. Curr Cancer Drug Targets.
25. Huang E, Bi Y, Jiang W, Luo X, Yang K, et al. (2012) Conditionally Immortalized Mouse Embryonic Fibroblasts Retain Proliferative Activity without Compromising Multipotent Differentiation Potential. PLoS One 7: e32428.
26. Wang J, Zhang H, Zhang W, Huang E, Wang N, et al. (2014) Bone Morphogenetic Protein-9 (BMP9) Effectively Induces Osteo/Odontoblastic Differentiation of the Reversibly Immortalized Stem Cells of Dental Apical Papilla. Stem Cells Dev 23: 1405–1416.
27. Wu N, Zhang H, Deng F, Li R, Zhang W, et al. (2014) Overexpression of Ad5 precursor terminal protein accelerates recombinant adenovirus packaging and amplification in HEK-293 packaging cells. Gene Ther 21: 629–637.

28. Feng T, Li Z, Jiang W, Breyer B, Zhou L, et al. (2002) Increased efficiency of cloning large DNA fragments using a lower copy number plasmid. Biotechniques 32: 992, 994, 996 passim.

29. Sharff KA, Song WX, Luo X, Tang N, Luo J, et al. (2009) Hey1 Basic Helix-Loop-Helix Protein Plays an Important Role in Mediating BMP9-induced Osteogenic Differentiation of Mesenchymal Progenitor Cells. J Biol Chem 284: 649–659.

30. Kang Q, Song WX, Luo Q, Tang N, Luo J, et al. (2009) A comprehensive analysis of the dual roles of BMPs in regulating adipogenic and osteogenic differentiation of mesenchymal progenitor cells. Stem Cells Dev 18: 545–559.

31. Yang R, Jiang M, Kumar SM, Xu T, Wang F, et al. (2011) Generation of melanocytes from induced pluripotent stem cells. J Invest Dermatol 131: 2458–2466.

32. Cheng H, Jiang W, Phillips FM, Haydon RC, Peng Y, et al. (2003) Osteogenic activity of the fourteen types of human bone morphogenetic proteins (BMPs). J Bone Joint Surg Am 85-A: 1544–1552.

33. Kang Q, Sun MH, Cheng H, Peng Y, Montag AG, et al. (2004) Characterization of the distinct orthotopic bone-forming activity of 14 BMPs using recombinant adenovirus-mediated gene delivery. Gene Ther 11: 1312–1320.

34. Luo J, Deng ZL, Luo X, Tang N, Song WX, et al. (2007) A protocol for rapid generation of recombinant adenoviruses using the AdEasy system. Nat Protoc 2: 1236–1247.

35. Kong Y, Zhang H, Chen X, Zhang W, Zhao C, et al. (2013) Destabilization of Heterologous Proteins Mediated by the GSK3beta Phosphorylation Domain of the beta-Catenin Protein. Cell Physiol Biochem 32: 1187–1199.

36. Liu X, Qin J, Luo Q, Bi Y, Zhu G, et al. (2013) Cross-talk between EGF and BMP9 signalling pathways regulates the osteogenic differentiation of mesenchymal stem cells. J Cell Mol Med.

37. Wang Y, Hong S, Li M, Zhang J, Bi Y, et al. (2013) Noggin resistance contributes to the potent osteogenic capability of BMP9 in mesenchymal stem cells. J Orthop Res 31: 1796–1803.

38. Gao Y, Huang E, Zhang H, Wang J, Wu N, et al. (2013) Crosstalk between Wnt/beta-Catenin and Estrogen Receptor Signaling Synergistically Promotes Osteogenic Differentiation of Mesenchymal Progenitor Cells. PLoS One 8: e82436.

39. Luo Q, Kang Q, Si W, Jiang W, Park JK, et al. (2004) Connective Tissue Growth Factor (CTGF) Is Regulated by Wnt and Bone Morphogenetic Proteins Signaling in Osteoblast Differentiation of Mesenchymal Stem Cells. J Biol Chem 279: 55958–55968.

40. Tang N, Song WX, Luo J, Luo X, Chen J, et al. (2009) BMP9-induced osteogenic differentiation of mesenchymal progenitors requires functional canonical Wnt/beta-catenin signaling. J Cell Mol Med 13: 2448–2464.

41. Zhang Y, Chen X, Qiao M, Zhang BQ, Wang N, et al. (2014) Bone morphogenetic protein 2 inhibits the proliferation and growth of human colorectal cancer cells. Oncol Rep.

42. Chen X, Luther G, Zhang W, Nan G, Wagner ER, et al. (2013) The E-F Hand Calcium-Binding Protein S100A4 Regulates the Proliferation, Survival and Differentiation Potential of Human Osteosarcoma Cells. Cell Physiol Biochem 32: 1083–1096.

43. Huang J, Bi Y, Zhu GH, He Y, Su Y, et al. (2009) Retinoic acid signalling induces the differentiation of mouse fetal liver-derived hepatic progenitor cells. Liver Int 29: 1569–1581.

44. Zhang W, Deng ZL, Chen L, Zuo GW, Luo Q, et al. (2010) Retinoic acids potentiate BMP9-induced osteogenic differentiation of mesenchymal progenitor cells. PLoS One 5: e11917.

45. Rastegar F, Gao JL, Shenaq D, Luo Q, Shi Q, et al. (2010) Lysophosphatidic acid acyltransferase beta (LPAATbeta) promotes the tumor growth of human osteosarcoma. PLoS One 5: e14182.

46. Su Y, Wagner ER, Luo Q, Huang J, Chen L, et al. (2011) Insulin-like growth factor binding protein 5 suppresses tumor growth and metastasis of human osteosarcoma. Oncogene 30: 3907–3917.

47. Huang E, Zhu G, Jiang W, Yang K, Gao Y, et al. (2012) Growth hormone synergizes with BMP9 in osteogenic differentiation by activating the JAK/STAT/IGF1 pathway in murine multilineage cells. J Bone Miner Res 27: 1566–1575.

48. Si W, Kang Q, Luu HH, Park JK, Luo Q, et al. (2006) CCN1/Cyr61 Is Regulated by the Canonical Wnt Signal and Plays an Important Role in Wnt3A-Induced Osteoblast Differentiation of Mesenchymal Stem Cells. Mol Cell Biol 26: 2955–2964.

49. Peng Y, Kang Q, Cheng H, Li X, Sun MH, et al. (2003) Transcriptional characterization of bone morphogenetic proteins (BMPs)-mediated osteogenic signaling. J Cell Biochem 90: 1149–1165.

50. Zhu GH, Huang J, Bi Y, Su Y, Tang Y, et al. (2009) Activation of RXR and RAR signaling promotes myogenic differentiation of myoblastic C2C12 cells. Differentiation 78 195–204.

51. Luo X, Sharff KA, Chen J, He TC, Luu HH (2008) S100A6 expression and function in human osteosarcoma. Clin Orthop Relat Res 466: 2060–2070.

52. Hu N, Jiang D, Huang E, Liu X, Li R, et al. (2013) BMP9-regulated angiogenic signaling plays an important role in the osteogenic differentiation of mesenchymal progenitor cells. J Cell Sci 126: 532–541.

53. Luo X, Chen J, Song WX, Tang N, Luo J, et al. (2008) Osteogenic BMPs promote tumor growth of human osteosarcomas that harbor differentiation defects. Lab Invest 88: 1264–1277.

54. Bi Y, Huang J, He Y, Zhu GH, Su Y, et al. (2009) Wnt antagonist SFRP3 inhibits the differentiation of mouse hepatic progenitor cells. J Cell Biochem 108: 295–303.

55. Luo J, Tang M, Huang J, He BC, Gao JL, et al. (2010) TGFbeta/BMP type I receptors ALK1 and ALK2 are essential for BMP9-induced osteogenic signaling in mesenchymal stem cells. J Biol Chem 285: 29588–29598.

56. Chen L, Jiang W, Huang J, He BC, Zuo GW, et al. (2010) Insulin-like growth factor 2 (IGF-2) potentiates BMP-9-induced osteogenic differentiation and bone formation. J Bone Miner Res 25: 2447–2459.

57. Zhang J, Weng Y, Liu X, Wang J, Zhang W, et al. (2013) Endoplasmic reticulum (ER) stress inducible factor cysteine-rich with EGF-like domains 2 (Creld2) is an important mediator of BMP9-regulated osteogenic differentiation of mesenchymal stem cells. PLoS One 8: e73086.

58. Yan D, Wiesmann M, Rohan M, Chan V, Jefferson AB, et al. (2001) Elevated expression of axin2 and hnkd mRNA provides evidence that Wnt/beta -catenin signaling is activated in human colon tumors. Proc Natl Acad Sci U S A 98: 14973–14978.

59. He TC, Sparks AB, Rago C, Hermeking H, Zawel L, et al. (1998) Identification of c-MYC as a target of the APC pathway [see comments]. Science 281: 1509–1512.

60. Kim JH, Liu X, Wang J, Chen X, Zhang H, et al. (2013) Wnt signaling in bone formation and its therapeutic potential for bone diseases. Ther Adv Musculoskelet Dis 5: 13–31.

61. Luu HH, Song WX, Luo X, Manning D, Luo J, et al. (2007) Distinct roles of bone morphogenetic proteins in osteogenic differentiation of mesenchymal stem cells. J Orthop Res 25: 665–677.

62. Conte C, Dastugue B, Vaury C (2002) Promoter competition as a mechanism of transcriptional interference mediated by retrotransposons. Embo J 21: 3908–3916.

63. Wodarz A, Nusse R (1998) Mechanisms of Wnt signaling in development. Annu Rev Cell Dev Biol 14: 59–88.

64. Deng ZL, Sharff KA, Tang N, Song WX, Luo J, et al. (2008) Regulation of osteogenic differentiation during skeletal development. Front Biosci 13: 2001–2021.

65. Tetsu O, McCormick F (1999) Beta-catenin regulates expression of cyclin D1 in colon carcinoma cells. Nature 398: 422–426.

66. He TC, Chan TA, Vogelstein B, Kinzler KW (1999) PPARdelta is an APC-regulated target of nonsteroidal anti-inflammatory drugs. Cell 99: 335–345.

67. Luo J, Chen J, Deng ZL, Luo X, Song WX, et al. (2007) Wnt signaling and human diseases: what are the therapeutic implications? Lab Invest 87: 97–103.

68. Lamplot JD, Qin J, Nan G, Wang J, Liu X, et al. (2013) BMP9 signaling in stem cell differentiation and osteogenesis. Am J Stem Cells 2: 1–21.

20

The Aggregation of Four Reconstructed Zygotes is the Limit to Improve the Developmental Competence of Cloned Equine Embryos

Andrés Gambini[1,2], Adrian De Stefano[1], Romina Jimena Bevacqua[1,2], Florencia Karlanian[1], Daniel Felipe Salamone[1,2]*

1 Laboratory of Animal Biotechnology, Faculty of Agriculture, University of Buenos Aires, Buenos Aires, Argentina, **2** National Institute of Scientific and Technological Research, Buenos Aires, Argentina

Abstract

Embryo aggregation has been demonstrated to improve cloning efficiency in mammals. However, since no more than three embryos have been used for aggregation, the effect of using a larger number of cloned zygotes is unknown. Therefore, the goal of the present study was to determine whether increased numbers of cloned aggregated zygotes results in improved *in vitro* and *in vivo* embryo development in the equine. Zona-free reconstructed embryos (ZFRE's) were cultured in the well of the well system in four different experimental groups: I. 1x, only one ZFRE per microwell; II. 3x, three per microwell; III. 4x, four per microwell; and IV. 5x, five ZFRE's per microwell. Embryo size was measured on day 7, after which blastocysts from each experimental group were either a) maintained in culture from day 8 until day 16 to follow their growth rates, b) fixed to measure DNA fragmentation using the TUNEL assay, or c) transferred to synchronized mares. A higher blastocyst rate was observed on day 7 in the 4x group than in the 5x group. Non-aggregated embryos were smaller on day 8 compared to those aggregated, but from then on the *in vitro* growth was not different among experimental groups. Apoptotic cells averaged 10% of total cells of day 8 blastocysts, independently of embryo aggregation. Only pregnancies resulting from the aggregation of up to four embryos per microwell went beyond the fifth month of gestation, and two of these pregnancies, derived from experimental groups 3x and 4x, resulted in live cloned foals. In summary, we showed that the *in vitro* and *in vivo* development of cloned zona-free embryos improved until the aggregation of four zygotes and declined when five reconstructed zygotes were aggregated.

Editor: Jason Glenn Knott, Michigan State University, United States of America

Funding: This work was partially funded by SIDUS Company. The funders had no role in study design, data collection and analysis, decision to publish, or preparation of the manuscript.

* Email: mailto:salamone@agro.uba.ar

Introduction

To date, many equine clones have been reported; however, cloning efficiency remains low [1–9]. Research is hampered in this species by the limited number of slaughterhouses and low recovery rates of oocytes by transvaginal aspiration.

As a means to improve cloning efficiency, the strategy of embryo aggregation has been applied in several species. These include the mouse [10], [11], bovine [12–14], pig [15] and horse [8]. These studies have reported benefits of embryo aggregation for *in vitro* and/or *in vivo* embryo development. In addition, embryo aggregation has been successfully used for chimera production [16–23], and to improve the establishment of parthenogenetic stem cells and the expression of imprinted genes [24].

The aggregation of two or three embryos at the onset of cloned embryo development could compensate for epigenetic defects of individual cells with different reprogramming status. This appears to be one reason for the improved developmental competence of aggregated embryos [10,25]. Furthermore, despite the fact that aggregated embryos are larger on day 7, after day 8 they are no different from non-aggregated embryos [8]. To elucidate this intriguing fact, mechanisms such as apoptosis must be studied. Apoptosis is a cellular mechanism that controls cell numbers during embryonic development and levels of apoptotic cells can be used as an indicator of embryo quality [26–30]. To date, apoptosis levels have been measured in the horse by the TUNEL assay for both ICSI embryos and for those produced *in vivo* [31], [32].

The aim of this work was to determine whether the number of aggregated zygotes changes the aggregation strategy efficiency. Up to five equine clones were aggregated and the *in vitro* and *in vivo* embryo development were evaluated. In addition, the effect of aggregation on embryo quality was measured by evaluating blastocyst size, DNA fragmentation levels, *in vitro* embryo growth beyond day 8 and the establishment of pregnancies and cloned foal production.

Materials and Methods

Chemicals

Except otherwise indicated, all chemicals were obtained from Sigma Chemicals Company (St. Louis, MO, USA).

Animal Welfare

All the research protocols were in accordance with the recommendations of the guidelines stated in the Guide for the Care and Use of Agricultural Animals in Agricultural Research and Teaching. The study design was approved by the Ethics and Animal Welfare Committee for the Faculty of Agriculture, University of Buenos Aires under number CEyBAFAUBA2014/1. All efforts were made to minimize animal suffering. All animals were housed at "Don Antonio" equine center in Buenos Aires, Argentina. Trained people provided daily care and feeding, and horses had permanent *ad-libitum* access to water. Recipient palpations, ultrasounds, hormone treatments and embryo transfer procedures were always performed by trained veterinarians.

Cell culture

Fibroblasts were obtained by culture of skin biopsies from an Argentinean Polo Pony (donor cell A) and a Show-jumping horse (donor cell B). They were cultured in Dulbecco's modified Eagle's Medium (DMEM; 11885, Gibco, Grand Island, NY, USA) supplemented with 10% fetal bovine serum (FBS; 10499-044, Gibco), 1% antibiotic–antimycotic (ATB; 15240-096, Gibco), and 1 μl/ml insulin-transferrin-selenium (ITS; 51300-044, Gibco) in 6.5% CO_2 in humidified air at 39°C. After establishment of the primary culture, fibroblasts were expanded, frozen in DMEM with 20% FBS and 10% DMSO, and stored in liquid nitrogen. Donor cells were induced into quiescence by being grown to confluence. Cells were trypsinized before use and resuspended in TALP-H with 10% FBS.

Oocyte collection and *in vitro* maturation

Slaughterhouse ovaries were collected and transported to the laboratory within 4–7 h, at 26–28°C. Equine oocyte recovery was performed by a combination of scraping and washing of all visible follicles using a syringe filled with DMEM/Nutrient Mixture F-12 medium (DMEM/F12; D8062), supplemented with 20 IU mL-1 heparin (H3149). Oocytes were matured for 24–26 h in 100 μl microdrops of bicarbonate-buffered TCM-199 (31100-035; Gibco) supplemented with 10% FBS, 2.5 μL/mL ITS, 1 mM sodium pyruvate (P2256), 100 mM cysteamine (M9768), 0.1 mg/mL of follicle-stimulating hormone (NIH-FSH-P1, Folltropin; Bioniche, Belleville, ON, Canada) and 1% ATB, under mineral oil (M8410). Maturation conditions were 6.5% CO_2 in humidified air at 39°C.

Cumulus and zona pellucida removal

Cumulus cells were removed by a combined treatment of pipeting oocytes in 0.05% Trypsin-EDTA (25300, Gibco) and vortexing them for 2 minutes in hyaluronidase [H4272; 1 mg/mL in Hepes-buffered Tyrodes medium containing albumin, lactate and pyruvate (TALP-H)]. Oocytes were individually observed under stereoscopic microscopy to confirm the presence of the first polar body.

In order to prepare the metaphase II oocytes for enucleation, the zona pellucida was removed by incubating oocytes for 3–6 min in 1.5 mg/ml pronase (P8811) in TALP-H on a warm plate. Zona-free oocytes (ZF-oocytes) were rinsed in TALP-H and placed in a microdrop of Synthetic Oviductal Fluid (SOF), supplemented with 2.5% FBS and 1% ATB, until enucleation.

Oocyte enucleation

Aspiration of the metaphase plate was performed in a microdrop of TALP-H containing 0.5 μg/ml of cytochalasin B (C6762). A blunt pipette was used for the aspiration, and a closed holding pipette to support the oocyte during the procedure. In order to observe the metaphase plate under UV light, ZF-oocytes were incubated (5 min), prior to enucleation, in a microdrop of SOF containing 1 μg/mL Hoechst bisbenzimide 33342 (H33342). Zona-free enucleated oocytes (ZFE-oocytes) were kept in a SOF microdrop until nuclear transfer.

Nuclear transfer and cloned embryo reconstruction

Zona-free enucleated oocytes were individually washed for a few seconds in 50 μl drops of 1 mg/ml phytohemagglutinin (L8754) dissolved in TCM-Hepes, and then dropped over a donor cell resting on the bottom of a 100 μl TALP-H drop; consequently these two structures were attached. Formed cell couplets were washed in fusion medium [0.3 M mannitol (M9546), 0.1 mM $MgSO_4$ (M7506), 0.05 mM $CaCl_2$ (C7902), 1 mg/ml polyvinyl alcohol (P8136)], and then fused in a fusion chamber containing 2 ml of warm fusion medium. A double direct current pulse of 1.2 kV/cm V, each pulse for 30 μs, 0.1 s apart was utilized for fusion. Couplets were individually placed in a 10 μl drop of SOF medium supplemented with 2.5% FBS and incubated under mineral oil, at 39°C in 5% CO_2 in air. Twenty minutes after the first round of fusion, non-fused couplets were re-fused.

Chemical activation

Two hours after the first round of fusion, zona-free reconstructed embryos (ZFRE's) were subjected to chemical activation. Chemical activation was achieved by a 4 min treatment in TALP-H containing 8.7 mM ionomycin (I24222; Invitrogen, Carlsbad, CA, USA) followed by a 4 h individual culture in a 5 μl drop of SOF supplemented with 1 mM 6-dimethylaminopurine (D2629) and 5 mg/ml cycloheximide (C7698).

In vitro embryo culture until day 8 and embryo aggregation

In vitro culture of ZFRE's was carried out in microwells containing 50 μl microdrops of DMEM/F12 medium under mineral oil. These microwells were produced using a heated glass capillary lightly pressed to the bottom of a 35 x 10 mm Petri dish. Four different experimental groups were set up according to the number of ZFRE's placed per each microwell: I. Group **1x**: one ZFRE per microwell (non aggregated embryos), II. Group **3x**: three ZFRE's per microwell, III. Group **4x**: four ZFRE's per microwell, IV. Group **5x**: five ZFRE's per microwell. Culture conditions were 5% O_2, 5% CO_2 and 90% N_2 in a humidified atmosphere at 38.5°C. Half of the medium was renewed on Day 3, with DMEM/F-12 HAM medium containing 10% FBS, and 1% ATB. A similar ratio of ZFRE's/culture medium was maintained for all experimental groups. Cleavage was assessed 72 h after activation, and rates of blastocyst formation and their diameter were recorded at Day 7 and Day 8 when the embryos were either fixed for TUNEL assay, maintained in *in vitro* culture or transferred to synchronized mares.

In vitro embryo culture beyond day 8

Fourteen derived B donor cell blastocysts from all experimental groups were kept in *in vitro* culture from day 8 until day 16–17 unless they collapsed earlier. One blastocyst from the 4x experimental group was found collapsed on day 15 and was not included for apoptosis analysis. On day 12, blastocysts were placed in a fresh 100 μl microdrop of DMEM/F12 medium containing 15% FBS and 1% ATB. Blastocyst diameters were measured daily using a millimeter eyepiece. At day 16, embryos were fixed for TUNEL assay.

Embryo fixing and TUNEL assay

DNA fragmentation was evaluated using the DeadEnd Fluorometric TUNEL System (Promega G3250, Madison, WI, USA). Embryos were fixed in 4% paraformaldehyde in DPBS, washed in BSA (A7906) solution (1 mg BSA/ml DPBS), permeabilized with 0.5% Triton X-100 in DPBS for 15 min at room temperature, and rinsed again in BSA solution. After three washes, embryos were incubated in the dark for 2 h at 39°C in a buffer consisting of equilibration buffer and a nucleotide mix containing fluorescein-dUTP and terminal deoxynucleotidyl transferase. Negative controls lacked the terminal deoxynucleotidyl transferase. The nuclei were counterstained with 0.5% propidium iodide for 30 min at room temperature. Embryos were washed in BSA solution and mounted on a glass slide in 70% v/v glycerol under a coverslip. Embryos were analyzed on a Nikon Confocal C.1 scanning laser microscope. An excitation wavelength of 488 nm was selected for detection of fluorescein-12-dUTP and a 544 nm wavelength to excite propidium iodide. Images of serial optical sections were recorded every 1.5–2 μm vertical step along the Z-axis of each embryo. Three-dimensional images were constructed using EZ-C1 3.9 software (Nikon Corporation, Japan). Total cell numbers and DNA-fragmented nuclei were counted manually for day 8 embryos. Due to the large number of cells of day 16 embryos, cells of five different areas of each day 16 blastocyst were counted using the ImageJ software (1.47 version, Wayne Rasband National Institutes of Health, USA).

Embryo transfer and production of cloned foals

Blastocyst transfers to recipients were performed during the breeding season. Mares aged 3 to 10 years were examined 2–3 times/week by transrectal ultrasound (5MHz linear probe, Aloka 500) to determine the phase of their estrous cycle. Prostaglandin F2 Alfa (Ciclase, Sintex, Buenos Aires, Argentina) and human chorionic gonadotropin (Ovusyn, Sintex, Buenos Aires, Argentina) were used to synchronize the day of ovulation. Transcervical embryo transfer was performed 5 to 7 days after ovulation, with one or two day 7 blastocysts. Blastocysts were transported in a 0.5 cc straw containing DMEM/F12, and the shipping container was held at 36°C for the 3 h transportation interval. Pregnancies were diagnosed by transrectal ultrasound 15 days after ovulation. At day 300 of gestation pregnant mares were moved to an equine hospital (KAWELL, Equine Rehabilitation Center, Solís, Argentina) where they were monitored until parturition.

DNA comparison

The cloned foals were confirmed by an external laboratory (Laboratorio de Genética Aplicada de la Sociedad Rural Argentina, ISAG code 84535). Twenty eight loci were compared using hair samples from each foal and its respective donor animal.

Statistical analysis

Differences among treatments in each experiment were determined using GraphPad Prism software version 5. Blastocyst rates, embryo size and pregnancy rates were analyzed by Chi-square or Fisher's exact test. TUNEL-positive cells were evaluated with the Kruskal-Wallis non parametric test and Dunn's post test. The effect of treatment on *in vitro* embryo growth rates was assessed by one-way within subjects (repeated measures) analysis of variance. Multiple observations of embryo growth rates in the same experimental units (embryo/days) constituted the within subject factor. *Post-hoc* pairwise comparisons of mean growing rates/days were performed by the Tukey Honestly Significant Differences test.

Results

In vitro embryo development of aggregated cloned equine embryos until day 8

A total of 765 ZFRE's were produced and cultured *in vitro* in four different experimental groups. Cleavage and blastocyst rates on day 7 and day 8 per embryo (microwell) and per ZFRE were recorded for all experimental groups (**Table 1**). A significant improvement of blastocyst rates per embryo was observed on day 7 when numbers of aggregated zygotes were up to 4/well. Furthermore, aggregation did not involve the use of additional oocytes to obtain blastocysts, since no significant differences from the control group were observed on day 7. On day 7, there were fewer blastocysts in the 5x group per number of ZFRE's compared to the 4x experimental group but no significant differences were found from the control group. The aggregation of four zygotes resulted in the best rate of *in vitro* embryo development, whereas when five reconstructed zygotes were aggregated embryo development rates decreased. Additional data are available in Table S1 and Table S2 showing the *in vitro* cloned equine embryo developmental competence per somatic donor cell. Embryo aggregation showed a similar effect between donor cells. Diameters of day 7 blastocysts are shown in **Table 2**. Cloned blastocyst size was smaller when no embryo aggregation was used. During zona-free embryo development some blastomeres were observed to be sloughed off into the microwell. This situation was observed in all experimental groups.

In vitro development of cloned blastocysts beyond day 8

In vitro growth rates of 14 cloned blastocysts (donor cell B) were determined daily. Growth patterns were similar between aggregated and non-aggregated groups. The number of embryos analyzed per group were: 1x: (n = 3), 3x: (n = 4), 4x: (n = 4) and 5x (n = 3). Each embryo within each experimental group derived from different replicates. Mean embryo sizes per day ±SD were: Day 8, 130.12 μm ±32.88; Day 9, 219.12 μm ±76.32; Day 11, 489.44 μm ±265.32; Day 12, 713.72 μm ±391.91; Day 13, 954.55 μm ±327.34; Day 14, 1664.87 μm ±713.54; Day 15, 2212.20 μm ±783.32 and day 16, 2677.18 μm ±977.43. **Figure 1** shows an equine zona-free cloned blastocyst from the 3x experimental group during *in vitro* embryo culture.

DNA fragmentation levels in cloned equine aggregated embryos on day 8 and day 16 by TUNEL assay

DNA fragmentation levels (mean ± SEM) of day 8 and day 16 cloned embryos of all experimental groups are shown in **Table 3**. There were no differences between groups, on day 8, in the levels of fragmented DNA, with an average of 10% positive TUNEL cells seen in all embryos (**Figure 2**). TUNEL-positive cells in day 8 embryos were: Group 1x (n = 6): 9.26%, 10.26%, 12.41%, 13.04%, 17.39% and 25.58%; Group 3x (n = 3): 8.97%, 9.43% and 10.29%; Group 4x (n = 5): 3.62%, 9.04%, 10.61%, 11.63% and 11.84%; and Group 5x (n = 2): 8.86% and 11.16%. There were no differences between groups on day 16 in the levels of fragmented DNA, with an average of 2.5% positive TUNEL cells seen in all cloned embryos (**Figure 3**). TUNEL-positive cell percentage of total cells counted per day 16 embryos and experimental group were: Group 1x (n = 3): 4.23%, 3.64% and

Table 1. Effects of equine cloned embryo aggregation on *in vitro* development until day 8.

Experimental groups	No. of ZFRE's (%)	No. of embryos (well)	No. of cleaved (%)	Blastocyst production Day 7 No.	% per Embryo	% per ZFRE	Day 8 No.	% per Embryo	% per ZFRE
1x	131	131	99 (75.57)	13	9.92[a]	9.92[ac]	20	15.27[a]	15.27
3x	228	76	193 (85.4)	26	34.21[b]	11.40[ac]	40	52.63[b]	17.54
4x	292	73	229 (78.16)	42	57.53[c]	14.38[bc]	52	71.23[b]	17.80
5x	115	23	90 (78.26)	7	30.43[b]	6.08[a]	15	65.22[b]	13.04
Total	765	303	611 (79.87)	88	29.04	11.50	127	41.91	16.60

Values with different superscripts in a column are significantly different (Chi-square Test P<0.05) (a, b, c). ZFRE's: Zona-free reconstructed embryos.

4.42%; Group 3x (n = 2): 3.35% and 0.33%; Group 4x (n = 3): 0.15%, 0.88% and 2.03%; and Group 5x (n = 3): 1.80%, 2.91% and 4.79%. Additional data are available in Figure S1.

In vivo development of aggregated cloned equine embryos

Embryo transfer, pregnancy and survival rates for all experimental groups are shown in **Table 4**. Early pregnancy rates were higher when aggregated embryos were transferred; however, no statistical differences were found in pregnancy rates between non-aggregated and aggregated groups. Only pregnancies resulting from the aggregation of up to four embryos per well survived beyond the fifth month of gestation. One of the pregnant mares from the 3x experimental group showed clinical signs of Equine Metabolic Syndrome, dying in the last month of gestation. The cloned fetus presented normal vital parameters until the death of the mare. The two cloned foals obtained in this study derived from experimental groups 3x and 4x (**Figure 4**). Their gestation times were normal.

Both foals needed neonatology assistance and they responded positively to treatment (oxygen, antibiotics and parental nutrition). The cloned foal derived from experimental group 3x presented a high degree of angular and flexural forelimb deformities.

Discussion

This study analyzed the effect of an increase in the numbers of aggregated zygotes on *in vitro* and *in vivo* cloned embryo development in the equine. Embryo aggregation has previously proved to enhance the efficiency of cloning. However, to date, studies on mammalian cloned embryo aggregation have focused on determining the effect of the aggregation of a maximum of three zygotes on *in vitro* and *in vivo* embryo production [8], [10–14], [25], [33–36].

In vitro embryo development of aggregated embryos until day 8

The aggregation of three and four embryos per microwell improved blastocyst rates on a per embryo basis. Unexpectedly, blastocyst development in the 5x experimental group at day 7 was not improved over that of the 4x experimental group, and blastocyst rates were similar to those obtained in the 3x experimental group. Furthermore, as we have previously demonstrated [8], embryo aggregation also enhanced blastocyst diameters on day 7. If the positive effects of embryo aggregation are due to an epigenetic compensation and/or an increase in embryo cell number [10], [13], [15], [25], then an increase in the number of aggregated zygotes should correlate to an improvement in *in vitro* embryo development. Some of the reported benefits of embryo aggregation are related to a normalization of gene expression and developmental potential [10], [11], [37], an increase in cell number [10], [12], [33] and a consequent improvement in blastocyst development rates [10], [13], [33]. Additionally, the lower results of the 5x experimental group indicate that these benefits are related to the number of aggregated zygotes. Possible mechanisms to explain this could be associated to an altered microenvironment inside the microwell, limited capability of the embryo to incorporate cells during embryo development or to alterations in cell cycle or cell death.

Each microwell provides a particular microenvironment reported to be beneficial for the embryo [38–40]. Embryo aggregation may induce modifications in this microenvironment leading to positive or negative effects depending on the number of reconstructed zygotes placed per microwell. In addition, the effect

Table 2. Effects of equine cloned embryo aggregation on *in vitro* embryo size at day 7.

Experimental groups	No. blastocyst	Blastocyst diameter				
		80–119 μm (%)	120–169 μm (%)	180–219 μm (%)	230–269 μm (%)	≥270 μm (%)
1x	13	8 (61.54)[a]	4 (30.77)[ac]	1 (7.69)[a]	0 (0)[a]	0 (0)[a]
3x	25	6 (24.00)[b]	4 (16.00)[a]	11 (44.00)[b]	3 (12.00)[a]	1 (4.00)[a]
4x	35	10 (28.57)[b]	7 (20.00)[a]	14 (40.00)[b]	3 (8.57)[a]	1(2.86)[a]
5x	7	0 (0)[b]	5 (71.43)[bc]	1 (14.29)[ab]	1 (14.29)[a]	0 (0)[a]
Total	80	26 (32.50)	20 (25.0)	25 (31.25)	7 (8.75)	2 (2.50)

Values with different superscripts in a column are significantly different (Fisher's exact test P<0.05) (*a, b*).

of embryo culture density has been reported to alter embryo development [40], [41]; furthermore in the bovine, the distance between individual embryos in culture has been shown to influence preimplantation development [42]. Therefore, placing an excessive number of embryos per microwell could negatively affect the microenvironment of the microwell and consequently the developmental capability of aggregated embryos. In addition, when aggregation is performed, the individual capability of each ZFRE to produce a blastocyst is lost. The maximum blastocyst rate per ZFRE when 5x aggregation is performed is 20%. As this maximum value was not reached, we do not consider this a reason

for the reduced developmental competence of this experimental group.

The high number of initial reconstructed zygotes of the 5x group could impact negatively on embryo development. It has been suggested that cell proliferation is a competitive process that allows recognition and elimination of defective cells during the early stages of development [43]. However, this situation could be altered with an excessive number of embryo cells. Furthermore, the developmental kinetics of *in vitro*-produced embryos is related to their developmental competence [44], and can be affected by the volume of the cytoplasm of the initial oocyte [45]. Even though

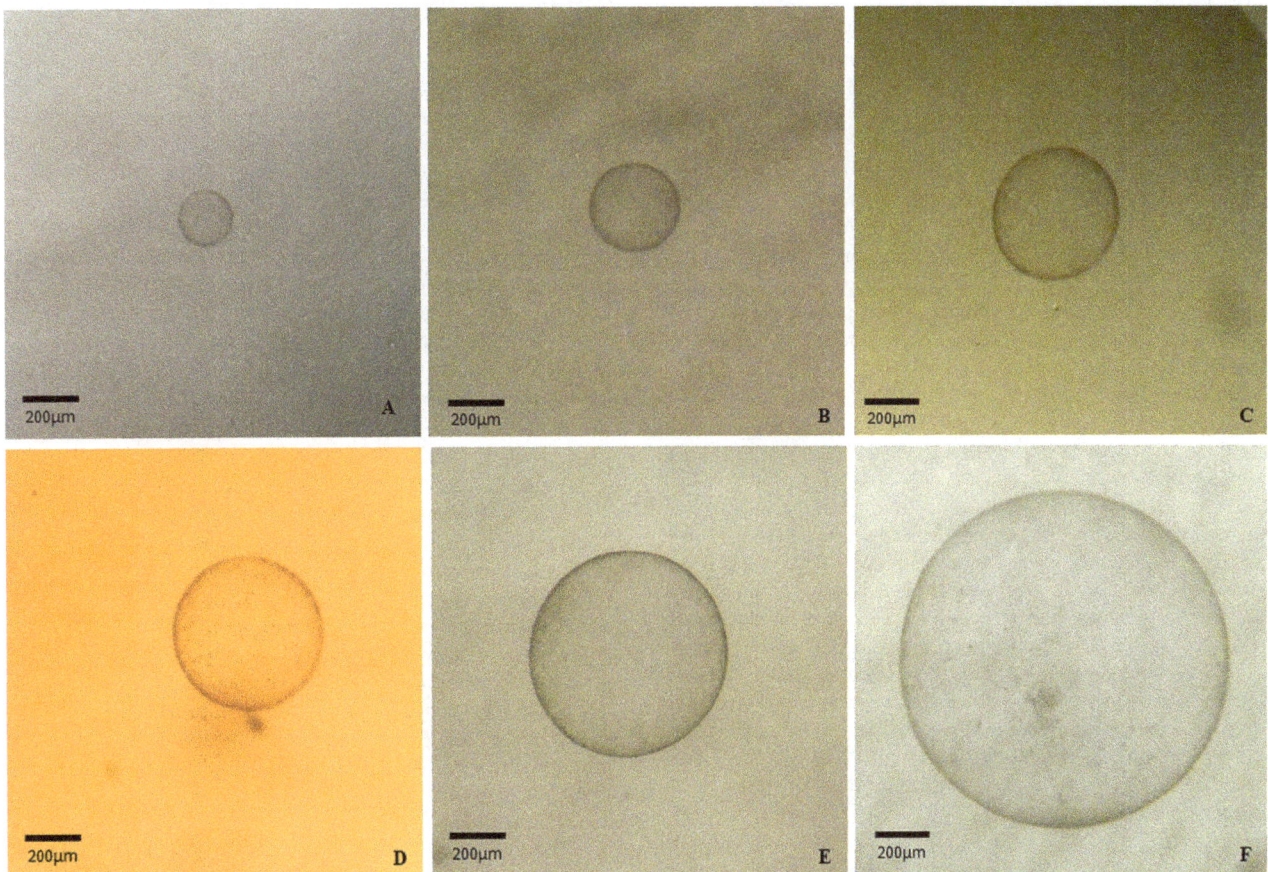

Figure 1. Photographs of an equine cloned aggregated blastocyst in *in vitro* embryo culture beyond day 8. An equine cloned zona free blastocyst placed in a 100 μl drop of DMEM/F12 medium derived from the experimental group 3x during *in vitro* embryo culture from day 8 until day 16. (**A**) Day 9, 195.21 μm. (**B**) Day 10, 314.28 μm. (**C**) Day 11 444.75 μm. (**D**) Day 12, 498.76 μm. (**E**) Day 13, 672.17 μm. (**F**) Day 14, 1127.98 μm.

Table 3. Evaluation of DNA fragmentation levels in equine aggregated and non-aggregated cloned blastocysts at day 8 and day 16.

Blastocysts	Experimental group	No.	TUNEL+ cells (Mean ±SEM)	Evaluated cells (Mean ±SEM)	TUNEL+/evaluated cells (Mean ±SEM)
Day 8	1x	6	26.83±3.99	217.3±51.92	14.64±2.45
	3x	3	21.80±5.70	228.4±52.88	9.37±0.25
	4x	5	39.20±8.38	427.8±65.28	9.34±1.51
	5x	2	25.00±1.00	252.0±19.00	10.01±1.15
Day 16	1x	3	38.67±15.30	963.7±383.6	4.09±0.23
	3x	2	58.00±54.00	2163±962.5	1.83±1.5
	4x	3	48.67±22.38	4618±1207	1.02±0.54
	5x	3	56.33±12.20	2038±788.0	3.16±0.87

No significant differences were detected within blastocyst day (Kruskal-Wallis test P<0.05).

embryo aggregation does not imply an increase in the cytoplasm-nucleus ratio, an increased embryo volume could also alter embryo developmental kinetics and the developmental competence of aggregated embryos. In addition, a study suggested that the length of the cell cycle can be regulated in the early post-implantation mouse chimeric embryo in order to compensate for increased pre-implantation cell numbers induced by aggregation [46].

In vitro embryo development beyond day 8

Day 8 cultured blastocysts expanded and their cell numbers increased; however, no statistical differences were observed among experimental groups. Previous studies in mice revealed that size regulation occurred in aggregated chimeric embryos during the early post-implantation stage [46]. Moreover, despite the abnormal proportions of ICM and trophectoderm in 4x aggregated chimeric mice blastocysts, the proportions of the tissues derived from them is already normal by day 5 [47]. Our observations together with those previously reported [8] support the notion that size regulation of aggregated cloned equine embryos occurs in the pre-implantation stages.

In vitro embryo development after day 8 allows the study of embryo developmental competence in a controlled environment avoiding the disadvantages associated with embryo transfer. Among domestic animals, in vitro development of embryos after reaching the blastocyst stage has been studied in the bovine [48–54] and porcine [51]. In the present study, the zona-free cloned blastocyst maintained its spherical shape (see **Figure 1**). It has been suggested that the embryo capsule is largely responsible for maintaining the spherical shape of the conceptus in the equine after day 6 [55]. Nevertheless, a capsule could not be clearly identified by microscopy in any experimental group. Thus, if present, it must be deficient. Alterations in early embryonic coats have been reported for equine ICSI [56] and cloned embryos [8] and also for rabbit [57]. Hence, improvements in in vitro culture conditions are necessary to allow for the normal formation of early embryonic coats in this species.

DNA fragmentation levels of aggregated cloned equine embryos

Regardless of the fact that aggregated embryos began their development with more cells, DNA fragmentation levels in cloned blastocysts on day 8 were the same among experimental groups, with an average of 10% of cells being apoptotic. Thus, at this stage of development, the beneficial effects of embryo aggregation would not be related to an anti-apoptotic phenomenon in the horse. This contrasts with recent observations in pigs that indicate that embryo aggregation has an anti-apoptotic effect due to fewer numbers of apoptotic cells [34]. Species differences in embryo physiology such as in the number of embryonic cells could be related to the effects of embryo aggregation. On the other hand, size regulation of aggregated chimeric mouse embryos was reported to be not related to cell death since more than 90% of the cells were synthesizing DNA and were presumably viable [46].

Apoptosis may have detrimental effects if either the number of apoptotic cells or the proportion of these cells are elevated [58]. In the equine, in vivo embryos recovered on day 6 did not show apoptotic cells, while 4% of cells had fragmented DNA in in vitro embryos produced by ICSI [32]. Consequently, the higher proportion of apoptotic cells observed in our study could be a reason for the lower developmental competence generally reported for in vitro-produced equine embryos. A very interesting observation was that day 16 embryos had a similar proportion of apoptotic cells to day 8 embryos independently of aggregation. Therefore, in vitro embryo culture in DMEM/F12 medium allows embryo cell proliferation without inducing apoptosis.

In vivo embryo development of aggregated embryos

In this study, viable pregnancies resulted from the aggregation of three and four reconstructed zygotes. In our previous publication, embryo aggregation increased pregnancy rates for cloned horses, being higher for 2x and even higher for 3x aggregations [8]. Embryo aggregation of up to three zygotes also improved pregnancy rates and in vivo embryo development in the mouse [10] and bovine [60]. Nevertheless, the aggregation of more than four cloned embryos did not improve pregnancy rates in the present study. These observations agree with suggestions previously reported [61] that the yield of live-born pups from chimeric aggregation of five or more embryos would be low. On the other hand, the abnormalities detected in the 3x cloned foals of the present study have been reported in 50% of cloned foals [59], indicating that angular and flexural limb deformities are not induced by zona removal or embryo aggregation.

In conclusion, the data presented in this paper indicate that cloned embryo aggregation in the equine results in increased blastocyst rates at day 7 for the aggregation of up to 4 zygotes.

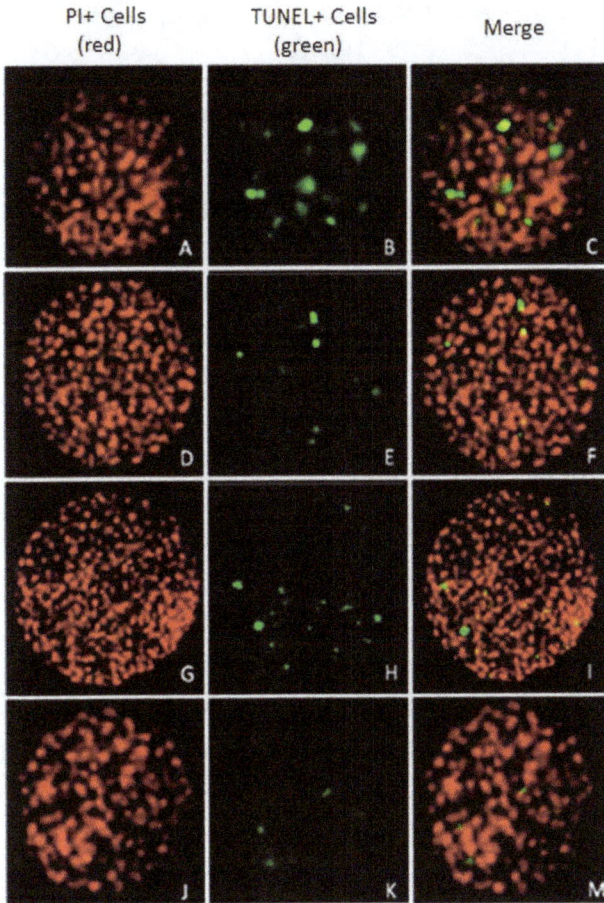

Figure 2. Photomicrographs of day 8 equine cloned embryo expression of TUNEL. (A, B, C) Day 8 Non-aggregated cloned equine embryo, 40x zoom. **(D, E, F)** Day 8 3x aggregated cloned embryo, 40x zoom. **(G, H, I)** Day 8 4x aggregated cloned embryo, 40x zoom. **(J, K, M)** Day 8 5x aggregated cloned embryo, 40x zoom.

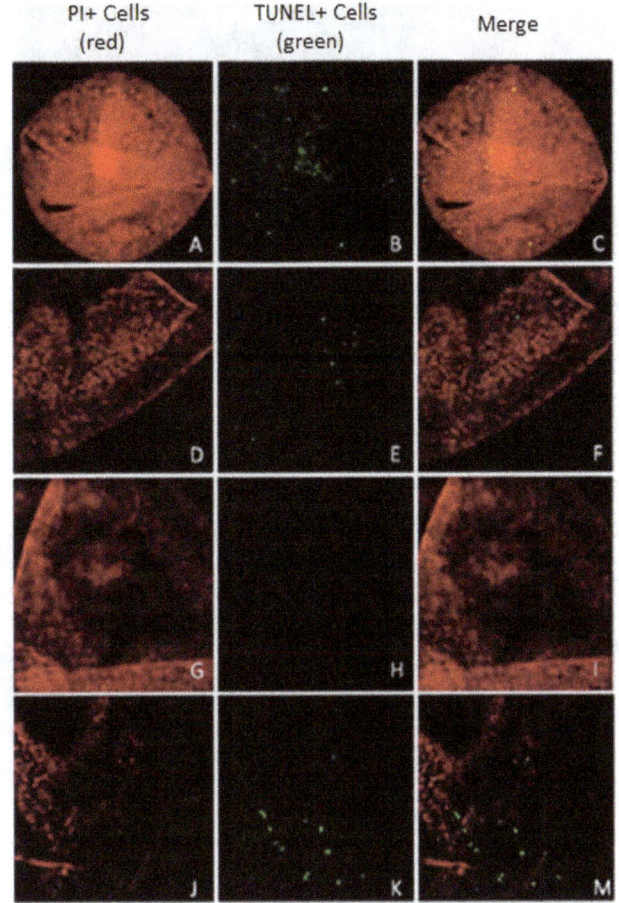

Figure 3. Photomicrographs of day 16 equine cloned embryo expression of TUNEL. (A, B, C) Day 16 Non-aggregated cloned equine embryo, 20x zoom. **(D, E, F)** Day 16 3x aggregated cloned embryo, 20x zoom. **(G, H, I)** Day 16 4x aggregated cloned embryo, 20x zoom. **(J, K, M)** Day 16 5x aggregated cloned embryo, 20x zoom.

Beyond four reconstructed zygotes, blastocyst rates do not continue to increase. Aggregated cloned embryos were initially larger, but *in vitro* embryo size compensated after day 8 as we have previously reported. A similar proportion of apoptotic cells was observed on day 8 in all experimental groups, and this phenomenon does not appear to be responsible for the observed compensation. Only aggregated embryos from groups 3x and 4x produced cloned offspring. This is the first report to show that the *in vitro* and *in vivo* development of cloned zona-free embryos can be improved when up to four zygotes are aggregated.

Table 4. Effects of equine cloned embryo aggregation on *in vivo* development.

Experimental Groups	No. of Recipient	Pregnant recipients (%)	Pregnancy Dynamic			No. offspring (%)
			1st month no. (%)	5th month no. (%)	8th month no. (%)	
1x	10	1 (10.00)	1 (100.00)	0 (0)	0 (0)	0 (0)
3x	17	3 (17.64)	2 (66.66)	2 (66.66)	2 (66.66)*	1 (33.33)
4x	11	2 (18.18)	1 (50.00)	1 (50.00)	1 (50.00)	1 (50.00)
5x	5	0 (0)	0 (0)	0 (0)	0 (0)	0 (0)
Total	44	6 (13.95)	4 (66.66)	3 (50.00)	3 (50.00)	2 (33.33)

No significant differences were found (Fisher's exact test P<0.05). *One pregnant mare died of Equine Metabolic Syndrome.

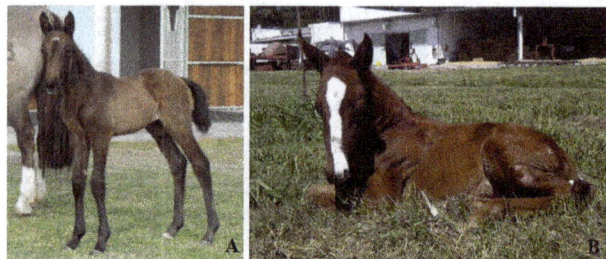

Figure 4. Photographs of equine cloned foals derived from aggregated embryos. (**A**) Equine cloned foal derived from 4x experimental group, born on the 18th of September, 2013. (**B**) Equine cloned foal derived from 3x experimental group, born on the 12th of January, 2013.

Supporting Information Legends

Figure S1   Scatter plot of TUNEL positive cells proportion in cloned equine blastocysts. (A) Day 8 blastocysts mean TUNEL-positive cells of groups 1x, 3x, 4x and 5x. (B) Day 16

blastocysts mean TUNEL-positive cells of groups 1x, 3x, 4x and 5x.
(TIFF)

Table S1   Effects of equine cloned embryo aggregation on *in vitro* development until day 8. Donor Cell A. (DOCX)

Table S2   Effects of equine cloned embryo aggregation on *in vitro* development until day 8. Donor Cell B. (DOCX)

Acknowledgments

The authors wish to thank Dr. Rafael Fernandez-Martin and DVM María Belén Rodriguez for reading and discussing the manuscript. Also thanks to Dr Elizabeth Crichton for English revision.

Author Contributions

Conceived and designed the experiments: AG DFS. Performed the experiments: AG AD FK RJB. Analyzed the data: AG DFS. Contributed reagents/materials/analysis tools: DFS. Wrote the paper: AG DFS.

References

1. Woods GL, White KL, Vanderwall DK, Li GP, Aston KI, et al. (2003) A mule cloned from fetal cells by nuclear transfer. Science 301: 1063.
2. Galli C, Lagutina I, Crotti G, Colleoni S, Turini P, et al. (2003) Pregnancy: a cloned horse born to its dam twin (letter). Nature 424: 635.
3. Lagutina I, Lazzari G, Duchi R, Colleoni S, Ponderato N, et al. (2005) Somatic cell nuclear transfer in horses: effect of oocyte morphology, embryo reconstruction method and donor cell type. Reproduction 130: 559–567.
4. Hinrichs K, Choi YH, Love CC, Chung YG, Varner DD (2006) Production of horse foals via direct injection of roscovitine-treated donor cells and activation by injection of sperm extract. Reproduction 131: 1063–1072.
5. Hinrichs K, Choi YH, Varner DD, Hartman DL (2007) Production of cloned horse foals using roscovitine-treated donor cells and activation with sperm extract and/or ionomycin. Reproduction 134: 319–325.
6. Lagutina I, Lazzari G, Duchi R, Turini P, Tessaro I, et al. (2007) Comparative aspects of somatic cell nuclear transfer with conventional and zona-free method in cattle, horse, pig and sheep. Theriogenology 67: 90–98.
7. Choi YH, Hartman DL, Fissore RA, Bedford-Guaus SJ, Hinrichs K (2009) Effect of sperm extract injection volume, injection of PLCzeta cRNA, and tissue cell line on efficiency of equine nuclear transfer. Cloning Stem Cells 11: 301–308.
8. Gambini A, Jarazo J, Olivera R, Salamone DF (2012) Equine cloning: *in vitro* and *in vivo* development of aggregated embryos. Biol Reprod.87: 15 1–9.
9. Choi YH, Norris JD, Velez IC, Jacobson CC, Hartman DL, et al. (2013) A viable foal obtained by equine somatic cell nuclear transfer using oocytes recovered from immature follicles of live mares. Theriogenology 79: 791–796.
10. Boiani M, Eckardt S, Leu NA, Scholer HR, McLaughlin KJ (2003) Pluripotency deficit in clones overcome by clone-clone aggregation: epigenetic complementation? EMBO J 22: 5304–5312.
11. Balbach ST, Esteves TC, Brink T, Gentile L, McLaughlin KJ, et al. (2010) Governing cell lineage formation in cloned mouse embryos. Dev Biol 343: 71–83.
12. Zhou W, Xiang T, Walker S, Abruzzese RV, Hwang E, et al. (2008) Aggregation of bovine cloned embryos at the four-cell stage stimulated gene expression and *in vitro* embryo development. Mol Reprod Dev 75: 1281–1289.
13. Ribeiro ES, Gerger RP, Ohlweiler LU, Ortigari I Jr, Mezzalira JC, et al. (2009) Developmental potential of bovine hand-made clone embryos reconstructed by aggregation or fusion with distinct cytoplasmic volumes. Cloning and Stem Cells 11: 377–386.
14. Akagi S, Yamaguchi D, Matsukawa K, Mizutani E, Hosoe M, et al. (2011) Developmental ability of somatic cell nuclear transferred embryos aggregated at the 8-cell stage or 16- to 32-cell stage in cattle. J Reprod Dev 57: 500–506.
15. Terashita Y, Sugimura S, Kudo Y, Amano R, Hiradate Y, et al. (2011) Improving the quality of miniature pig somatic cell nuclear transfer blastocysts: aggregation of SCNT embryos at the four-cell stage. Reprod Domest Anim 46: 189–196.
16. Hillman N, Sherman MI, Graham C (1972) The effect of spatial arrangement on cell determination during mouse development. J Embryol Exp Morphol 28: 263–278.
17. Boediono A, Suzuki T, Li LY, Godke RA (1999) Offspring born from chimeras reconstructed from parthenogenetic and in vitro fertilized bovine embryos. Mol Reprod Dev 53: 159–170.
18. Lee SG, Park CH, Choi DH, Kim HS, Ka HH, et al. (2007) *In vitro* development and cell allocation of porcine blastocysts derived by aggregation of *in vitro* fertilized embryos. Mol Reprod Dev 74: 1436–1445.
19. Yang F, Hao R, Kessler B, Brem G, Wolf E, et al. (2007) Rabbit somatic cell cloning: effects of donor cell type, histone acetylation status and chimeric embryo complementation. Reproduction 133: 219–230.
20. Ohtsuka M, Miura H, Gurumurthy CB, Kimura M, Inoko H, et al. (2012) Fluorescent transgenic mice suitable for multi-color aggregation chimera studies. Cell Tissue Res 350: 251–260.
21. He W, Kong Q, Shi Y, Xie B, Jiao M, et al. (2013) Generation and developmental characteristics of porcine tetraploid embryos and tetraploid/diploid chimeric embryos. Genomics Proteomics Bioinformatics 11: 327–333.
22. Hiriart MI, Bevacqua RJ, Canel NG, Fernández-Martín R, Salamone DF (2013) Production of chimeric embryos by aggregation of bovine egfp eight-cell stage blastomeres with two-cell fused and asynchronic embryos. Theriogenology 80: 357–364.
23. Nakano K, Watanabe M, Matsunari H, Matsuda T, Honda K, et al. (2013) Generating porcine chimeras using inner cell mass cells and parthenogenetic preimplantation embryos. PLoS One 8: e61900.
24. Shan ZY, Wu YS, Shen XH, Li X, Xue Y, et al. (2012) Aggregation of pre-implantation embryos improves establishment of parthenogenetic stem cells and expression of imprinted genes. Dev Growth Differ 54: 481–488.
25. Eckardt S, McLaughlin KJ (2004) Interpretation of reprogramming to predict the success of somatic cell cloning. Anim Reprod Sci 83: 97–108.
26. Hardy K (1997) Cell death in the mammalian blastocyst. Mol Hum Reprod 3: 919–925.
27. Hardy K, Stark J, Winston RM (2003) Maintenance of the inner cell mass in human blastocysts from fragmented embryos. Biol Reprod 68: 1165–1169.
28. Hao Y, Lai L, Mao J, Im GS, Bonk A, et al. (2003) Apoptosis and *in vitro* development of preimplantation porcine embryos derived *in vitro* or by nuclear transfer. Biol Reprod 69: 501–507.
29. Maddox-Hyttell P, Gjorret JO, Vajta G, Alexopoulos NI, Lewis I, et al. (2003) Morphological assessment of preimplantation embryo quality in cattle. Reproduction 61: 103–116.
30. Melka MG, Rings F, Hölker M, Tholen E, Havlicek V, et al. (2010) Expression of apoptosis regulatory genes and incidence of apoptosis in different morphological quality groups of *in vitro*-produced bovine pre-implantation embryos. Reprod Domest Anim 45: 915–921.
31. Moussa M, Tremoleda JL, Duchamp G, Bruyas JF, Colenbrander B, et al. (2004) Evaluation of viability and apoptosis in horse embryos stored under different conditions at 5 degrees C. Theriogenology 61: 921–932.
32. Pomar FJ, Teerds KJ, Kidson A, Colenbrander B, Tharasanit T, et al. (2005) Differences in the incidence of apoptosis between *in vivo* and *in vitro* produced blastocysts of farm animal species: a comparative study. Theriogenology 63: 2254–2268.
33. Tecirlioglu RT, Cooney MA, Lewis IM, Korfiatis NA, Hodgson R, et al. (2005) Comparison of two approaches to nuclear transfer in the bovine: hand-made cloning with modifications and the conventional nuclear transfer technique. Reprod Fertil Dev 17: 573–585.
34. Misica-Turner PM, Oback FC, Eichenlaub M, Wells DN, Oback B (2007) Aggregating embryonic but not somatic nuclear transfer embryos increases cloning efficiency in cattle. Biol Reprod 76: 268–278.

35. Oback B (2008) Climbing mount efficiency-small steps, not giant leaps towards higher cloning success in farm animals. Reprod Domest Anim 43: 407–416.

36. Siriboon C, Tu CF, Kere M, Liu MS, Chang HJ, et al. (2014) Production of viable cloned miniature pigs by aggregation of handmade cloned embryos at the 4-cell stage. Reprod Fertil Dev 26: 395–406.

37. Kurosaka S, Eckardt S, Ealy AD, McLaughlin KJ (2007) Regulation of blastocyst stage gene expression and outgrowth interferon tau activity of somatic cell clone aggregates. Cloning Stem Cells 9: 630–641.

38. Vajta G, Peura TT, Holm P, Paldi A, Greve T, et al. (2000) Method for culture of zona-included or zona-free embryos: the Well of the Well (WOW) system. Mol Reprod Dev 55: 256–264.

39. Taka M, Iwayama H, Fukui Y (2005) Effect of the Well of the Well (WOW) system on in vitro culture for porcine embryos after intracytoplasmic sperm injection. J Reprod Dev 51: 533–537.

40. Hoelker M, Rings F, Lund Q, Ghanem N, Phatsara C, et al. (2009) Effect of the microenvironment and embryo density on developmental characteristics and gene expression profile of bovine preimplantative embryos cultured in vitro. Reproduction 137: 415–425.

41. Sananmuang T, Phutikanit N, Nguyen C, Manee-In S, Techakumphu M, et al. (2013) In vitro culture of feline embryos increases stress-induced heat shock protein 70 and apoptotic related genes. J Reprod Dev 59: 180–188.

42. Gopichandran N, Leese HJ (2006) The effect of paracrine/autocrine interactions on the in vitro culture of bovine preimplantation embryos. Reproduction 131: 269–277.

43. Sancho M, Di-Gregorio A, George N, Pozzi S, Sánchez JM, et al. (2013) Competitive interactions eliminate unfit embryonic stem cells at the onset of differentiation. Dev Cell 26: 19–30.

44. Balbach ST, Esteves TC, Houghton FD, Siatkowski M, Pfeiffer MJ, et al. (2012) Nuclear reprogramming: kinetics of cell cycle and metabolic progression as determinants of success. PLoS One 7: e35322.

45. Li J, Li R, Villemoes K, Liu Y, Purup S, et al. (2013) Developmental potential and kinetics of pig embryos with different cytoplasmic volume. Zygote 15: 1–11.

46. Lewis NE, Rossant J (1982) Mechanism of size regulation in mouse embryo aggregates. J Embryol Exp Morphol 72: 169–81.

47. Rands GF (1986) Size regulation in the mouse embryo. I. The development of quadruple aggregates. J Embryol Exp Morphol 94: 139–148.

48. Bertolini M, Beam SW, Shim H, Bertolini LR, Moyer AL, et al. (2002) Growth, development, and gene expression by in vivo and in vitro-produced Day 7 and 16 embryos. Mol Reprod Dev 63: 318–328.

49. Vajta G, Hyttel P, Trounson AO (2000) Post-hatching development of in vitro produced bovine embryos on agar and collagen gels. Anim Reprod Sci 60-61: 208.

50. Brandão DO, Maddox-Hyttel P, Løvendahl P, Rumpf R, Stringfellow D, et al. (2004) Post hatching development: a novel system for extended in vitro culture of bovine embryos. Biol Reprod 71: 2048–2055.

51. Vejlsted M, Du Y, Vajta G, Maddox-Hyttel P (2006) Post-hatching development of the porcine and bovine embryo–defining criteria for expected development in vivo and in vitro. Theriogenology 65: 153–165.

52. Alexopoulos NI, French AJ (2009) The prevalence of embryonic remnants following the recovery of post-hatching bovine embryos produced in vitro or by somatic cell nuclear transfer. Anim Reprod Sci 114: 43–53.

53. Machado GM, Ferreira AR, Guardieiro MM, Bastos MR, Carvalho JO, et al. (2013) Morphology, sex ratio and gene expression of day 14 in vivo and in vitro bovine embryos. Reprod Fertil Dev 25: 600–608.

54. Machado GM, Ferreira AR, Pivato I, Fidelis A, Spricigo JF, et al. (2013) Post-hatching development of in vitro bovine embryos from day 7 to 14 in vivo versus in vitro. Mol Reprod Dev 80: 936–947.

55. Allen WR, Stewart F (2001) Equine placentation. Reprod Fertil Dev 13: 623–634.

56. Tremoleda JL, Stout TAE, Lagutina I, Lazzari G, Bevers MM, et al. (2003) Effects of in vitro production on horse embryo morphology, cytoskeletal characteristics, and blastocyst capsule formation. Biol Reprod 69: 1895–1906.

57. Fischer B, Mootz U, Denker HW, Lambertz M, Beier HM (1991) The dynamic structure of rabbit blastocyst coverings. III. Transformation of coverings under non-physiological developmental conditions. Anat Embryol (Berl) 183: 17–27.

58. Levy RR, Cordonier H, Czyba JC, Goerin JF (2001) Apoptosis in preimplantation mammalian embryo and genetics. Int J Anat Embryol 106: 101–108.

59. Johnson AK, Clark-Price SC, Choi YH, Hartman DL, Hinrichs K (2010) Physical and clinicopathologic findings in foals derived by use of somatic cell nuclear transfer: 14 cases (2004-2008). J Am Vet Med Assoc 236: 983–990.

60. Pedersen HG, Schmidt M, Sangild PT, Strobech L, Vajta G, et al. (2005) Clinical experience with embryos produced by handmade cloning: work in progress. Mol Cell Endocrinol 234: 137–143.

61. Petters RM, Mettus RV (1984) Survival rate to term of chimeric morulae produced by aggregation of five to nine embryos in the mouse, Mus musculus. Theriogenology 22: 167–174.

PERMISSIONS

The contributors of this book come from diverse backgrounds, making this book a truly international effort. This book will bring forth new frontiers with its revolutionizing research information and detailed analysis of the nascent developments around the world.

We would like to thank all the contributing authors for lending their expertise to make the book truly unique. They have played a crucial role in the development of this book. Without their invaluable contributions this book wouldn't have been possible. They have made vital efforts to compile up to date information on the varied aspects of this subject to make this book a valuable addition to the collection of many professionals and students.

This book was conceptualized with the vision of imparting up-to-date information and advanced data in this field. To ensure the same, a matchless editorial board was set up. Every individual on the board went through rigorous rounds of assessment to prove their worth. After which they invested a large part of their time researching and compiling the most relevant data for our readers.

The editorial board has been involved in producing this book since its inception. They have spent rigorous hours researching and exploring the diverse topics which have resulted in the successful publishing of this book. They have passed on their knowledge of decades through this book. To expedite this challenging task, the publisher supported the team at every step. A small team of assistant editors was also appointed to further simplify the editing procedure and attain best results for the readers.

Apart from the editorial board, the designing team has also invested a significant amount of their time in understanding the subject and creating the most relevant covers. They scrutinized every image to scout for the most suitable representation of the subject and create an appropriate cover for the book.

The publishing team has been an ardent support to the editorial, designing and production team. Their endless efforts to recruit the best for this project, has resulted in the accomplishment of this book. They are a veteran in the field of academics and their pool of knowledge is as vast as their experience in printing. Their expertise and guidance has proved useful at every step. Their uncompromising quality standards have made this book an exceptional effort. Their encouragement from time to time has been an inspiration for everyone.

The publisher and the editorial board hope that this book will prove to be a valuable piece of knowledge for researchers, students, practitioners and scholars across the globe.

LIST OF CONTRIBUTORS

Laura Alcazar-Fuoli, Timothy Cairns, Paul Bowyer and Elaine Bignell
Manchester Fungal Infection Group, Institute for Inflammation and Repair, Faculty of Medicine and Human Sciences, Manchester Academic Health Science Centre, The University of Manchester, Manchester, United Kingdom

Jordi F. Lopez
Department of Environmental Chemistry, Institute of Environmental Assessment and Water Research (IDÆA), Consejo Superior de Investigaciones Científicas, c/Jordi Girona, Barcelona, Spain

Bozo Zonja and Sandra Pérez
Water and Soil Quality Research Group, Institute of Environmental Assessment and Water Research (IDÆA), Consejo Superior de Investigaciones Cientı´ficas, c /Jordi G irona, B arcelona, Spain

Damiá Barceló
Water and Soil Quality Research Group, Institute of Environmental Assessment and Water Research (IDÆA), Consejo Superior de Investigaciones Cientı´ficas, c/Jordi Girona, Barcelona, Spain
Catalan Institute of Water Research, ICRA, C/Emili Grahit, 101, edifici H2O Parc Científic i Tecnoló gic de la Universitat de Girona, Girona, Spain

Yasuhiro Igarashi
Biotechnology Research Center, Toyama Prefectural University 5180 Kurokawa, Imizu, Toyama, Japan

Lien Tembuyser, Véronique Tack and Elisabeth M. C. Dequeker
Department of Public Health and Primary Care, Biomedical Quality Assurance Research Unit, KU Leuven – University of Leuven, Leuven, Belgium

Karen Zwaenepoel and Patrick Pauwels
Department of Pathology, Antwerp University Hospital, Edegem, Belgium

Keith Miller
UK NEQAS ICC & ISH, London, United Kingdom

Lukas Bubendorf
Institute for Pathology, Basel University Hospital, Basel, Switzerland

Keith Kerr
Department of Pathology, Aberdeen Royal Infirmary, Aberdeen, United Kingdom

Ed Schuuring
Department of Pathology and Medical Biology, University of Groningen, Groningen, the Netherlands

Erik Thunnissen
Department of Pathology, VU University Medical Center, Amsterdam, The Netherlands

Chrysostomos Tornari, Emily R. Towers, Jonathan E. Gale, Sally J. Dawson
UCL Ear Institute, University College London, London, United Kingdom

Ming-Shian Tsai
Department of Surgery, E-DA Hospital, Kaohsiung, Taiwan
The School of Medicine for Post-Baccalaureate, I-Shou University, Kaohsiung, Taiwan

Po-Huang Lee
Department of Surgery, E-DA Hospital, Kaohsiung, Taiwan
Department of Surgery, National Taiwan University Hospital, National Taiwan University College of Medicine, Taipei, Taiwan

Yu-Chun Lin, Shih-Che Huang and Ying-Hsien Kao
Department of Medical Research, E-DA Hospital, Kaohsiung, Taiwan

Cheuk-Kwan Sun
Department of Medical Education, E-DA Hospital, Kaohsiung, Taiwan

Akira Shimamoto, Harunobu Kagawa, Kazumasa Zensho, Yukihiro Sera and Hidetoshi Tahara
Department of Cellular and Molecular Biology, Graduate School of Biomedical & Health Sciences, Hiroshima University, Hiroshima, Japan

Yasuhiro Kazuki and Mitsuo Oshimura
Department of Biomedical Science, Institute of Regenerative Medicine and Biofunction, Graduate School of Medical Science, Tottori University, Yonago, Japan

Mitsuhiko Osaki
Department of Biomedical Science, Institute of Regenerative Medicine and Biofunction, Graduate School of Medical Science, Tottori University, Yonago, Japan
Division of Pathological Biochemistry, Faculty of Medicine, Tottori University, Yonago, Japan

Yasuhito Ishigaki
Medical Research Institute, Kanazawa Medical University, Kahoku, Ishikawa, Japan

Kanya Hamasaki and Yoshiaki Kodama
Department of Genetics, Radiation Effects Research Foundation, Hiroshima, Japan

Shinsuke Yuasa and Keiichi Fukuda
Department of Cardiology, Keio University School of Medicine, Tokyo, Japan

Kyotaro Hirashima and Hiroyuki Seimiya
Division of Molecular Biotherapy, The Cancer Chemotherapy Center, Japanese Foundation For Cancer Research, Tokyo, Japan

Hirofumi Koyama and Takahiko Shimizu
Department of Advanced Aging Medicine, Chiba University Graduate School of Medicine, Chiba, Japan

Minoru Takemoto and Koutaro Yokote
Department of Clinical Cell Biology and Medicine, Chiba University Graduate School of Medicine, Chiba, Japan

Makoto Goto
Division of Orthopedic Surgery & Rheumatology, Tokyo Women's Medical University Medical Center East, Tokyo, Japan

Giandomenico Turchiano, Maria Carmela Latella and Alessandra Recchia
Center for Regenerative Medicine, Department of Life Sciences, University of Modena and Reggio Emilia, Modena, Italy

Fulvio Mavilio
Center for Regenerative Medicine, Department of Life Sciences, University of Modena and Reggio Emilia, Modena, Italy
Genethon, Evry, France

Andreas Gogol-Döring
German Centre for Integrative Biodiversity Research (iDiv) Halle-Jena-Leipzig, Leipzig, Germany
Institute of Computer Science, Martin Luther University Halle-Wittenberg, Halle, Germany

Claudia Cattoglio
Howard Hughes Medical Institute, Department of Molecular and Cell Biology, University of California, Berkeley, Berkeley, California, United States of America

Zsuzsanna Izsvák
Max Delbruck Center for Molecular Medicine, Berlin, Germany

Zoltán Ivics
Division of Medical Biotechnology, Paul Ehrlich Institute, Langen, Germany

Theodoros Goulas, Anna Cuppari, Raquel Garcia-Castellanos, Joan L. Arolas and F. Xavier Gomis-Rüth
Proteolysis Lab, Molecular Biology Institute of Barcelona, CSIC, Barcelona Science Park, Helix Building, Barcelona, Spain

Scott Snipas
Sanford-Burnham Medical Research Institute, La Jolla, California, United States of America

Rudi Glockshuber
Institute of Molecular Biology and Biophysics, Department of Biology, Zurich, Switzerland

Jun Uetake
Transdisciplinary Research Integration Center, Minato-ku, Tokyo, Japan
National Institute of Polar Research, Tachikawa, Tokyo, Japan

Sota Tanaka, Hideaki Motoyama and Satoshi Imura
Faculty of Science, Chiba University, Chiba, Chiba, Japan

Kosuke Hara
Graduate School of Science, Kyoto University, Kyoto, Japan

Yukiko Tanabe
Institute for Advanced Study, Waseda University, Shinjuku-ku, Tokyo, Japan

Denis Samyn
Department of Mechanical Engineering, Nagaoka University of Technology, Nagaoka, Nigata, Japan

Shiro Kohshima
Wildlife Research Center, Kyoto University, Kyoto, Kyoto, Japan

Jialin Li and Nan Li and Song Qin
Key Laboratory of Coastal Biology and Bioresource Utilization, Yantai Institute of Coastal Zone Research, Chinese Academy of Sciences, Yantai, China

Shuxian Yu and Yinchu Wang
Key Laboratory of Coastal Biology and Bioresource Utilization, Yantai Institute of Coastal Zone Research, Chinese Academy of Sciences, Yantai, China,
Graduate University of Chinese Academy of Sciences, Beijing, China

Fuchao Li
Key Laboratory of Experimental Marine Biology, Institute of Oceanology, Chinese Academy of Sciences, Qingdao, China

Tao Zou
Key Laboratory of Coastal Environmental Processes and Ecological
Remediation, Yantai Institute of Coastal Zone Research, Chinese Academy of Sciences, Yantai, China

Guangyi Wang
Tianjin University Center for Marine Environmental Ecology, School of Environmental Sciences and Engineering, Tianjin University, Tianjin, China, Department of Microbiology, University of Hawaii at Manoa, Honolulu, Hawaii, United States of America

Xiaoyun Huang
College of Biological Science and Technology, Fuzhou University, Fuzhou 350108, P.R. China

Guozeng Wang, Juan Lin and Xiu Yun Ye
College of Biological Science and Technology, Fuzhou University, Fuzhou 350108, P.R. China
National Engineering Laboratory for High-efficiency Enzyme Expression,
Fuzhou 350002, P. R. China

Tzi Bun Ng
School of Biomedical Sciences, Faculty of Medicine, The Chinese University of Hong Kong, Shatin, New Territories, Hong Kong, China

Paula Bustamante and Omar Orellana
Programa de Biología Celular y Molecular, ICBM, Facultad de Medicina, Universidad de Chile, Santiago, Chile

Mario Tello
Centro de Biotecnología Acuícola, Departamento de Biología, Facultad de Químicay Biología, Universidad de Santiago de Chile, Santiago, Chile

Naoki Tanigawa, Toshitsugu Fujita and Hodaka Fujii
Chromatin Biochemistry Research Group, Combined Program on Microbiology and Immunology, Research Institute for Microbial Diseases, Osaka University, Suita, Osaka, Japan

Anna Podnos, Ismat Khatri and Zhiqi Chen
University Health Network, Toronto General Hospital, Toronto, Canada

Reginald M. Gorczynski
University Health Network, Toronto General Hospital, Toronto, Canada
Department of Immunology, Faculty of Medicine, University of Toronto, and Institute of Medical Science, University of Toronto, Toronto, Ontario, Canada

Nuray Erin
Department of Medical Pharmacology, Akdeniz University, School of Medicine, Antalya, Turkey

Paola Genevini, Giulia Papiani, Annamaria Ruggiano and Francesca Navone
Institute of Neuroscience, Consiglio Nazionale delle Ricerche, and Department of Medical Biotechnology and Translational Medicine (BIOMETRA), Universitá degli Studi di Milano, Milano, Italy

Nica Borgese
Institute of Neuroscience, Consiglio Nazionale delle Ricerche, and Department of Medical Biotechnology and Translational Medicine (BIOMETRA), Universitá degli Studi di Milano, Milano, Italy, Department of Health Science, Magna Graecia University of Catanzaro, Catanzaro, Italy

Lavinia Cantoni
Department of Molecular Biochemistry and Pharmacology, Istituto di Ricerche Farmacologiche "Mario Negri", Milan, Italy

Yahai Lu
College of Resources and Environmental Sciences, China Agricultural University, Beijing, 100193, China

Lei Cheng
College of Resources and Environmental Sciences, China Agricultural University, Beijing, 100193, China Key Laboratory of Development and Application of Rural
Renewable Energy, Biogas Institute of Ministry of Agriculture, Chengdu, 610041, China

Qiang Li and Hui Zhang
Key Laboratory of Development and Application of Rural Renewable Energy, Biogas Institute of Ministry of Agriculture, Chengdu, 610041, China

Shengbao Shi and Jianfa Chen
State Key Laboratory of Petroleum Resources and Prospecting, China University of Petroleum (Beijing), Beijing, 102200, China

Michael Kacik
Faculty of Medicine, Philipps-University Marburg & Medical Center I, Clemenshospital/University Hospital of University Münster, 48153 Münster, Germany

Aida Oliván-Viguera
Aragon Institute of Health Sciences I+CS/IIS, 50009 Zaragoza, Spain

Ralf Kö hler
Aragon Institute of Health Sciences I+CS/IIS, 50009 Zaragoza, Spain
Fundación Agencia Aragonesa para la Investigación y Desarrollo (ARAID), 50018 Zaragoza, Spain

Claudia Skamel
Campus Technologies Freiburg (CTF) GmbH, Agency for Technology Transfer at the University and University Medical Center Freiburg, Freiburg, Germany

Stephen G. Aller
Department of Pharmacology and Toxicology and Center for Structural Biology, University of Alabama at Birmingham, Birmingham, Alabama, United States of America

Alain Bopda Waffo
Department of Biological Sciences, Alabama State University, Montgomery, Alabama, United States of America

Philipp Sundermann, Michael Tachezy, Jakob R. Izbicki and Maximilian Bockhorn
Department of General, Visceral and Thoracic Surgery, University Medical Center Hamburg-Eppendorf, University of Hamburg, Hamburg, Germany

Florian Gebauer
Department of General, Visceral and Thoracic Surgery, University Medical Center Hamburg-Eppendorf, University of Hamburg, Hamburg, Germany
Institute of Anatomy and Experimental Morphology and University Cancer Center Hamburg (UCCH), University Medical-Center Hamburg-Eppendorf, Hamburg, Germany

Daniel Wicklein, Jennifer Horst, Hanna Maar and Udo Schumacher
Institute of Anatomy and Experimental Morphology and University Cancer Center Hamburg (UCCH), University Medical-Center Hamburg-Eppendorf, Hamburg, Germany

Thomas Streichert
Institute of Clinical Chemistry, University Medical-Center Hamburg-Eppendorf, Hamburg, Germany

Tian Yang
Department of Cell Biology, Third Military Medical University, Chongqing, 400038, China

Fang Deng
Department of Cell Biology, Third Military Medical University, Chongqing, 400038, China
Molecular Oncology Laboratory, Department of Orthopaedic Surgery and Rehabilitation Medicine, The University of Chicago Medical Center, Chicago, IL, 60637, United States of America

Xiang Chen, Sahitya Denduluri, Melissa Li, Nisha Geng, Guolin Zhou, Hue H. Luu and Rex C. Haydon
Molecular Oncology Laboratory, Department of Orthopaedic Surgery and
Rehabilitation Medicine, The University of Chicago Medical Center, Chicago, IL, 60637, United States of America

Zhan Liao and Tong-Chuan He
Molecular Oncology Laboratory, Department of Orthopaedic Surgery and
Rehabilitation Medicine, The University of Chicago Medical Center, Chicago, IL, 60637, United States of America
Department of Orthopaedic Surgery, the Affiliated Xiang-Ya Hospital of Central South University, Changsha, 410008, China

Zhengjian Yan, Zhongliang Wang, Youlin Deng, Jing Wang, Qian Zhang, Qiang Wei and Penghui Zhang
Molecular Oncology Laboratory, Department of Orthopaedic Surgery and
Rehabilitation Medicine, The University of Chicago Medical Center, Chicago, IL, 60637, United States of America
Ministry of Education Key Laboratory of Diagnostic Medicine, and the Affiliated Hospitals of Chongqing Medical University, Chongqing, 400016, China

Zhonglin Zhang and Ruifang Li
Molecular Oncology Laboratory, Department of Orthopaedic Surgery and
Rehabilitation Medicine, The University of Chicago Medical Center, Chicago, IL, 60637, United States of America
Department of Surgery, the Affiliated Zhongnan Hospital of Wuhan University, Wuhan, 430071, China

Jixing Ye
Molecular Oncology Laboratory, Department of Orthopaedic Surgery and
Rehabilitation Medicine, The University of Chicago Medical Center, Chicago, IL, 60637, United States of America

School of Bioengineering, Chongqing University, Chongqing, 400044, China

Min Qiao
Molecular Oncology Laboratory, Department of Orthopaedic Surgery and
Rehabilitation Medicine, The University of Chicago Medical Center, Chicago, IL, 60637, United States of America
Ministry of Education Key Laboratory of Diagnostic Medicine, and the Affiliated Hospitals of Chongqing Medical University, Chongqing, 400016, China
Department of Orthopaedic Surgery, the Affiliated Xiang-Ya Hospital of Central South University, Changsha, 410008, China

Lianggong Zhao
Molecular Oncology Laboratory, Department of Orthopaedic Surgery and
Rehabilitation Medicine, The University of Chicago Medical Center, Chicago, IL, 60637, United States of America
Department of Orthopaedic Surgery, the Second Affiliated Hospital of Lanzhou University, Lanzhou, Gansu, 730000, China

Russell R. Reid
Molecular Oncology Laboratory, Department of Orthopaedic Surgery and
Rehabilitation Medicine, The University of Chicago Medical Center, Chicago, IL, 60637, United States of America
The Laboratory of Craniofacial Biology, Department of Surgery, The University of Chicago Medical Center, Chicago, IL, 60637, United States of America

Adrian De Stefano and Florencia Karlanian
Laboratory of Animal Biotechnology, Faculty of Agriculture, University of Buenos Aires, Buenos Aires, Argentina

Andrés Gambini, Romina Jimena Bevacqua and Daniel Felipe Salamone
National Institute of Scientific and Technological Research, Buenos Aires, Argentina

Index